ALGEBRA AND TRIGONOMETRY

GORDON FULLER

Professor of Mathematics Emeritus   *Texas Tech University*

ALGEBRA AND TRIGONOMETRY

McGRAW-HILL BOOK COMPANY   *New York   St. Louis   San Francisco*
Düsseldorf   Johannesburg   Kuala Lumpur   London   Mexico   Montreal
New Delhi   Panama   Rio de Janeiro   Singapore   Sydney   Toronto

This book was set in Laurel
by Graphic Services, Inc.,
printed on permanent paper
by Halliday Lithograph Corporation,
and bound by The Maple Press Company.
The designer was J. E. O'Connor;
the drawings were done by AYXA, Inc.
The editors were E. M. Millman and Joan A. DeMattia.
Peter D. Guilmette supervised production.

ALGEBRA AND TRIGONOMETRY

Library of Congress Catalog Card Number 74-116663

07-022605-9

89101112 HDHD 798765

PREFACE

This book has been planned for college students who are not sufficiently prepared in algebra and trigonometry to begin the study of analytic geometry and calculus. The treatment has been somewhat "modernized" through the procedures, terminology, and the emphasis on the basic concepts. This approach, together with precise definitions, makes possible a judicious amount of rigor. And this, in turn, facilitates the mastery of the essential techniques.

The exposition throughout is stated in clear and simple language. Each new topic is amply illustrated with solved problems which have been carefully planned to illustrate the principles involved, and also to help obviate any troublesome points. In fact, the organization and exposition are meant to be such that the material will be largely self-teaching.

The concept of a set is introduced in the first chapter. The treatment, though brief, suffices for the use of the concept in the chapters which follow. Set notation and set operations appear at various places where they tend to promote understanding and help with simplifications.

The field axioms of real numbers and certain theorems based on the axioms are treated in Chapter 2. This discussion makes way for justifying the fundamental operations on algebraic expressions. Although students at this stage have been using the axioms and theorems, many do not realize that the theorems stem from the axioms. The dependence of the theorems on the axioms has been emphasized by frequent reference to the axioms in proving the theorems.

The coverage of this material will, of course, depend on the class at hand. As a minimum accomplishment, however, the students, by understanding both the basic nature of the axioms and the statements contained in the theorems, should begin to envision elementary algebra as being a logical structure.

In Chapters 3 through 7 we take up a number of algebraic topics. These include the fundamental operations on algebraic expressions, factoring, functions and relations, exponents and radicals, and linear equations. Although the treatment is quite elementary, many freshmen need to study this part rather thoroughly, whereas for others a rapid review may be sufficient. Functions and relations are defined as sets of ordered pairs of numbers with the distinction between the two concepts clearly pointed out. A careful discussion is given of the operations on equations which lead to equivalent equations.

We turn next to topics of trigonometry. Here we show the close tie between the trigonometric functions of angles and the trigonometric functions of real numbers. Because of this close relation, we have the option in the discussions which follow of considering the domains of the trigonometric functions as sets of angles or as sets of real numbers. Hence, separate treatments of the two types of functions are not necessary.

We introduce certain right-triangle problems using a four-place table of natural trigonometric functions. The data in most of the problems consist of two and three significant figures. Hence the necessary computations are not tedious, and this permits the student to do enough problems in a short time to grasp the ideas involved and to appreciate this practical aspect of trigonometry. Here, and throughout the book, the results of worked-out computational problems and the answers to such problems are consistent with the accuracy of the given data.

The basic identities of trigonometry are handled in careful detail; the derivations are carried out with full explanations and are illustrated with numerous well-chosen examples. Suggestions are listed for proving other identities. The distance formula, derived earlier, is employed in establishing the formula for the cosine of the difference of two angles (or numbers).

Complex numbers are discussed quite adequately in Chapter 11. We have chosen the rectangular form $a + bi$ and the trigonometric form

$r(\cos\theta + i \sin\theta)$ in the treatment. Attention is given to the fact that complex numbers obey the field axioms of real numbers.

The next six chapters deal with algebraic material. Among the topics in this part are the order axioms of real numbers and certain theorems based on the axioms. Included in the discussion are inequalities of one and two variables and systems of inequalities. Mathematical induction is introduced and applied in proving the binomial theorem. Several of the elementary properties of equations are treated.

The study of analytic trigonometry is completed in Chapters 18 and 19. Careful discussion is given to the line-segment representation of the values of the trigonometric functions. This representation, stemming directly from the definitions of the functions, furnishes an easy way of visualizing the variation of the values of the functions as the angle increases from 0 to 2π. The graphs of the trigonometric functions are treated with ample explanations. In addition to the graphs of the trigonometric functions of x, the graphs of the functions of bx and $bx + c$, where b and c are constants, are discussed. Also included are the sums of functions of these types. Sets of real numbers are chosen as the domains in all the graphs.

Trigonometric equations and inverse trigonometric functions are taken up next; they are presented with clear explanations and illustrative examples.

A chapter on logarithms precedes the study of the oblique triangle. The theoretical aspects of logarithms are discussed quite thoroughly. The law of sines, the law of cosines, the law of tangents, and the half-angle formulas are derived and used in solving triangles. In some classes, at least, it will be necessary to omit portions of this material, particularly the use of the law of tangents and the half-angle formulas. However, a satisfactory coverage can be made in a very few lessons by assigning problems which call for accuracy to only two or three significant figures. This will enable the student to do enough problems in a short time to gain an appreciation of the methods and principles involved. In addition, the experience in deriving the various formulas, or understanding the derivations, is of value to the student.

The concluding chapter deals with an introduction to matrices and the most important elementary properties of determinants. The expansion of a determinant by cofactors and a practical method for evaluating determinants are discussed. Cramer's rule is employed in solving systems of linear equations.

As with the text proper, meticulous care has been used in planning the exercises, each of which has an abundance of problems. Answers to the odd-numbered problems are included in the text, and answers to the even-numbered ones are given in an instructor's manual. All the necessary numerical tables are bound with the book.

The book is flexible enough to accommodate students with varying degrees of preparation. Suitable material may be selected for any group whose needs would best be served by three class meetings a week for two semesters. In most instances, however, four or five class meetings a week for one semester would likely be preferable.

Much of the algebra material of this book is taken from *College Algebra,* third edition, by Gordon Fuller (1969). The use of this material, by permission of Van Nostrand Reinhold Company, has facilitated the preparation of this book, and we are grateful to them for their kindness and generosity.

Gordon Fuller

CONTENTS

ALGEBRA AND TRIGONOMETRY

1

SETS AND OPERATIONS

1-1 SETS

The idea of a set is a fundamental concept in mathematics. Through repeated use of the word, we have come to think of a set as a collection of objects or other entities. Thus, as examples, we understand the meaning of a set of china, a set of chairs, a set of books, a set of smugglers, and on and on. In fact, the meaning of a set is so definitely and firmly fixed in our minds that any attempt to define the idea would be futile.

Each object of a set is called a *member* or *element* of the set. As with the word "set," we take "member" and "element" as undefined terms. A set may be specified by listing its elements, if possible, or by a description such that it can be determined if a given object is, or is not, a member of the set.

Many kinds of sets occur in mathematics. We mention a few examples of such sets, using a capital letter to stand for each set.

$A =$ the set of angles having measures between 0° and 30°
$B =$ the set of vowels in the English alphabet
$C =$ the set of positive integers less than 6†
$D =$ the set of all numbers between 2 and 3
$E =$ the set of all positive integers
$F =$ the set of positive odd integers less than 50

We note that the elements of B, C, and F can be listed. But we cannot list the elements of A, D, and E. If the elements can be listed from the first through the last, we say the set is *finite;* otherwise, the set is *infinite.*

It is customary to specify a set by enclosing a description of the set within braces. If a set is finite, however, it may be simpler just to list the elements. We may specify the above sets with the following notation.

$$\begin{aligned} A &= \{\theta | 0° < \theta < 30°\} \\ B &= \{a, e, i, o, u\} \\ C &= \{1, 2, 3, 4, 5\} \\ D &= \{x | 2 < x < 3\} \\ E &= \{x | x \text{ is a positive integer}\} \\ F &= \{1, 3, 5, \ldots, 49\} \end{aligned}$$

The Greek letter θ in set A stands for an arbitrary member of the set and the symbols $<$ and $>$ mean, respectively, "is less than" and "is greater than." The vertical bars in A, D, and E may be read "such that." Thus, in words, "A is the set of angles θ such that θ is greater than 0° and less than 30°." The three dots in set F indicate that the missing odd integers are members of the set. With this explanation, we see that the members of each set from A to F are definitely determined.

We use the symbol $\in$ to show that a given object is a member of a set. Thus, referring to sets A and D above, we write

$$\theta \in A \qquad \text{and} \qquad x \in D$$

The symbol $\in$ may be read either as "is a member of" or as "belongs to." The symbol $\notin$ is read "is not a member of." Thus, $25° \in A$ but $35° \notin A$. The letters θ and x, as used here, are variables, as in the following definition.

† We assume the existence of the numbers 1, 2, 3, 4, and so on. This set of numbers is called the *counting numbers*, the *natural numbers*, or the *positive integers.*

Definition 1-1 A symbol, usually a letter, which may stand for any member of a specified set of objects is called a *variable*. If the set has only one member the symbol is called a *constant*.

1-2 RELATED SETS

In this section we shall point out certain ways in which two sets may be related.

Definition 1-2 Two sets A and B are said to be *equal* ($A = B$) if each element of set A is an element of set B and each element of set B is an element of set A.

The sets $A = \{x, y, z\}$ and $B = \{y, z, x\}$, for example, are equal.

Definition 1-3 If it is possible to pair each element of a set A with exactly one element of a set B and each element of B with exactly one element of A, then we say the elements of the sets can be arranged in a *one-to-one correspondence*.

Definition 1-4 Two sets A and B are said to be *equivalent* ($A \leftrightarrow B$) if their elements can be put into a one-to-one correspondence.

According to this definition the infinite set of positive integers and the infinite set of positive even integers are equivalent. The equivalence of the sets $A = \{1, 2, 3, \ldots\}$ and $B = \{2, 4, 6, \ldots\}$ becomes evident from the following method of pairing the elements:

$$(1, 2), (2, 4), (3, 6), (4, 8), (5, 10), \ldots$$

Definition 1-5 If each element of a set A is an element of set B, then A is called a *subset* of B. If A is a subset of B and if B has one or more elements not belonging to A, then A is a *proper* subset of B.

To indicate that A is a subset of B, we write $A \subseteq B$, which is read "A is contained in B" or "B contains A." This notation tells us simply that every element of A is an element of B, but it does not tell us whether or not A is a proper subset. If, however, A is a proper subset, we may emphasize that relationship by writing $A \subset B$.

The following four subsets are some, but not all, of the subsets of $\{a, b, c, d\}$.

$$\{a\} \subset \{a, b, c, d\}$$
$$\{b, c\} \subset \{a, b, c, d\}$$
$$\{a, c, d\} \subset \{a, b, c, d\}$$
$$\{a, b, c, d\} \subseteq \{a, b, c, d\}$$

The fourth subset illustrates that *a set is a subset of itself;* this is a consequence of the very definition of a subset.

Thus far, we have mentioned sets which contain one or more elements. Conditions may be specified, however, such that a set has no element. For example, the set of blondes in a group of five brunettes has no element. A set which contains no element is called the *null,* or *empty,* set. The symbol $\varnothing$ is used to denote the empty set. Thus,

$$\{x|x \text{ is a square circle}\} = \varnothing$$
$$\{x|x \text{ is an integer between 5 and 6}\} = \varnothing$$

Since $\varnothing$ has no element, we may say, and we do say, that each element of $\varnothing$ belongs to any set A, and therefore $\varnothing \subseteq A$. In other words, the empty set is a subset of every set.

EXERCISE 1-1

Enclose within braces (a) a list of the elements of each set and (b) a description of the set.

1. The positive integers less than 5.
2. The positive even integers between 1 and 11.
3. The days of the week.
4. The months of the year having 30 days.
5. The first three presidents of the United States.
6. The counting numbers which are divisible by 3 and less than 13.
7. Books of the Pentateuch.

Tell if the two sets are equal, equivalent, or if one is a subset of the other.

8. $A = \{4, 7, 11\}$, $B = \{7, 4, 11\}$

9. $C = \{a, e, i, o, u\}$, $D = \{u, v, x, y\}$

10. $C = \{x|x \text{ is a number between 1 and 20 and divisible by 4}\}$.
$D = \{t|t \text{ is a positive integer less than 20}\}$.

11. $S = \{x|x \text{ is a lady president of the United States}\}$.
$T = \{x|x \text{ is a positive integer less than 1}\}$.

12. Given $\{5, 7, 8\} = \{x, 8, 7\}$, find x.

13. Find a set of values for a, b, c such that $\{a, b, c\} = \{0, 1, 4\}$.

14. Write all the subsets of **(a)** $\{1\}$, **(b)** $\{u, v\}$.

15. Write the eight subsets of $\{1, 2, 3\}$.

16. If $E = \{2, 5, 8, 11\}$ and $F = \{8, 11, 14, 2\}$, tell which of the following is true:
(a) $8 \in E$ **(b)** $8 \subset F$ **(c)** $8 \subset E$
(d) $\varnothing \in F$ **(e)** $\varnothing \subset F$ **(f)** $\{8\} \in E$
(g) $\{8\} \subset E$ **(h)** $E \leftrightarrow F$
(i) $\{2, 5, 8\}$ is a proper subset of E.

17. If $A \subseteq B$ and $B \subseteq A$, show that A is (or is not) equal to B.

18. By using definitions concerning sets in Sec. 1-2, show that the following are true for sets A, B, and C:
(a) $A = A$ (equality is reflexive).
(b) If $A = B$, then $B = A$ (equality is symmetric).
(c) If $A = B$ and $B = C$, then $A = C$ (equality is transitive).

1-3 OPERATIONS ON SETS

We know that new sets, called *subsets*, can be obtained from a given set. In this section we shall consider operations by which new sets can be obtained from two given sets.

Definition 1-6 The set consisting of the totality of elements under consideration in a particular discussion is called the *universal set*. The universal set is commonly denoted by U.

Definition 1-7 The set of all elements which belong to a given universal set U and do not belong to a given subset A of U is called the *complement* of A.

Denoting the complement of a set A by A', we may write

$$A' = \{x \mid x \in U \quad \text{and} \quad x \notin A\}$$

Suppose, for example, that we have under consideration the sets $A = \{a, b, c\}$ and $B = \{c, d, f, g\}$. Then we could choose $U = \{a, b, c, d, f, g\}$ and have

$$A' = \{d, f, g\} \qquad \text{and} \qquad B' = \{a, b\}$$

Definition 1-8 The *union* of two sets A and B, denoted by $A \cup B$, is defined as the set composed of all elements which belong to A or to B or to both A and B.

It follows from this definition that if $x \in A$, then $x \in A \cup B$; and if $x \in B$, then $x \in A \cup B$. Hence, we have

$$A \cup B = \{x \mid x \in A \text{ or } x \in B\}$$

Definition 1-9 The *intersection* of two sets A and B, denoted by $A \cap B$, is defined as the set consisting of all elements that belong both to A and to B.

We express this definition symbolically by writing

$$A \cap B = \{x \mid x \in A \text{ and } x \in B\}$$

EXAMPLE 1 If $A = \{u, v, w, x, y\}$ and $B = \{a, v, w, y\}$, then

$$A \cup B = \{u, v, w, x, y, a\}$$

and

$$A \cap B = \{v, w, y\}$$

EXAMPLE 2 If $C = \{5, 7, 11\}$ and $D = \{\text{sun, moon}\}$, then

$$C \cup D = \{5, 7, 11, \text{sun, moon}\}$$

and

$$C \cap D = \varnothing$$

Since C and D have no common element, their intersection is the empty set $\varnothing$.

Definition 1-10 Two sets which have no common element are called *disjoint sets*.

Figures 1-1 and 1-2 show illustrations of the union and intersection of two sets. We let A stand for the set of all points inside the larger circle, and B the set of all points inside the smaller circle. Drawings of this kind are called Venn diagrams in honor of the English logician John Venn (1834–1883).

Definition 1-11 If x and y denote two objects (alike or different) with x specified as the first object and y the second object, then the symbol (x, y) is called an *ordered pair*.

Definition 1-12 If X and Y are sets, the set of all possible ordered pairs (x, y) such that $x \in X$ and $y \in Y$ is called the *product set*, or *Cartesian product*, of X and Y; or, in symbols,

$$X \times Y = \{(x, y) | x \in X \quad \text{and} \quad y \in Y\}$$

EXAMPLE 3 If $X = \{a, b\}$ and $Y = \{1, 2, 3\}$, then the product set is

$$X \times Y = \{(a, 1), (a, 2), (a, 3), (b, 1), (b, 2), (b, 3)\}$$

EXAMPLE 4 If $S = \{(a, b)\}$, then

$$S \times S = \{(a, a), (a, b), (b, a), (b, b)\}$$

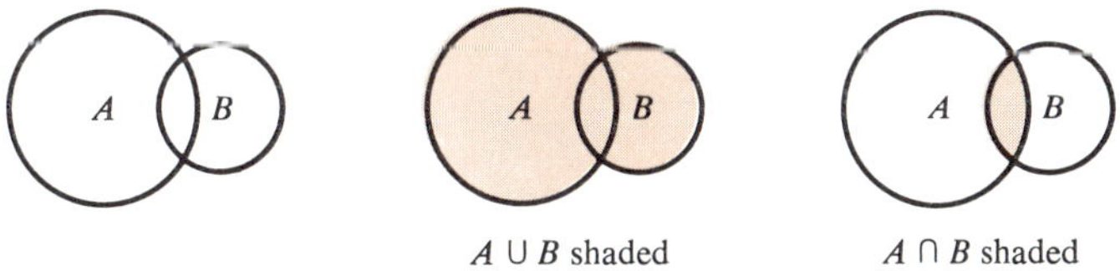

Figure 1-1 $A \cup B$ shaded $A \cap B$ shaded

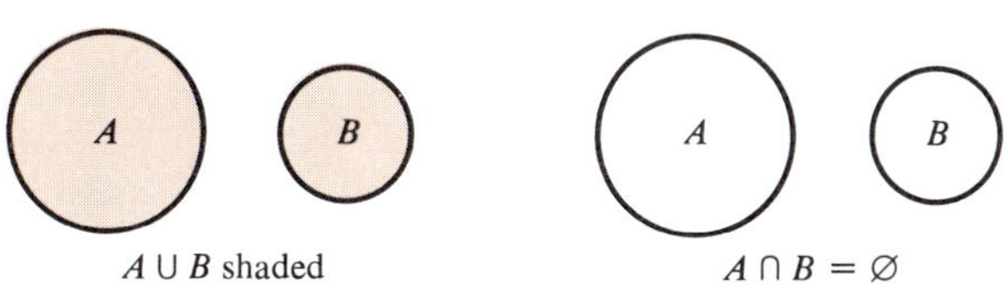

Figure 1-2 $A \cup B$ shaded $A \cap B = \varnothing$

EXERCISE 1-2

If $A = \{0, 1, 2, 3, 4, 5\}$, $B = \{0, 1, 3, 4\}$, and $C = \{4, 5, 6, 7\}$, enclose within braces the elements of the following sets.

1. $A \cup B$ **2.** $A \cap B$ **3.** $A \cup C$

4. $B \cap B$ **5.** $C \cup C$ **6.** $A \cup \varnothing$

7. $B \cap \varnothing$ **8.** $A \cup (B \cup C)$ **9.** $A \cup (B \cap C)$

10. $A \cap (B \cap C)$

11. $A \cap (B \cup C)$

12. $A \cup (B \cap \varnothing)$

If $A = \{p, q, r, s\}$, $B = \{r, s, t, u\}$, and $C = \{t, u, v, w\}$, determine if the following equations are true by examining each member of the equations.

13. $A \cap (B \cup C) = (A \cap B) \cup (A \cap C)$

14. $A \cup (B \cap C) = (A \cup B) \cap (A \cup C)$

15. $A \cap (\varnothing \cup C) = (A \cup B) \cap (A \cup C)$

Prove that if A and B are any sets, then each of the following equations is true.

16. $A \cup A = A$ **17.** $A \cap A = A$

18. $A \cap B = B \cap A$ **19.** $A \cup B = B \cup A$

Draw Venn diagrams to illustrate the following theorems.

20. *Theorem:* If $A \subset B$, then $B \cup A = B$.

21. *Theorem:* If $B \cup A = B$, then $A \subset B$.

22. *Theorem:* If $A \cap B = A$, then $A \subset B$.

23. *Theorem:* If $A \subset B$, then $A \cap B = A$.

24. If $X = \{1\}$ and $Y = \{2, 3\}$, find $X \times Y$ and $Y \times X$.

25. If $P = \{1, 2\}$ and $Q = \{2, 3\}$, find $P \times Q$ and $Q \times P$.

26. If $S = \{a, b, c\}$ and $T = \{1, 2\}$, find $S \times T$.

2

THE REAL NUMBERS

2-1 A DEDUCTIVE SYSTEM

In this chapter we will establish the working rules of algebra. It is interesting to see how directly and efficiently these rules may be established by the deductive method.

We recall that a theorem of high school plane geometry consists of a hypothesis and a conclusion, and that the proof of the theorem is accomplished by a process of logical reasoning which shows that the conclusion is a consequence of the hypothesis. Among the theorems which are established by such a process, there must be a first theorem. The proof of the first theorem clearly cannot rest on other theorems; the proof must involve an unproved statement or statements. The unproved statements are called *axioms* or *postulates;* these are assumed to be true.

There are also numerous definitions of terms in plane geometry. The first of a succession of definitions must describe a term with words whose meanings are assumed to be known. Such words are called *undefined terms.* The meaning of the words "point" and "line," for example, are classed as undefined terms.

The undefined terms and axioms form the basis (starting point) or foundation for proving theorems. With the proof of each theorem, however, the basis for proving other theorems is enlarged. Since theorems stem from the undefined terms and axioms, plane geometry is called a *deductive,* or *axiomatic, system.*

Algebra, like geometry and many other areas of mathematics, is a deductive system. This system is based on a set of axioms and undefined terms which we shall introduce and study in this chapter. In the treatment we shall consider a set of undefined elements called *real numbers,* and indicate the set by $R = \{a, b, c, \ldots\}$. The set will have two basic undefined operations called *addition* and *multiplication.* The symbol $+$ denotes the operation of addition and the symbols $\cdot$ and $\times$ denote multiplication. Thus, if a and b stand for real numbers, we write $a + b$ for addition and $a \cdot b$, $a \times b$, or just ab to indicate multiplication. Symbols grouped together are also used to indicate multiplication. Thus, $(a)(b)$, $a(b)$, and $(a)b$ all mean multiplication. The result of addition is called the *sum,* and the result of multiplication, the *product.*

Later, we shall introduce numbers that are different from real numbers. In the meantime, however, we shall frequently, for brevity, use the word *number* for real number.

The equals sign in the number system denotes the relation of identity. Thus, $a = b$ means that a and b are symbols for the same number. For example, we may write IV $= 4$ and $\frac{1}{2} = 0.5$. A statement of equality such as $a = b$ is called an *equation.*

For the equality relation, we assume the following properties for all elements $a, b, c, \ldots$ of R.

Axiom E-1 $a = a$ (*reflexive* axiom).

Axiom E-2 If $a = b$, then $b = a$ (*symmetric* axiom).

Axiom E-3 If $a = b$ and $b = c$, then $a = c$ (*transitive* axiom).

Axiom E-4 If $a = b$ and $c = d$, then $a + c = b + d$ (*additive* axiom).

Axiom E-5 If $a = b$ and $c = d$, then $ac = bd$ (*multiplicative* axiom).

The last three properties are described, respectively, by the statements:

1. Quantities equal to the same quantity are equal to each other.
2. If equals are added to equals, then the sums are equal.
3. If equals are multiplied by equals, then the products are equal.

The five axioms seem to be quite reasonable. The first axiom, stating that a number is equal to itself, appears to be acceptable. If in the equation $a = b$, we replace a by its equal b and b by its equal a, we obtain $b = a$. Hence, Axiom E-2 is really an agreement that any quantity may be replaced by an equal quantity; this is called the *principle of substitution*. Axioms E-3, E-4, and E-5 likewise follow from this principle. To check Axiom E-4, we start with the sum $a + c$. Then, replacing a by its equal b and c by its equal d, we have $a + c = b + d$.

If in Axioms E-4 and E-5 we use $c = c$ instead of $c = d$, we see that $a + c = b + c$ and $ac = bc$. These results give rise to the familiar rules:

1. The same number may be added to both sides of an equation.
2. Both sides of an equation may be multiplied by the same number.

2-2 FIELD AXIOMS OF REAL NUMBERS

Here we shall state six axioms, called *axioms of a field,* for the set of real numbers R. Later (Sec. 13-1), additional axioms will be introduced to complete our study of real numbers. In the following discussions the grouping symbols ( ) and [ ] indicate that the enclosed numbers are to be considered as a single quantity.

Axiom 1 *The closure laws:* For any $a, b \in R$,

$$a + b \in R \qquad \text{and} \qquad a \cdot b \in R$$

This axiom simply states that the operations of addition and multiplication on two numbers which belong to R yield numbers which also belong to R. That is, the sum $a + b$ and the product $a \cdot b$ are real numbers. Hence, the set R is said to be *closed* under these operations.

Axiom 2 *The commutative laws:* For any $a, b \in R$,

$$a + b = b + a \qquad \text{and} \qquad a \cdot b = b \cdot a$$

The commutative axiom tells us that the sum or product of two numbers does not depend on the order in which the addition or multiplication is performed. Thus, $3 + 7 = 7 + 3$ and $3 \cdot 7 = 7 \cdot 3$.

Axiom 3 *The associative laws:* For any $a, b, c \in R$,

$$(a + b) + c = a + (b + c) \qquad \text{and} \qquad (ab)c = a(bc)$$

The associative axiom tells us that the sum or product of any three numbers is independent of the way in which the numbers are grouped for these operations. In fact, Axioms 1 to 3 permit us to add or multiply any number of real numbers in any chosen order. For two of the several possible ways of finding the product of four numbers, we have

$$\begin{aligned} 2 \cdot 3 \cdot 4 \cdot 5 &= (2 \cdot 3) \cdot 4 \cdot 5 = (6 \cdot 4) \cdot 5 = 24 \cdot 5 = 120 \\ &= 2 \cdot (3 \cdot 4) \cdot 5 = 2 \cdot (12 \cdot 5) = 2 \cdot 60 = 120 \end{aligned}$$

Axiom 4 *The distributive law:* For any $a, b, c \in R$,

$$a(b + c) = ab + ac$$

The distributive axiom states that a particular product is equal to a sum, and, conversely, the sum is equal to a product since equality is symmetric. The name of this axiom seems appropriate because the multiplier *a is distributed to each element of the sum* $(b + c)$. The distributive law, as we shall see, has important consequences; it is basic to the structure of elementary algebra.

A consequence of Axiom 4 is that the product of a number and the sum of three or more numbers is the same as the sum of the products of the first number by each of the numbers forming the sum. Thus, for the sum of three numbers, we have

$$\begin{aligned} a(b + c + d) &= a[b + (c + d)] && \text{Axiom 3} \\ &= ab + a(c + d) && \text{Axiom 4} \\ &= ab + (ac + ad) && \text{Axiom 4} \\ &= ab + ac + ad && \text{Axiom 3} \end{aligned}$$

Axiom 5 *The identity elements:* For any $a \in R$, there exists a real number called "zero," denoted by 0, such that

$$a + 0 = a$$

For any $a \in R$, there exists a real nonzero number called *one*, denoted by 1, such that

$$a \cdot 1 = a$$

The number 0 is called the *identity element for addition*, and the number 1 is called the *identity element for multiplication.* Since 0 and 1 belong to R, we have, by Axiom 2,

$$a + 0 = 0 + a = a$$

and

$$a \cdot 1 = 1 \cdot a = a$$

Axiom 6 *The inverse elements:* For any $a \in R$, there exists a real number, denoted by $-a$, such that

$$a + (-a) = 0$$

For any nonzero number $a \in R$, there exists a real number, denoted by either

$$\frac{1}{a} \quad \text{or} \quad 1/a$$

such that

$$a \cdot \frac{1}{a} = 1$$

The number $-a$ is called the *additive inverse of a*, the *negative of a*, or *minus a*. The number $1/a$ is called the *multiplicative inverse of a* or the *reciprocal of a.*

By the commutative axiom, we have

$$a + (-a) = (-a) + a = 0$$

$$a \cdot \frac{1}{a} = \frac{1}{a} \cdot a = 1$$

Hence, we see that a is the additive inverse of $-a$, and $a \neq 0$ is the multiplicative inverse of $1/a$. (The symbol $\neq$ is read "not equal to.")

EXERCISE 2-1

Naming the appropriate axiom or axioms, explain why each statement 1 through 14 is true.

1. $5 + 7 = 7 + 5$

2. $8 + 0 = 8$

3. $0 + 8 = 8$

4. $(-2) \cdot 1 = -2$

5. $1 \cdot 6 = 6$

6. $4 + (-4) = 0$

7. $-9 + [-(-9)] = 0$

8. $(-ab) + (ab) = 0$

9. $-7\left(\frac{1}{-7}\right) = 1$

10. $7\left(\frac{1}{7}\right) = 1$

11. $4(3 + 5) = 4 \cdot 3 + 4 \cdot 5$

12. $(3 + 5)4 = 4 \cdot 3 + 4 \cdot 5$

13. $x + y = 1 \cdot x + 1 \cdot y$

14. $0 + 0 = 0$

Find the additive inverse and also the multiplicative inverse of each of the following numbers.

15. 3

16. -3

17. 1

18. $\frac{3}{4}$

19. $-a$

20. $(4x + y)$

21. $\dfrac{5x}{2y}$

22. $-\dfrac{3x}{4y}$

23. $-(2x + 3y)$

24. Assuming that $d \neq 0$, give the value of x if

(a) $dx = d$

(b) $\frac{1}{d}(x) = 1$

Show that the following statements are true.

25. $(b + c)a = ba + ca$

26. $(ab)c = (ca)b$

27. $(b + c + d)a = ab + ac + ad$

28. $(a + 4) + b = 4 + (a + b)$

29. $-a + (a + b) = b$

30. $\frac{1}{b}(ab) = a$

2-3 CERTAIN CONSEQUENCES OF THE FIELD AXIOMS*

We shall next establish some important theorems which follow from the field axioms. The theorems represent basic properties of the set of real numbers and will be applied in later work. Most of the steps in the proofs of the theorems will be justified by referring to the appropriate axioms or previously established theorems. In cases in which references are missing, the student should supply the reasons. The student may also establish the theorems which are stated without proofs.

Theorem 2-1 If, for any $a, b, c \in R$, $a + c = b + c$, then $a = b$.

Proof:

$a + c = b + c$	Given equation
$(a + c) + (-c) = (b + c) + (-c)$	Adding $-c$ to both sides
$a + [c + (-c)] = b + [c + (-c)]$	Axiom 3
$a + 0 = b + 0$	Axiom 6
$a = b$	Axiom 5

Theorem 2-2 If $a, b, c \in R$, $c \neq 0$, and if $ac = bc$, then $a = b$.

To prove the theorem, start by multiplying both sides of $ac = bc$ by $1/c$.

Theorems 2-1 and 2-2 are sometimes called, respectively, the *cancellation law of addition* and the *cancellation law of multiplication.*

Theorem 2-3 The additive inverse of any $a \in R$ is unique.

Proof: Let us assume that some number b is an additive inverse of a. Thus,

* The way in which the material in this chapter is covered will vary from class to class. As a minimum accomplishment, however, the students should know the axioms, understand the statements contained in the theorems, and be able to follow at least a few of the proofs.

$a + b = 0$	Axiom 6
$(a + b) + (-a) = 0 + (-a)$	Axiom E-4
$(b + a) + (-a) = -a + 0$	Axiom 2
$b + [a + (-a)] = -a + 0$	Axiom 3
$b + 0 = -a + 0$	Axiom 6
$b = -a$	Axiom 5

Hence, any additive inverse of a is equal to $-a$.

Theorem 2-4 For any nonzero number $a \in R$, the multiplicative inverse is unique.

Theorem 2-5 The identity element for addition (0) is unique, and the identity element for multiplication (1) is unique.

Theorem 2-6 For any $a \in R$, $-(-a) = a$.

Proof:

$(-a) + [-(-a)] = 0$	Axiom 6

and also

$a + (-a) = 0$	Axiom 6
$(-a) + a = 0$	Axiom 2

These equations reveal that $-(-a)$ and a are both additive inverses of $-a$. Hence by Theorem 2-3, we conclude that $-(-a) = a$.

Theorem 2-7 For any $a \in R$, $a \cdot 0 = 0$.

Proof:

$0 + 0 = 0$	Axiom 5
$(0 + 0)a = 0 \cdot a$	Axiom E-5
$a(0 + 0) = a \cdot 0$	Axiom 2
$a \cdot 0 + a \cdot 0 = a \cdot 0 + 0$	Axioms 4 and 5
$a \cdot 0 + a \cdot 0 = 0 + a \cdot 0$	Axiom 2
$a \cdot 0 = 0$	Theorem 2-1

Theorem 2-8 If $a, b \in R$, and $ab = 0$, then either $a = 0$ or $b = 0$.

Proof: If $a = 0$, the theorem is certainly verified. Hence, we assume that $a \neq 0$. Then $1/a \in R$ and

$$\frac{1}{a}(ab) = \frac{1}{a}(0)$$

$$= 0 \qquad \text{Theorem 2-7}$$

But

$$\frac{1}{a}(ab) = b\left(a \cdot \frac{1}{a}\right) \qquad \text{Axioms 2 and 3}$$

$$= b \cdot 1 \qquad \text{Axiom 6}$$

$$= b \qquad \text{Axiom 5}$$

We have found that

$$b = \frac{1}{a}(ab) \qquad \text{and} \qquad \frac{1}{a}(ab) = 0$$

Accordingly, we conclude, by Axiom E-3, that $b = 0$.

The two preceding theorems permit us to state that *the product of two numbers is equal to zero if, and only if, at least one of the numbers is zero.*

Theorem 2-9 If $a, b \in R$, then $(-a)b = -(ab)$.

Proof:

$$ab + (-a)b = ba + b(-a) \qquad \text{Axiom 2}$$

$$= b[a + (-a)] \qquad \text{Axiom 4}$$

$$= b \cdot 0 \qquad \text{Axiom 6}$$

$$= 0 \qquad \text{Theorem 2-7}$$

Hence, $(-a)b$ is the additive inverse of ab. Since $-(ab)$ is also the additive inverse of ab, by Theorem 2-3, we have $(-a)b = -(ab)$.

Corollary $(-a)b = a(-b) = -(ab)$ and $-1(b) = -b$

Theorem 2-10 For any $a, b \in R$, $(-a)(-b) = ab$.

This theorem follows quite readily from Theorems 2-9 and 2-6, and we leave the proof to the student. We remark that if $a = b = -1$, then $(-1)(-1) = 1$.

Definition 2-1 The operation of subtraction is defined by the equation

$$a - b = a + (-b)$$

We call $a - b$ the "difference of a and b" or "b subtracted from a" or "a minus b." It is important to note that the symbol $(-)$, called the *minus sign*, is used in two different ways in the equation defining subtraction. The minus sign in $a - b$ denotes the operation of subtraction, and $-b$ in $a + (-b)$ is the negative of b. That is, we obtain the quantity $a - b$ by adding a and the negative of b.

Theorem 2-11 If $a, b, c \in R$, then $a(b - c) = ab - ac$.

Proof:

$a(b - c) = a[b + (-c)]$	Definition 2-1
$= ab + a(-c)$	Axiom 4
$= ab + [-(ac)]$	Theorem 2-9 (Corollary)
$= ab - ac$	Definition 2-1

Definition 2-2 The operation of division is defined by the equation

$$\frac{a}{b} = a\left(\frac{1}{b}\right)$$

where b is different from zero.

We read the fraction a/b as "a divided by b"; a is called the *numerator*, and b the *denominator*. The result of a division is called the *quotient*. We defined subtraction in terms of addition, and here we define division in terms of multiplication; that is, a divided by b is equal to a multiplied by the reciprocal of b.

The following properties of a nonzero number a can now be easily established:

$$\frac{a}{a} = 1 \qquad \frac{a}{1} = a \qquad \frac{0}{a} = 0$$

We suggest that the student make the proofs.

Theorem 2-12 For any $a, b, c, d \in R$, except $b = d = 0$,

$$\frac{a}{b} \cdot \frac{c}{d} = \frac{ac}{bd}$$

Proof:

$$\frac{a}{b} \cdot \frac{c}{d} = a\left(\frac{1}{b}\right)c\left(\frac{1}{d}\right) \qquad \text{Definition 2-2}$$

$$= a\left(\frac{1}{b}\right)c\left(\frac{1}{d}\right)bd\left(\frac{1}{bd}\right) \qquad \text{Axioms 6 and 5}$$

$$= \left(ac \cdot \frac{1}{bd}\right) \cdot b\left(\frac{1}{b}\right) \cdot d\left(\frac{1}{d}\right) \qquad \text{Axioms 2 and 3}$$

$$= \left(ac \cdot \frac{1}{bd}\right) \cdot 1 \cdot 1 \qquad \text{Axiom 6}$$

$$= ac \cdot \frac{1}{bd} \qquad \text{Axiom 5}$$

$$= \frac{ac}{bd} \qquad \text{Definition 2-2}$$

Corollary

$$\frac{ac}{bc} = \frac{a}{b}$$

This result follows at once by taking $d = c$ in Theorem 2-12.

Reading this corollary from left to right and from right to left, we have justifications for the following rules.

1. The numerator and denominator of a fraction may be divided by any nonzero number without changing the value of the fraction.
2. The numerator and denominator of a fraction may be multiplied by any nonzero number without changing the value of the fraction.

Theorem 2-13 For any $a, b, c \in R$, $c \neq 0$,

$$\frac{a}{c} + \frac{b}{c} = \frac{a + b}{c}$$

This theorem is an immediate consequence of the definition of division and the distributive axiom. The theorem may be extended to show that the sum of three or more fractions with the same denominator is equal to the sum of the numerators divided by the common denominator.

Theorem 2-14 For any $a, b, c, d \in R$, with $b \neq 0$ and $d \neq 0$,

$$\frac{a}{b} + \frac{c}{d} = \frac{ad + bc}{bd}\dagger$$

Proof:

$$\frac{a}{b} + \frac{c}{d} = \frac{ad}{bd} + \frac{cb}{db} \qquad \text{Theorem 2-12 (Corollary)}$$

$$= \frac{ad + bc}{bd} \qquad \text{Theorem 2-13}$$

For convenience, we sometimes express the quotient of two fractions, for example, a/b divided by c/d, by the notation

$$\frac{a}{b} \div \frac{c}{d} \qquad \text{or by} \qquad \frac{a/b}{c/d}$$

Theorem 2-15 For any $a, b, c, d \in R$, with b, c, and d nonzero numbers,

$$\frac{a}{b} \div \frac{c}{d} = \frac{a}{b} \cdot \frac{d}{c}$$

Proof:

$$\frac{a}{b} \div \frac{c}{d} = \frac{a/b}{c/d}$$

$$= \frac{(a/b)(d/c)}{(c/d)(d/c)} \qquad \text{Theorem 2-12 (Corollary)}$$

$$= \frac{(a/b)(d/c)}{1} \qquad \text{Theorem 2-12}$$

$$= \frac{a}{b} \cdot \frac{d}{c}$$

From this theorem, we state the familiar rule for division.

† Note that this is the pattern you learned in elementary algebra for adding two fractions with unlike denominators.

Rule To divide one fraction by another fraction, multiply the dividend by the inverted divisor.

EXERCISE 2-2

1. Let a be a nonzero number and use the equations $a \cdot 0 = 0$ and $a \cdot 1 = a$ to point out why $1 \neq 0$.
2. Let $a = -1$ in the distributive axiom and show that $-(b + c) = -b - c$.
3. Show that $-(a - b) = -a + b$ and that $-(a + b - c - d) = -a - b + c + d$.

Show that the following statements are correct. The letters stand for any real numbers except that no denominator shall be equal to 0.

4. $-0 \cdot 0 = 0$
5. If $a + b = 0$, then $b = -a$.
6. If $x + c = a$, then $x = a - c$.
7. If $a = b$, then $-a = -b$.
8. If $\dfrac{a}{b} = x$, then $a = bx$.
9. If $x = y$, then $x/y = 1$.
10. If $\dfrac{a}{b} = \dfrac{c}{d}$, then $ad = bc$.
11. If $ad = bc$, then $a/b = c/d$.
12. Prove that the identity element for addition is unique.
13. Prove Theorems 2-10 and 2-13.

2-4 RATIONAL AND IRRATIONAL NUMBERS

Thus far in our study of real numbers, we have not given explicit attention to any particular subsets of the set R. We shall now discuss two special subsets of the real number system. We continue to assume the existence of the natural numbers 1, 2, 3, . . . , which are also called the counting numbers or the positive integers. Then, by the additive inverse axiom, we have the negatives of the natural numbers $-1, -2, -3, \ldots$. The natural numbers, their negatives, and zero constitute the integers of the real number system. The set of integers give rise to a new set of numbers as here defined.

Definition 2-3 Any number which can be expressed as the quotient of two integers (division by zero excluded) is called a *rational number.*

According to this definition, numbers like $\frac{17}{3}$, $-\frac{8}{5}$, and 5.28 are rational numbers. Any integer can be represented by itself divided by 1, and is therefore a rational number. There are real numbers, however, which are not equal to the quotient of two integers. These numbers, constituting one important subset of the real number system, are called *irrational numbers.*

A rational number can be represented decimally by the process of long division. Thus, as examples, $\frac{11}{4} = 2.75$ and $\frac{3}{11} = 0.2727\cdots$. The first of these decimals is said to terminate, but the second is nonterminating because the digits 27 are repeated indefinitely. These examples illustrate the fact that the decimal representation of a rational number will be either a terminating or a nonterminating decimal. A *repeating* decimal,* as we shall learn in Chap. 14, can be expressed readily as the quotient of two integers. We conclude then that an irrational number cannot be represented by a terminating or a repeating decimal.

As an illustration of the existence of irrational numbers, we shall prove that $\sqrt{2}$ (the number whose square is 2) is irrational. First, however, we need to make a preliminary discussion.

An integer is called an *even integer* or an *odd integer* depending on whether it can be expressed as $2n$ or $2n + 1$, where n is an integer. Thus, $10 = 2(5)$ is an even integer and $-7 = 2(-4) + 1$ is an odd integer. Since any integer p is either even or odd, there is some integer n such that

$$p = 2n \qquad \text{or} \qquad p = 2n + 1$$

Hence,

$$p^2 = 4n^2 = 2(2n^2) \quad \text{or} \quad p^2 = 4n^2 + 4n + 1 = 2(2n^2 + 2n) + 1$$

Examining these equations, we discover that the square of an even integer is an even integer and the square of an odd integer is an

* A repeating decimal is a decimal which from some stage on consists of an endless repetition of a group of one or more digits.

odd integer. Accordingly, an integer p is an even integer if p^2 is an even integer, and is an odd integer if p^2 is an odd integer. We are now ready to prove the following theorem.

Theorem 2-16 The number $\sqrt{2}$ is an irrational number.

Proof: To establish the theorem we need to show that $\sqrt{2}$ is not equal to the quotient of two integers. We proceed by supposing that there is some rational number $p/q = \sqrt{2}$. Furthermore, we may use Theorem 2-12 (Corollary) to put p/q in lowest terms; then p and q are not both divisible by any integers except 1 and -1. Then squaring, we get

$$\frac{p^2}{q^2} = 2 \qquad \text{and} \qquad p^2 = 2q^2$$

The last equation reveals that p^2 is divisible by 2, and consequently $p = 2n$ for some integer n. Then, substituting, we obtain

$$4n^2 = 2q^2 \qquad \text{and} \qquad 2n^2 = q^2$$

The second equation here shows that q^2 is divisible by 2, which makes q an even integer. Thus on the basis of our assumption that the $\sqrt{2}$ is rational, we have shown that p and q are both even integers. This is contrary to our agreement that the fraction p/q is in lowest terms. This means that our assumption was wrong, and we conclude, therefore, that the number $\sqrt{2}$ is not a rational number.

In an analogous fashion we could prove that numbers like $\sqrt{3}$, $\sqrt{6}$, and $\sqrt{10}$ are irrational numbers. As an exercise, the student might make the proof for one of these numbers.

We state the following properties of real numbers.

Property 1 The sum, difference, and product of any two integers is an integer.

Property 2 The sum, difference, and product of any two rational numbers is a rational number. The quotient obtained by dividing a rational number by a nonzero rational number is a rational number.

We accept Property 1 as a postulate; we do not prove the property. Property 2 follows from Theorems 2-12, 2-14, 2-15, and Property 1 if we let the letter symbols stand for integers.

2-5 THE REAL NUMBER AXIS

The representation of all real numbers by points on a line is a basic concept of mathematics. To establish this representation, we first choose a direction on a line as positive (to the right in Fig. 2-1) and select a point of the line, which we call the *origin*, to represent the number zero. Next, we choose some convenient unit of length and mark points at distances 1, 2, 3, . . . units to the right of the origin to represent the natural numbers. In the same way, we locate points to the left of the origin to represent the negatives of the natural numbers. We now have a one-to-one correspondence between the set of integers and the points of the line thus far located.

The point corresponding to any nonintegral rational number can (at least theoretically) be determined. Let p and q be natural numbers, with p/q not an integer. And let a line segment of length 1 be divided into q equal parts, a process discussed in plane geometry. Then p/q has its point at a distance p of these smaller units to the right of the origin, and the negative of p/q has its point the same distance to the left of the origin.

The points corresponding to some, but not all, irrational numbers can be determined. Figures 2-2 and 2-3 exhibit line segments of lengths $\sqrt{2}$ and $\sqrt{3}$ and open the way for finding the points corresponding to these numbers. Each side of the right triangle in Fig. 2-2 is of unit length and consequently the length of the

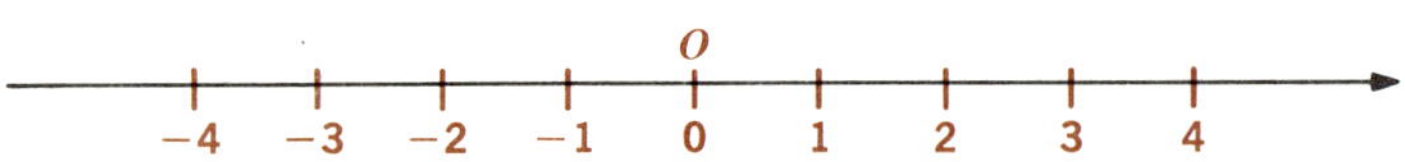

Figure 2-1

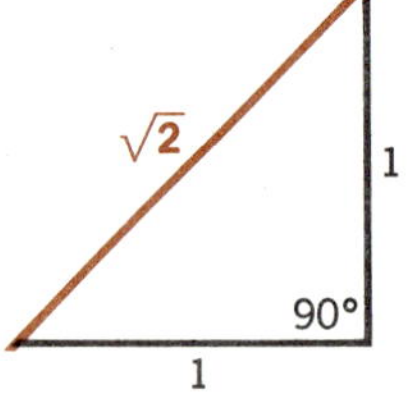

Figure 2-2

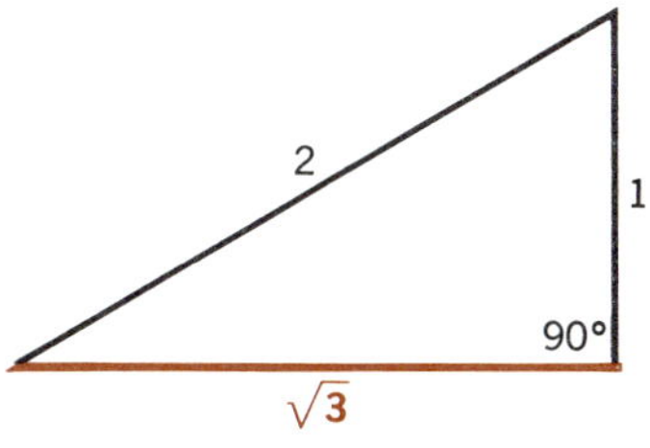

Figure 2-3

hypotenuse, according to the Pythagorean theorem,* is equal to $\sqrt{2}$. This shows that $\sqrt{2}$, which we have already proved cannot be a rational number, actually exists. The length of the hypotenuse in Fig. 2-3 is 2, and the lengths of the sides of the right angle are 1 and $\sqrt{3}$. These values, we note, satisfy the Pythagorean theorem.

Although line segments of lengths equal to some irrational numbers cannot be constructed geometrically, we assume that every irrational number has a corresponding point on the line of Fig. 2-1 at a distance from the origin equal to the number. In summary, then, we assume that every real number corresponds to one point of the line, and, conversely, every point of the line corresponds to one real number.

The line of Fig. 2-1, with its points corresponding to real numbers, is called a *real number axis* or a *real number scale.* The number corresponding to a point on the line is called the *coordinate of the point.* For convenience, we shall sometimes speak of a point as being a number, and vice versa. For example, we may say "the point 5" when we mean "the number 5," and "the number 5" when we mean "the point 5."

Thus far in this chapter, we have spoken of real numbers and their negatives, but have not mentioned positive numbers and negative numbers. Now, however, we arbitrarily define the numbers corresponding to points on the number axis to the right of the origin as *positive numbers* and those to the left of the origin as *negative numbers.* We note that the positive numbers correspond to points in the chosen positive direction from the origin, and the negative numbers correspond to points in the opposite or negative direction from the origin. In view of this circumstance, we shall consider the coordinate of a point on a number axis to be a *directed distance* from the origin—positive if in the positive direction from the origin, and negative if in the negative direction.

* The Pythagorean theorem states that the square of the length of the hypotenuse of a right triangle is equal to the sum of the squares of the lengths of the remaining sides.

Figure 2-4

The one-to-one correspondence between the set of real numbers and the points of a line furnish a geometrical representation of the *order* property of real numbers, a topic which we shall study in Chap. 13. Consider a pair of numbers a and b and their corresponding points on a number axis. Either the two points coincide or do not coincide. If the points coincide, we say the numbers a and b are equal. If the point a is in the negative direction from b (to the left in Fig. 2-4), we say a is less than b. If point a is in the positive direction from point b, we say that a is greater than b. We express these three possibilities by writing

$$a < b \qquad a = b \qquad a > b$$

where the symbols $<$ and $>$ respectively mean "is less than" and "is greater than." We assume, then, that for any two real numbers one and only one of these relations is true. We note that when $b = 0$, these relations become

$$a < 0 \qquad a = 0 \qquad a > 0$$

Hence, any real number is negative, equal to zero, or positive—one of the three.

Our discussion of real numbers in this section and the previous one enable us to make the following observations: The set of rational numbers and the set of irrational numbers are proper subsets of the system of real numbers. The rational numbers contain the set of integers as a subset and the nonintegral rational numbers as a subset. And, finally, the integers consist of positive integers, zero, and negative integers. This makeup of the system of real numbers is pictured in the accompanying diagram.

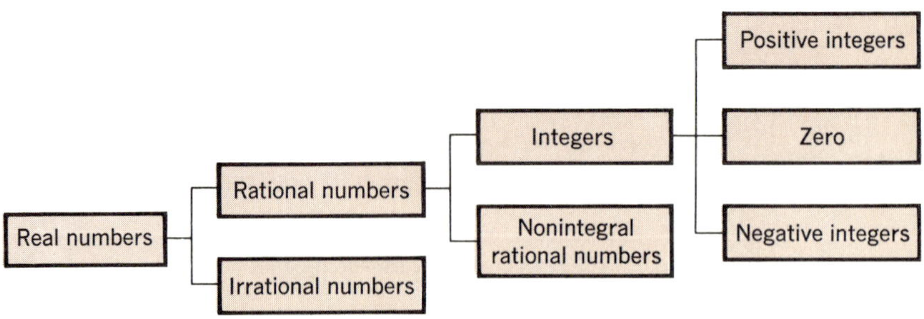

EXERCISE 2-3

Show that each of the following numbers can be expressed as the quotient of two integers and therefore is rational.

1. 0.38 **2.** $5\frac{7}{8}$ **3.** -11.62 **4.** 0

Express each of the following rational numbers decimally.

5. $\frac{5}{6}$ **6.** $\frac{5}{9}$ **7.** $\frac{34}{11}$

8. $\frac{21}{37}$ **9.** $\frac{1}{7}$ **10.** $\frac{14}{13}$

11. Let p and q be positive integers with $p < q$. In dividing p by q, what is the maximum number of remainders in the steps of the process until a remainder is repeated? Noting that each remainder is an integer less than q, can you conclude that the quotient is a repeating decimal?

12. Prove that p is divisible by 3 if p^2 is divisible by 3. Then prove that $\sqrt{3}$ is irrational.

13. Prove that $\sqrt{5}$ is irrational.

Place the appropriate symbol "<" or ">" between each pair of numbers. Also locate the points on a number axis corresponding to the pair of numbers.

14. 5 7 **15.** 4 9 **16.** 6 5

17. 8 2 **18.** 6 -3 **19.** 3 -3

20. -2 -1 **21.** -5 -2

3

OPERATIONS ON ALGEBRAIC EXPRESSIONS

3-1 SOME DEFINITIONS

The field axioms and theorems of the previous chapter furnish the basis for the operations which we shall perform in this chapter. However, it is also necessary to first introduce certain ideas and definitions.

Addition, subtraction, multiplication, and division are called the *fundamental operations of algebra.*

Each of two or more numbers which are multiplied together to form a product is called a *factor* of the product. In the product $7xy$, for example, 7, x, and y are the factors, with x and y being the *literal factors.* Usually a numerical factor, such as the 7, is called the *coefficient* of the literal factors. But, more generally, any factor, or factors, may be regarded as the coefficient of the other factors. Thus in $7xy$, $7x$ is the coefficient of y, and x is the coefficient of $7y$.

Either a single number, or two or more numbers combined with some or all of the fundamental operations of algebra using the appropriate signs is called an *algebraic expression.* An algebraic expression may be separated into parts by plus and minus signs. Each part with the sign preceding it is called a *term.* Frequently, the sign of the first term of an algebraic expression is omitted; in all such cases the plus sign is understood.

EXAMPLES

$3a$ is an expression of one term.

$2x - 3y$ is an expression with the terms $+2x$ and $-3y$.

$mn - pq + 7$ is an expression with the terms mn, $-pq$, and $+7$.

An expression of one term is called a *monomial;* an expression of two terms is called a *binomial;* and an expression of three terms is called a *trinomial.* The word *multinomial* is used to indicate an expression of two or more terms.

The terms in an algebraic expression with the same literal factor, or factors, are called *like terms.* The expression $3xy - 3y - 2xy + 4x$, for example, has $3xy$ and $-2xy$ as like terms.

Frequently, the concept of the absolute value of a real number is of particular significance. Relative to this concept, we have the following definition.

Definition 3-1 The absolute value of a real number a, denoted by $|a|$, is the real number such that

$$|a| = a \text{ when } a \text{ is positive or zero}$$

$$|a| = -a \text{ when } a \text{ is negative}$$

According to the definition, the absolute value of every nonzero number is positive and the absolute value of zero is zero. Thus,

$$|4| = 4 \qquad |-4| = -(-4) = 4 \qquad |0| = 0$$

3-2 ADDITION AND SUBTRACTION

In arithmetic, positive numbers are added, but in algebra addition is performed on both positive and negative numbers. Addition in this wider system of numbers is sometimes called *algebraic addition.*

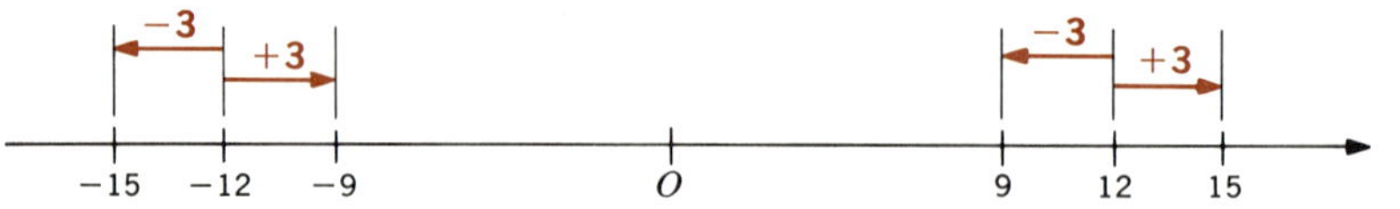

Figure 3-1

EXAMPLE 1

$$12 + 3 = 15$$
$$12 - 3 = 9$$
$$-12 + 3 = -(12 - 3) = -9$$
$$-12 - 3 = -(12 + 3) = -15$$

These results are obtained graphically in Fig. 3-1. We note, for example, that the point corresponding to 15 is 3 units to the right of the point corresponding to 12. All points located in this manner by two positive numbers are to the right of the origin, and this motivates us to *assume* that the sum of two positive numbers is positive. The results of the four additions are illustrations of the following laws of signs for the addition of numbers.

Law 1 The sum of two numbers of the same sign (both positive or both negative) is equal to the sum of their absolute values preceded by the common sign.

Law 2 The sum of two numbers of different signs is obtained by subtracting the smaller absolute value from the larger absolute value and taking the sign of the number of larger absolute value.

EXAMPLE 2 Find the sum of $4xy$ and $-7xy$.

Solution:

$$\begin{aligned} 4xy + (-7xy) &= 4xy - 7xy && \text{Definition 2-1} \\ &= (4 - 7)xy && \text{Distributive axiom} \\ &= -3xy && \text{Law 2} \end{aligned}$$

EXAMPLE 3 Combine like terms of the expression

$$12a + 7ab - 5 - 8a + 4ab - a + 3$$

Solution: Grouping like terms, we have

$$\begin{aligned}(12a - 8a - a) + (7ab + 4ab) + (-5 + 3) \quad & \text{Associative axiom} \\ = (12 - 8 - 1)a + (7 + 4)ab + (-5 + 3) \quad & \text{Distributive axiom} \\ = 3a + 11ab - 2 \quad & \end{aligned}$$

EXAMPLE 4 From $4x + 7y - 4z$ subtract $-6x + 5y + 3z$.

Solution: We may obtain the difference by finding the sum of the first given expression (the minuend) and the additive inverse of the second expression (the subtrahend). We get the additive inverse of the subtrahend by changing the sign of each of its terms. Thus, we have

$$\begin{array}{rl} 4x + 7y - 4z & \\ 6x - 5y - 3z & \\ \hline 10x + 2y - 7z & \quad \text{Sum} \end{array}$$

Alternatively, we may mentally change the signs of the terms of the subtrahend, and write

$$\begin{array}{rl} 4x + 7y - 4z & \quad \text{Minuend} \\ -6x + 5y + 3z & \quad \text{Subtrahend} \\ \hline 10x + 2y - 7z & \quad \text{Difference} \end{array}$$

In performing algebraic operations, it is sometimes necessary to place parentheses (or other grouping symbols) around an expression or to remove parentheses from an expression. The following examples illustrate the procedure for introducing and for removing parentheses when a plus sign precedes the parentheses and also when a minus sign precedes the parentheses.

EXAMPLE 5

$$\begin{aligned}+(4x - 3y + 5) &= 1 \cdot (4x - 3y + 5) \quad && \text{Identity element} \\ &= 1 \cdot (4x) + 1(-3y) + 1 \cdot 5 \quad && \text{Distributive axiom} \\ &= 4x - 3y + 5 \quad && \text{Identity element}\end{aligned}$$

EXAMPLE 6

$$-(4x - 3y + 5)$$

$= -(4x - 3y + 5) \cdot 1$	Identity element
$= -1 \cdot (4x - 3y + 5)$	Theorem 2-9 (Corollary)
$= (-1)(4x) + (-1)(-3y) + (-1)5$	Distributive axiom
$= -4x + 3y - 5$	Theorems 2-9 and 2-10

As these illustrations show, parentheses preceded by a plus sign may be placed around, or removed from, an expression if no change is made in the signs of the terms of the expression. If a minus sign precedes the parentheses, the signs of all terms must be changed.

Frequently, one or more sets of grouping symbols are contained within another. To remove such multiple sets, it is advisable to begin with the innermost symbols.

EXAMPLE 7

$$\begin{aligned} &\{5x - [4x - (2x - y) - 3y] + 6y\} \\ &\quad = \{5x - [4x - 2x + y - 3y] + 6y\} \\ &\quad = \{5x - [2x - 2y] + 6y\} \\ &\quad = \{5x - 2x + 2y + 6y\} \\ &\quad = 3x + 8y \end{aligned}$$

EXERCISE 3-1

Perform the indicated operations.

1. $14 - 6 - 2$ **2.** $7 - 4 + 3$ **3.** $3 - 7 - 4$

4. $9 - 5 - 8$ **5.** $6 - 13 + 7$ **6.** $2 + 8 - 5$

7. $|7| + |-3| - |-2|$ **8.** $|9 - 6| + |4 - 10|$

9. $|7| - |4| - |-1|$ **10.** $|8 - 4| - |-6|$

11. $|-5| + |-6| - |-2|$ **12.** $|5| - |0| + |-3|$

13. $3y - 5y - 7y$ **14.** $m - 3m + 4m$

15. $-x - 7x - 4x$ **16.** $8a + 3a - 9a$

17. $6x - 4x - 8x$ **18.** $mn - 6mn + 2mn$

Find the sum of the two expressions in problems 19 through 23. Then subtract the second expression from the first.

19. $\begin{array}{r} 2x + 8y - 7 \\ 3x + 7y + 4 \\ \hline \end{array}$ **20.** $\begin{array}{r} 4a + 8b - 10c \\ -6a \qquad\quad + 8c \\ \hline \end{array}$

21. $\begin{array}{l} 9x + 9y \\ 2x - 7y + 6z \\ \hline \end{array}$ **22.** $\begin{array}{r} 2x + 5xy + 3yz + 4a \\ 2x - 6xy - 5yz - 9a \\ \hline \end{array}$

23. $\begin{array}{l} 4x \qquad\quad - 3yz + 10a \\ 2x + xy \qquad\quad - 11a \\ \hline \end{array}$

In problems 24 through 27, add the three expressions in each problem, then subtract the third expression from the sum of the first and second.

24. $6a - 10b + 12c$; $-13a + 3b + 9c$; $8a - 8b - 13c$

25. $3xy + 5yz - 2x$; $3x - 4xy + 6yz$; $2yz - 3x + 4xy$

26. $5ab - 7ac + 2bc$; $3bc - 4ab + 6ac$; $7ab + 10ac - 3bc$

27. $3rs - 8rt + 10st$; $4rs + 8st - 3r$; $5rt - 6st - 5s$

Remove the grouping symbols and simplify by combining like terms.

28. $(-4y) - (2y) - (-5y)$

29. $(5y) - (7y) + (-6y)$

30. $3a - (b - c) + (2a - b - 3c)$

31. $3x + (y - 4) - (4x - 3)$

32. $(2x - 3y) - (3x - 4y + 3)$

33. $(x - 2y) - (3x - 4y) - (-x)$

34. $a - (b + c - 4) + (a + c + 2)$

35. $1 - [a + 2b - (3 - a) + 3]$

36. $7a - [(a + 3b) - (2a - 2b)]$

37. $-[x + (3 - x) - (3 + 4x)]$

38. $\{6 - [3 - (6 - 8) - 4] - 8\}$

39. $[(-10 + 4) - 6] - [-(4 - 8) - 7]$

40. $\{[(4 - 2) + (10 + 4)] - (-13 + 8) - 3\}$

41. $2xy - \{-(4xy + 4x) + [2y - (2x - xy - x + 2xy)]\}$

42. $-\{a - 3ab + 2b - [2a - 4ab + 5b - (2a - 3b) + 4]\}$

43. $5 + \{2x + 3y - [3 - x - 9y + (3y - 6x + 3) - 2] - (8x - y)\}$

44. $-\{4 - (7x - 8y + 3) - [(3x - 2y + 6) - (x + 2y) - 5] + (-6x + y)\}$

45. $7a - \{(-4ab - b + 10) - [-(4a - 6ab - 5) + (-7 + b + 5)]\}$

Place parentheses preceded by a plus sign around the last three terms of each expression. Then let a minus sign precede the parentheses.

46. $3 - 4x - 5y + 6z$
47. $a + 3b - 2c - 4$
48. $c - xy + 4 - yz$
49. $2 + 3 - p + q$
50. $2x - x - 4y - 3z$
51. $r - 2s + 3t - st$

3-3 MULTIPLICATION

In products involving positive and negative numbers, we need to know the sign of the product. To determine this, we first *assume* that the product of two positive numbers is positive. Then according to Theorems 2-9 and 2-10, we have

$$-a(b) = -(ab)$$

$$(-a)(-b) = ab$$

These equations, being true for any real numbers a and b, are certainly true when a and b are positive numbers. If a and b are positive, $-a$ and $-b$ are negative, then the first equation reveals that the product of a negative number and a positive number is negative, and the second equation reveals that the product of two negative numbers is positive. So we have a justification for the following law of signs for products.

Law The product of two numbers of like signs is positive. The product of two numbers of unlike signs is negative.

EXAMPLE 1

$$(-3)(-4)(5) = (12)(5) = 60$$
$$(-3)(4)(5) = (-12)(5) = -60$$
$$(-3)(-4)(-5) = (12)(-5) = -60$$

The same factor often occurs two, three, or more times in a product. Although the product may be expressed by writing the factor the proper number of times, it is more convenient to use a shorthand notation. Thus,

$$a^2 = a \cdot a$$
$$a^3 = a \cdot a \cdot a$$
$$a^4 = a \cdot a \cdot a \cdot a$$

In general, we have the following definition.

Definition 3-2 If a is any number and n is a positive integer, then a^n denotes the product of n factors each of which is a; that is,

$$a^n = a \cdot a \cdot a \cdot a \cdots a \qquad (n \text{ factors})$$

The quantity a^n is called the nth *power of a*. The number a is the base and n is the *exponent*. The first power of a number is usually expressed without an exponent. Thus, by definition, $a^1 = a$. The second and third powers a^2 and a^3 are called "a square" and "a cube," respectively.

Theorem 3-1 If a is a real number and m and n are positive integers, then

$$a^m a^n = a^{m+n}$$

Proof: The quantity a^m means the product of m factors each of which is a, and a^n means a is a factor n times. In all, a occurs as a factor $m + n$ times, which we exhibit by writing

$$a^m a^n = (\overbrace{a \cdot a \cdot a \cdot a \cdots a}^{m \text{ factors}})(\overbrace{a \cdot a \cdot a \cdot a \cdots a}^{n \text{ factors}}) = a^{m+n}$$

As illustrations of this law of exponents, we have

$$3^2 \cdot 3^4 = 3^{2+4} = 3^6$$

$$a^2a^5a^3 = a^{2+5+3} = a^{10}$$

$$(3a^2b^3)(2ab^4) = 3 \cdot 2 \cdot a^2 \cdot a^1 \cdot b^3 \cdot b^4 = 6a^3b^7$$

EXAMPLE 2 Multiply $2x^2 - 3xy + 2y^2$ by $3y^3$.

Solution: Using the distributive axiom for multiplication and Theorem 3-1, we get

$$\begin{aligned} 3y^3(2x^2 - 3xy + 2y^2) &= 3y^3(2x^2) + 3y^3(-3xy) + 3y^3(2y^2) \\ &= 6x^2y^3 - 9xy^4 + 6y^5 \end{aligned}$$

The intermediate step is written to emphasize the process. The usual procedure is to write only the final result.

The following theorems may be established by repeated applications of the distributive axiom.

Theorem 3-2 The product of two multinomials is the sum of all results obtained by multiplying all the terms of one multinomial by each term of the other.

EXAMPLE 3 Multiply $3x^2 - xy + 2y^2$ by $x^2 + 3xy - y^2$.

Solution: We obtain the product by multiplying the first multinomial, in turn, by x^2, $3xy$, and $-y^2$, and then adding the separate results. For convenience in adding, we arrange the work so that like terms are in columns.

$$\begin{array}{rrrrrl} & 3x^2 & -\ xy & +\ 2y^2 & & \\ & x^2 & +\ 3xy & -\ y^2 & & \\ \hline 3x^4 & -\ x^3y & +\ 2x^2y^2 & & & \text{Multiplying by } x^2 \\ & +\ 9x^3y & -\ 3x^2y^2 & +\ 6xy^3 & & \text{Multiplying by } 3xy \\ & & -\ 3x^2y^2 & +\ xy^3 & -\ 2y^4 & \text{Multiplying by } -y^2 \\ \hline 3x^4 & +\ 8x^3y & -\ 4x^2y^2 & +\ 7xy^3 & -\ 2y^4 & \text{Adding} \end{array}$$

EXERCISE 3-2

Multiply as indicated and collect like terms.

1. $(5)(-4)(-3)$ **2.** $(2)(-1)(8)$

3. $(-6)(-4)(-2)$ **4.** $a^3 \cdot a^4$

5. $x \cdot x^4 \cdot x^3$

6. $3x^2 \cdot x^3 \cdot x^4$

7. $4y^9 \cdot y^{10} \cdot y^{11}$

8. $(3xy)(2xy)$

9. $(-4a^2b)(3ab^2)$

10. $(-2a^3)(-3a)$

11. $(xy)(-2y^2)(3x)$

12. $(-xy)(2y)(-4x)$

13. $(-ab)(-ab^2)(b^2)$

14. $(-x^2y)(-2y)(-4x)$

15. $(x^2y)(-xy^3)(2xy)$

16. $3x(3x - 2y)$

17. $-3a(2a - 3b)$

18. $5m(2 - 4m)$

19. $-6a^2(4a - 3)$

20. $xy(2x^2 - 3y^2)$

21. $2x^2y(2xy^2 - y)$

22. $(x + 2y)(x - 2y)$

23. $(a - b)(2a + b)$

24. $(3x + y)(2x - 2y)$

25. $(2x + 3)(3x - 4)$

26. $(a + 2b)(3a - b)$

27. $(m - n)(3m - 2n)$

28. $3x^2(5x^2 - 7x + 2)$

29. $2ab(a^2 - 3ab + 3ab^3)$

30. $-3xy^2(2x + 4xy - 4y + 2)$

31. $7xy(2y - 3xy + y^2 - 5)$

32. $-x^2y^2(x^3 + 2xy + 3x^2y^3)$

33. $-2mn(2 - 5m + 6mn - 3n)$

34. $(2a - 3)(4a^2 + 6a + 9)$

35. $(x + y)(x^2 - xy + y^2)$

36. $(m + 5n)(m^2 - 5mn + 25n^2)$

37. $(y + 2x)(y^2 - xy - x^2)$

38. $(c - 4a)(2c^2 - 5ac - a^2)$

39. $(a \quad 3b)(3a^2 - 6ab + 4b^2)$

40. $(b^2 + 2)(2b^2 - 3b + 4)$

41. $(x^2 - 7)(3x^2 - 5x + 6)$

42. $(x + y + 2)(x - y + 2)$

43. $(2x + y - 3)(3x - y - 3)$

44. $(a^2 + 3a + 2)(a^2 - a - 4)$

45. $(2x^2 - x + 7)(3x^2 + x - 4)$

46. $(x + 1)(x - 2)(x + 3)$

47. $(2x - 1)(x + 3)(x + 4)$

48. $(3a - 2)(a + 2)(2a - 1)$

49. $(4 - x)(4 + 2x)(2 - x)$

3-4 DIVISION

In this section, we shall consider the problem of dividing one algebraic expression by another. First, however, let us learn how to obtain the sign of the quotient by dividing one number by another. If c is the quotient of a number a divided by a nonzero number b, then

$$\frac{a}{b} = c \qquad \text{and} \qquad a = bc$$

The last equation tells us that the quotient c is positive if a and b are both positive or both negative, and that c is negative if a and b have unlike signs. Accordingly, we have the following law of signs for division.

Law The quotient of two numbers of like signs is positive. The quotient of two numbers of unlike signs is negative.

In defining division (Definition 2-2), we rejected zero as a divisor. The reason for rejecting zero becomes evident when we try to use zero as a divisor. To divide a number a (different from zero) by zero requires that we find a number c such that $a = 0 \cdot c$. This is impossible, of course, because the product of zero and any number is equal to zero. The equation would be satisfied, however, if the dividend a is also zero. But in this case the quotient c could be any number whatever since $0 = 0 \cdot c$. Hence, division by zero is never allowed.

To prepare the way for dividing a multinomial by a monomial, we need to establish another law of exponents.

Theorem 3-3 If m and n are positive integers and $a \neq 0 \in R$, then

$$\begin{aligned} \frac{a^m}{a^n} &= a^{m-n} && \text{if} \quad m > n \\ &= 1 && \text{if} \quad m = n \\ &= \frac{1}{a^{n-m}} && \text{if} \quad m < n \end{aligned}$$

Proof: If $m > n$, we have by Theorem 3-1

$$\begin{aligned}\frac{a^m}{a^n} &= \frac{a^n \cdot a^{m-n}}{a^n}\\ &= \frac{a^n}{a^n}(a^{m-n})\\ &= a^{m-n}\end{aligned}$$

If $m = n$, we have from the definition of division (Definition 2-2) and Axiom 6

$$\frac{a^m}{a^n} = \frac{a^m}{a^m} = a^m\left(\frac{1}{a^m}\right) = 1$$

We leave the proof of the last part of the theorem to the student.
As illustrations of the theorem, we write

$$\frac{y^5}{y^3} = y^{5-3} = y^2 \qquad \text{and} \qquad \frac{9x^3}{-3x^5} = -\frac{3}{x^2}$$

EXAMPLE 1 Divide $6x^2y - 3xy^2 + 12x$ by $3x$.

Solution: Theorem 2-13 permits us to divide each term of the multinomial by the monomial. Hence,

$$\frac{6x^2y - 3xy^2 + 12x}{3x} = \frac{6x^2y}{3x} - \frac{3xy^2}{3x} + \frac{12x}{3x} = 2xy - y^2 + 4$$

We next consider the problem of division when the dividend and divisor are multinomials. We shall, however, restrict the multinomials to a special kind called *polynomials*.

An algebraic expression in which each term involving x is of the form kx^n, with n a positive integer and k free of x, is called a *polynomial in x*. The polynomial may have one or more terms not involving x. According to this definition

$$3x + 4 \qquad \tfrac{5}{4}x^3 - \tfrac{1}{3}x^2 - x - 1 \qquad 3x^4 - 5x^2 + 7x$$

are polynomials in x.

The *degree* of each term of a polynomial in x is the exponent of x in that term, and the degree of the polynomial is defined as the greatest exponent of x in the polynomial. For example, the terms of

$$7x^4 + 6x^3 - x^2 - 5x + 3$$

are respectively of degrees four, three, two, one, and zero. Note that we regard the term 3 to be of degree zero.

We may have polynomials in one, two, three, or more letters. The degree of a term with respect to a set of letters is the sum of the exponents of those letters in the term. And the greatest sum thus obtained among the terms of the polynomial determines the degree of the polynomial with respect to the set of letters. The expression

$$2x^3y^2z - 3x^2y + xy - z + 5$$

is a polynomial of degree three in x, of degree two in y, of degree one in z, of degree five in x and y, of degree four in x and z, of degree three in y and z, and of degree six in x, y, and z.

We now describe a method of dividing one polynomial by another. The procedure is quite similar to that of long division in arithmetic.

1. Arrange the dividend and divisor in order of descending powers of a common letter, leaving a gap for any missing power of the letter in the dividend.
2. Divide the first term of the dividend by the first term of the divisor. This gives the first term of the quotient.
3. Multiply the divisor by the first term of the quotient and subtract the result from the dividend.
4. Consider the remainder thus obtained as a new dividend and repeat steps 2 and 3 to find the second term of the quotient and the next remainder.
5. Continue this process until a remainder is obtained which is zero or is of lower degree in the common letter than the degree of the divisor.

EXAMPLE 2 Divide $17x - 6x^2 - 12$ by $2x - 3$.

Solution: We write the dividend and divisor in order of descending powers of x and carry out the steps listed above.

$$\begin{array}{lrl} & -3x + 4 & \text{Quotient} \\ \text{Divisor} \quad 2x - 3 \overline{)} & -6x^2 + 17x - 12 & \text{Dividend} \\ & -6x^2 + 9x & \\ \hline & 8x - 12 & \\ & 8x - 12 & \\ \hline & 0 & \text{Remainder} \end{array}$$

Hence, we have

$$\frac{-6x^2 + 17x - 12}{2x - 3} = -3x + 4$$

EXAMPLE 3 Divide $6a^4 - 29a^2 - 21a - 10$ by $2a^2 - 4a - 3$.

Solution: We carry out the steps listed above using an arrangement which facilitates the operations.

$$\begin{array}{llrrrrl}
 & & 3a^2 + & 6a & + 2 & & \text{Quotient} \\
\text{Divisor} & 2a^2 - 4a - 3 \,\big) & 6a^4 & & - 29a^2 & - 21a - 10 & \text{Dividend} \\
 & & 6a^4 - & 12a^3 & - 9a^2 & & \\
 & & & 12a^3 & - 20a^2 & - 21a & \\
 & & & 12a^3 & - 24a^2 & - 18a & \\
 & & & & 4a^2 & - 3a - 10 & \\
 & & & & 4a^2 & - 8a - 6 & \\
 & & & & & 5a - 4 & \text{Remainder}
\end{array}$$

Hence,

$$\frac{6a^4 - 29a^2 - 21a - 10}{2a^2 - 4a - 3} = 3a^2 + 6a + 2 + \frac{5a - 4}{2a^2 - 4a - 3}$$

We see that the complete quotient consists of a polynomial plus a fraction. In a division of this kind the polynomial itself is often referred to as the *quotient*—but sometimes, more accurately, as the *partial quotient.* Using these descriptive words, we may check the result of a division by the relation

$$\text{Dividend} = (\text{partial quotient})(\text{divisor}) + \text{remainder}$$

If the remainder is zero, the dividend is equal to the product of the complete quotient and the divisor.

EXERCISE 3-3

Perform the indicated divisions.

1. $\dfrac{6x^4}{2x^2}$ 2. $\dfrac{-25x^3y^2}{-5xy}$ 3. $\dfrac{12ab}{-4}$

4. $\dfrac{a^3b^4}{a^2b}$ 5. $\dfrac{14a^4b^3}{-7a^3b}$ 6. $\dfrac{-6a^2b^3c^2}{-3abc}$

7. $\dfrac{21x^2 - 14x^3}{7x^2}$ 8. $\dfrac{9a^2b^3 - 15ab}{3ab}$ 9. $\dfrac{5x^2y^3 - 10x^4y^5}{5x^2y^3}$

10. $\dfrac{mn^2 - m^2n - mn}{mn}$ 11. $\dfrac{5m^2 - 10m - 5m^3}{5m}$

12. $\dfrac{8m^3 - 4m^2 + 4m}{-4m}$ 13. $\dfrac{x^2 + 7x + 6}{x + 1}$

14. $\dfrac{x^2 - 8x + 7}{x - 1}$ 15. $\dfrac{x^2 + 2x - 3}{x + 3}$

16. $\dfrac{12x^2 - x - 19}{3x - 4}$ 17. $\dfrac{8x^2 + 10x - 5}{4x - 1}$

18. $\dfrac{10x^2 - 21x - 9}{2x - 5}$ 19. $\dfrac{a^3 - b^3}{a - b}$

20. $\dfrac{a^3 + b^3}{a + b}$ 21. $\dfrac{x^3 - 2x^2 + 96}{x + 4}$

22. $\dfrac{2x^3 - 9x - 30}{x - 3}$ 23. $\dfrac{2x^3 - 20x - 5}{2x + 6}$

24. $\dfrac{2x^3 - 6x^2 + 7}{2x - 4}$ 25. $(4x^3 - 4x^2 - x - 3) \div (2x - 3)$

26. $(12x^3 + 5x^2 + 13x + 8) \div (3x + 2)$

27. $(3x^4 - 5x^3 + 12x^2 - 8x + 8) \div (x^2 - 2x + 4)$

28. $(4x^4 + x^3 - 4x^2 + 6x - 3) \div (x^2 + x - 1)$

29. $(2x^4 + x^3y - 4x^2y^2 + 11xy^3 - 4y^4) \div (2x^2 - 3xy + 4y^2)$

30. $(4x^4 - 2x^3 - 8x^2 + 3x + 1) \div (2x - 3)$

31. $(10x^4 - 23x^3 - 18x^2 + 29x - 7) \div (5x - 4)$

32. $(x^5 + x^4 + x^3 + 3x^2 + 10x + 6) \div (x^2 + x + 2)$

33. $(2x^5 - 5x^4 + 6x^3 + 4x^2 - 11x + 4) \div (2x^2 - 3x + 1)$

3-5 SPECIAL PRODUCTS

In many problems certain kinds of algebraic expressions occur over and over as factors to be multiplied. Consequently, it is worthwhile to learn to write the products speedily. We shall list some formulas for multiplying which will greatly condense the procedure used in Sec. 3-3. Each formula may be verified by carrying out the multiplication indicated on the left side. The student should memorize the formulas and apply them until proficiency is acquired in obtaining products of factors of these types, and, conversely, in supplying the factors from the products. It is important to understand that the letters in the formulas may stand for any quantity.

Product of the sum and difference of two numbers:

$$(x + y)(x - y) = x^2 - y^2 \tag{1}$$

Square of a binomial:

$$(x + y)^2 = x^2 + 2xy + y^2 \tag{2}$$

$$(x - y)^2 = x^2 - 2xy + y^2 \tag{3}$$

Product of two binomials with like terms:

$$(ax + by)(cx + dy) = acx^2 + (ad + bc)xy + bdy^2 \tag{4}$$

The cube of a binomial:

$$(x + y)^3 = x^3 + 3x^2y + 3xy^2 + y^3 \tag{5}$$

$$(x - y)^3 = x^3 - 3x^2y + 3xy^2 - y^3 \tag{6}$$

With the exception of formula (4), the student might find it helpful to express the formulas in words. We suggest the following for formulas (1) and (3), respectively.

The product of the sum and difference of two numbers is equal to the square of the first minus the square of the second.

The square of the difference of two numbers is equal to the square of the first minus twice the product of the numbers plus the square of the second.

EXAMPLE 1 Perform the multiplication $(4a + 3b)(4a - 3b)$.

Solution: We apply formula (1) with $x = 4a$ and $y = 3b$. Thus,

$$(4a + 3b)(4a - 3b) = (4a)^2 - (3b)^2 = 16a^2 - 9b^2$$

EXAMPLE 2 Find the product $(x - 2y + 3z)(x - 2y - 3z)$.

Solution: Grouping terms and applying formulas (1) and (3), we have

$$\begin{aligned}[(x - 2y) + 3z][(x - 2y) - 3z] &= (x - 2y)^2 - (3z)^2 \\ &= x^2 - 4xy + 4y^2 - 9z^2\end{aligned}$$

EXAMPLE 3 Write by inspection the product of $3x - 4y$ and $2x + 5y$.

Solution: We write the binomials and describe a procedure. The product of the first terms of the binomials is $(3x)(2x) = 6x^2$, and of the second terms $(-4y)(5y) = -20y^2$.

$$\underbrace{(3x - \overbrace{4y)(2x}^{-8xy} + 5y)}_{15xy} \qquad -8xy + 15xy = 7xy$$

As indicated in the diagram, the product of the two inner terms is $(-4y)(2x) = -8xy$ and the product of the two outer terms is $(3x)(5y) = 15xy$, which gives $7xy$ as the sum of these like terms. Hence, we have

$$(3x - 4y)(2x + 5y) = 6x^2 + 7xy - 20y^2$$

The middle term of this product is called the *cross product.* Being the sum of two terms, it is the least simple to find. However, the idea is to determine mentally each term of a product of this kind and to write the result at once.

EXAMPLE 4 Find the cube of $5m - 2n$.

Solution: We may either use formula (6) with $x = 5m$ and $y = 2n$ or, alternatively, use formula (5) with $x = 5m$ and $y = -2n$. Using the latter, we have

$$\begin{aligned}(5m - 2n)^3 &= (5m)^3 + 3(5m)^2(-2n) + 3(5m)(-2n)^2 + (-2n)^3 \\ &= 125m^3 - 150m^2n + 60mn^2 - 8n^3\end{aligned}$$

EXAMPLE 5 Find the square of $2x - y + 3z$.

Solution: Treating $2x - y$ as one term and using formulas (2) and (3), we have

$$\begin{aligned}[(2x - y) + 3z]^2 &= (2x - y)^2 + 2(2x - y)(3z) + (3z)^2 \\ &= 4x^2 - 4xy + y^2 + 12xz - 6yz + 9z^2\end{aligned}$$

EXERCISE 3-4

Write the products by inspection.

1. $(x + 3)(x - 3)$
2. $(4 + x)(4 - x)$
3. $(2x + 3y)(2x - 3y)$
4. $(x^2 + 2y)(x^2 - 2y)$
5. $(5 + 3a^2)(5 - 3a^2)$
6. $(a + bc)(a - bc)$
7. $(a - 4)^2$
8. $(a + 2b)^2$
9. $(3x - 2y)^2$
10. $(2x^2 + y^2)^2$
11. $(x^3 - 2y^3)^2$
12. $(xy^2 + 3)^2$
13. $[(x - 1) + y]^2$
14. $(2 - x - y)^2$
15. $(3x - 4y + 2)^2$
16. $(x^2 + x - 1)^2$
17. $(x + 5)(x + 2)$
18. $(x - 7)(x + 4)$
19. $(x - 5)(2x + 1)$
20. $(3x - 1)(2x + 3)$
21. $(4a + 1)(2a - 3)$
22. $(7a + 2b)(3a - 4b)$
23. $(5a + 2b)(3a - 4b)$
24. $(3a^2 + 2)(4a^2 - 2)$
25. $(2r^2 + 3s^2)(3r^2 - 2s^2)$
26. $(x + 1)^3$
27. $(x - 3)^3$
28. $(2x - y)^3$
29. $(5 - x)^3$
30. $(3x - 2y)^3$
31. $(x^2 + y^2)^3$
32. $(mn - 3y)^3$
33. $(3m - 4xy)^3$
34. $(x - y + z)^2$
35. $(x + 2y - z)^2$
36. $(a - b - 2c)^2$
37. $(a^2 + b^2 + 1)^2$
38. $(x^2 - x + 1)^2$
39. $(2x + 3y + 2z)^2$
40. $[(x - y) - 2]^3$
41. $[(2x - y) - 1]^3$
42. $(x - 2y + 3)^3$

Treat the factors in each problem as the sum and difference of two quantities and find the product.

43. $(a - b + 1)(a - b - 1)$
44. $(x + y + 3)(x + y - 3)$
45. $(1 - 2m + n)(1 - 2m - n)$

46. $(x - y + 1)(x + y - 1)$

47. $(x + 5y - 1)(x - 5y + 1)$

48. $(x^2 + 1 - x)(x^2 - 1 + x)$

49. $(x + y + z + 1)(x + y - z - 1)$

50. $(2x + y - 2z + 3)(2x + y + 2z - 3)$

4

FACTORING AND OPERATIONS ON FRACTIONS

4-1 SIMPLE TYPES OF FACTORING

Much of Chap. 3 dealt with finding the product of factors. In this chapter we shall consider the inverse problem of finding the factors of products. This process, called *factoring*, has important applications. Factoring at this stage is usually restricted to expressing polynomials with rational coefficients as the product of factors where each factor is also a polynomial with rational coefficients. A polynomial is said to be completely factored when none of its factors can be factored. Each factor is then expressible only as 1 times itself or minus 1 times its negative and consequently is called a *prime factor*.

In this section, we shall take up polynomials whose factors are quite easily determined.

Type 1: Common factors. If each term of a polynomial has a common factor, the distributive law can be applied to express the polynomial as the product of two factors, one of which is the common factor.

EXAMPLE 1

$$4x^4y^3 - 6x^3y^3 + 2x^3y^2 = 2x^3y^2(2xy - 3y + 1)$$

EXAMPLE 2

$$a(x + y) + b(x + y) = (x + y)(a + b)$$

Type 2: The difference of two squares. By interchanging the members of the special product formula $(x + y)(x - y) = x^2 - y^2$, we obtain the factoring formula

$$x^2 - y^2 = (x + y)(x - y)$$

EXAMPLE 1

$$9x^2 - 4y^2 = (3x)^2 - (2y)^2 = (3x + 2y)(3x - 2y)$$

EXAMPLE 2

$$16ax^2 - 4ay^2 = 4a(4x^2 - y^2) = 4a(2x + y)(2x - y)$$

Note: As here illustrated, the removal of a common factor, if any, should be the first step in factoring.

EXAMPLE 3

$$(2a + b)^2 - (c - d)^2 = (2a + b + c - d)(2a + b - c + d)$$

Type 3: Perfect-square trinomial. The formulas for the square of a binomial (Sec. 3-5) yield the factoring formulas

$$x^2 + 2xy + y^2 = (x + y)^2$$
$$x^2 - 2xy + y^2 = (x - y)^2$$

EXAMPLE 1

$$\begin{aligned} 9x^2 + 30x + 25 &= (3x)^2 + 2(3x)5 + (5)^2 \\ &= (3x + 5)^2 \end{aligned}$$

EXAMPLE 2

$$4x^2 - 12xy + 9y^2 = (2x)^2 - 2(2x)(3y) + (3y)^2$$
$$= (2x - 3y)^2$$

Type 4: The sum and difference of two cubes. By multiplying, we may verify the formulas

$$x^3 + y^3 = (x + y)(x^2 - xy + y^2)$$
$$x^3 - y^3 = (x - y)(x^2 + xy + y^2)$$

EXAMPLE 1

$$64x^3 + 27y^3 = (4x)^3 + (3y)^3$$
$$= (4x + 3y)(16x^2 - 12xy + 9y^2)$$

EXAMPLE 2

$$54x^4y - 2xy^4 = 2xy(27x^3 - y^3)$$
$$= 2xy(3x - y)(9x^2 + 3xy + y^2)$$

EXERCISE 4-1

Factor each expression completely.

1. $x^3 + x^2y - 3xy^3$

2. $6x^3y - 2xy^3 + 10xy$

3. $3x^2 + 3xy - 6xy^2$

4. $4(x + 2) + a(x + 2)$

5. $m(2a + b) - n(2a + b)$

6. $3x(1 - y) + 2a(1 - y)$

7. $x^2 - 9y^2$

8. $x^2 - 16$

9. $1 - 25y^2$

10. $9a^2 - 4b^2c^2$

11. $2ax^2 - 32ay^2$

12. $4m^3 - 16m$

13. $(a + 1)^2 - b^2$

14. $(2a - b)^2 - 9c^2$

15. $9x^2y^2 - (z - 1)^2$

16. $x^2 + 8x + 16$

17. $x^2 - 4x + 4$

18. $4x^2 - 12x + 9$

19. $9a^2 + 6ab + b^2$

20. $25a^2x^2 - 10ax + 1$

21. $3x^2 - 18xy + 27y^2$

22. $2ax^2 + 24ax + 72a$

23. $x^3 + 8y^3$

24. $64a^3 - b^3$

25. $8b^3 - 1$

26. $8p^3q^3 + 27$

27. $5xy^4 - 5x^4y$

28. $a^4 - 27ab^3$

29. $8(x - 1)^3 + y^3$

30. $(x - 2y)^3 - 8y^3$

31. $54x^6y - 2x^3y^4$

32. $x^4 - 256$

33. $1 - 81y^4$

34. $625a^4 - b^4$

35. $(x^3)^2 - (y^3)^2$

36. $x^6 - 64y^6$

37. $64x^6 + y^6$

38. $3a^6b^2 - 3a^2b^2$

39. $162a^5b^2 - 2ab^2$

40. $y^6 - 729x^6$

41. $x^4 - 8x^2 + 16$

42. $x^4 - 18x^2 + 81$

43. $(2x - 3y)^2 + 10(2x - 3y) + 25$

44. $(a + 2b)^2 - 6(a + 2b) + 9$

45. $(16x^2 - 24xy + 9y^2) - 1$

46. $(x^4)^2 - (y^4)^2$

47. $x^8 - 1$

48. $(x^4)^2 - (y^2)^2$

49. $x^8 - 16$

50. $x^{12} - y^{12}$

4-2 GENERAL SECOND-DEGREE TRINOMIAL

Some of the trinomials of the preceding section were expressible as the squares of binomials. We shall now consider trinomials which are equal to the product of two distinct binomials. To obtain a formula for this purpose, we reverse the members of formula (4) in Sec. 3-5, and get

$$acx^2 + (ad + bc)xy + bdy^2 = (ax + by)(cx + dy)$$

To factor a trinomial of this kind, we must find rational values for a, b, c, and d from the known values ac, $ad + bc$, and bd. If such values exist, we say the trinomial is factorable. Although factoring for this case is less simple than factoring perfect-square trinomials, the following examples show a procedure.

EXAMPLE 1 Factor the trinomial $x^2 - 2x - 15$.

Solution: If x is the first term of each factor, the product will contain x^2. The product of the second terms must be -15. This value is the product of each of the pairs

$$1 \text{ and } -15, \qquad -1 \text{ and } 15, \qquad 3 \text{ and } -5, \qquad -3 \text{ and } 5$$

Observing the middle term of the given trinomial, we see that the numbers whose product is -15 must have the sum -2. Clearly, only the pair 3 and -5 satisfy these conditions. Hence,

$$x^2 - 2x - 15 = (x + 3)(x - 5)$$

EXAMPLE 2 Factor the expression $10x^2 + 11xy - 6y^2$.

Solution: Using positive coefficients, we see that the pair x and $10x$ and $2x$ and $5x$ have $10x^2$ as the product. The product of the second terms must be $-6y^2$. This product is obtainable from each of the pairs

$$y \text{ and } -6y, \qquad -y \text{ and } 6y, \qquad 2y \text{ and } -3y, \qquad -2y \text{ and } 3y$$

The sum of the cross products of the factors (the first term of each factor times the second term of the other) must yield $11xy$. By trying each of the pairs for the first terms with each of the pairs for the second terms, we may determine that $2x + 3y$ and $5x - 2y$ yield the correct cross product. Hence,

$$10x^2 + 11xy - 6y^2 = (2x + 3y)(5x - 2y)$$

We chose positive coefficients for the first terms of the factors. We could, of course, use $-2x$ and $-5x$ for the first terms and come out with $-2x - 3y$ and $-5x + 2y$ as the factors.

EXERCISE 4-2

Factor each expression completely.

1. $x^2 + 9x + 18$	**2.** $x^2 - 6x + 8$
3. $x^2 + 9x + 20$	**4.** $x^2 + x - 12$
5. $x^2 - 2x - 15$	**6.** $x^2 - 5x - 14$
7. $a^2 + 4ab - 21b^2$	**8.** $a^2 + 5ab - 36b^2$
9. $a^2 + 11ab + 18b^2$	**10.** $2a^2 - a - 3$
11. $3a^2 - 4a - 4$	**12.** $4a^2 - 23a - 6$
13. $10x^2 + 13x - 3$	**14.** $4x^2 - 4x - 15$

15. $24 + 10x - 21x^2$

16. $21 + 25x - 4x^2$

17. $3x^2y^2 + 11xy + 8$

18. $6x^2 + 11xy - 7y^2$

19. $12x^3 + 11x^2 + 2x$

20. $x^4 - 13x^3 + 42x^2$

21. $4x^2y^2 + 5xy^3 - 6y^4$

22. $m^4 - 3m^2 - 4$

23. $m^4 + 8m^2 + 12$

24. $m^4 + 7m^2 + 12$

25. $6m^4 - 5m^2 - 6$

26. $2m^4 - m^2 - 1$

27. $4m^4 + 17m^2 + 4$

28. $3m^4 - 10m^2 - 8$

29. $4m^4 - 37m^2 + 9$

30. $9m^4 + 13m^2 + 4$

31. $x^6 + 7x^3 - 8$

32. $8x^6 + 7x^3 - 1$

33. $x^6 + 28x^3 + 27$

34. $(x - 4y)^2 + 3(x - 4y) - 10$

35. $(x + y)^2 - (x + y) - 2$

36. $12x^3y - 27x^2y^2 + 6xy^3$

37. $4x^4y - 2x^3y^2 - 20x^2y^3$

4-3 FACTORING BY GROUPING

Some polynomials whose terms have no common factor can be separated into groups of terms which have a common factor. And some polynomials not in the forms of the difference of two squares can be so expressed by a proper grouping of terms.

EXAMPLE 1

$$\begin{aligned} ax + ay + 5x + 5y &= a(x + y) + 5(x + y) \\ &= (a + 5)(x + y) \end{aligned}$$

EXAMPLE 2

$$\begin{aligned} x^3 + y^3 - 4x - 4y &= (x + y)(x^2 - xy + y^2) - 4(x + y) \\ &= (x + y)(x^2 - xy + y^2 - 4) \end{aligned}$$

EXAMPLE 3

$$\begin{aligned} 9 - 4x^2 + 12xy - 9y^2 &= 9 - (4x^2 - 12xy + 9y^2) \\ &= 9 - (2x - 3y)^2 \\ &= (3 + 2x - 3y)(3 - 2x + 3y) \end{aligned}$$

EXAMPLE 4 Factor the expression $x^4 - 6x^2y^2 + 25y^4$.

Solution: The given expression would be a perfect square if the middle term were $10x^2y^2$. By adding $16x^2y^2$, we obtain the perfect square. If we add $16x^2y^2$, we must subtract $16x^2y^2$. Thus, we have

$$\begin{aligned} x^4 - 6x^2y^2 + 25y^4 &= x^4 + 10x^2y^2 + 25y^4 - 16x^2y^2 \\ &= (x^2 + 5y^2)^2 - (4xy)^2 \\ &= (x^2 + 5y^2 + 4xy)(x^2 + 5y^2 - 4xy) \end{aligned}$$

The device of adding and subtracting a term works only if the expression to be factored becomes a perfect square by the addition, not the subtraction, of a term, and only if the term added, and also subtracted, is a perfect square.

EXERCISE 4-3

Factor each expression completely.

1. $4x - 12 + xy - 3y$
2. $ax + 4x + ay + 4y$
3. $ax + 2cx - ay - 2cy$
4. $12xy + 3cx + 9cy + 4x^2$
5. $3xy - ay + 9ax - 3a^2$
6. $x^3 - 3x - 2x^2 + 6$
7. $4 + 2m + 4m^2 + 2m^3$
8. $4m^3 - 3m^2 - 4m + 3$
9. $x^2 - 2x + 1 - ax + a$
10. $x^2 + 4xy + 4y^2 + 4x + 8y$
11. $x^2 - y^2 - 5x - 5y$
12. $x^2 - 4y^2 - 6x + 12y$
13. $x^2 - 2xy + y^2 - 1$
14. $x^2 - y^2 - 10x + 25$
15. $a^2 - 4b^2 + 12b - 9$
16. $25 - a^2 + 2ab - b^2$
17. $x^2 + 2x + 1 - y^2 + 2yz - z^2$
18. $x^2 - 4x + 4 - 4y^2 + 12yz - 9z^2$
19. $x^2 + 2xy + y^2 - a^2 - 2ab - b^2$
20. $x^2 + 2x - y^2 - 2y$
21. $x^2 + 8x - y^2 - 4y + 12$
22. $x^3 - y^3 + x - y$
23. $a^3 - b^3 - a^2 + 2ab - b^2$
24. $x^4 + 4$
25. $64a^4 + 1$
26. $x^4 - 3x^2 + 1$

27. $a^4 + a^2b^2 + b^4$

28. $x^4 + 11x^2y^2 + 36y^4$

29. $a^4 + 7a^2b^2 + 16b^4$

30. $4a^4 + 7a^2b^2 + 16b^4$

31. $9x^4 - 37x^2 + 4$

32. $25x^4 - 5x^2 + 4$

4-4 ALGEBRAIC FRACTIONS

In Chap. 2, we defined a fraction and proved some theorems which provide the basis for certain operations on fractions. The fraction, defined there, involved literal numbers. Fractions of this kind are sometimes called *algebraic fractions* to distinguish them from the fractions of arithmetic.

We may think of a fraction as having three signs associated with it: the sign preceding the fraction, the sign of the numerator, and the sign of the denominator. If $c = -1$, Theorem 2-12 (Corollary) becomes $-a/-b = a/b$. Hence, we may reverse the signs of the numerator and denominator of a fraction without changing the value of the fraction. And if we replace b by $1/b$ in Theorem 2-9, we find that the value of a fraction is not changed by reversing the sign of the numerator and the sign of the fraction. So we may write

$$\frac{a}{b} = \frac{-a}{-b} = -\frac{-a}{b} = -\frac{a}{-b} \qquad b \neq 0$$

As these equations show, any two of the three signs associated with a fraction may be reversed without altering the value of the fraction.

If the numerator and denominator of a fraction are in factored form, the signs of the terms of any two of the factors, or the signs of the terms of one of the factors and the sign of the fraction, may be changed without altering the value of the fraction. We leave the proof of this statement to the student.

In dealing with algebraic fractions, we shall agree that the literal number or numbers in a denominator may not have values which make the denominator equal to zero.

4-5 REDUCTION TO LOWEST TERMS

A fraction is said to be in *lowest terms* or in *simplest form* if the numerator and denominator have no common factor except ± 1. Hence, we may determine if a fraction is in lowest terms by expressing the numerator and denominator as products of their prime factors. Any factor appearing in both the numerator and denominator can then readily be removed by division.

EXAMPLE 1 Reduce the fraction $\dfrac{x^2 - x - 6}{x^2 - 7x + 12}$ to lowest terms.

Solution:

$$\frac{x^2 - x - 6}{x^2 - 7x + 12} = \frac{(x - 3)(x + 2)}{(x - 3)(x - 4)} = \frac{x + 2}{x - 4}$$

The common factor is removed by dividing the numerator and denominator by $x - 3$.

EXAMPLE 2 Reduce the following fraction to lowest terms:

$$\frac{x^2 - 9x + 20}{16 - x^2}$$

Solution:

$$\begin{aligned}\frac{x^2 - 9x + 20}{16 - x^2} &= \frac{(x - 4)(x - 5)}{(4 - x)(4 + x)} \\ &= \frac{(x - 4)(5 - x)}{(x - 4)(4 + x)} \\ &= \frac{5 - x}{4 + x}\end{aligned}$$

Observe that we obtained a common factor in the numerator and denominator by reversing the signs of the terms in a factor of the numerator and in a factor of the denominator.

The process of eliminating a common factor from the numerator and denominator of a fraction is called *multiplicative canceling.* This operation is sometimes used erroneously by striking out equal quantities which are not factors. For example, the fraction

$$\frac{(2x + y) + 6}{(2x + y)(3x + 4y)}$$

has $2x + y$ as a factor of the denominator but $2x + y$ is a term of the numerator, not a factor. Striking out the equal quantities here would mean that the denominator is divided by $2x + y$ and that $2x + y$ is subtracted from the numerator. This is a serious error and should be avoided. As given, the fraction is in lowest terms.

EXERCISE 4-4

Reduce each fraction to lowest terms.

1. $\dfrac{20x^2y}{60xy^2}$
2. $\dfrac{18a^2bc^2}{22ab^2c}$
3. $\dfrac{27a^4b^6}{9a^6b}$
4. $\dfrac{3x}{x^2 - 2x}$
5. $\dfrac{ax - 4x}{a^2 - 4a}$
6. $\dfrac{2x + 3}{4x^2 - 9}$
7. $\dfrac{(x + 2)^2}{x^2 - 4}$
8. $\dfrac{a^2 + ab}{ab + b^2}$
9. $\dfrac{x - 3}{3x - x^2}$
10. $\dfrac{6y - 6x}{3x - 3y}$
11. $\dfrac{(a - b)^2}{b^2 - a^2}$
12. $\dfrac{3x^2 - 15xy}{2x^2 - 10xy}$
13. $\dfrac{3a^2bc - 9a^2b}{6abc - 18ab}$
14. $\dfrac{x^2y - xy^3}{5x^3y - 5x^2y^3}$
15. $\dfrac{2a^2 + ab - b^2}{a^2 - b^2}$
16. $\dfrac{3x^2 + x - 2}{3x^2 - 11x + 6}$
17. $\dfrac{1 - 16x^2}{4x^2 + 7x - 2}$
18. $\dfrac{3 + 5x + 2x^2}{3 + 2x - x^2}$
19. $\dfrac{9 - 9x + 2x^2}{9 - 6x + x^2}$
20. $\dfrac{a^3 + b^3}{a^2 - ab + b^2}$
21. $\dfrac{a^3 - b^3}{a^2 - 2ab + b^2}$
22. $\dfrac{x^3 + x^2y + xy^2}{x^3 - y^3}$
23. $\dfrac{x^2 - y^2}{x^2 - y^2 + x + y}$
24. $\dfrac{4a^2 - b^2 + 2a - b}{4a^2 - b^2}$
25. $\dfrac{y^3 + y^2 + y + 1}{y^3 - y^2 + y - 1}$
26. $\dfrac{a^3 + 2a^2 - a - 2}{a^3 - 2a^2 - a + 2}$
27. $\dfrac{c^3 - c^2 + 2c - 2}{c^3 + 3c^2 + 2c + 6}$

4-6 MULTIPLICATION AND DIVISION OF FRACTIONS

Theorems 2-12 and 2-15 furnish procedures for multiplying and dividing fractions, and are expressed by the equations

$$\frac{a}{b}\cdot\frac{c}{d} = \frac{ac}{bd} \qquad b, d \neq 0$$

$$\frac{a}{b} \div \frac{c}{d} = \frac{a}{b}\cdot\frac{d}{c} \qquad b, c, d \neq 0$$

These theorems permit us to make the following statements:

1. The product of two or more fractions is equal to the product of the numerators divided by the product of the denominators.
2. The quotient of two fractions is equal to the dividend multiplied by the inverted divisor.

It is usually desirable to reduce the product, or the quotient, of fractions to lowest terms. That being the case, the best procedure is to write each fraction in factored form. Common factors of the numerator and denominator can then easily be removed by division.

EXAMPLE 1 Multiply

$$\frac{2x^2 + x - 3}{x^2 - 1} \quad \text{by} \quad \frac{x^2 + 2x + 1}{3x^2 + 5x + 2}$$

Solution: Factoring the numerators and the denominators, we have

$$\frac{(2x + 3)(x - 1)}{(x + 1)(x - 1)} \cdot \frac{(x + 1)(x + 1)}{(3x + 2)(x + 1)} = \frac{2x + 3}{3x + 2}$$

EXAMPLE 2 Find the product of

$$(x^2 - y^2) \cdot \frac{3x^2 + xy - 4y^2}{2x^2 + 3xy + y^2} \cdot \frac{2x + y}{(x - y)^2}$$

Solution: We write $x^2 - y^2$ in fractional form and factor all numerators and denominators. Thus, we obtain

$$\frac{(x + y)(x - y)}{1} \cdot \frac{(3x + 4y)(x - y)}{(2x + y)(x + y)} \cdot \frac{2x + y}{(x - y)^2}$$
$$= \frac{(x + y)(x - y)^2(2x + y)(3x + 4y)}{(x + y)(x - y)^2(2x + y)} = 3x + 4y$$

EXAMPLE 3 Divide

$$\frac{a^2 - 9}{a^3 - 1} \quad \text{by} \quad \frac{a^2 - 2a - 3}{a^2 + a + 1}$$

Solution: To find the quotient, we invert the divisor and then multiply. Thus, we get

$$\frac{a^2 - 9}{a^3 - 1} \cdot \frac{a^2 + a + 1}{a^2 - 2a - 3} = \frac{(a + 3)(a - 3)}{(a - 1)(a^2 + a + 1)} \cdot \frac{a^2 + a + 1}{(a - 3)(a + 1)}$$

$$= \frac{a + 3}{(a + 1)(a - 1)}$$

EXAMPLE 4 Perform the indicated operations:

$$\frac{6x^2 - xy - y^2}{3x^2 + 7xy + 2y^2} \cdot \frac{x^2 + 5xy + 6y^2}{2x^2 + xy - y^2} \div \frac{x^2 - 9y^2}{x^2 - y^2}$$

Solution: Inverting the divisor and factoring the numerators and denominators, we obtain

$$\frac{(2x - y)(3x + y)}{(3x + y)(x + 2y)} \cdot \frac{(x + 2y)(x + 3y)}{(2x - y)(x + y)} \cdot \frac{(x + y)(x - y)}{(x + 3y)(x - 3y)}$$

$$= \frac{(2x - y)(3x + y)(x + 2y)(x + 3y)(x + y)(x - y)}{(2x - y)(3x + y)(x + 2y)(x + 3y)(x + y)(x - 3y)}$$

$$= \frac{x - y}{x - 3y}$$

EXERCISE 4-5

Perform the indicated operations and express each result in lowest terms.

1. $\dfrac{30y}{7x} \cdot \dfrac{14x}{15y}$ **2.** $\dfrac{3a^2b}{4x^2} \cdot \dfrac{2x}{9a^3b}$ **3.** $4ab \cdot \dfrac{3}{8ab^2}$

4. $\dfrac{48x^2}{13y^2} \cdot \dfrac{39y}{64x^2}$ **5.** $\dfrac{34x}{9ab^2} \cdot \dfrac{27ab}{17x^2}$ **6.** $\dfrac{a^3x^4}{4by} \cdot \dfrac{8b^2y}{5a^4x^3}$

7. $\dfrac{a^2}{14c} \div \dfrac{9a}{7bc}$ **8.** $\dfrac{6z^3}{5x^2y} \div \dfrac{3z^3}{10xy^3}$ **9.** $\dfrac{8x^3y^3}{3} \div 64x^3$

10. $\dfrac{x + 1}{y + 1} \cdot \dfrac{y^2 - 1}{x^2 + x}$ **11.** $\dfrac{2b + 2a}{2y - x} \cdot \dfrac{x - 2y}{a + b}$

12. $\dfrac{a^2 - 1}{1 - x^2} \cdot \dfrac{1 - x^3}{1 - a}$ **13.** $\dfrac{2a - 2b}{y - x} \div \dfrac{a - b}{x - y}$

14. $(x-6) \div \frac{2x-12}{3x}$

15. $\frac{2x+4y}{x+y} \div \frac{x+2y}{x^2-y^2}$

16. $\frac{a+b}{x+y} \cdot \frac{x^3+y^3}{a^3+b^3}$

17. $\frac{2+x}{2-x} \cdot \frac{8-x^3}{4-x^2}$

18. $\frac{a-b}{2x-y} \cdot \frac{4x^2-y^2}{a^3-b^3}$

19. $\frac{x^2+7x-30}{x^2-x-12} \cdot \frac{24-2x-x^2}{30-7x-x^2}$

20. $\frac{x^2-2xy-3y^2}{x^2-6xy+9y^2} \cdot \frac{x^2-9y^2}{7x+21y}$

21. $\frac{3x^2-7x+2}{x^2-49} \cdot \frac{2ax-14a}{3bx-b}$

22. $\frac{a^3+8}{2a-1} \cdot \frac{2a^2-5a+2}{a^2-4}$

23. $\frac{x^2-x-2}{x^2-x-6} \div \frac{x^2+8x+7}{x^2-6x+9}$

24. $\frac{9y^2-16x^2}{3x^2+3xy} \div \frac{3y^2-xy-4x^2}{2xy+2y^2}$

25. $\frac{a^3-27}{a^2+3a+2} \div \frac{a^2+3a+9}{a^3+2a^2-a-2}$

26. $\frac{2x-1}{x^2+x+1} \div \frac{4x^2-1}{x^4+x^2+1}$

27. $\frac{1-x}{1+x} \div \left(\frac{x^2}{1-x^2} \div \frac{2x}{1-x}\right)$

28. $\left(\frac{1-x}{1+x} \div \frac{x^2}{1-x^2}\right) \div \frac{2x}{1-x}$

29. $\frac{4x^2-8x+3}{x^2+x-12} \cdot \frac{x^2+7x+12}{2x^2+3x-9} \cdot \frac{2x^2-5x-3}{4x^2-1}$

30. $\frac{x^2-6x+9}{6x^2+35x+36} \cdot \frac{x^2+37x+36}{x+1} \cdot \frac{6x^2-19x+3}{(x-3)^2}$

31. $\frac{3a^2+20a+25}{10a^2+a-2} \cdot \frac{8a^2-2a-3}{3a^2-a-10} \div \frac{12a^2+11a-15}{5a^2+8a-4}$

32. $\frac{x^2-xy-12y^2}{x^2-6xy-7y^2} \cdot \frac{x^2+7xy+6y^2}{x^2+4xy-12y^2} \div \frac{x^2-xy-12y^2}{x^2-9xy+14y^2}$

33. $\frac{x^2-4y^2}{3x^2+13xy-10y^2} \cdot \frac{4x^2+2xy-2y^2}{3x^2+4xy+y^2} \div \frac{2x^2-5xy+2y^2}{9x^2-12xy+4y^2}$

4-7 ADDITION OF FRACTIONS

The sum of two or more fractions with the same denominator is equal to the fraction whose numerator is the algebraic sum of the numerators and whose denominator is the common denominator (Theorem 2-13). The result obtained in this manner is usually called *addition*, even though a minus sign precedes some of the fractions. To add fractions that have different denominators, it is necessary to first express the fractions as equivalent fractions with a common denominator (Theorem 2-14). In view of this circumstance, we shall introduce the idea of a common multiple of polynomials.

A polynomial which is exactly divisible by two or more polynomials is called a *common multiple* of the latter polynomials. The common multiple determined in the following way is called the *lowest common multiple* (LCM):

1. Factor each of the given polynomials.
2. Form a product by using all the different prime factors of the polynomials, giving each prime factor the largest exponent which that factor has in any of the polynomials.

EXAMPLE 1 Find the LCM of the polynomials

$$6x^2 - 3x - 3, \qquad 7x^2 + 14x + 7, \qquad 12x^2 - 12$$

Solution: Factoring each polynomial, we have

$$\begin{aligned} 6x^2 - 3x - 3 &= 3(2x^2 - x - 1) = 3(2x + 1)(x - 1) \\ 7x^2 + 14x + 7 &= 7(x^2 + 2x + 1) = 7(x + 1)^2 \\ 12x^2 - 12 &= 12(x^2 - 1) = 2^2 \cdot 3(x + 1)(x - 1) \end{aligned}$$

Selecting the proper power of each prime factor, we get

$$2^2 \cdot 3 \cdot 7(2x + 1)(x - 1)(x + 1)^2 = 84(2x + 1)(x - 1)(x + 1)^2$$

which is the lowest common multiple of the given polynomials.

The LCM may be expressed in factored form or the actual multiplication may be carried out. The factored form, however, is usually more convenient.

In adding fractions, we shall use the LCM of the various denominators for the common denominator. It is then called the *lowest common denominator* (LCD). Having found the LCD, we obtain the sum of the fractions by the following operations:

1. Divide the LCD by the denominator of each fraction. Then multiply the numerator and denominator of the fraction by the quotient thus obtained.
2. Add the resulting numerators, being careful to reverse the signs of all terms of the numerator of any fraction preceded by a minus sign. This gives the numerator of the sum of the fractions; the denominator is the common denominator.

EXAMPLE 2 Add the fractions

$$\frac{3}{(x+3)(x-4)}+\frac{2}{(2x-1)(4-x)}$$

Solution: The factor $4 - x$ in the second denominator is the negative of the factor $x - 4$ in the first denominator. To handle this situation, we reverse the signs of the terms of $4 - x$ and change the sign preceding the second fraction from plus to minus. Then $(x + 3)(x - 4)(2x - 1)$ is the LCD of the resulting fractions. Hence, we obtain

$$\begin{aligned}
&\frac{3}{(x+3)(x-4)}-\frac{2}{(2x-1)(x-4)}\\
&\quad=\frac{3(2x-1)}{(x+3)(x-4)(2x-1)}-\frac{2(x+3)}{(x+3)(x-4)(2x-1)}\\
&\quad=\frac{6x-3-2x-6}{(x+3)(x-4)(2x-1)}\\
&\quad=\frac{4x-9}{(x+3)(x-4)(2x-1)}
\end{aligned}$$

EXAMPLE 3 Add the fractions

$$\frac{3x-1}{(x-1)(x+1)}-\frac{x+3}{(x+1)(x+2)}-\frac{1}{x+2}$$

Solution: We see at once that $(x - 1)(x + 1)(x + 2)$ is the LCD, and so we multiply the numerators, respectively, by $x + 2$, $x - 1$, and $(x - 1)(x + 1)$ and place the products, preceded by the proper signs, over the common denominator. Then we proceed with the simplification process to obtain the sum of the given fractions in lowest terms. Using these steps, we get

$$\frac{3x-1}{(x-1)(x+1)} - \frac{x+3}{(x+1)(x+2)} - \frac{1}{x+2}$$

$$= \frac{(3x-1)(x+2) - (x+3)(x-1) - (x+1)(x-1)}{(x-1)(x+1)(x+2)}$$

$$= \frac{3x^2+5x-2-x^2-2x+3-x^2+1}{(x-1)(x+1)(x+2)}$$

$$= \frac{x^2+3x+2}{(x-1)(x+1)(x+2)}$$

$$= \frac{(x+1)(x+2)}{(x-1)(x+1)(x+2)}$$

$$= \frac{1}{x-1}$$

EXERCISE 4-6

Find the LCM and leave each result in factored form except for the numerical coefficient.

1. $20x^2y, \quad 18xy^2, \quad 30xy^3$ **2.** $36xy^2, \quad 60x^3, \quad 84y^3z^2$

3. $4x^2-8x, \quad 3x^2+6x, \quad x^2-4$

4. $6x^3-2x^2, \quad 9x+3, \quad 9x^2-1$

5. $2a^2+5ab+2b^2, \quad 6a^2+ab-b^2, \quad 3a^2+5ab-2b^2$

6. $a^2-ab-2b^2, \quad 2a^2-7ab+6b^2, \quad 2a^2-ab-3b^2$

Combine into a single fraction in lowest terms.

7. $\dfrac{3a+4b}{3c}+\dfrac{2a-3b}{3c}$ **8.** $\dfrac{5a-2b}{a-6}-\dfrac{a+b}{a-6}$

9. $\dfrac{a-2}{3-a}+\dfrac{a+2}{a-3}$ **10.** $\dfrac{7x}{25}+\dfrac{x}{5}-\dfrac{x}{10}$

11. $\dfrac{1}{a}+\dfrac{1}{b}+\dfrac{1}{c}$ **12.** $\dfrac{1}{2a}+\dfrac{1}{a}+\dfrac{1}{3}$

13. $\dfrac{x}{x-5}-\dfrac{2}{x}$ **14.** $\dfrac{2a}{a-5}+\dfrac{5}{a-2}$

15. $\dfrac{3}{x+3} - \dfrac{4}{x-9}$

16. $\dfrac{4}{x+2y} + \dfrac{3}{x-y}$

17. $\dfrac{3}{4-a^2} + \dfrac{2}{a-2}$

18. $\dfrac{x+2}{3x-12} - \dfrac{2x-1}{4-x}$

19. $\dfrac{1}{2b+y} + \dfrac{2b-y}{1}$

20. $x + 2y - \dfrac{x^2}{x-3y}$

21. $2x - 4 + \dfrac{x+3}{x^2+1}$

22. $\dfrac{4a}{a+6} - \dfrac{2a}{a-3} + 2$

23. $\dfrac{a+2}{a-5} - \dfrac{4a}{a+3} + 3$

24. $\dfrac{3}{a-2} + \dfrac{2}{a-3} + \dfrac{1}{a+1}$

25. $\dfrac{3}{1-a} - \dfrac{2}{2-a} + \dfrac{2}{3-a}$

26. $\dfrac{a-5}{2-a} - \dfrac{a+4}{1-a} + \dfrac{a+3}{1+a}$

27. $\dfrac{m+5}{m-2} + \dfrac{m+3}{m+2} - \dfrac{m-4}{m-1}$

28. $\dfrac{3m}{m+1} + \dfrac{2m}{m-1} - \dfrac{4m^2}{m^2-1}$

29. $\dfrac{4m}{2m-1} + \dfrac{6m}{2m+1} - \dfrac{16m^2}{4m^2-1}$

30. $\dfrac{3m}{3m+1} + \dfrac{2m}{3m-1} - \dfrac{12m^2}{9m^2-1}$

31. $\dfrac{2m^2-7m}{9-4m^2} + \dfrac{m-1}{2m+3} - \dfrac{m+1}{2m-3}$

32. $\dfrac{2m}{n^2-m^2} + \dfrac{1}{m-n} - \dfrac{1}{m+n}$

33. $\dfrac{5}{x^2-x+1} - \dfrac{x^2-1}{x^3+1}$

34. $\dfrac{3m+n}{m^2-n^2} - \dfrac{1}{m+n} - \dfrac{2n}{m^2-mn}$

35. $\dfrac{2a+1}{a^2+a-2} + \dfrac{3a-2}{a^2-4} - \dfrac{2a-3}{a^2-3a+2}$

36. $\dfrac{2x^2-5y^2}{x^2+xy-2y^2} - \dfrac{x-2y}{x-y} + \dfrac{x-2y}{x+2y}$

37. $\dfrac{5x+10}{x^2+3x-4}+\dfrac{2}{x^2+7x+12}+\dfrac{x+4}{x^2+2x-3}$

38. $\dfrac{5x}{2x^2-x-3}+\dfrac{1}{2x^2+3x+1}+\dfrac{x+2}{4x^2-4x-3}$

39. $\dfrac{2}{6+a-2a^2}+\dfrac{1}{4a^2+4a-3}+\dfrac{3}{2a^2+a-1}$

40. $\dfrac{y-14}{3y^2-7y-6}+\dfrac{2y-4}{3y^2+8y+4}+\dfrac{y+5}{y^2-y-6}$

41. $\dfrac{2x+7}{2x^2+5x-7}-\dfrac{x-2}{2x^2+3x-14}-\dfrac{x-1}{x^2-3x+2}$

4-8 COMPLEX FRACTIONS

Thus far in this chapter, we have considered simple fractions—fractions whose numerators and denominators are nonfractional. We now introduce fractions which contain a fraction in the numerator or denominator, or both. A fraction of this kind is called a *complex fraction.*

To simplify a complex fraction means to express it as an equivalent fraction that has no fraction in either the numerator or denominator. The simplification may be accomplished by either of the following methods. Sometimes one of these methods is more convenient than the other:

1. Reduce the numerator and denominator separately to simple fractions and then divide.
2. Multiply the numerator and denominator of the main fraction by an expression such that the numerator and denominator of the new fraction will be nonfractional.

EXAMPLE 1 Simplify the complex fraction

$$\frac{\dfrac{1}{x+1}-\dfrac{1}{x+2}}{\dfrac{1}{x+2}-\dfrac{1}{x+3}}$$

Solution: We add the fractions in the numerator and add the fractions in the denominator, and then divide. Thus,

$$\frac{\dfrac{1}{x+1}-\dfrac{1}{x+2}}{\dfrac{1}{x+2}-\dfrac{1}{x+3}}=\frac{\dfrac{(x+2)-(x+1)}{(x+1)(x+2)}}{\dfrac{(x+3)-(x+2)}{(x+2)(x+3)}}$$

$$=\frac{\dfrac{1}{(x+1)(x+2)}}{\dfrac{1}{(x+2)(x+3)}}$$

$$=\frac{x+3}{x+1}$$

EXAMPLE 2 Simplify the complex fraction

$$\frac{\dfrac{8}{x-2}+4}{\dfrac{x+1}{x-2}-\dfrac{x-1}{x+2}}$$

Solution: Adding the fractions in the numerator and the denominator separately, we have

$$\frac{\dfrac{8}{x-2}+4}{\dfrac{x+1}{x-2}-\dfrac{x-1}{x+2}}=\frac{\dfrac{8+4x-8}{x-2}}{\dfrac{x^2+3x+2-(x^2-3x+2)}{(x-2)(x+2)}}$$

$$=\frac{4x}{x-2}\cdot\frac{(x-2)(x+2)}{6x}=\frac{2(x+2)}{3}$$

We can also simplify the given fraction by multiplying the numerator and denominator by the LCM of the three simple fractions which occur. Thus, we have

$$\frac{(x-2)(x+2)\left(\dfrac{8}{x-2}+4\right)}{(x-2)(x+2)\left(\dfrac{x+1}{x-2}-\dfrac{x-1}{x+2}\right)}$$

$$=\frac{8(x+2)+4(x-2)(x+2)}{(x+2)(x+1)-(x-2)(x-1)}$$

$$=\frac{8x+16+4(x^2-4)}{x^2+3x+2-(x^2-3x+2)}$$

$$=\frac{4x^2+8x}{6x}=\frac{2(x+2)}{3}$$

EXERCISE 4-7

Simplify the complex fractions.

1. $\dfrac{\frac{2}{3} - \frac{3}{5}}{\frac{1}{3} + \frac{7}{10}}$ 2. $\dfrac{1 - \frac{2}{21}}{\frac{4}{7} + \frac{1}{3}}$ 3. $\dfrac{2 - \frac{5}{9}}{3 - \frac{5}{6}}$ 4. $\dfrac{\dfrac{2}{x} - \dfrac{1}{2}}{\dfrac{2}{x} + \dfrac{1}{2}}$

5. $\dfrac{\dfrac{ac}{bd} + 1}{\dfrac{a}{b} - \dfrac{c}{d}}$ 6. $\dfrac{\dfrac{x}{3} - \dfrac{3}{4}}{\dfrac{x}{4} - \dfrac{4}{3}}$ 7. $\dfrac{1 - \dfrac{3}{x}}{1 + \dfrac{3}{x}}$ 8. $\dfrac{x + 1 - \dfrac{2}{x}}{x - 1 + \dfrac{2}{x}}$

9. $\dfrac{\dfrac{2}{x} - 2x}{\dfrac{2}{x} - x - 1}$ 10. $\dfrac{2 - \dfrac{x - 1}{x}}{4 - \dfrac{4x - 3}{x}}$ 11. $\dfrac{1 + \dfrac{1}{x} - \dfrac{2}{x^2}}{1 + \dfrac{3}{x} - \dfrac{4}{x^2}}$

12. $\dfrac{x - \dfrac{x^2 + 1}{2x}}{\dfrac{2}{x} - \dfrac{2}{x^2}}$ 13. $\dfrac{\dfrac{1}{x} - \dfrac{5}{x + 1}}{\dfrac{3}{x} - \dfrac{2}{x + 1}}$ 14. $\dfrac{x + \dfrac{x - 1}{x + 1}}{x - \dfrac{x - 1}{x + 1}}$

15. $\dfrac{\dfrac{1}{x + 3} + \dfrac{1}{x - 3}}{\dfrac{4}{x + 3} - \dfrac{2}{x - 3}}$ 16. $\dfrac{\dfrac{1}{2x - 1} + \dfrac{1}{3x - 1}}{\dfrac{1}{2x - 1} - \dfrac{2}{3x - 1}}$

17. $\dfrac{\dfrac{a + b}{a - b} - \dfrac{a - b}{a + b}}{\dfrac{a}{a - b} - \dfrac{a}{a + b}}$ 18. $\dfrac{\dfrac{x + y}{x - y} - \dfrac{x - y}{x + y}}{\dfrac{x - y}{x + y} + \dfrac{1}{x^2 - y^2}}$

19. $\dfrac{3 - \dfrac{x - 6}{x^2 - 6x + 8}}{2 + \dfrac{x + 7}{x^2 - 2x - 8}}$ 20. $\dfrac{2 + \dfrac{x + 6}{x^2 - x - 6}}{1 - \dfrac{x - 6}{x^2 - x - 6}}$

21. $\dfrac{1 + \dfrac{6}{x^2 + x - 6}}{1 + \dfrac{9}{x^2 + 2x - 8}}$ 22. $\dfrac{\dfrac{6x}{x + 2y} - 2}{\dfrac{3x}{x - 2y} + 3}$

23. $\dfrac{1 - \dfrac{6d}{2c + 3d}}{2 + \dfrac{d}{c - 2d}}$ 24. $\dfrac{\dfrac{2x}{y - x} + 2}{\dfrac{x^2}{x^2 - y^2} - 1}$

25. $\dfrac{1}{1 - \dfrac{2}{1 + (2/x)}}$ 26. $\dfrac{x}{x - \dfrac{1}{2 - (1/x)}}$

5

FUNCTIONS AND RELATIONS

5-1 RECTANGULAR COORDINATES

In Sec. 2-5, we discussed a scheme for putting the system of real numbers into a one-to-one correspondence with the points of a line. We now develop a plan for putting the points of a plane into a one to one correspondence with a set of pairs of real numbers. The device for this association of points and number pairs is called a *rectangular Cartesian coordinate system.* This coordinate system was introduced in 1637 by René Descartes, a French mathematician and philosopher. The Cartesian and other coordinate systems open the way for reducing geometric problems to arithmetic ones and consequently have contributed greatly to the advancement of mathematics and its applications in various areas of learning.

We draw a horizontal line and a vertical line meeting at O (Fig. 5-1). The point O is called the *origin;* the horizontal line OX, the x *axis;* and the vertical line OY, the y *axis.* The x axis and the y axis, taken together, are called the *coordinate axes,* and the

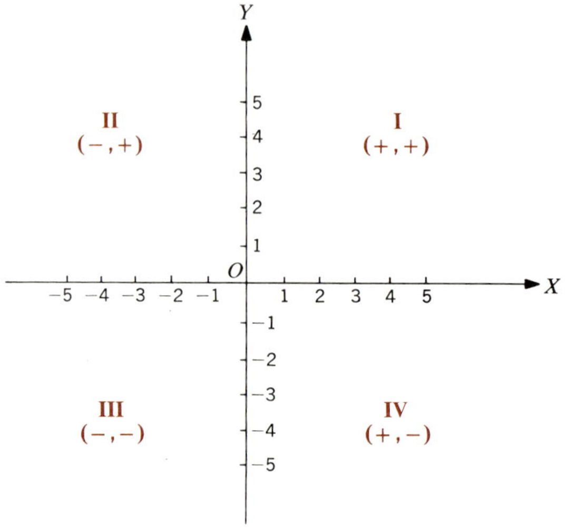

Figure 5-1

plane determined by the coordinate axes is called the *coordinate plane.* With a convenient unit of length, we make a real-number scale on each coordinate axis, letting the origin be the zero point. The positive direction is chosen to the right on the x axis and upward on the y axis, as indicated by the arrowheads in the figure.

If P is a point of the coordinate plane, we define the distances of the point from the coordinate axes to be *directed distances.* That is, the distance from the y axis is positive if P is to the right of the y axis and negative if P is to the left, and the distance from the x axis is positive if P is above the x axis and negative if P is below the x axis. Each point P of the plane has associated with it a pair of numbers called *coordinates,* which are defined as follows.

Definition 5-1 The x *coordinate,* or *abscissa,* of a point P is the directed distance from the y axis to the point. The y *coordinate,* or *ordinate,* of a point P is the directed distance from the x axis to the point.

A point whose abscissa is x and whose ordinate is y is designated by (x, y), in that order, the abscissa always coming first. Hence, the coordinates of a point are called an *ordered pair* of numbers. Although a pair of coordinates determine a point, the coordinates themselves are often referred to as a point.

A point of given coordinates is *plotted* by measuring the proper distances from the axes and marking the point thus located. For example, if the coordinates of a point are $(-4, 3)$, the abscissa -4 means the point is 4 units to the left of the y axis and the ordinate 3 (plus sign understood) means the point is 3 units above the x axis. Consequently, we locate the point by going from the origin 4 units to the left along the x axis and then 3 units upward parallel to the y axis (Fig. 5-2).

The coordinate axes divide their plane into four parts, called *quadrants*, which are numbered I to IV in Fig. 5-1. The coordinates of a point in the first quadrant are both positive, which is indicated in the figure by $(+, +)$. The signs of the coordinates in each of the other quadrants are similarly indicated.

We assume that to any pair of real numbers (coordinates) there corresponds one definite point. Conversely, we assume that to each point of the plane there corresponds one definite pair of coordinates. This relation of points on a plane and pairs of real numbers is called a *one-to-one correspondence.*

The cross product of two sets of real numbers A and B is defined to be the set of all ordered pairs (x, y) such that x is a member of A and y is a member of B. Using this definition and letting $A = B = R$, we may denote the set of number pairs corresponding to the points of a coordinate plane by the notation

$$R \times R = \{(x, y) | x \in R, y \in R\}$$

5-2 RELATIONS AND FUNCTIONS

The concepts of relations and functions, which we shall discuss in this chapter, are fundamental in many areas of mathematics. We begin our discussion of these topics by defining a relation.

Definition 5-2 A *relation* is a set of ordered pairs of numbers. The set of first elements of the ordered pairs is called the *domain* of the relation and the set of second elements is called the *range* of the relation.

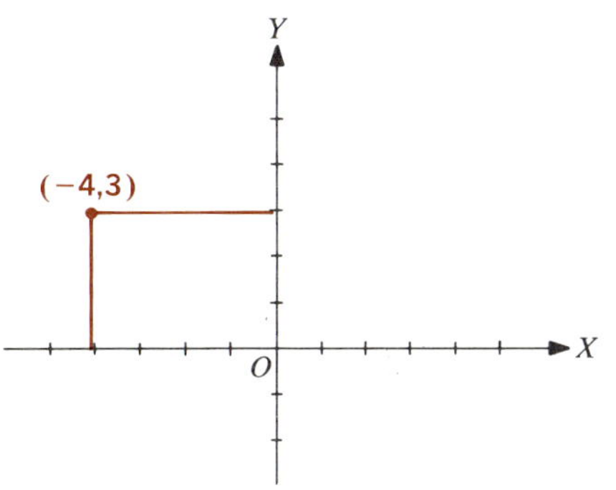

Figure 5-2

EXAMPLE 1 The set of number pairs

$$\{(-1, -1), (0, 1), (1, 1), (1, 2), (3, 0)\}$$

defines a relation. The set $\{-1, 0, 1, 3\}$ is the domain and the set $\{-1, 1, 2, 0\}$ is the range. The points of the coordinate plane determined by the number pairs of this relation are plotted in Fig. 5-3.

EXAMPLE 2. Relations are frequently determined by a given domain and an equation which provides a rule for computing the second element of each number pair from the first element. To illustrate, let the second element of each number pair (x, y) be given by the equation $y = 3x + 1$, and let the domain be the set of real numbers R. These conditions definitely define the relation. The set of number pairs comprising the relation can, with the specified domain, be expressed in the form

$$\{(x, y) | y = 3x + 1\}$$

This relation consists of infinitely many number pairs, and the elements therefore cannot be listed. We can, of course, find y for any particular value of x. Thus, the number pairs for $x = -2$, 0, and 3 are

$$(-2, -5), (0, 1), (3, 10)$$

In this chapter, we shall mainly be concerned with a special kind of relation which is defined as follows:

Definition 5-3 Let X and Y denote sets of real numbers, and let (x, y) denote an ordered pair of numbers such that to each $x \in X$ there is a unique $y \in Y$. The totality of number pairs thus determined is called a *function*. The set X is called the *domain* of the function and the set Y is called the *range* of the function.

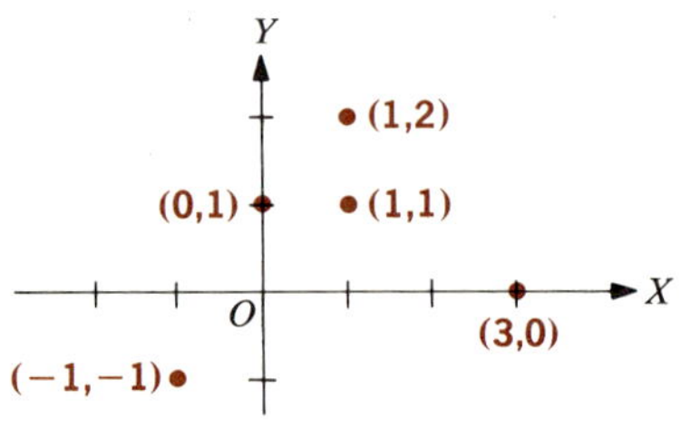

Figure 5-3

We emphasize that this definition requires that for any value x of the domain there is one and only one corresponding value of y. A relation does not have this stipulation. Consequently, a relation may have more than one value of the range corresponding to a value of the domain. Thus, we see that a function is a relation but a relation is not necessarily a function. Consider, for example, the sets

$$\{(x, y)|y = \sqrt{x}\} \qquad \text{and} \qquad \{(x, y)|y = \pm\sqrt{x}\}$$

with the domain of each set given by $X = \{1, 2, 3\}$. The first set has a unique value of y for each value of x, and therefore is both a function and a relation. The second set, having two values of y for each value of x, is a relation but not a function. We note too that the set of number pairs in Example 1 is a relation but not a function, whereas the set of number pairs in Example 2 is a relation and also a function.

The second element of a number pair of a function is called the *value of the function* corresponding to the value of the first element. If we let f stand for the set of number pairs comprising a function and let x stand for any element of the domain, then the notation $f(x)$ denotes the value of the function at x. Although $f(x)$ stands for a value of a function, the symbol is sometimes said, somewhat loosely, to be a "function of x." That is, the second element of an ordered pair of a function, depending for its value on the first element, is called a *function of* the first element. We warn that $f(x)$ in this context is *not a product*—not f times x. We also remark that letters other than f and x may be used in this situation. Thus, if h stands for a function and s for an element of the domain, then $h(s)$ stands for the value of the function at s.

Suppose, for example, that we define a function by a given domain and an equation such as $y = x^2 - 2x$ or, alternatively, $f(x) = x^2 - 2x$. We may then say that y, or $f(x)$, is a function of x and compute the value corresponding to any x value of the domain. Thus, replacing x by 3 in the form

$$f(x) = x^2 - 2x$$

we find

$$f(3) = 3^2 - 2(3) = 3$$

Similarly,

$$f(-2) = (-2)^2 - 2(-2) = 8$$

$$f(5) = 5^2 - 2(5) = 15$$

Hence, (3, 3), (−2, 8), (5, 15) are three number pairs of the function in question.

EXAMPLE 3 If $h(s) = 3s^2 + 4$, find $h(3t)$ and $h(t + 1)$.

Solution: We replace s first by $3t$ and then by $t + 1$ and get

$$h(3t) = 3(3t)^2 + 4 = 27t^2 + 4$$
$$h(t + 1) = 3(t + 1)^2 + 4 = 3t^2 + 6t + 7$$

Although a function or a relation may be completely defined by an equation and a specified domain, often an equation is given with no mention of the domain. In all such cases, it is to be understood that the domain consists of all real numbers for which the equation yields real numbers for the values of the function or relation.

EXAMPLE 4 Find the domain and range of the function defined by $y = \sqrt{9 - x^2}$.

Solution: A square root of a negative number is not a real number. This is true because the square of either a positive or a negative number is positive. So the domain of this function consists of the set of all numbers from -3 to 3, inclusive. Any value of $x < -3$ or of $x > 3$ makes $9 - x^2$ negative and does not belong to the domain. The range consists of the set of all numbers from 0 to 3, inclusive. We use the symbol $\leq$ (read "is less than or equal to") and express these descriptions of the domain and range by

$$X = \{x | -3 \leq x \leq 3\} \qquad Y = \{y | 0 \leq y \leq 3\}$$

EXERCISE 5-1

List the elements of the domain and the elements of the range of each set of ordered number pairs. Tell which are functions.

1. $\{(3, 4), (-3, 4), (3, -4), (-3, -4)\}$
2. $\{(-1, 1), (0, 0), (1, 1), (2, 8)\}$
3. $\{(0, 4), (1, -1), (2, 2), (3, 5)\}$
4. $\{(2, -4), (3, 5), (2, 4), (3, 6)\}$
5. $\{(-3, 0), (-2, 1), (0, 3), (0, 4)\}$
6. $\{(-1, 0), (0, 1), (3, 2), (8, 3)\}$

Find the set of ordered pairs of numbers (x, y) which are defined by the given equation and the given domain X.

7. $y = 2x - 3, \quad X = \{-1, 0, 2, 3\}$
8. $y = x^2 + 1, \quad X = \{-2, 0, 1, 2\}$
9. $y = x + 1, \quad X = \{-1, 0, 3, 8, 15\}$
10. $y = 1 - x^2, \quad X = \{-2, -1, 0, 1, 2\}$
11. If $f(x) = x - 1$, find $f(0), \quad f(-1), \quad f(2), \quad f(4), \quad f(6)$
12. If $f(x) = 4 - 3x$, find $f(1), \quad f(-1), \quad f(2), \quad f(\frac{2}{3}), \quad f(-\frac{1}{3})$
13. If $g(t) = t^2 - 2t + 2$, find $g(-2), \quad g(-1), \quad g(4), \quad g(2)$
14. If $F(s) = 2s^2 - 3s - 5$, find $F(1), \quad F(3), \quad F(-2), \quad F(-1)$
15. If $F(x) = (x - 1)/(x - 2)$, find $F(0), \quad F(1), \quad F(-3), \quad F(4)$
16. If $h(t) = (3t - 1)/t$, find $h(2), \quad h(-1), \quad h(\frac{1}{3}), \quad h(3)$
17. If $F(x) = x^2 + 1$, find $F(s^2), \quad F(x - 2), \quad [F(x - 1)]^2$
18. If $g(t) = t^2$, find $g(t + h), \quad g(t + h) - g(t)$
19. If $F(t) = t^2 - 3t$, find $F(t + h), \quad F(t + h) - F(t)$

Determine the domain and range of the function defined by each of the following equations.

20. $f(x) = x^2 - 4$
21. $f(x) = \sqrt{3x}$
22. $f(x) = 1/(x - 2)$
23. $f(x) = \sqrt{16 - x^2}$
24. $f(x) = \sqrt{x^2 - 16}$
25. $f(x) = \sqrt{x^2 + 1}$

5-3 GRAPHS OF FUNCTIONS AND RELATIONS

Consider the function defined by the equation

$$y = f(x) = x^2 - 3x - 2$$

If a value is assigned to x, the corresponding value of $f(x)$ may be computed. Several values of x and the corresponding values of $y = f(x)$ are shown in the table. These number pairs furnish a picture of the relation of x and y. A better representation is obtained, however, by plotting each value of x and the corresponding value of y as the abscissa and ordinate of a point, and then drawing a smooth curve through the points thus determined.

This process is called *graphing the function*, and the curve is called the *graph* or *locus* of the function.

x	-1.5	-1	0	1	2	3	4	4.5
$y = f(x)$	4.75	2	-2	-4	-4	-2	2	4.75

The plotted points (Fig. 5-4) extend from $x = -1.5$ to $x = 4.5$. Points farther to the left and right, as well as any number of intermediate points, could be located. But the plotted points show about where the intermediate points would lie. Hence, we can use a few points to draw a curve which is reasonably accurate. The exact graph satisfies the following definition.

Definition 5-4 The graph of a function (or relation) consists of the set of all points whose coordinates are the ordered pairs of the function (or relation). If the function (or relation) is defined by an equation, the graph of the equation is the same as the graph of the function (or relation).

The point or points of the graph of a function which are on the x axis are of special interest. The abscissa of such a point makes the value of y, or the function, equal to zero and is called a *zero of the function*. We estimate the abscissas of the points where the graph in Fig. 5-4 cuts the x axis to be $x = -0.6$ and $x = 3.6$. These values are approximations to the zeros of the function represented by the graph.

EXAMPLE 1 Draw the graph of the function defined by $y = 1 - 2x$.

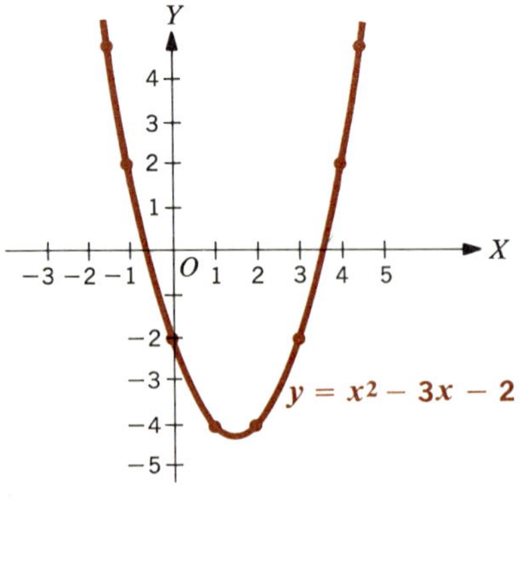

Figure 5-4

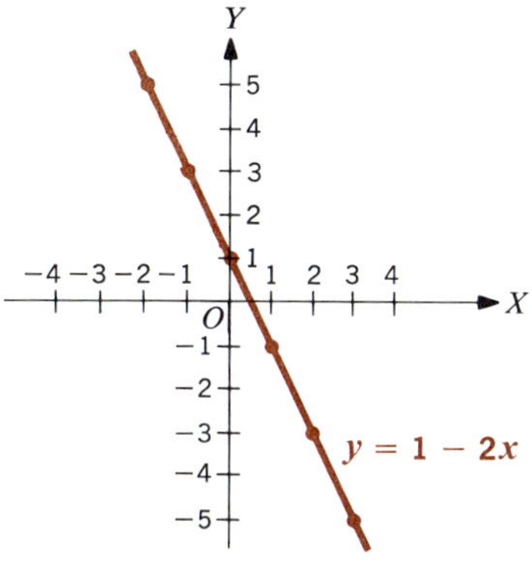

Figure 5-5

Solution: We assign certain values to x and obtain the corresponding values of y, as shown in the table. The tabulated values enable us to plot some points of the graph and draw a curve through them (Fig. 5-5). The graph is a straight line and shows the zero of the function to be $x = \frac{1}{2}$.

x	-2	-1	0	1	2	3
y	5	3	1	-1	-3	-5

EXAMPLE 2 Draw the graph of $y = |x - 2|$.

Solution: Recalling the definition of the absolute value of a number (Definition 3-1), we prepare a table of values and sketch the graph shown in Fig. 5-6.

x	-2	0	2	4	6
y	4	2	0	2	4

EXAMPLE 3 Construct the graph of the relation defined by $y^2 = 4x$.

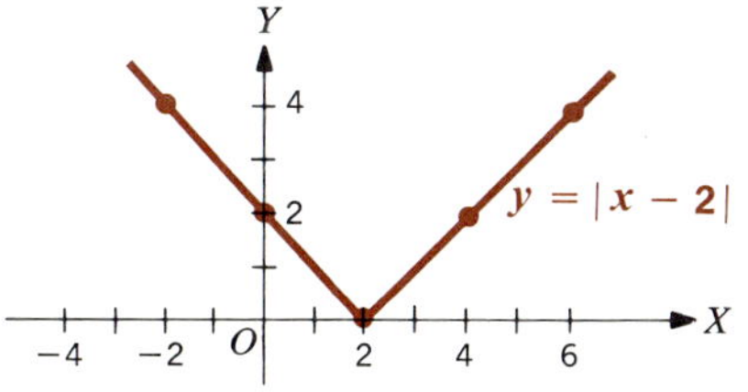

Figure 5-6

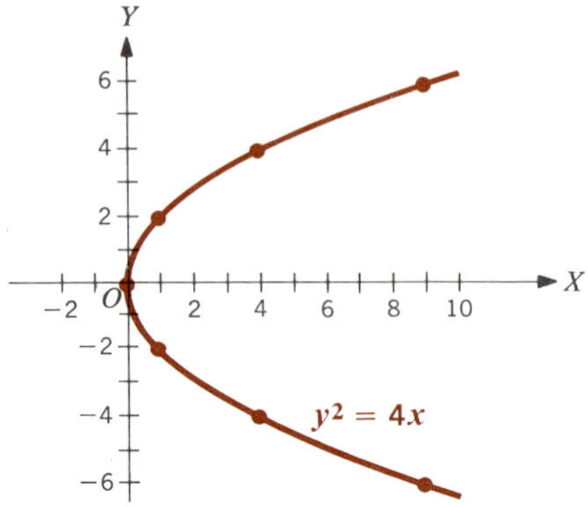

Figure 5-7

Solution: Solving for y gives $y = \pm 2\sqrt{x}$. We draw the graph (Fig. 5-7) by use of the following table.

x	0	1	4	9
y	0	± 2	± 4	± 6

EXERCISE 5-2

Draw the graph of the function defined by each of the following equations. Give the zeros of the functions.

1. $y = x - 2$ **2.** $y = 2x + 4$ **3.** $y = 1 - x$

4. $y = 3x - 5$ **5.** $y = \frac{1}{2}x$ **6.** $y = 1 - \frac{1}{3}x$

7. $y = 1 - |x|$ **8.** $y = |x|$ **9.** $y = |x + 2|$

Draw the graph of the function defined by each of the following equations and give (or estimate) its zeros. Use all the integral values of x in the indicated domain and enough fractional values to determine the shape of the curve.

10. $y = x^2$ from $x = -3$ to $x = 3$

11. $y = x^2 - 4x$ from $x = -1$ to $x = 5$

12. $y = 2x^2 - 9$ from $x = -3$ to $x = 3$

13. $y = -x^2 + 3x + 3$ from $x = -2$ to $x = 5$

14. $y = 3x^2 - 6x - 5$ from $x = -1$ to $x = 3$

15. $y = x^3 - 2x^2$ from $x = -1$ to $x = 3$

16. $y = x^3 + 2x^2$ from $x = -3$ to $x = 1$

Draw the graphs of the relations defined by the following equations. (See Table 1 in the Appendix for square roots of numbers.)

17. $y^2 = x$ from $x = 0$ to $x = 9$

18. $y^2 = 9x$ from $x = 0$ to $x = 9$

19. $y = 25 - x^2$ from $x = -5$ to $x = 5$

20. $y = 16 - x^2$ from $x = -4$ to $x = 4$

5-4 THE DISTANCE FORMULA AND THE CIRCLE

The distance between two points of the coordinate plane, or the length of the line segment connecting them, can be determined from the coordinates of the points. Let $P_1(x_1, y_1)$ and $P_2(x_2, y_2)$ be the coordinates of any two points of the plane. Then a line through P_1 parallel to the x axis and a line through P_2 parallel to the y axis intersect at a point P_3 whose abscissa is x_2 and whose ordinate is y_1. Referring to Fig. 5-8, or Fig. 5-9, we express the distances between P_1 and P_3 and between P_2 and P_3, respectively, by

$$P_1P_3 = |x_2 - x_1| \qquad \text{and} \qquad P_2P_3 = |y_2 - y_1|$$

Designating the distance between P_1 and P_2 by d and applying the Pythagorean theorem (see problem 28, Exercise 5-3), we have

$$d = \sqrt{|x_2 - x_1|^2 + |y_2 - y_1|^2}$$

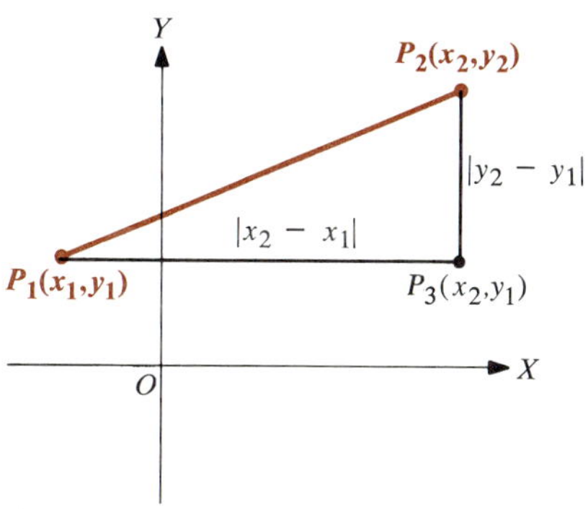

Figure 5-8

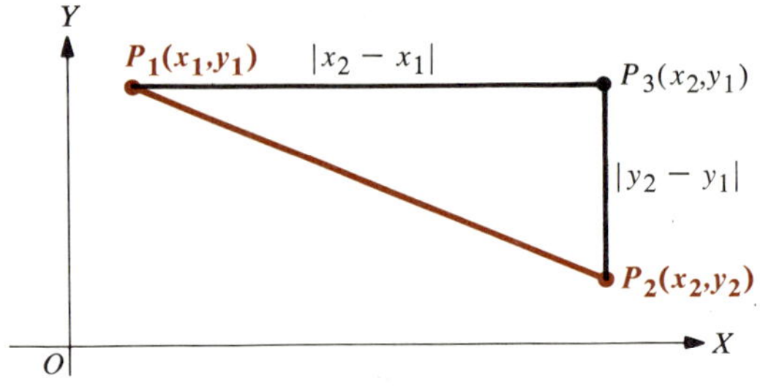

Figure 5-9

Finally, noting that $|x_2 - x_1|^2 = (x_2 - x_1)^2$ and $|y_2 - y_1|^2 = (y_2 - y_1)^2$, we express the formula for the distance between two points in the form

$$d = \sqrt{(x_2 - x_1)^2 + (y_2 - y_1)^2} \tag{1}$$

In applying this formula, either point may be designated by (x_1, y_1) and the other by (x_2, y_2). This results because the two differences involved are squared.

EXAMPLE 1 Find the lengths of the sides of the triangle (Fig. 5-10) with vertices at $A(-2, -3)$, $B(6, 1)$, and $C(-2, 5)$.

Solution:

$$\begin{aligned}
AB &= \sqrt{[(6 - (-2)]^2 + [1 - (-3)]^2} \\
&= \sqrt{64 + 16} = \sqrt{80} = 4\sqrt{5} \\
BC &= \sqrt{(-2 - 6)^2 + (5 - 1)^2} \\
&= \sqrt{64 + 16} = \sqrt{80} = 4\sqrt{5} \\
CA &= \sqrt{[-2 - (-2)]^2 + (5 + 3)^2} \\
&= \sqrt{0 + 64} = 8
\end{aligned}$$

Two of the sides have equal lengths and the triangle is isosceles.

The distance formula (1) has numerous important uses in mathematical situations. We shall now apply the formula to find the equation of a circle. We recall, of course, that a circle is the locus of all points in a plane that are equally distant from a fixed point of the plane. The fixed point is called the *center*, and the distance from the center to a point of the circle is called the *radius*.

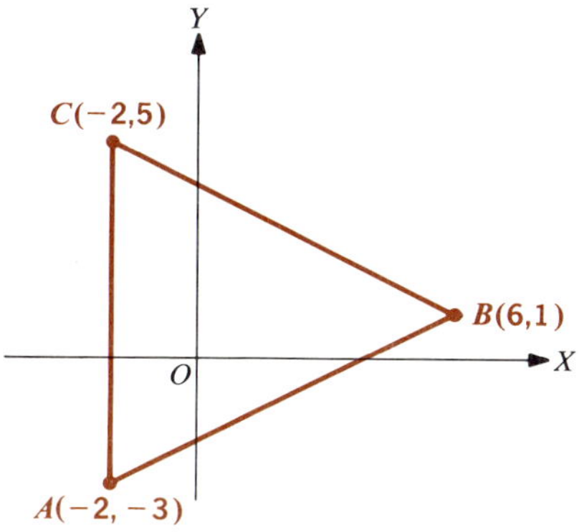

Figure 5-10

If we let $C(h, k)$ be the coordinates of the center of a circle of radius r (Fig. 5-11), then any point $P(x, y)$ of the circle must satisfy the condition

$$\sqrt{(x - h)^2 + (y - k)^2} = r$$

or, squaring,

$$(x - h)^2 + (y - k)^2 = r^2 \tag{2}$$

Equation (2) is the general equation of a circle. The circle consists of the set of all points whose coordinates satisfy the equation. For the special case in which the center is at the origin and the radius is equal to 1, the equation becomes

$$x^2 + y^2 = 1$$

We shall make considerable use of this circle in Chap. 8.

EXAMPLE 2 Find the equation of the circle which passes through the points (0, 0), (4, 0), and (0, −4).

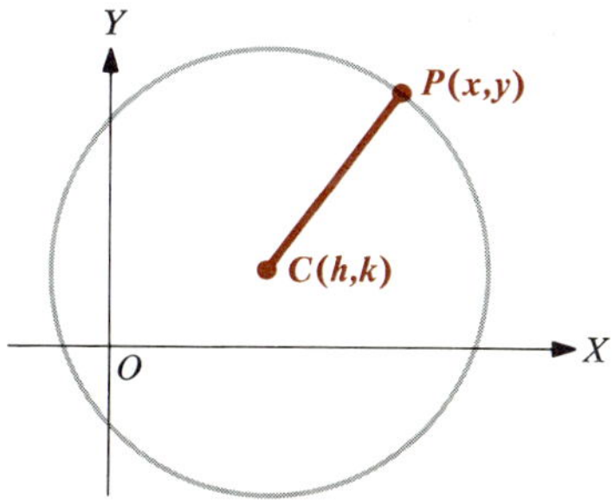

Figure 5-11

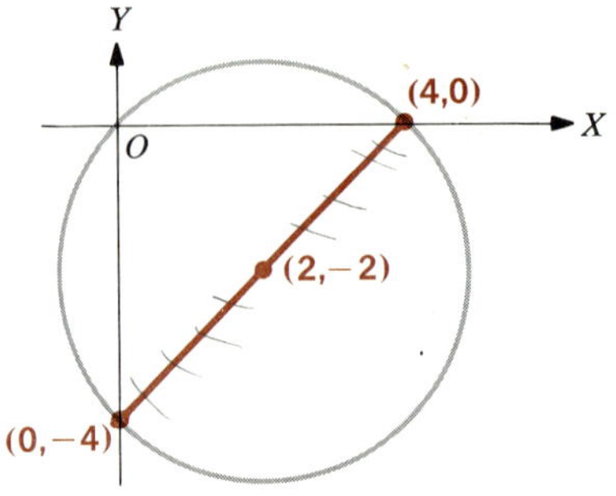

Figure 5-12

Solution: These points are vertices of a right triangle (Fig. 5-12). Then, recalling that the midpoint of the hypotenuse of a right triangle is equidistant from the three vertices, we see that $(2, -2)$ are the coordinates of the center of the circle. The distance from this point to each of the vertices is $\sqrt{8}$. Hence, taking $r^2 = 8$, we have the equation

$$(x - 2)^2 + (y + 2)^2 = 8$$

EXERCISE 5-3

Find the distance between the pair of points in each problem 1 through 6.

1. $(3, 1)$, $(7, 4)$ **2.** $(4, -3)$, $(-8, 2)$

3. $(-3, -2)$, $(0, 1)$ **4.** $(5, -12)$, $(0, 0)$

5. $(-4, 0)$, $(0, 3)$ **6.** $(7, 2)$, $(4, -1)$

In each problem 7 through 10 draw the triangle having the given vertices and find the lengths of the sides.

7. $A(-1, 1)$, $B(-1, 4)$, $C(3, 4)$

8. $A(2, -1)$, $B(4, 2)$, $C(5, 0)$

9. $A(0, 0)$, $B(5, -2)$, $C(-3, 2)$

10. $A(0, -3)$, $B(3, 0)$, $C(0, -4)$

Determine, by the distance formula, if the three points in each problem 11 through 16 are the vertices of a scalene triangle (no equal sides), an isosceles triangle (two equal sides), or a right triangle (see problem 28), or if the three points are on a straight line (the distance between two of the points equal to the sum of the other distances).

11. $A(6, 1)$, $B(2, -4)$, $C(-2, 1)$
12. $A(1, 4)$, $B(10, 6)$, $C(2, 2)$
13. $A(-2, 1)$, $B(5, -2)$, $C(3, 3)$
14. $A(1, -1)$, $B(1, -7)$, $C(8, 3)$
15. $A(2, 1)$, $B(-1, 2)$, $C(5, 0)$
16. $A(3, 2)$, $B(0, 0)$, $C(9, 6)$

In each problem 17 through 27 write the equation of the circle which satisfies the given conditions.

17. Center $(-6, 2)$, radius 5
18. Center $(4, 3)$, radius 5
19. Center $(0, 0)$, radius 6
20. Center $(-4, -3)$, radius 7
21. The line segment joining $A(0, 0)$ and $B(6, 0)$ is a diameter.
22. The line segment joining $A(2, 3)$ and $B(2, -3)$ is a diameter.
23. The center is at $(2, 4)$, and the circle passes through $(-1, -1)$.
24. The center is at $(1, -3)$, and the circle passes through $(-3, 5)$.
25. The circle is tangent to the x axis, and the center is at $(-4, 1)$.
26. The circle passes through the points $(0, 0)$, $(2, 0)$, and $(0, 2)$.
27. The circle passes through the points $(1, 1)$, $(5, 1)$, and $(1, 5)$.
28. The Pythagorean theorem states that the square of the length of the hypotenuse of a right triangle is equal to the sum of the squares of the lengths of the remaining two sides. Use Fig. 5-13 to prove that $c^2 = a^2 + b^2$. *Hint:* The area of the outer square is given by

$$A = (a + b)^2 \qquad \text{and also by} \qquad A = c^2 + 4(\tfrac{1}{2}ab)$$

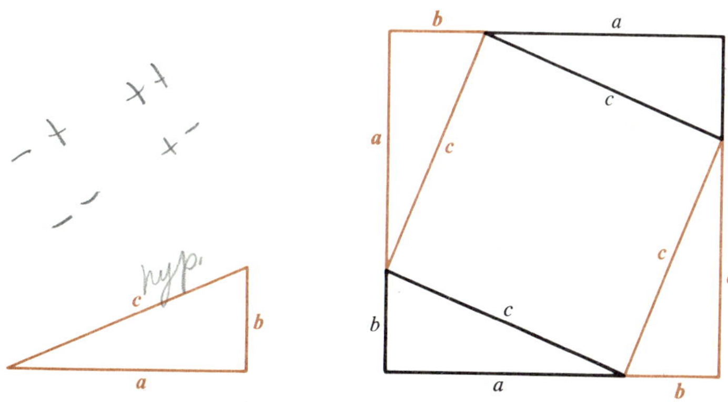

Figure 5-13

6

LINEAR EQUATIONS

6-1 CONDITIONAL EQUATIONS AND IDENTITIES

A statement that two quantities are equal is called an *equation.* The customary way of writing an equation is to place the symbol = (read "is equal to") between the equal quantities. An equation then has two members, the left member and the right member. Equations usually have one or more letters which are regarded as *variables.* Numbers which, when substituted for the variables, make the two members of the equation equal are said to *satisfy* or be a *solution* of the equation. The totality of solutions is called the *solution set.*

EXAMPLES

$$x - 3 = 4 \tag{1}$$

$$x^2 - 2x - 15 = 0 \tag{2}$$

$$(x - 2)^2 = x^2 - 4x + 4 \tag{3}$$

$$\frac{1}{x-1} - \frac{1}{x} = \frac{1}{x(x-1)} \tag{4}$$

The first equation is satisfied when $x = 7$ and for no other number. The second equation is satisfied for $x = -3$ and $x = 5$; any other value for x makes the two members of the equation unequal. The right member of Eq. (3) is the result of the operation indicated on the left side. Clearly, then, the equation is satisfied when x is replaced by any real number. The right member of Eq. (4) is the sum of the fractions of the left member, and the equation is true for all values of x except $x = 0$ and $x = 1$. Either of these values produces a zero in a denominator on the left and the denominator on the right. Consequently, we say that each member of the equation is *undefined* for these values of x.

The two types of equations which we have illustrated are named in accordance with the following definitions.

Definition 6-1 An equation which is satisfied by some, but not all, of the values of the variables for which the members of the equation are defined is called a *conditional equation*.

Definition 6-2 An equation which is satisfied by all the values of the variables for which the members of the equation are defined is called an *identity*.

According to these definitions, Eqs. (1) and (2) are conditional equations, and Eqs. (3) and (4) are identities. The solution sets of the four equations are, respectively,

$$\{7\}, \qquad \{-3, 5\}, \qquad \{x|x \in R\}, \qquad \{x|x \in R, x \neq 0 \text{ or } 1\}$$

In Chap. 4, we equated expressions to the product of their factors. The equations thus formed are identities. In this chapter, we take up conditional equations, first studying equations that have only one variable. A solution of an equation in one variable is called a *root* of the equation. To *solve* an equation of this kind means to find all of its roots. Before solving equations, however, we must discuss certain operations that are required in the process.

6-2 OPERATIONS ON EQUATIONS

Two equations are said to be *equivalent* if their solution sets are the same. The equations $x - 3 = 2$ and $x + 1 = 6$, for example, are equivalent; each has the one root 5. Neither of these equations, however, is equivalent to $(x - 5)(x - 1) = 0$. The values $x = 1$ and $x = 5$ satisfy the last equation. But $x = 1$ does not satisfy the other equations.

In the process of finding the solutions of an equation it is usually necessary to perform operations, starting with the given equation, which lead to other equations. It is important, therefore, to know if the new equations are equivalent to the original equation. The following theorems pertain to this question.

Theorem 6-1 If the same polynomial is added to both members of an equation, an equivalent equation is obtained.

Proof: Suppose we let $f(x)$ and $g(x)$ stand for the members of an equation and let $h(x)$ denote a polynomial in x. Then we need to show that the equations

$$f(x) = g(x) \tag{1}$$

and

$$f(x) + h(x) = g(x) + h(x) \tag{2}$$

are equivalent. If some number r is a root of Eq. (1), then $f(r) = g(r)$. Now r is also a root of Eq. (2) if

$$f(r) + h(r) = g(r) + h(r)$$

Since $f(r)$, $g(r)$, and $h(r)$ are constants and $f(r) = g(r)$, we know, by the additive axiom E-4, Sec. 2-1, that this equation is true. Consequently, any root of Eq. (1) is a root of Eq. (2).

Conversely, if r is a root of Eq. (2), then the equation

$$f(r) + h(r) = g(r) + h(r)$$

is true, and Theorem 2-1 requires that $f(r) = g(r)$. Hence, any root of Eq. (2) is a root of Eq. (1).

We now have shown that any root of Eq. (1) is a root of Eq. (2), and that any root of Eq. (2) is a root of Eq. (1). This establishes the equivalence of the equations and completes the proof of the theorem.

Theorem 6-2 If both members of an equation are multiplied by or divided by the same nonzero constant, an equivalent equation is obtained.

We leave the proof of this theorem to the student.

Theorem 6-2 applies to multiplication or division by a constant. If, however, both members of an equation are multiplied by or divided by an expression which contains a variable, the resulting equation will in some cases not be equivalent to the original one. Thus, for example, multiplying both members of $4x - 8 = 12$ by x gives $4x^2 - 8x = 12x$. The new equation is satisfied by $x = 5$ and also by $x = 0$; whereas the first equation is satisfied by $x = 5$ but not by $x = 0$. Conversely, if we start with $4x^2 - 5x = 12x$ and divide by x, a root is lost rather than gained. Notice, however, that each of these operations, yielding nonequivalent equations, involves the variable.

A root of the new equation which is obtained by an operation such as that illustrated here, and which is not a root of the original equation is called an *extraneous root.* An extraneous root can be detected by substituting solutions of the new equation in the given equation. If all the roots are desired, an operation which leads to an equation with fewer roots than the original equation should be avoided.

6-3 LINEAR EQUATIONS IN ONE VARIABLE

An equation of the first degree in its variable, or variables, is called a *linear equation.* In this section, we shall consider equations of the form, or reducible to the form,

$$ax + b = 0 \tag{1}$$

where x is a variable and a and b are specified constants except that a is different from zero. The equation, being of the first degree in x, is a linear equation. We solve the equation in the following way:

$$ax + b = 0 \qquad \text{Given equation}$$

$$ax = -b \qquad \text{Adding } -b \text{ to both sides}$$

$$x = -\frac{b}{a} \qquad \text{Dividing by } a$$

Notice that here each of the two operations yields an equivalent equation, and therefore $-b/a$ is a root of the given equation. Furthermore, the replacement of x by any other value would make a false equation. Hence, the solution set $\{-b/a\}$ consists of a single element. Because of this restriction on the variable in an equation of the form (1), the symbol x is sometimes called an *unknown.*

EXAMPLE 1 Solve the equation $9x - 2 = 3x - 11$.

Solution:

$$9x - 2 = 3x - 11 \qquad \text{Given equation}$$

$$9x - 2 + 2 - 3x = 3x - 11 + 2 - 3x \qquad \text{Adding } 2 - 3x$$

$$6x = -9 \qquad \text{Collecting terms}$$

$$x = -1.5 \qquad \text{Dividing by 6}$$

Check: We substitute -1.5 for x in each member of the given equation and find

$$9(-1.5) - 2 = -13.5 - 2 = -15.5 \qquad \text{For left member}$$

$$3(-1.5) - 11 = -4.5 - 11 = -15.5 \qquad \text{For right member}$$

The value $x = -1.5$ is a root because it makes the members of the equation equal, and therefore the solution set is $\{-1.5\}$.

EXAMPLE 2 Solve the equation $3x + 4 = 5ax + 2c$ for x.

Solution:

$$3x + 4 = 5ax + 2c \qquad \text{Given equation}$$

$$3x + 4 - 5ax - 4 = 5ax + 2c - 5ax - 4 \qquad \text{Adding } -5ax - 4$$

$$3x - 5ax = 2c - 4 \qquad \text{Collecting terms}$$

$$(3 - 5a)x = 2c - 4 \qquad \text{Factoring}$$

$$x = \frac{2c - 4}{3 - 5a} \qquad \text{Dividing by } 3 - 5a$$

This result may be checked by substituting in the given equation. Hence, the solution set is $\{(2c - 4)/(3 - 5a)\}$.

An equation involving simple fractions may be reduced to nonfractional form by multiplying both members by the LCM of the denominators. This process is called *clearing of fractions.* If none of the denominators contains the variable, the new equation is equivalent to the original equation. If, however, the variable appears in any denominator, the new equation may have roots in addition to those of the original equation. It is essential, therefore, that all roots of the new equation be checked in the given equation.

EXAMPLE 3 Solve the equation

$$\frac{4}{x+1} - \frac{2}{x-1} = \frac{x-5}{x^2-1}$$

Solution: Multiplying both members of the equation by $(x+1)(x-1)$, the LCM of the denominators gives

$$\begin{aligned} 4(x-1) - 2(x+1) &= x - 5 \\ 4x - 4 - 2x - 2 &= x - 5 && \text{Simplifying} \\ 2x - 6 &= x - 5 && \text{Collecting terms} \\ x &= 1 && \text{Adding } 6 - x \end{aligned}$$

The value $x = 1$ is a root of the equation obtained by clearing the given equation of fractions. This value, however, makes two of the denominators of the given fraction equal to zero, and consequently must be rejected. Hence, the given equation has no solution.

EXAMPLE 4 Solve the equation

$$\frac{x^2-2}{2x^2+x-1} = \frac{x}{x+1} - \frac{x-1}{2x-1}$$

Solution: Multiplying both members of the equation by $(x+1)(2x-1)$, the LCM of the denominators, we have

$$\begin{aligned} x^2 - 2 &= x(2x-1) - (x-1)(x+1) \\ x^2 - 2 &= 2x^2 - x - x^2 + 1 && \text{Simplifying} \\ x &= 3 && \text{Adding } 2 - x^2 + x \end{aligned}$$

Check: Replacing x by 3 in the original equation gives

$$\frac{9-2}{18+3-1} = \frac{7}{20} \qquad \text{For left member}$$

$$\frac{3}{3+1} - \frac{3-1}{6-1} = \frac{7}{20} \qquad \text{For right member}$$

The value $x = 3$ satisfies the given equation, and therefore the solution set is $\{3\}$.

EXERCISE 6-1

Solve each equation for x. Reject any value for x which makes a denominator equal to zero.

1. $3x + 4 = 5x + 6$
2. $x - 11 = 5x + 5$
3. $2(x + 4) + 4 = 3(4 - 5x)$
4. $5(3 + 2x) + 2(x - 7) + 6 = 0$
5. $3(5x + 20) + 8 = 4x + 13$
6. $(x - 1)^2 + x + 9 = (x + 1)^2$
7. $(3x - 1)(x - 1) = 3x^2 + 12$
8. $(x - 1)(x + 2) = (x - 4)(x + 3)$
9. $x(9 - 2a) = 3 + 5a$
10. $3 + ax - 4x - 18 = 0$
11. $3(x - a) = 5 - 2(ax - 4)$
12. $4x - 3a = 5ax + 7b$
13. $6 - ax = a - 2bx - 5$
14. $a(x - 1) - b(x - 2) = a - 4b$
15. $cx - bx + 3a = 3x + 7$
16. $2a(x + 4) + 3b(x - 1) = 3x + c$
17. $\dfrac{x + 2}{3} = \dfrac{3x + 4}{4} - \dfrac{5x}{8}$
18. $\dfrac{2x + 1}{21} - \dfrac{x}{7} = \dfrac{x - 1}{3} + \dfrac{1}{3}$
19. $\dfrac{x + 1}{6} - \dfrac{x - 5}{3} = \dfrac{2x + 7}{4}$
20. $\dfrac{4 - 3x}{4} - \dfrac{3 - 4x}{5} = \dfrac{5 - 2x}{2}$
21. $\dfrac{2}{x} - \dfrac{11}{10} = \dfrac{1}{x} + \dfrac{2}{5}$

22. $\dfrac{5}{7x} - \dfrac{3}{14} = \dfrac{7}{2x} + 3$

23. $\dfrac{x^2 - 1}{x^2 - 4} = \dfrac{x}{x - 2} + \dfrac{2}{x + 2}$

24. $\dfrac{4x^2}{x^2 - 1} + \dfrac{3}{x + 1} = \dfrac{4x}{x - 1}$

25. $\dfrac{4}{x^2 - 1} - \dfrac{3}{x + 1} = \dfrac{2}{x - 1}$

26. $\dfrac{1}{x^2 - x} = \dfrac{2}{x - 1} - \dfrac{1}{x}$

27. $\dfrac{3x + 5}{x^2 + 5x} + \dfrac{x + 4}{x + 5} = \dfrac{x + 1}{x}$

28. $\dfrac{7}{6x + 3a} - \dfrac{2}{3x} = \dfrac{1}{2x + a}$

29. $\dfrac{2}{x - 2} + \dfrac{3}{x + 2} = \dfrac{4x - 4}{x^2 - 4}$

30. $\dfrac{x - 3}{x} - \dfrac{x + 4}{x - 5} = \dfrac{11x + 10}{x(5 - x)}$

6-4 WORD PROBLEMS

The equations in Exercise 6-1 are either linear or expressible linearly by clearing of fractions. We next consider word problems which lead to equations of the same type. The equation in each problem is to be found from the described relations among known and unknown numbers. We make the following suggestions for attacking such problems.

1. Read and reread the problem until what is stated is clear, getting well in mind the given numbers and the unknown numbers (numbers to be found). Often a diagram will be helpful in analyzing a problem.
2. Express each unknown number in terms of a single letter.
3. Find the quantities, involving the given numbers and the unknown numbers, which are equal. Then form an equation.
4. Solve the equation and check the result.

EXAMPLE 1 The length of a rectangle exceeds its width by 3 feet. If each dimension were increased by 2 feet, the area would be increased by 30 square feet. Find the original dimensions.

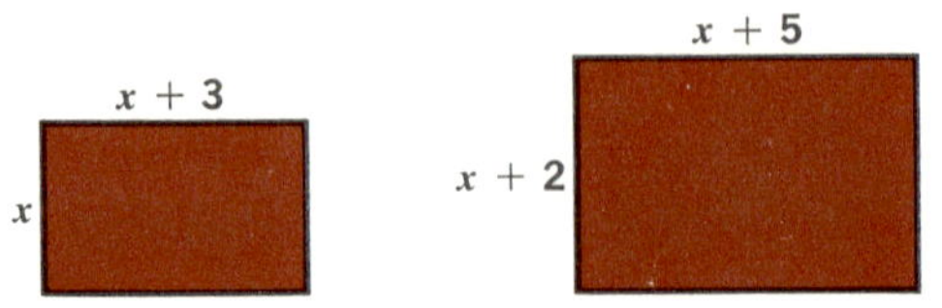

Figure 6-1

Solution: If x feet by $x + 3$ feet are the original dimensions, then $x + 2$ feet by $x + 5$ feet are the new dimensions (Fig. 6-1). Since the increase in area is 30 square feet, we have

$$\begin{aligned}(x + 2)(x + 5) - x(x + 3) &= 30\\ x^2 + 7x + 10 - x^2 - 3x &= 30\\ 4x &= 20\\ x &= 5\end{aligned}$$

The original dimensions are 5 feet by 8 feet and the new dimensions are 7 feet by 10 feet. The areas, then, are 40 square feet and 70 square feet, which yield a difference of 30 square feet.

EXAMPLE 2 A can do a certain task in 12 hours, B in 15 hours, and C in 18 hours. How long will it take to do the task if A and B work 2 hours and then A and C finish the task?

Solution: The fractional parts of the task done in 1 hour by A, B, and C are $\frac{1}{12}$, $\frac{1}{15}$, $\frac{1}{18}$, respectively. If we let x be the total number of hours required to do the task, then A works x hours, B works 2 hours, and C works $x - 2$ hours. Hence,

$$\begin{aligned}\frac{x}{12} &= \text{part of task done by } A\\ \frac{2}{15} &= \text{part done by } B\\ \frac{x - 2}{18} &= \text{part done by } C\end{aligned}$$

The whole task is completed by the sum of these parts, and therefore

$$\frac{x}{12} + \frac{2}{15} + \frac{x-2}{18} = 1$$

$$15x + 24 + 10x - 20 = 180 \qquad \text{Multiplying by 180}$$

$$25x = 176 \qquad \text{Collecting terms}$$

$$x = 7\tfrac{1}{25} \text{ hours} \qquad \text{Dividing by 25}$$

EXAMPLE 3 How many ounces of pure silver must be added to 18 ounces of 60% pure silver to make an alloy that is 76% pure silver?

Solution: The necessary equation may be set up by equating the amount of silver in the separate parts to the amount of silver in the mixture. Letting x stand for the number of ounces of silver to be added, we arrange the data thus:

x ounces:	100% pure = x oz silver
18 ounces:	60% pure = 10.8 oz silver
$(x + 18)$ ounces:	76% pure = $0.76(x + 18)$ oz silver

The x ounces of silver plus the original 10.8 ounces constitute the amount of silver in the mixture, and therefore

$$x + 10.8 = 0.76x + 13.68$$

$$0.24x = 2.88$$

$$x = 12 \text{ oz}$$

EXERCISE 6-2

1. The sum of three numbers is 138. The second number is 6 more than the smallest and the third number is 12 more than the smallest. Find the numbers.

2. Three numbers are so related that the second number is 2 more than the first number, and the third number is 4 more than the first. Find the numbers if the sum of their squares is 140 more than three times the square of the smallest number.

3. The length of a rectangle exceeds its width by 4 units. If each dimension were increased by 3 units, the area would be in-

creased by 81 square units. Find the dimensions of the rectangle.

4. The altitude of a triangle is five-thirds the length of its base. If the altitude were increased by 4 feet and the base decreased by 2 feet, the area would be unchanged. Find the length of the base and altitude.

5. A can paint a house in 12 days and B can paint the house in 15 days. How long will it take to paint the house if both men work?

6. A man can do a certain task in 6 days, another man can do the task in 8 days, and a boy can do the task in 10 days. How long will it take to do the task if all three work?

7. A can do a certain job in 8 hours, B can do the job in 12 hours, and C can do the job in 16 hours. How long will it take to do the job if A and B work 1 hour and B and C finish the job?

8. A tank can be filled by one pipe in 9 hours and by another pipe in 12 hours. Starting empty, how long will it take to fill the tank, with both pipes operating, if a third pipe is removing water at the rate per hour equal to one-sixth the capacity of the tank?

9. How many ounces of pure silver must be added to 100 ounces of 40% pure silver to make an alloy that is 65% pure silver?

10. A perfumer wishes to blend perfume valued at \$4.10 an ounce with perfume worth \$2.50 an ounce to obtain a mixture of 80 ounces worth \$3.00 an ounce. How much of the \$4.10 perfume should he use?

11. An alloy of silver and gold weighs 15 ounces in air and 14 ounces in water. Assuming that silver loses one-tenth of its weight in water and that gold loses one-nineteenth of its weight, how many ounces of each metal are in the alloy?

12. The denominator of a fraction exceeds the numerator by 2. If 3 is subtracted from the numerator and 6 is subtracted from the denominator, the sum of the original fraction and the new fraction is equal to 2. Find the original fraction.

13. The numerator of a fraction is 4 less than the denominator. If the numerator is doubled and the denominator diminished by 2, the sum of the original fraction and the new one is 3. Find the original fraction.

14. The numerator of a fraction is 2 less than the denominator. If the numerator is multiplied by 3 and the denominator decreased by 4, the sum of the original fraction and the new fraction is 4. Find the original fraction.

15. A man loans \$10,000, part at 4.75% annual interest and the rest at 5.5%. Find the amount of each loan if the total annual income is \$520.

16. A man loans \$6,250 at one annual interest rate and \$7,500 at a 1% greater rate. The \$7,500 loan earns \$137.50 more per year than the \$6,250 loan earns. Find the interest rates.

17. A man invests \$3,000 at a certain interest rate and \$4,000 at a $\frac{1}{2}$% greater rate. His annual income from the two investments is \$352.50. Find the interest rates.

18. In college algebra a student earns an average grade of 83 on his homework and 73 on his hour tests. What final examination grade will give him a total average of 80 if homework counts one-tenth, hour tests six-tenths, and the final examination three-tenths?

19. A person having Federal Medical Insurance pays \$36 per year in premiums. This medical insurance will pay 80% of covered medical expenses which remain after the insured pays the first \$50. Now suppose that the insured has another policy which will pay 20% of the medical expense plus an amount equal to 20% of the amount the government pays. If the premium for this policy is \$84 per year, find the amount of a medical bill which the insured may incur during a year so that he will recover exactly the medical expense plus the cost of the premiums on the two policies.

The following formulas will be helpful in the problems 20 through 24. If a body travels at a constant speed for a given time, the relation among the distance d, the speed s, and the time t is expressed by the formulas:

$$d = st, \qquad s = \frac{d}{t}, \qquad t = \frac{d}{s},$$

20. A man drove 220 miles in $3\frac{1}{2}$ hours. Part of the trip was at 60 miles per hour and the rest at 65 miles per hour. Find the time spent at the lower speed.

21. Town A is 11 miles west of town B. A man walks from A to B at the rate of 3 miles per hour, and another man,

starting at the same time, walks from B to A at the rate of 4 miles per hour. Find the times after starting that the men are 2 miles apart.

22. Two men starting at a point on a 1-mile circular racetrack walk in opposite directions and meet in 6 minutes. But if they walk in the same direction, the faster walker gains one lap in 30 minutes. Find the speed of each man in miles per hour.

23. The speed of a motorboat in still water is 16 miles per hour. Find the speed of a river's current if the motorboat goes 5 miles downstream in the same time required to go 3 miles upstream. *Hint:* The speed downstream is the speed in still water plus the rate of flow of the river, and the speed upstream is the speed in still water minus the rate of flow of the river.

24. A motorboat goes 3 miles upstream in the same time required to go 5 miles downstream. If the rate of flow of the river is 3 miles per hour, find the speed of the motorboat in still water.

6-5 LINEAR EQUATIONS IN TWO VARIABLES

An equation of the form

$$ax + by + c = 0 \tag{1}$$

where a, b, and c are constants with a and b not both zero, is called a *linear equation* in the variables x and y. A pair of values, one for x and one for y, which satisfy the equation is called a *solution* of the equation. We have already solved equations of the form (1) with $b = 0$. Now we assume that $b \neq 0$ and express the equation in the equivalent form (Sec. 6-2)

$$y = -\frac{a}{b}x - \frac{c}{b}$$

This equation reveals that there is a unique value of y corresponding to any arbitrary value of x. Hence, we say there are infinitely many solutions of the equation. The totality of solutions is called the *solution set* of the equation. If we let A stand for the solution set, we may express the set as

$$A = \left\{(x, y) \mid y = -\frac{a}{b}x - \frac{c}{b}\right\}$$

or, equivalently,

$$A = \{(x, y) \mid ax + by + c = 0\}$$

Referring to Definition 5-4, we note that the graph of a linear equation in x and y is the set of all points (x, y) in the plane whose coordinates satisfy the equation. And it is proved in analytic geometry that the graph is a straight line. The name linear equation is derived from this. The function defined by the equation is quite appropriately called a *linear function.* Clearly, the graph of this kind of equation (or function) can be drawn from two of its points.

Two linear equations in the same two variables are said to constitute a *system* of equations. We represent such a system by the equations

$$\begin{aligned} a_1x + b_1y &= c_1 \\ a_2x + b_2y &= c_2 \end{aligned} \tag{2}$$

We have seen how to find the solution set of a linear equation in one variable, and we shall now define the solution set of a pair of equations of the form in Eqs. (2).

Definition 6-3 If we let the solution sets of the separate equations be denoted by

$$A = \{(x, y) \mid a_1x + b_1y = c_1\} \quad \text{and} \quad B = \{(x, y) \mid a_2x + b_2y = c_2\}$$

then the solution set of the system of equations is defined to be the intersection $A \cap B$ of the sets A and B.

We shall consider the problem of solving systems of the form as in Eqs. (2). We begin by using a graphical method. Since the graph of a linear equation in x and y is a straight line, we shall illustrate each of the following cases:

1. The graphs are intersecting lines.
2. The graphs are parallel lines.
3. The graphs are coincident lines.

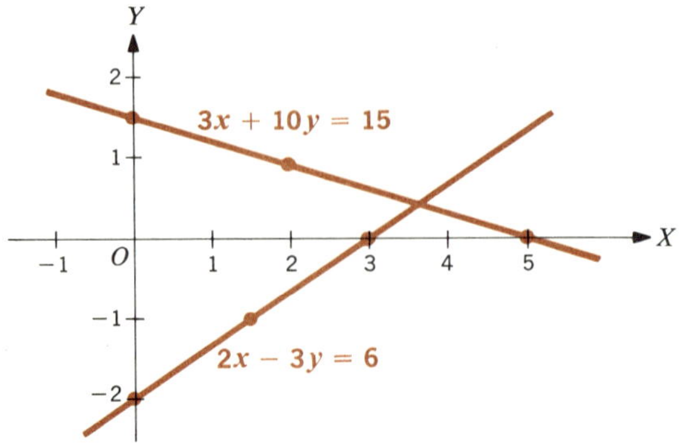

Figure 6-2

EXAMPLE 1 Graphically, solve the system

$$2x - 3y = 6$$
$$3x + 10y = 15$$

Solution: We prepare a table of corresponding values of x and y as in Sec. 5-3. Although two number pairs are sufficient for drawing a line, we suggest a third pair to serve as a check against errors.

$2x - 3y = 6$:

x	0	3	1.5
y	−2	0	−1

$3x + 10y = 15$:

x	0	5	2
y	1.5	0	0.9

The lines intersect as shown in Fig. 6-2. We estimate the coordinates of the intersection point as $x = 3.6$ and $y = 0.4$. These values are approximately correct. They satisfy the first equation, but make the left side of the second equation 14.8 instead of 15. So we take $\{(3.6, 0.4)\}$ as an approximate solution.

EXAMPLE 2 Construct the graph of the system of equations

$$3x + 4y = 5$$
$$6x + 8y = -2$$

Solution: We make a table of corresponding values for each equation.

$3x + 4y = 5$:

x	-1	3
y	2	-1

$6x + 8y = -2$:

x	1	-3
y	-1	2

The graph (Fig. 6-3) consists of two lines which appear to be parallel. If the lines are parallel, there is no pair of numbers for x and y which satisfy both equations. However, independent of the graph, we can show that there is no solution. Thus, multiplying both members of the first equation by 2, which yields an equivalent equation (Theorem 6-2), we may write the system as

$$6x + 8y = 10$$
$$6x + 8y = -2$$

A pair of numbers which makes $6x + 8y$ equal to 10 would certainly not make $6x + 8y = -2$. Hence, the lines of the graph have no common point, and consequently the given system of equations has no solution. We may express this condition by saying that the solution set is the null set $\varnothing$ (Sec. 1-3).

EXAMPLE 3 Finally, we examine the system

$$3x + 4y = 5$$
$$6x + 8y = 10$$

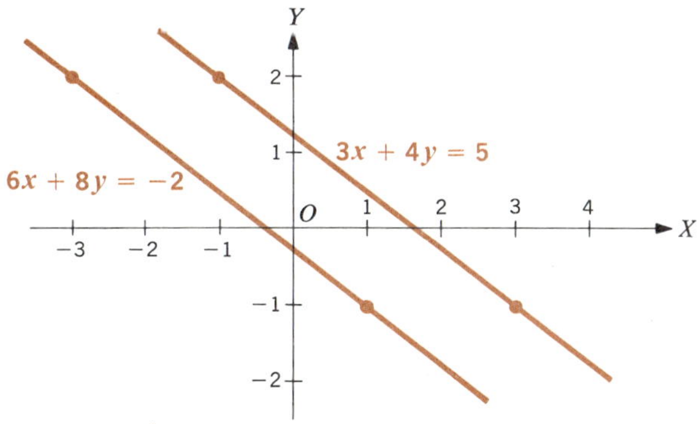

Figure 6-3

If we multiply the members of the first equation by 2, we obtain the second equation. This means that the graphs of the given equations are coincident lines and that every pair of values which satisfies one of the equations satisfies the other. Hence, we express the solution set by $\{(x, y)|3x + 4y = 5\}$.

Summarizing, we have illustrated the following kinds of solution sets of a system of the form in Eqs. (2):

1. The solution set is a single pair of numbers. In this case the graphs of the equations intersect, in a single point, and the equations are said to be *consistent.*
2. The solution set is the null set $\varnothing$. The graphs of the equations are distinct parallel lines, and the equations are said to be *inconsistent.*
3. The solution set of one equation is the same as the solution set of the other, and the equations are said to be *dependent.* The graphs of the equations coincide.

We observe that, in effect, a pair of dependent equations is really only one equation. Consequently the solution set consists of infinitely many pairs of numbers. In the other possibilities, consistent and inconsistent equations, the values of x and y for a solution are held to a single pair or no solution exists. Because of this restriction on x and y, we shall frequently refer to these letters in a system of the form (2) as *unknowns* instead of variables.

EXERCISE 6-3

Draw the graphs of the equations in each of the following systems. Estimate, to one decimal place, the coordinates of the intersection point of each consistent system. For each of the other systems tell if the equations are inconsistent or dependent.

1. $x + y = 6$, $x - y = 2$

2. $2x - 3y = 2$, $x + 2y = -6$

3. $3x + 2y = 6$, $2x - 3y = 6$

4. $y + 4x = 3$, $2y + 8x = 6$

5. $7x + 3y = 3$, $14x + 6y = 6$

6. $4x - 6y = 12$, $2x - 3y = -6$

7. $4x + 4y = 9$, $x - y = 0$

8. $x + 5y = 1$, $2x + 3y = -2$

9. $3x - 4y = 5$
$2x - 3y = 3$

10. $5x + 2y = 10$
$5x + 2y = -10$

11. $4x - 6y = 15$
$6x - 9y = 20$

12. $8x - 6y = 12$
$12x - 9y = 18$

13. $5x - 4y = 8$
$3x + 5y = -15$

14. $4x + 5y = -15$
$7x + 10y = -25$

15. $9x - y = 4$
$x - 9y = 3$

6-6 SOLUTION BY ALGEBRAIC METHODS

Next, we discuss two different types of algebraic methods for solving a system of linear equations in two unknowns. In each case the solution will be obtained through operations which reduce the given system to an equivalent system in which x is equal to a constant and y is equal to a constant.

Type A: Elimination by addition or subtraction. We apply this method in solving the system.

$$4x + 11y = 6 \tag{1}$$

$$5x - 3y = -26 \tag{2}$$

Solution: If we multiply the members of Eq. (1) by 5 and those of Eq. (2) by 4, we obtain the equivalent system

$$20x + 55y = 30 \tag{3}$$

$$20x - 12y = -104 \tag{4}$$

We now subtract the corresponding members of Eq. (4) from Eq. (3) to eliminate x. Thus, subtracting equals from equals, we get

$$67y = 134$$

$$y = 2$$

If we substitute 2 for y in either of the given equations, we find $x = -4$. We check by substituting $x = -4$, $y = 2$ in the left members of the given equations.

Check:

$$4(-4) + 11(2) = -16 + 22 = 6$$
$$5(-4) - 3(2) = -20 - 6 = -26$$

Hence, the ordered pair of numbers $(-4, 2)$ satisfies both equations, and $\{(-4, 2)\}$ is the solution set.

As illustrated in this example, the following steps are involved in the addition or subtraction method:

1. Multiply the members of the equations by properly chosen constants so as to obtain new equations with the coefficients of one of the unknowns the same in absolute value.
2. Then, by adding or subtracting, eliminate that unknown.
3. Solve the resulting equation for its unknown.
4. Substitute the number thus found in either of the original equations and find the other unknown.

Step 1 should be omitted if the coefficients of one of the unknowns in the given equations have the same absolute value.

Type B: Elimination by substitution. We use this method in solving the system

$$5x - 3y = 16 \qquad (1)$$
$$4x - 5y = 5 \qquad (2)$$

Solving Eq. (1) for y, we get

$$y = \frac{5x - 16}{3}$$

By substituting this expression for y in Eq. (2), we obtain a linear equation in x. Thus,

$$4x - 5\left(\frac{5x - 16}{3}\right) = 5$$

Multiplying by 3 gives

$$12x - 25x + 80 = 15$$
$$-13x = -65$$
$$x = 5$$

Then

$$y = \frac{5x - 16}{3} = \frac{25 - 16}{3} = 3$$

The solution set is $\{(5, 3)\}$.

If either of the above methods of solution is applied to an inconsistent or a dependent system of equations, both unknowns will be eliminated simultaneously. If the elimination process leads to an equation whose members are different constants, the equations are inconsistent. If the resulting members are the same constant, the equations are dependent.

EXERCISE 6-4

Solve each pair of equations, if found consistent, by the addition or subtraction method. Then check your solution by using the substitution method.

1. $x - 3y = 3$, $x + y = 1$ **2.** $4x - 3y = -2$, $2x + 3y = 2$ **3.** $3x + 4y = 1$, $5x + 2y = 11$

4. $4x - 3y = 3$, $8x - 6y = 6$ **5.** $3x + 15y = 2$, $2x + 10y = 3$ **6.** $4x - 2y = -3$, $2x + 4y = 9$

Solve each system for x and y by either elimination process and check your results.

7. $5y + 5x = 18$, $y - x = 0$ **8.** $5x + y = 6$, $3x + 2y = -2$

9. $4x - 3y = 5$, $3x - 2y = 3$ **10.** $7x + 3y = 4$, $4x + 2y = 3$

11. $3x - 4y = 7$, $5x - 5y = 8$ **12.** $5x + 7y = -2$, $4x + 6y = -3$

13. $3x + 2y = -3$, $2x - 5y = 17$ **14.** $2x - 7y = 17$, $4x - 5y = 25$

15. $3x + 4y = 7$, $2x + 3y = 6$ **16.** $12x - 12y = 13$, $4x + 6y = 1$

17. $2x - 4y = -3$, $4x + 2y = 9$ **18.** $ax - 3by = c$, $2ax + 2by = 9c$

19. $3x + ay = 4$, $2x + ay = 2$ **20.** $x - y = 2a$, $x + y = 2b$

21. $ax - by = 2b$
$3ax + 4by = 7a - b$

22. $\dfrac{x-2}{2x+2} = \dfrac{y+3}{2y-3}$
$\dfrac{x-1}{x+2} = \dfrac{3y+8}{3y-2}$

23. $\dfrac{2x-3}{y+7} = \dfrac{2x-1}{y+5}$
$\dfrac{y+3}{x+5} = \dfrac{5y-14}{5x-4}$

24. $\dfrac{x+8}{y+8} = \dfrac{2x-5}{2y-3}$
$\dfrac{y+7}{x+8} = \dfrac{y-5}{x-6}$

Solve each system for $1/x$ and $1/y$ and, from the results, obtain x and y.

25. $\dfrac{1}{x} + \dfrac{2}{y} = 3$
$\dfrac{1}{x} - \dfrac{3}{y} = 2$

26. $\dfrac{2}{x} - \dfrac{3}{y} = 1$
$\dfrac{4}{x} - \dfrac{3}{y} = 4$

27. $\dfrac{5}{x} + \dfrac{3}{y} = 7$
$\dfrac{4}{x} + \dfrac{3}{y} = 2$

28. $\dfrac{5}{3x} - \dfrac{1}{y} = \dfrac{2}{3}$
$\dfrac{3}{5x} - \dfrac{1}{2y} = \dfrac{1}{10}$

29. $\dfrac{5}{9x} - \dfrac{2}{3y} = 3$
$\dfrac{1}{3x} - \dfrac{3}{4y} = \dfrac{5}{2}$

30. $\dfrac{3}{2x} - \dfrac{1}{3y} = \dfrac{3}{2}$
$\dfrac{1}{2x} - \dfrac{2}{3y} = \dfrac{2}{3}$

31. Show that the equations of the system

$$a_1x + b_1y = c_1$$
$$a_2x + b_2y = c_2$$

are dependent if

$$\frac{a_1}{a_2} = \frac{b_1}{b_2} = \frac{c_1}{c_2}$$

Hint: Set each of the three fractions equal to a constant k. Show also that the equations are inconsistent if

$$\frac{a_1}{a_2} = \frac{b_1}{b_2} \neq \frac{c_1}{c_2}$$

6-7 WORD PROBLEMS LEADING TO SYSTEMS OF EQUATIONS

There are many kinds of verbal problems involving two or more unknowns which can be solved by setting up a system of equations from the known data. The reasoning by which the equations are obtained is much the same as for the equations of one unknown in Sec. 6-4.

EXAMPLE 1 If the length of a certain rectangle were increased by 2 feet and the width decreased by 1 foot, the area would be decreased by 2 square feet. But if the length were decreased by 2 feet and the width increased by 2 feet, the area would be increased by 16 square feet. Find the dimensions of the rectangle.

Solution: Three rectangles are involved in this problem. If we let x feet be the length and y feet the width of the first rectangle, the dimensions of the second rectangle are $x + 2$ feet by $y - 1$ feet and the dimensions of the third rectangle are $x - 2$ feet by $y + 2$ feet. Since the area of the first rectangle exceeds the area of the second rectangle by 2 square feet and the area of the third rectangle exceeds the area of the first rectangle by 16 square feet, we can form the equations

$$xy - (x + 2)(y - 1) = 2$$
$$(x - 2)(y + 2) - xy = 16$$

Simplifying these equations, we obtain the equivalent system

$$x - 2y = 0$$
$$x - y = 10$$

whose solution is $x = 20$ feet, $y = 10$ feet.

EXAMPLE 2 A man rows 4 miles upstream in a river and back in $2\frac{1}{2}$ hours. If his rate downstream is 4 times his rate upstream, find his rowing rate in still water and the rate of the current.

Solution: If we let x miles per hour stand for his rate in still water and y miles per hour stand for the rate of the current, then the rate upstream is $x - y$ and the rate downstream is $x + y$. Since time is equal to distance divided by rate, $4/(x - y)$ is the rate upstream and $4/(x + y)$ is the rate downstream. From the preceding information, we have the system of equations

$$x + y = 4(x - y)$$
$$\frac{4}{x - y} + \frac{4}{x + y} = \frac{5}{2}$$

By solving the first equation for y, say, and substituting the result in the second equation, it may be determined that $x = 5$. This value for x in the first equation gives $y = 3$. Hence, the rowing rate in still water is 5 miles per hour and the rate of the current is 3 miles per hour.

Skip

EXERCISE 6-5

Find the solution of each of the following problems by first setting up a system of two equations in two unknowns.

1. A cattleman sold 40 calves and 60 sheep to a buyer for \$3,040; and, pricing the animals in the same way, he sold 50 calves and 70 sheep to another buyer for \$3,690. Find the price per head of each kind of animal.

2. If the numerator and denominator of a certain fraction are each decreased by 3, the value of the new fraction is $\frac{1}{2}$. But if the numerator of the original fraction is increased by 1 and the denominator decreased by 1, the resulting fraction is equal to $\frac{3}{4}$. Find the original fraction.

3. A man has two investments, one paying 5% annual interest and the other 6%. The total annual income from the investments is \$305. If the interest rates were interchanged, the total annual income would be \$300. Find the amount of each investment.

4. If the length of a certain rectangle were increased by 3 feet and the width decreased by 2 feet, the area would be decreased by 12 square feet. If the length were decreased by 2 feet and the width increased by 2 feet, the area would be increased by 12 square feet. Find the dimensions of the rectangle.

5. A and B can paint a certain house if A works 12 days and B works 24 days, or they can paint the house if A works 18 days and B works 16 days. How long would it take each working alone to paint the house?

6. A man and a boy perform a task in 2 days. They completed one-half of the task the first day with the man working 8 hours and the boy working 10 hours. On the next day they finished the task with the man working 4 hours and the boy working 15 hours. Find how many hours it would take for each to do the task working alone.

7. The sum of the reciprocals of two numbers is 13. Three times the reciprocal of one of the numbers is 4 more than twice the reciprocal of the other number. Find the two numbers.

8. A certain two-digit number is equal to 4 times the sum of its digits. If 36 were added to the number, the digits would be reversed. Find the number. *Hint:* If x is the tens digit and y the units digit, the value of the number is $10x + y$.

9. An airplane travels 360 miles with the wind in 1 hour and 20 minutes and returns against the wind in 2 hours and 15 minutes. Find the speed of the airplane in still air and the speed of the wind.

10. A boat goes 6 miles downstream in a river in 30 minutes and returns upstream in 45 minutes. Find the speed of the boat in still water and the speed of the current.

11. A man rows 12 miles upstream in a river and back in $7\frac{1}{2}$ hours. He can row 1 mile upstream in the same time that he can row 4 miles downstream. Find the speed of the boat in still water and the speed of the current.

12. A laborer was employed to work for a daily wage of \$30 and pay \$4 per day for board. He was idle on certain days and received no wage for that time, but was charged for food every day. His net pay at the end of the period was \$764. Had his daily wage been \$36 and his board \$6, the net pay would have been \$876. How many days did he work and how many days was he idle?

13. The sum of the three angles of a triangle is 180°. The sum of two of the angles is equal to the third angle and the difference of the two angles is equal to two-thirds of the third angle. Find the angles.

14. To provide change for a day's business, a merchant buys \$75 worth of half-dollars, quarters, dimes, and nickels. Altogether there are 360 coins, and the number of half-dollars is the same as the number of quarters, and the number of dimes is

the same as the number of nickels. Find the number of coins of each kind.

15. A man makes three investments totaling \$12,000 with interest rates of $2\frac{1}{2}\%$, 3%, and $3\frac{1}{2}\%$ per year. The total annual income is \$370, and the income from the 3% investment is \$10 less than the combined income from the other investments. Find the amount of each investment.

6-8 LINEAR EQUATIONS IN THREE UNKNOWNS

We next consider the problem of solving a system of three linear equations in three unknowns. The algebraic methods which were used for solving a system of equations in two unknowns will serve in an analogous fashion to yield the solution for the case of three unknowns.

A system of three linear equations in three unknowns has one solution, no solution, or infinitely many solutions. This same situation was discussed in the case of two unknowns. But we shall consider systems of three unknowns which have one and only one solution.

EXAMPLE Solve the system of equations

$$3x + 2y + 2z = 8 \tag{1}$$

$$2x + 3y + 4z = 11 \tag{2}$$

$$4x - 4y - 3z = 8 \tag{3}$$

Solution: Selecting y as the first unknown to be eliminated, we multiply the members of Eq. (1) by 2, and use Eq. (3) unchanged and obtain

$$6x + 4y + 4z = 16$$

$$4x - 4y - 3z = 8$$

Adding, we have

$$10x + z = 24$$

Next, we multiply the members of Eq. (1) by 3, and the members of Eq. (2) by 2. This gives

$$9x + 6y + 6z = 24$$
$$4x + 6y + 8z = 22$$

Subtracting, we get

$$5x - 2z = 2$$

We now have the equations $10x + z = 24$ and $5x - 2z = 2$. The solution of these equations is $x = 2$, $z = 4$. Now x and z may be replaced by these values in any of the given equations to yield $y = -3$. Hence, $x = 2$, $y = -3$, $z = 4$ is the solution of the given system. We use $\{(2, -3, 4)\}$ to denote the solution set.

The following steps may be employed to solve a system of linear equations of three unknowns:

1. Choose a pair of equations from the three given equations and eliminate one of the unknowns.
2. Select one of the same pair and the unused equation and eliminate the same unknown.
3. Solve the equations resulting from steps 1 and 2 for the two unknowns.
4. Substitute these values in one of the given equations and obtain the remaining unknown.

EXERCISE 6-6

Solve the following systems of equations for x, y, and z or u, v, and w. Check your results.

1. $3x + 4y + z = 1$
   $x + y - 3z = 3$
   $3x + 2y - 2z = 0$

2. $2x + y - 3z = 9$
   $3x - y - 2z = 11$
   $x + y - z = 2$

3. $4u - v - 2w = 10$
   $3u - 2v - w = 5$
   $2u - 3v - 2w = 10$

4. $3u - v - 3w = 9$
   $4u - v - 2w = 13$
   $7u - 2v + w = 24$

5. $4x - 2y - 3z = 3$
   $3x + 6y - 4z = 3$
   $x - 2y - z = 1$

6. $2x + 3y + 4z = 8$
   $3x + 2y + 3z = 8$
   $2x - 2y - z = 3$

7. $3x + 2y - 5z = 28$
   $4x - 5y - 4z = 9$
   $5x + 4y + 3z = 26$

8. $5x + 7y + 9z = 14$
   $7x - 3y + 4z = 11$
   $3x - 4y - 8z = -5$

9. $$\begin{aligned} 5x + 4y + 3z &= 2 \\ 4x + 2y - 2z &= -1 \\ 6x + 4y - 2z &= -1 \end{aligned}$$

10. $$\begin{aligned} u + v &= 1 \\ u - w &= -1 \\ v - w &= -6 \end{aligned}$$

11. $$\begin{aligned} ax + by &= 2 \\ ax + 2cz &= 3 \\ by + 3cz &= 2 \end{aligned}$$

12. $$\begin{aligned} 4u + 9v &= 4 \\ u + 6w &= 4 \\ 3v - 4w &= -1 \end{aligned}$$

13. $$\begin{aligned} u - v - w &= 2a \\ u + v - w &= 2b \\ u - v + w &= 2c \end{aligned}$$

14. $$\begin{aligned} ax - by - cz &= 6 \\ ax + by - cz &= 8 \\ ax + by + cz &= 0 \end{aligned}$$

15. $$\begin{aligned} ax + 2by + 4cz &= 5 \\ 2ax - 3by + 2cz &= 6 \\ 3ax - 4by + 3cz &= 8 \end{aligned}$$

16. $$\begin{aligned} ax - by + az &= a^2 \\ x - y + z &= a \\ bx + ay - az &= -b^2 \end{aligned}$$

17. $$\begin{aligned} ax - by + a^2z &= b^2 \\ bx + ay - b^2z &= b^2 \\ x - y + az &= b \end{aligned}$$

18. $$\begin{aligned} 2x + 3y &= 2 \\ x + 2z &= 1 \\ 3y + 8z &= 3 \end{aligned}$$

19. $$\begin{aligned} \frac{1}{x} + \frac{2}{y} - \frac{1}{z} &= \frac{4}{3} \\ \frac{2}{x} + \frac{3}{y} - \frac{2}{z} &= \frac{5}{3} \\ \frac{3}{x} + \frac{4}{y} - \frac{6}{z} &= 3 \end{aligned}$$

20. $$\begin{aligned} \frac{2}{x} - \frac{2}{y} - \frac{3}{z} &= 4 \\ \frac{1}{x} - \frac{4}{y} + \frac{4}{z} &= 0 \\ \frac{3}{x} - \frac{3}{y} - \frac{2}{z} &= 1 \end{aligned}$$

21. $$\begin{aligned} \frac{2}{x} - \frac{5}{y} - \frac{3}{z} &= -2 \\ \frac{3}{x} + \frac{4}{y} - \frac{6}{z} &= 9 \\ \frac{5}{x} - \frac{3}{y} + \frac{3}{z} &= 1 \end{aligned}$$

6-9 RATIO AND PROPORTION

Two numbers may be compared in magnitude by dividing one by the other. The quotient is called the *ratio* of one number to the other. If x and y ($\neq 0$) stand for two numbers, the ratio of x to y may be expressed in any of the forms

$$\frac{x}{y} \qquad \text{or} \qquad x/y, \qquad x \div y, \qquad x:y$$

The fractional form (x/y) is usually preferable when algebraic operations are to be performed. The form $x:y$ (read "x is to y") uses the colon to indicate division.

Two like quantities may be compared in magnitude by first expressing them in terms of the same unit of measure. For example, the ratio of 16 hours to 3 days is

$$\tfrac{16}{72} = \tfrac{2}{9}$$

Hence, the ratio of 16 hours to 3 days is 2 to 9. The same information is conveyed by saying the ratio of 3 days to 16 hours is 9 to 2. Often the magnitudes of unlike things are expressed as a ratio. For example, if the cost of a city lot of 50 feet frontage is \$3,000, we use the ratio 3000/50 in finding the cost per foot of frontage.

A statement of equality of two ratios is called a *proportion*. A proportion may be expressed in any of the forms

$$\frac{a}{b} = \frac{c}{d}, \qquad a \div b = c \div d, \qquad a:b = c:d$$

The form $a:b = c:d$ is read "a is to b as c is to d." The numbers a and d are called the *extremes* and b and c the *means* of the proportion. The number d is called the *fourth* proportional to a, b, and c. If the means are equal ($b = c$), the proportion becomes

$$\frac{a}{b} = \frac{b}{d}$$

In this special case b is called the *mean proportional* to a and d, and d is the *third proportional* to a and b.

EXAMPLE 1 Find the third proportional to 4 and 7.

Solution: To form the proportion, we observe that each of the means is 7. Then letting x stand for the third proportional, we have

$$4:7 = 7:x \qquad \text{or} \qquad \frac{4}{7} = \frac{7}{x}$$

hence, clearing of fractions, $4x = 49$ and $x = 12.25$.

EXAMPLE 2 A triangle has sides of lengths 24, 34, and 44 inches. If the length of the shortest side of a similar triangle is 16 inches, find the length of the longest side.

Solution: We recall that the corresponding sides of similar figures are proportional. Then, if x inches denotes the length of the longest side, we have

$$\frac{x}{44} = \frac{16}{24}$$

$$24x = (44)(16)$$

$$x = 29\tfrac{1}{3} \text{ inches}$$

Sometimes, we express the relative magnitudes of three numbers in the form of a ratio. Suppose, for example, that the ratio of x to y is 2 to 3 and that the ratio of y to z is 3 to 7. Then we say that x is to y is to z as 2 is to 3 is to 7, and write

$$x:y:z = 2:3:7$$

Proportions occur in many areas of mathematics and other sciences. The following theorems express useful properties of proportions. The values of a, b, c, and d are assumed to be such that no denominator is equal to zero.

Theorem 6-3 If $\dfrac{a}{b} = \dfrac{c}{d}$, then $ad = bc$.

Theorem 6-4 If $\dfrac{a}{b} = \dfrac{c}{d}$, then $\dfrac{a}{c} = \dfrac{b}{d}$.

Theorem 6-5 If $\dfrac{a}{b} = \dfrac{c}{d}$, then $\dfrac{b}{a} = \dfrac{d}{c}$.

Theorem 6-6 If $\dfrac{a}{b} = \dfrac{c}{d}$, then $\dfrac{a+b}{b} = \dfrac{c+d}{d}$.

Theorem 6-7 If $\dfrac{a}{b} = \dfrac{c}{d}$, then $\dfrac{a-b}{b} = \dfrac{c-d}{d}$.

Theorem 6-8 If $\dfrac{a}{b} = \dfrac{c}{d}$, then $\dfrac{a+b}{a-b} = \dfrac{c+d}{c-d}$.

Theorem 6-3 tells us that the *product of the extremes of a proportion is equal to the product of the means.* The theorem may be established by multiplying both members of the proportion by bd. Theorem 6-5 may be proved by dividing the members of $bc = ad$ by ac. Theorem 6-6 may be proved by first adding 1 to both members of the proportion, and Theorem 6-8 may readily be obtained from Theorems 6-6 and 6-7.

EXERCISE 6-7

Express each ratio as a fraction in lowest terms.

1. 4 feet to 10 inches **2.** 15¢ to \$1.20

3. 8 hours to 3 days **4.** 6 quarts to 2 gallons

5. \$1.60 to 3 quarters **6.** 80 square inches to 2 square feet

Find the fourth proportional to each set of numbers, and the third proportional to the first two numbers.

7. 6, 3, 4 **8.** 5, 6, −2

9. −9, 11, −3 **10.** 3, 1, 9

Find the mean proportional to each pair of numbers.

11. 3, 12 **12.** 28, 7 **13.** 6, 24 **14.** 25, 9

Find the unknown or unknowns in each problem.

15. $4:2 = x:5$ **16.** $9:6 = 7:x$ **17.** $x:9 = 4:x$

18. $(2 + x):3 = 5:2$ **19.** $5:x = x:20$

20. $2:x = 3:(x + 3)$ **21.** $x:2:3 = 4:5:y$

22. $2:3:x = y:5:7$ **23.** $x:y:3 = 3:2:5$

24. The locations of two school buildings, as shown on a city map, are 5 inches apart. If the map scale is 3 inches to 7 miles, find the distance between the two buildings.

25. Divide 144 into two parts which have the ratio 4:5.

26. The sides of a triangle are 7, 9, and 10 inches long. The longest side of a similar triangle is 8 inches. Find the lengths of the other sides of the triangle.

27. Three men do a piece of work for \$123. If this sum is to be divided among the men in the ratio 3:4:5, find the three resulting amounts.

6-10 VARIATION

In our discussion of ratios thus far we have dealt with fixed quantities. We next consider the ratio of two variables which are so related that their ratio is constant, or, equivalently, one variable is a constant times the other variable. Variables of this kind are plentiful in science, business, and many everyday situations.

If a variable y is equal to a constant times another variable x, then y is said to *vary directly as* x. The relation may be written as

$$y = kx$$

The fixed number k is called the *constant of proportionality,* or the *constant of variation.* The expressions

y varies directly as x
y varies as x
y is directly proportional to x
y is proportional to x

are all used with the same meaning.

The formula for the circumference of a circle $C = 2\pi r$ shows that the circumference varies directly as the radius and that 2π is the constant of variation. The formula for the area of a circle $A = \pi r^2$ shows that the area varies directly as the square of the radius and that π is the constant of proportionality.

Suppose that z is always equal to a constant times the product of two variables x and y; that is

$$z = kxy$$

We describe this relation by saying that *z varies jointly as x and y.*

The area of a triangle varies jointly as the base and altitude. The formula $A = \frac{1}{2}bh$ shows the constant of variation to be $\frac{1}{2}$.

EXAMPLE 1 The weight of a sphere is proportional to the cube of its radius. A certain sphere of radius 3 inches weighs 18 pounds. Find the weight of a sphere of radius 6 inches which is made of the same kind of material.

Solution: We let W pounds stand for the weight and r inches for the radius and write

$$W = kr^3$$

To determine k, we substitute 18 for W and 3 for r. Thus, we find

$$18 = 27k \qquad \text{and} \qquad k = \tfrac{2}{3}$$

Then for all spheres of this material, we have

$$W = \frac{2r^3}{3}$$

Hence, the weight of the sphere of radius 6 inches is

$$W = \frac{2(6)^3}{3} = 144 \text{ pounds}$$

If a variable y is always equal to a constant times the reciprocal of another variable x, then y is said to *vary inversely as* x, or *y is inversely proportional to* x. This relation is expressed by the equation

$$y = \frac{k}{x}$$

Boyle's law is an example of inverse variation. According to this law, the volume of a confined gas varies inversely with the pressure if the temperature remains constant. Thus, if V denotes the volume and p the pressure, then $V = k/p$.

The different kinds of variation may occur simultaneously in problems. In the relation

$$z = \frac{kx}{y}$$

z varies directly as x and inversely as y. The relation

$$W = \frac{kxy^3}{z^2}$$

means W varies jointly as xy^3 and inversely as z^2.

EXAMPLE 2 The maximum safe load for a horizontal beam supported at both ends varies jointly as the width and the square of the depth and inversely as the length. If a beam 2 inches wide, 6 inches deep, and 10 feet long can safely bear a load up to 1500 pounds, find the maximum safe load for a beam of the same material which is 3 inches wide, 9 inches deep, and 15 feet long.

Solution: Using L = load, w = width, d = depth, l = length, we write the law of the beam as

$$L = \frac{kwd^2}{l}$$

If we substitute $L = 1500$, $w = 2$, $d = 6$, and $l = 10$, we find

$$1500 = \frac{k(2)(6)^2}{10} \qquad \text{and} \qquad k = \frac{625}{3}$$

With this value for k, the formula becomes

$$L = \frac{625wd^2}{3l}$$

where L is in pounds, w and d are in inches, and l is in feet. Then the maximum safe load for the beam which has $w = 3$, $d = 9$, and $l = 15$ is given by

$$L = \frac{625(3)(9)^2}{3(15)} = 3375 \text{ pounds}$$

EXERCISE 6-8

Express the statements 1 through 5 as equations, using k for the constant of variation.

1. The surface area S of a sphere varies as the square of the radius r.
2. The area A of a rectangle varies jointly as the base b and altitude h.
3. The volume V of a circular cylinder is proportional to its altitude h and the square of its radius r.
4. The attraction F between two bodies varies inversely as the square of the distance s between them.
5. Neglecting air resistance, the distance s which a body falls from rest varies directly as the square of the time t.
6. The weight of a sphere is proportional to the cube of its radius. A certain sphere of radius 4 inches weighs 24 pounds. Find the weight of a sphere of radius 6 inches which is made of the same kind of material.

7. The force required to stretch a spring is proportional to the elongation. If 24 pounds stretches a spring 3 inches, find the force required to stretch the spring
(**a**) 2 inches, (**b**) $5\frac{1}{2}$ inches.

8. The amount of paint required to paint a circular floor varies as the square of the radius. If it takes 3 quarts for a floor of radius 10 feet, find the amount required for a floor of radius 16 feet.

9. A constant force moves a body in a line on a smooth surface. The distance which the body moves varies jointly as the force and the square of the time the force acts. If the distance is 48 feet when the force is 8 pounds and the time 2 seconds, find the distance when the force is 6 pounds and the time 3 seconds.

10. The volume of a confined gas varies inversely as the pressure if the temperature remains unchanged. If the volume is 720 cubic inches when the pressure is 9 pounds per square inch, find the volume when the pressure is
(**a**) 6 pounds, (**b**) 12 pounds, (**c**) 15 pounds.

11. The weight of a body above the earth's surface varies inversely as the square of the distance from the center of the earth. A body weighing 64 pounds at the surface (4000 miles from the center) would have what weight at a distance above the surface of
(**a**) 1000 miles, (**b**) 2000 miles, (**c**) 4000 miles?

12. The time required for a crew of men to do a job is inversely proportional to the product of the number of men and the number of hours they work per day. If a crew of 9 men working 10 hours a day do a job in 14 days, find how long it would take 12 men working 7 hours per day to do the job.

13. The volume of gas in a container varies directly as its absolute temperature and inversely as the pressure. If the volume is 600 cubic inches when the absolute temperature is 300° and the pressure 15 pounds per square inch, find the volume when the absolute temperature is 150° and the pressure is 60 pounds per square inch.

14. The electrical resistance of a wire varies directly as the length and inversely as the square of its radius. The resistance of a certain piece of wire of length 100 feet and radius 0.02 inch is 300 ohms. Find the resistance of a wire of the same material 2500 feet long and radius 0.05 inch.

15. The maximum safe load for a beam 16 feet long, 4 inches wide, and 5 inches deep is 4000 pounds. Find the maximum safe load for a beam of the same material which is 12 feet long, 3 inches wide, and 9 inches deep. (See Example 2, Sec. 6-10.)

16. The maximum safe load for a certain beam 3 inches wide, 8 inches deep, and 12 feet long is 2000 pounds. Find the maximum safe load for a beam of the same material which is 2 inches wide, 6 inches deep, and 10 feet long.

17. The kinetic energy of a moving body is proportional to the square of its velocity. Compare the kinetic energy of a car going 30 miles per hour with the same car going 40 miles per hour.

18. The illumination on a page is proportional to the wattage of the lamp and inversely proportional to the square of the distance from the lamp. Compare the illumination furnished by a 100-watt lamp at a distance of 3 feet with that of a 75-watt lamp at a distance of 2 feet.

19. Find the wattage of a lamp which will give the same amount of light at a distance of 15 feet as a 75-watt lamp will give at a distance of 5 feet. (See problem 18.)

20. Compare the maximum safe load for a beam 18 feet long, 4 inches wide, and 8 inches deep with that of a beam of like material 12 feet long, 3 inches wide, and 12 inches deep.

21. Compare the maximum safe load for a beam 15 feet long, 3 inches wide, and 5 inches deep with that of the same beam when it is turned so that the 3-inch dimension is the depth and the 5-inch dimension is the width.

7

EXPONENTS AND RADICALS

7-1 LAWS OF EXPONENTS

In Sec. 3-3, we defined a positive integral exponent. Accordingly a^n means the product formed by taking a as a factor n times. That is,

$$a^n = a \cdot a \cdot a \cdots a \quad (n \text{ factors})$$

The quantity a^n is called *the* n*th power of the base* a and n is called the *exponent.*

In this chapter we shall extend the definition of exponents to include all rational numbers. Before taking up new exponents, however, we state five laws of exponents and prove that the laws hold for positive integral exponents. The bases a and b in the statement of the laws may be any real numbers which do not make a denominator equal to zero. The proofs of Law 1 and Law 2 are given in Secs. 3-3 and 3-4.

Law 1

$$a^m a^n = a^{m+n}$$

Law 2

$$\frac{a^m}{a^n} = a^{m-n} \qquad \text{if } m > n$$
$$= 1 \qquad \text{if } m = n$$
$$= \frac{1}{a^{n-m}} \qquad \text{if } m < n$$

Law 3

$$(a^m)^n = a^{mn}$$

Proof: By definition, $(a^m)^n$ means a^m is to be taken as a factor n times. But each a^m has a as a factor m times. Hence, in all, a appears as a factor mn times, giving a^{mn}.

Law 4

$$(ab)^m = a^m b^m$$

Proof: By definition $(ab)^m$ means the product obtained by taking ab as a factor m times. Hence, the factor a occurs m times and the factor b occurs m times. By the commutative and associative axioms, we may rearrange the factors so that all the a factors come first and all the b factors follow. Thus, we may write $(ab)^m = a^m b^m$.

Law 5

$$\left(\frac{a}{b}\right)^m = \frac{a^m}{b^m} \qquad \text{if } b \neq 0$$

Proof: By definition, $(a/b)^m$ means the product of m factors each of which is the fraction a/b. Recalling the definition of the products of fractions, we have a^m for the numerator of the product of the fractions and b^m for the denominator.

EXAMPLES

$$(x^3)^2 = x^{3 \cdot 2} = x^6 \qquad (4x)^3 = 4^3 x^3 = 64x^3$$

$$\left(\frac{2a^2}{3b}\right)^4 = \frac{2^4 a^8}{3^4 b^4} = \frac{16a^8}{81b^4} \qquad \frac{a^{5m}}{a^{3m}} = a^{5m-3m} = a^{2m}$$

7-2 ZERO AND NEGATIVE INTEGRAL EXPONENTS

We have defined positive integral exponents and have established five laws of exponents which apply to them. Our next step is to extend the idea of exponents to include zero and the negative integers. The new exponents are to be defined in such a way that they obey the five laws of exponents.

First, let us determine what meaning should be given to the symbol a^0. If Law 2 is to hold when $m = n$, we have

$$\frac{a^n}{a^n} = a^{n-n} = a^0 \qquad (a \neq 0)$$

This division yields a zero exponent. But any nonzero number divided by itself has 1 as a quotient. This leads us to define the zero exponent as follows:

Definition 7-1 If a is a nonzero number, then

$$a^0 = 1$$

Next, in a similar way, we determine what meaning is to be given to a^{-n} when $-n$ is a negative integer. If Law 1 holds when $m = -n$, then

$$a^{-n}a^n = a^0 = 1$$

Dividing both members of this equation by a^n, we have

$$a^{-n} = \frac{1}{a^n} \qquad (a \neq 0)$$

Hence, we make the following definition:

Definition 7-2 If n is a positive integer and $a \neq 0$, then

$$a^{-n} = \frac{1}{a^n}$$

The definitions of a^0 and a^{-n} were reached, in each case, by applying one of the laws of exponents. However, it would be easy to show that all the laws are satisfied by the definitions.

EXAMPLE 1

$$(-3)^2(-3)^{-2} = (-3)^{2-2} = (-3)^0 = 1$$

EXAMPLE 2

$$\left(\frac{a}{b}\right)^{-3} = \frac{a^{-3}}{b^{-3}} = \frac{1/a^3}{1/b^3} = \frac{b^3}{a^3} = \left(\frac{b}{a}\right)^3$$

EXAMPLE 3

$$\frac{(x^2y^{-2})^3(x^{-2}y)^2}{(xy^{-1})^6} = \frac{(x^6y^{-6})(x^{-4}y^2)}{x^6y^{-6}} = \frac{y^2}{x^4}$$

EXAMPLE 4

$$\frac{ax^{-2}y^{-3}}{bz^{-3}} = \frac{a/x^2y^3}{b/z^3} = \frac{az^3}{bx^2y^3}$$

As illustrated by the last example, a factor of the numerator of a fraction may be moved to the denominator, or vice versa, if the sign of the exponent is changed. Terms of the numerator or denominator, however, cannot be handled in this manner. Thus,

$$\frac{1}{x^{-2} + y^{-2}} = \frac{1}{\dfrac{1}{x^2} + \dfrac{1}{y^2}} = \frac{x^2y^2}{x^2 + y^2}$$

and not equal to $x^2 + y^2$.

EXERCISE 7-1

Find the value of each of the following expressions, using the laws of exponents.

1. $2^2 \cdot 2^4$ **2.** $(2^2)^3$ **3.** 5^{-1}

4. $(-6)^0$ **5.** $(ab)^0$ **6.** 4^{-3}

7. $4^2 \cdot 4^{-3}$ **8.** $(4^{-2})^2$ **9.** $(2 \cdot 3)^{-1}$

10. $(3^{-2})^2$ **11.** $(-5)^{-2}$ **12.** $(3 \cdot 7)^0$

13. $(\frac{1}{7})^{-1}$ **14.** $(\frac{2}{5})^{-1}$ **15.** $(3 \cdot 7^0)^{-2}$

16. $\dfrac{5^7}{5^3}$ **17.** $\dfrac{8^5}{4^5}$ **18.** $\dfrac{27^2}{9^2}$

Simplify each expression by performing the indicated operations and leaving the result without zero or negative exponents.

19. $(3xy)^{-2}(6xy^3)$ **20.** $(x^{-2}y^2)^{-1}(x^3y^0)^2$

21. $(a^{-3}b)(a^{-1}b^{-1})^{-1}$ **22.** $\dfrac{a^{-1}b^{-2}}{a^{-2}b^{-1}}$

23. $\dfrac{2^{-1}a^{-4}b}{3^{-1}a^{-2}b^2}$ **24.** $\dfrac{5^{-2}a^3b^2}{10^{-1}a^0b^{-2}}$

25. $\dfrac{24a^3b^4}{8ab^3}$ **26.** $\dfrac{14a^7b^{-2}}{7a^4b^{-3}}$

27. $\dfrac{32x^2y^4z^5}{24x^5y^2z^6}$ **28.** $\dfrac{3x^3y^{-3}z^2}{6x^{-4}y^{-2}z}$

29. $\dfrac{5c^5}{3a^3b^4} \cdot \dfrac{9a^6}{10bc^4}$ **30.** $\dfrac{10a^{-2}bc}{7ab^{-1}c} \cdot \dfrac{14ab^2c^{-4}}{5ab^4c^{-3}}$

31. $\left(\dfrac{3yz}{7x^3}\right)^2\left(\dfrac{7x^2}{z}\right)^2$ **32.** $\dfrac{(8x^3y^6)^3}{(4x^4y^5)^3}$

33. $\left(\dfrac{p^{-3}q^{-2}}{2^0r^{-1}}\right)^{-2}$ **34.** $\dfrac{(5x^2y)^3}{(10xy^3)^2}$

35. $\left(\dfrac{p^2q^4}{pq^{-1}}\right)^{-3}$ **36.** $\left(\dfrac{25p^3q^{-2}}{5r^3}\right)^{-2}$

37. $\dfrac{2^{-1} - 3^{-1}}{2^{-1}}$ **38.** $\dfrac{2^{-1} - 1^{-1}}{2^{-1} + 1^{-1}}$

39. $\dfrac{3^{-1} + 5^{-1}}{3^{-1}}$ **40.** $\dfrac{x^{-1} + y^{-1}}{x^{-2}y^{-2}}$

41. $\dfrac{x^{-2}y^{-2}}{x^{-1} - y^{-1}}$ **42.** $\dfrac{y^{-2}}{x^{-2} + y^{-2}}$

43. $\dfrac{x^{-3} - y^{-3}}{x^{-3} + y^{-3}}$ **44.** $\dfrac{x^{-1}y^2 - x^2y^{-1}}{x^{-1} - y^{-1}}$

7-3 FRACTIONAL EXPONENTS

In this section, we shall extend the idea of exponents to include all rational numbers. Before introducing fractional exponents, however, we need to consider the following definition.

Definition 7-3 If a and b are two numbers such that the nth power of a (n a positive integer) is equal to b, then a is called an nth root of b; that is, if $a^n = b$, then a is an nth root of b.

According to this definition, the equations

$$2^2 = 4 \qquad (-2)^2 = 4 \qquad 3^3 = 27 \qquad (-3)^3 = -27$$

show that $+2$ and -2 are square roots of 4, that 3 is a cube root of 27, and -3 is a cube root of -27. Since 4 has two square roots, we might raise the question as to how many nth roots a number has. Although the proof comes later (Theorem 11-5), we state now that any nonzero number has two square roots, three cube roots, four fourth roots, and so on. But some of these roots involve a new kind of number. This new number, which we shall introduce later, is not a real number. We see at once that a negative number has no real even root (square, fourth, sixth, and so on). This is true because an even power of a positive or a negative number is positive. We wish now, however, to give attention to real numbers only. Hence, we shall defer consideration of numbers and roots of numbers which are not real.

In connection with roots of numbers, we make the following statements but do not prove them.

1. A positive number has exactly two real even roots, one positive and the other negative.
2. A positive or negative number has exactly one real odd root, the sign of the root being the same as the sign of the number.
3. A negative number has no real even root.

If n is a positive even integer, the positive nth root of a positive number a is called the *principal* nth *root of* a. When n is odd, the real nth root of a positive or negative number a is called the principal nth root. The principal nth root of a number is denoted by $\sqrt[n]{a}$. The symbol $\sqrt[n]{a}$ is called a *radical*, a is called the *radicand*, and n is called the *index*, or *order*, of the radical. We are excluding from consideration the case in which the radicand is negative and the index is an even number.

The following equations are illustrations of principal roots of numbers:

$$\sqrt{36} = 6 \qquad \sqrt[3]{8} = 2 \qquad \sqrt[4]{81} = 3 \qquad \sqrt[5]{-32} = -2$$

We note that -6 is a square root of 36 and -3 is a fourth root of 81, but neither is a principal root. The negative of the principal nth root of a number a is denoted by $-\sqrt[n]{a}$. Thus $-\sqrt[4]{81} = -3$.

We are now ready to consider exponents of the form m/n, where m is either a positive or negative integer and n is a positive integer. First, we take $m = 1$ and seek an interpretation of $a^{1/n}$. If Law 3 is to hold, we have

$$(a^{1/n})^n = a^{n/n} = a$$

This equation shows that the nth power of $a^{1/n}$ is equal to a, or that $a^{1/n}$ is an nth root of a. Specifying this root as the principal nth of a, by definition, we have

$$a^{1/n} = \sqrt[n]{a}$$

In this definition, a may be any real number when n is odd, but we exclude negative values for a when n is even.

Applying Law 3 again, with the integer $m \neq 1$, we have

$$a^{m/n} = (a^m)^{1/n} = \sqrt[n]{a^m}$$

and also

$$a^{m/n} = (a^{1/n})^m = (\sqrt[n]{a})^m$$

Summarizing, we have the following definition:

Definition 7-4 If m/n is a rational number with n positive, then

$$a^{m/n} = \sqrt[n]{a^m} = (\sqrt[n]{a})^m$$

The form $\sqrt[n]{a^m}$ means the principal nth root of a^m, and the form $(\sqrt[n]{a})^m$ means the mth power of the principal nth root of a. In each form the denominator n of the exponent indicates a root and the numerator m indicates a power. We note again, however, that n here stands for any positive integer and m for any positive or negative integer.

We began our study of exponents by defining positive integral exponents and establishing the five laws of operations. Then we extended the definitions to include all rational numbers. Although complete proofs of the laws of exponents were made only for the positive integral exponents, we can show that the laws hold for all rational exponents.

Assuming that Law 3 holds for rational exponents, we can show that a fractional exponent can be reduced to lowest terms. Thus, if m, n, and c are integers, n and c not zero, we have

$$a^{cm/cn} = (a^{m/n})^{c/c} = a^{m/n}$$

EXAMPLE 1

$$8^{2/3} = (\sqrt[3]{8})^2 = 2^2 = 4$$
$$8^{2/3} = \sqrt[3]{8^2} = \sqrt[3]{64} = 4$$

EXAMPLE 2

$$16^{-3/4} = \frac{1}{(\sqrt[4]{16})^3} = \frac{1}{2^3} = \frac{1}{8}$$

EXAMPLE 3

$$(-243)^{3/5} = (\sqrt[5]{-243})^3 = (-3)^3 = -27$$

EXAMPLE 4

$$(x^{1/4}y^{5/3})(x^{3/4}y^{1/3}) = x^{1/4+3/4}y^{5/3+1/3} = xy^2$$

EXAMPLE 5

$$\left(\frac{9a^{-4/3}}{b^{-2}c^{2/3}}\right)^{-1/2} = \frac{9^{-1/2}a^{2/3}}{bc^{-1/3}} = \frac{a^{2/3}c^{1/3}}{3b}$$

EXERCISE 7-2

Find the rational value of each expression.

1. $27^{1/3}$ **2.** $49^{1/2}$ **3.** $25^{3/2}$

4. $32^{3/5}$ **5.** $64^{5/6}$ **6.** $9^{-1/2}$

7. $27^{-2/3}$ **8.** $(-8)^{4/3}$ **9.** $(-64)^{2/3}$

10. $(\frac{8}{27})^{-2/3}$ **11.** $(-32)^{-2/5}$ **12.** $(\frac{81}{16})^{-3/4}$

13. $(3^{1/4})^4$ **14.** $(25^{-1/4})^{-2}$ **15.** $(16^{-1/3})^{3/2}$

Simplify each expression, leaving the results without zero or negative exponents.

16. $a^{1/3}a^{5/3}$ **17.** $a^{3/5}a^{-4/5}$ **18.** $a^{-4/3}a^{-2/3}$

19. $3^{3/2} \div 3^{2/3}$ **20.** $a^{1/3} \div a^{1/6}$ **21.** $a^{1/4} \div a^{-3/4}$

22. $(2x^{-2/3})^3$ **23.** $(3x^{-1/2})^4$ **24.** $(4^{1/2}x^{3/2})^4$

25. $(3a^{-3/8})^{-4}$ **26.** $(a^{-2}b^{-2/3})^{-4}$ **27.** $(3a^{-5/3})^{-3}$

28. $(a^{1/2} + 1)(a^{1/2} - 1)$ **29.** $(a^{1/2} + b^{1/2})^2$

30. $(a^{3/2} + 2)(a^{3/2} - 2)$ **31.** $\dfrac{x^{-1/2}y^{4/3}}{x^{3/2}y^{-2/3}}$

32. $\dfrac{36^{1/2}x^{-1}y^{-2}}{6^{1/2}x^{-2}y^{-1/3}}$ **33.** $\dfrac{9^{5/2}x^{3/4}y^{5/4}}{81^{1/2}x^{-1/4}y^{-1/2}}$

34. $\dfrac{8^{1/3}x^{-1/3}y^{5/3}}{4^{3/2}x^{1/2}y^{2/3}}$ **35.** $\dfrac{9^{3/2}x^{-5/2}y^{1/2}}{4^{5/2}x^{-1}y^{2/3}}$

36. $\dfrac{16^{1/2}x^{-1/4}y^{1/2}}{27^{1/3}x^{3/4}y^{2/3}}$ **37.** $\left(\dfrac{x^2y^{-4/5}}{x^0y^{-3/5}}\right)^{-10}$

38. $\left(\dfrac{3^{1/3}x^{-2/3}y^{1/2}}{2^{-1/3}x^{1/2}y^{3/2}}\right)^6$ **39.** $\dfrac{16^{1/2}x^{-1/4}y^{1/2}}{27^{1/3}x^{3/4}y^{2/3}}$

7-4 LAWS OF RADICALS

From the laws of exponents certain useful laws of radicals may be had. We list here six laws of radicals which are immediate consequences of the corresponding laws of exponents appearing in the column to the right. In these formulas we restrict c, m, and n to positive integers, and restrict a and b to the extent that a zero must not appear in a denominator and radicals of even order must have positive radicands.

Law 1: $\sqrt[n]{a^n} = (\sqrt[n]{a})^n = a$ $\qquad (a^n)^{1/n} = (a^{1/n})^n = a$

Law 2: $\sqrt[n]{ab} = \sqrt[n]{a}\sqrt[n]{b}$ $\qquad (ab)^{1/n} = a^{1/n}b^{1/n}$

Law 3: $\sqrt[n]{\dfrac{a}{b}} = \dfrac{\sqrt[n]{a}}{\sqrt[n]{b}}$ $\qquad \left(\dfrac{a}{b}\right)^{1/n} = \dfrac{a^{1/n}}{b^{1/n}}$

Law 4: $\sqrt[cn]{a^{cm}} = \sqrt[n]{a^m}$ $\qquad a^{cm/cn} = a^{m/n}$

Law 5: $\sqrt[n]{\sqrt[m]{a}} = \sqrt[nm]{a}$ $\qquad (a^{1/m})^{1/n} = a^{1/mn}$

Law 6: $\sqrt[n]{a^m}\sqrt[q]{a^p} = \sqrt[nq]{a^{mq+np}}$ $\qquad a^{m/n}a^{p/q} = a^{(mq+np)/nq}$

These laws may be employed in making some of the needed changes in radicals, the most common of which are

1. Removing factors from the radicand.
2. Making the radicand nonfractional.
3. Expressing a radical as one of lower order.
4. Bringing an outside factor inside the radical sign.

A radical is said to be in *simplest form* when operations 1, 2, and 3 are fully carried out. Operation 2 is called *rationalizing the denominator.*

In the illustrative examples and the exercises to follow, we shall assume that all literal numbers are positive.

EXAMPLE 1 Simplify the radical $\sqrt{75x^2y^3}$.

Solution:

$$\sqrt{75x^2y^3} = \sqrt{25x^2y^2(3y)} = 5xy\sqrt{3y}$$

EXAMPLE 2 Rationalize the denominators of the radicals

$$\sqrt{\frac{3}{5}} \quad \text{and} \quad \sqrt[4]{\frac{b}{2x^3}}$$

Solution:

$$\sqrt{\frac{3}{5}} = \sqrt{\frac{3 \cdot 5}{5 \cdot 5}} = \frac{\sqrt{15}}{5} = \frac{1}{5}\sqrt{15}$$

$$\sqrt[4]{\frac{b}{2x^3}} = \sqrt[4]{\frac{b \cdot 8x}{2x^3 \cdot 8x}} = \frac{\sqrt[4]{8bx}}{2x} = \frac{1}{2x}\sqrt[4]{8bx}$$

EXAMPLE 3 Simplify the radicals $\sqrt[4]{16a^2}$ and $\sqrt[6]{27a^3b^9}$.

Solution:

$$\sqrt[4]{16a^2} = \sqrt[4]{(4a)^2} = \sqrt{4a} = 2\sqrt{a}$$

$$\sqrt[6]{27a^3b^9} = \sqrt[6]{(3ab^3)^3} = \sqrt{3ab^3} = b\sqrt{3ab}$$

EXAMPLE 4 Bring the coefficient of the radical

$$2x\sqrt[3]{1-\frac{1}{4x^2}}$$

inside the radical sign.

Solution:

$$2x\sqrt[3]{1-\frac{1}{4x^2}} = \sqrt[3]{8x^3\left(1-\frac{1}{4x^2}\right)} = \sqrt[3]{8x^3-2x}$$

EXAMPLE 5 Express the repeated radical $\sqrt[4]{2^3\sqrt{3}}$ as a single radical.

Solution:

$$\sqrt[4]{2^3\cdot 3^{1/2}} = (2^3\cdot 3^{1/2})^{1/4} = 2^{3/4}\cdot 3^{1/8} = \sqrt[8]{2^6\cdot 3}$$

EXERCISE 7-3

Express each of the following radicals in simplest form.

1. $\sqrt{20}$	**2.** $\sqrt[3]{-16}$	**3.** $\sqrt{18a^2}$
4. $\sqrt{12a^4b^2}$	**5.** $\sqrt[3]{32x^2y^4}$	**6.** $\sqrt{100x^7y^7}$
7. $\sqrt[3]{27x^5y^4}$	**8.** $\sqrt[4]{32x^6y^3}$	**9.** $\sqrt[5]{32x^8}$
10. $\sqrt{\frac{2}{5}}$	**11.** $\sqrt{\frac{5}{3a}}$	**12.** $\sqrt{\frac{7}{8}}$
13. $\sqrt[3]{\frac{64}{9}}$	**14.** $\sqrt[3]{-\frac{2}{3}}$	**15.** $\sqrt{\frac{3}{2x^2}}$
16. $\sqrt[4]{\frac{5}{27}}$	**17.** $\sqrt{\frac{2}{3x^5}}$	**18.** $\sqrt[4]{\frac{x^4y^2}{3z}}$
19. $\sqrt[3]{\frac{3x}{2y}}$	**20.** $\sqrt[3]{\frac{-3x^4}{16y^4}}$	**21.** $\sqrt{\frac{1}{3x}}$
22. $\sqrt[4]{\frac{a}{4b^3}}$	**23.** $\sqrt[3]{\frac{x^2}{y^4z^4}}$	**24.** $\sqrt[4]{36}$
25. $\sqrt[4]{9a^6}$	**26.** $\sqrt[4]{25a^2}$	**27.** $\sqrt[6]{27x^9y^6}$

28. $\sqrt[6]{16x^4y^8}$ 29. $\sqrt[6]{\dfrac{25}{x^2}}$ 30. $\sqrt[4]{\dfrac{x^2}{y^6}}$

31. $\sqrt[6]{\dfrac{x^2 - 2x + 1}{9x^2}}$ 32. $\sqrt[4]{\dfrac{4x^6}{x^2 + 2x + 1}}$ 33. $\sqrt[6]{\dfrac{(a + 1)^3}{b^3c^3}}$

Inside the radical sign, bring the coefficient to the proper power.

34. $3\sqrt{2}$ 35. $5x\sqrt{y}$

36. $2x\sqrt{\dfrac{3 - x}{4x^2}}$ 37. $x\sqrt[3]{\dfrac{4 - x}{x^2}}$

38. $2a\sqrt{\dfrac{1}{4} - \dfrac{1}{a^2}}$ 39. $\dfrac{a}{b}\sqrt{\dfrac{2b^3}{a}}$

40. $\dfrac{3a}{x}\sqrt[3]{\dfrac{2x^5}{27a^2}}$ 41. $\dfrac{3a}{x^2}\sqrt[3]{\dfrac{5x^8}{9a^2}}$

42. $(2 + x)\sqrt[4]{\dfrac{2 - x}{(2 + x)^3}}$

Express each repeated radical as a single radical.

43. $\sqrt[3]{\sqrt{2}}$ 44. $\sqrt[3]{\sqrt{a^5}}$ 45. $\sqrt{3\sqrt[3]{9}}$

46. $\sqrt[3]{5\sqrt{3}}$ 47. $\sqrt[4]{2^3\sqrt{5}}$ 48. $\sqrt[4]{8\sqrt{8}}$

7-5 ADDITION AND SUBTRACTION OF RADICALS

Radicals of the same order and the same radicand are called *like radicals.* An algebraic sum of like radicals may be expressed as a single radical by use of the distributive law. Unlike radicals often become like radicals when they are simplified. Radicals not expressible as like radicals may have addition and subtraction indicated by the proper signs; they cannot be written as a single radical. Thus, the pair of radicals $\sqrt{ab}$ and $\sqrt[3]{ab}$, as well as the pair $\sqrt{ab}$ and $\sqrt{2ab}$, cannot be combined into a single radical. The radicals of the first pair are of different orders and those of the second pair are of different radicands.

EXAMPLE 1

$$\begin{aligned} 2\sqrt{50} - 5\sqrt{\tfrac{1}{2}} + \sqrt[4]{64} &= 2\sqrt{5^2 \cdot 2} - 5\sqrt{\tfrac{2}{4}} + \sqrt[4]{2^4 \cdot 2^2} \\ &= 10\sqrt{2} - \tfrac{5}{2}\sqrt{2} + 2\sqrt{2} \\ &= (10 - \tfrac{5}{2} + 2)\sqrt{2} \\ &= \tfrac{19}{2}\sqrt{2} \end{aligned}$$

EXAMPLE 2

$$\sqrt[3]{24a^4b} + \sqrt[3]{81ab^4} - \sqrt[3]{9a^4b^4}$$
$$= \sqrt[3]{2^3a^3 \cdot 3ab} + \sqrt[3]{3^3b^3 \cdot 3ab} - \sqrt[3]{a^3b^3 \cdot 9ab}$$
$$= 2a\sqrt[3]{3ab} + 3b\sqrt[3]{3ab} - ab\sqrt[3]{9ab}$$
$$= (2a + 3b)\sqrt[3]{3ab} - ab\sqrt[3]{9ab}$$

The unlike radicals $\sqrt[3]{3ab}$ and $\sqrt[3]{9ab}$ cannot be expressed as a single radical.

EXERCISE 7-4

Simplify the radicals in each of the following problems and then combine all like terms.

1. $\sqrt{18} + \sqrt{32} - \sqrt{50}$
2. $\sqrt{12} - \sqrt{27} - \sqrt{75}$
3. $\sqrt{112} + 3\sqrt{63} - \sqrt{28}$
4. $2\sqrt{75} - \sqrt{20} + 4\sqrt{12}$
5. $\sqrt{63} - \sqrt{50} + \sqrt{28}$
6. $\sqrt{\frac{1}{2}} - 3\sqrt{2} + \sqrt{\frac{1}{8}}$
7. $\sqrt{12} + \sqrt{\frac{1}{3}} + 3\sqrt{\frac{1}{27}}$
8. $\sqrt{60} - 3\sqrt{\frac{5}{3}} + 5\sqrt{\frac{3}{5}}$
9. $\sqrt[3]{54} + \sqrt[3]{40} - 3\sqrt[3]{16}$
10. $\sqrt[3]{81} + \sqrt[3]{54} - \sqrt[3]{16}$
11. $3\sqrt{18x} - \sqrt{8x^3} - \sqrt{2x}$
12. $\sqrt{16x^3y} - \sqrt{4xy^3} + \sqrt{xy}$
13. $3\sqrt{75x^6y^3} - \sqrt{12x^4y} - \sqrt{3x^2y}$
14. $\sqrt[3]{54ab^4} - \sqrt[3]{16a^4b^4} - \sqrt[3]{2ab}$
15. $\sqrt[3]{81a^2b^6} - \sqrt[3]{24a^5b^3} - \sqrt[3]{3a^2}$
16. $\sqrt[3]{8xy^2} + \sqrt[3]{27x^4y^5} + \sqrt[3]{27x^7y^5}$
17. $\sqrt[3]{16ab^2} + \sqrt[3]{2a^4b^2} - \sqrt[3]{2ab^2}$
18. $\sqrt[4]{25a^2} + \sqrt{20a^3} + \sqrt{45a}$
19. $\sqrt[6]{27a^3} - \sqrt[4]{9a^2} - \sqrt{3a^2}$
20. $6\sqrt[6]{8a^3} + 4\sqrt[4]{4a^6} + 3\sqrt{2a}$
21. $\sqrt{\dfrac{1}{2x}} + \sqrt{\dfrac{2}{x}} - \sqrt{\dfrac{x}{2}}$
22. $\sqrt[4]{x^2y^2} + \sqrt{\dfrac{x}{y}} + \sqrt{\dfrac{y}{x}} - \sqrt{\dfrac{1}{xy}}$
23. $\sqrt{\dfrac{a^2}{a^2 - b^2}} + \sqrt{\dfrac{a - b}{a + b}} - \sqrt{\dfrac{a + b}{a - b}}$

24. $\sqrt{\dfrac{4a}{3b}} - \dfrac{1}{2}\sqrt{\dfrac{25b}{a}} + \dfrac{1}{3}\sqrt{\dfrac{75a}{b}}$

25. $\sqrt[8]{\dfrac{16a^4}{b^4}} + \sqrt[4]{\dfrac{324a^2}{b^2}} + \sqrt{\dfrac{16a^4}{b^4}}$

26. $\sqrt[15]{\dfrac{x^5y^5}{z^5}} - \sqrt[9]{\dfrac{x^3y^3}{512z^3}} + \sqrt[6]{\dfrac{x^2y^2}{64z^2}}$

7-6 MULTIPLICATION AND DIVISION OF RADICALS

Two radicals of the same order can be multiplied by the use of Law 2, Sec. 7-4. To multiply radicals of different orders, it is first necessary to express them as radicals of the same order. The order of the new radicals should be the LCM of the orders of the original radicals. The order of a radical can be raised (Law 4, Sec. 7-4) by multiplying the order of the radical and the exponent of the radicand by the same positive integer greater than 1.

We note, however, that fractional exponents are defined in terms of radicals, and, conversely, a radical can be expressed in terms of fractional exponents. In view of this circumstance, operations on radicals can be carried out by the use of fractional exponents. In fact, the conversion to fractional exponents is preferable when complicated radicals are involved.

EXAMPLE 1 Find the product of the radicals

$$5\sqrt[3]{4a^2b} \qquad \text{and} \qquad 2\sqrt[3]{3ab^4}$$

Solution: The radicals are of the same order, and, by direct multiplication, we get

$$\begin{aligned} 5\sqrt[3]{4a^2b}\cdot 2\sqrt[3]{3ab^4} &= 5\cdot 2\sqrt[3]{4a^2b\cdot 3ab^4} \\ &= 10\sqrt[3]{12a^3b^3b^2} = 10ab\sqrt[3]{12b^2} \end{aligned}$$

EXAMPLE 2 Find the product of $2\sqrt{3} + 3\sqrt{5}$ and $\sqrt{3} - 4\sqrt{5}$.

Solution: The binomials have like radicals, and we multiply in the usual way. Thus,

$$\begin{aligned} (2\sqrt{3} + 3\sqrt{5})(\sqrt{3} - 4\sqrt{5}) &= 2\cdot 3 - 8\sqrt{15} + 3\sqrt{15} - 12\cdot 5 \\ &= -54 - 5\sqrt{15} \end{aligned}$$

EXAMPLE 3 Find the product of $5\sqrt[3]{2}$ and $4\sqrt{3}$.

Solution: We first express each radical as a radical of order 6, the LCM of the orders of the given radicals. Thus,

$$5\sqrt[3]{2} \cdot 4\sqrt{3} = 5\sqrt[6]{2^2} \cdot 4\sqrt[6]{3^3} = 20\sqrt[6]{2^2 \cdot 3^3} = 20\sqrt[6]{108}$$

Alternatively, using fractional exponents, we have

$$\begin{aligned} 5\sqrt[3]{2} \cdot 4\sqrt{3} &= 5 \cdot 2^{1/3} \cdot 4 \cdot 3^{1/2} \\ &= 20 \cdot 2^{2/6} \cdot 3^{3/6} = 20\sqrt[6]{2^2 \cdot 3^3} = 20\sqrt[6]{108} \end{aligned}$$

Law 3, Sec. 7-4, shows how to express the quotient of two radicals of the same order as a single radical. As in multiplication, however, radicals of different orders must first be converted to radicals of the same order. We shall consider the division of two radicals to be fully carried out when the quotient has no radical in the denominator and the radical in the numerator, if any, is expressed in simplest form. The process of ridding a denominator of a radical or radicals is called *rationalizing the denominator.*

EXAMPLE 4 Divide $2\sqrt{14x^3y}$ by $3\sqrt{7xy^3}$ and simplify the result.

Solution:

$$\frac{2\sqrt{14x^3y}}{3\sqrt{7xy^3}} = \frac{2}{3}\sqrt{\frac{14x^3y}{7xy^3}} = \frac{2}{3}\sqrt{\frac{2x^2}{y^2}} = \frac{2x}{3y}\sqrt{2}$$

EXAMPLE 5 Divide $3\sqrt[3]{5}$ by $6\sqrt{2}$ and simplify the result.

Solution:

$$\frac{3\sqrt[3]{5}}{6\sqrt{2}} = \frac{3\sqrt[3]{5} \cdot \sqrt{2}}{6\sqrt{2} \cdot \sqrt{2}} = \frac{3\sqrt[6]{5^2} \cdot \sqrt[6]{2^3}}{12} = \frac{3\sqrt[6]{5^2 \cdot 2^3}}{12} = \frac{1}{4}\sqrt[6]{200}$$

When the divisor is a binomial which has a radical in one or both terms, we rationalize the denominator by multiplying the dividend and divisor by a properly chosen expression. For this purpose we observe that if a and b are rational numbers, the product of $\sqrt{a} + \sqrt{b}$ and $\sqrt{a} - \sqrt{b}$ is the rational quantity $a - b$. Hence, a rationalizing factor is obtained by changing the sign of one term of the divisor.

EXAMPLE 6 Divide $(\sqrt{5} - 3\sqrt{2})$ by $(3\sqrt{5} + 2\sqrt{2})$.

Solution:

$$\frac{\sqrt{5} - 3\sqrt{2}}{3\sqrt{5} + 2\sqrt{2}} = \frac{\sqrt{5} - 3\sqrt{2}}{3\sqrt{5} + 2\sqrt{2}} \cdot \frac{3\sqrt{5} - 2\sqrt{2}}{3\sqrt{5} - 2\sqrt{2}}$$

$$= \frac{15 - 2\sqrt{10} - 9\sqrt{10} + 12}{45 - 8}$$

$$= \frac{27 - 11\sqrt{10}}{37}$$

EXERCISE 7-5

Multiply as indicated and simplify the results.

1. $\sqrt{3} \cdot \sqrt{7}$
2. $2\sqrt{13} \cdot 3\sqrt{26}$
3. $3\sqrt{6} \cdot \sqrt{12}$
4. $\sqrt{2} \cdot \sqrt{11} \cdot \sqrt{22}$
5. $\sqrt{3} \cdot \sqrt{5} \cdot \sqrt{45}$
6. $\sqrt{\frac{2}{5}} \cdot \sqrt{\frac{10}{3}} \cdot \sqrt{\frac{3}{2}}$
7. $\sqrt{18x^2y} \cdot \sqrt{3xy^4}$
8. $\sqrt{5xy^2} \cdot \sqrt{15x^3y^3}$
9. $(3\sqrt{x + y})^2$
10. $\sqrt[3]{6} \cdot \sqrt[3]{9}$
11. $\sqrt[3]{3a^2} \cdot \sqrt[3]{4a}$
12. $\sqrt[3]{4a^2} \cdot \sqrt[3]{16ab}$
13. $\sqrt[3]{3} \cdot \sqrt{2}$
14. $\sqrt[3]{5} \cdot \sqrt{2}$
15. $\sqrt[4]{2} \cdot \sqrt{8}$
16. $\sqrt{x} \cdot \sqrt[3]{x} \cdot \sqrt[4]{x}$
17. $\sqrt{3} \cdot \sqrt[3]{2} \cdot \sqrt[3]{4}$
18. $\sqrt[4]{4} \cdot \sqrt[3]{3} \cdot \sqrt{2}$
19. $(\sqrt{2} - \sqrt{3})(\sqrt{2} + \sqrt{3})$
20. $(2\sqrt{3} + 5)(2\sqrt{3} - 3)$
21. $(4\sqrt{3} + 3\sqrt{2})(\sqrt{3} - 2\sqrt{2})$
22. $\sqrt{6 + 4\sqrt{2}} \cdot \sqrt{6 - 4\sqrt{2}}$
23. $\sqrt{5 - \sqrt{a}} \cdot \sqrt{5 + \sqrt{a}}$
24. Find the value $x^2 - 6x + 7$ if $x = 3 - \sqrt{2}$.
25. Find the value of $2x^2 + x + 1$ if $x = \sqrt{2} + 1$.
26. Find the value of $x^2 + x - 5$ if $x = \sqrt{3} - \sqrt{2}$.

Perform the divisions and express each result in simplest form.

27. $\sqrt{7} \div \sqrt{63}$

28. $\sqrt{33} \div \sqrt{11}$

29. $\sqrt{28x} \div \sqrt{7x^2}$

30. $\sqrt{15a^4} \div \sqrt{5a}$

31. $2\sqrt{5x^3} \div \sqrt{20x}$

32. $\sqrt[3]{4} \div \sqrt[3]{108}$

33. $\sqrt[3]{21a^2} \div \sqrt[3]{7a^4}$

34. $\sqrt[3]{4x} \div \sqrt[3]{15x^4}$

35. $\sqrt{5} \div \sqrt[3]{5}$

36. $\sqrt{33} \div \sqrt[3]{11}$

37. $\sqrt[3]{a^2b} \div \sqrt{ab^2}$

38. $\sqrt[4]{24x^3y} \div \sqrt[4]{8xy^3}$

39. $\sqrt{x^2y} \div \sqrt[4]{xy^2}$

40. $\dfrac{\sqrt{6} - \sqrt{3}}{\sqrt{3}}$

41. $\dfrac{\sqrt{15} - \sqrt{35}}{\sqrt{5}}$

42. $\dfrac{1}{3 + \sqrt{5}}$

43. $\dfrac{1}{\sqrt{3} + \sqrt{2}}$

44. $\dfrac{\sqrt{5} - \sqrt{3}}{\sqrt{5} + \sqrt{3}}$

45. $\dfrac{\sqrt{2}}{5 - \sqrt{2}}$

46. $\dfrac{\sqrt{7} - \sqrt{5}}{\sqrt{7} + \sqrt{5}}$

47. $\dfrac{\sqrt{6} + 2\sqrt{3}}{2\sqrt{6} - \sqrt{3}}$

48. $\dfrac{4\sqrt{5} + \sqrt{3}}{2\sqrt{5} - \sqrt{3}}$

49. $\dfrac{3\sqrt{7} - 2\sqrt{2}}{2\sqrt{7} + 3\sqrt{2}}$

8

THE TRIGONOMETRIC FUNCTIONS

8-1 TRIGONOMETRY

In this chapter, we shall treat topics which belong to trigonometry, a branch of science which had its beginnings centuries ago. As early as the second century B.C., the Greek astronomer Hipparchus made a creditable advance toward the founding of this science by collecting and extending some of the basic ideas. Since then, and particularly in modern times, the knowledge and applications of trigonometry have increased tremendously. The earlier use of trigonometry consisted largely of computing unknown sides and angles of triangles in dealing with problems in land measurement, astronomy, and navigation. Although applications of this kind have continued through the centuries, this phase of trigonometry has been surpassed in importance by the development of other aspects of the subject. The methods and concepts of trigonometry, apart from their triangle-solving uses, have contributed to the advancement of numerous branches of mathematics. In particular, trigonometric analysis has been of great importance in the develop-

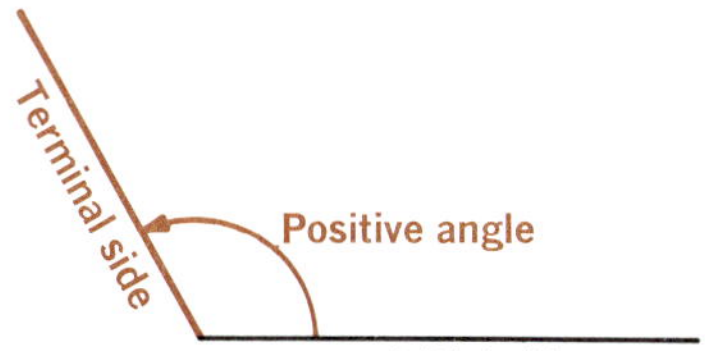

Figure 8-1

ment of the physical and engineering sciences. As the student progresses, he will discover that trigonometry is vital in the structure and applications of mathematics.

8-2 ANGLES AND UNITS OF MEASURE

The concept of an angle as introduced in elementary plane geometry is the figure formed by two half lines emanating from a common point.* The half lines are called the *sides* of the angle and the common point the *vertex.* The measure of the angle is taken as positive and equal to the amount of rotation, less than one revolution, required to bring a half line from the position of one side to the position of the other side. In trigonometry, however, the idea of an angle is enlarged to include angles of positive and negative measures of any magnitude.

In Fig. 8-1, we label one of the half lines the *initial side* and the other the *terminal side.* The curved arrow indicates a *counterclockwise* rotation from the initial side to the terminal side. For this direction of rotation, we define the angle and also the measure of the angle to be *positive.* In Fig. 8-2 the rotation from the initial side to the terminal side is *clockwise,* and we define the angle and its measure to be *negative.*

A half line may be rotated from coincidence with the initial side of an angle, by any chosen amount, to serve as the terminal side. Accordingly, we may visualize an angle as being formed in this way.

* A point O of a line together with the part of the line extending in one direction from O is called a *half line.*

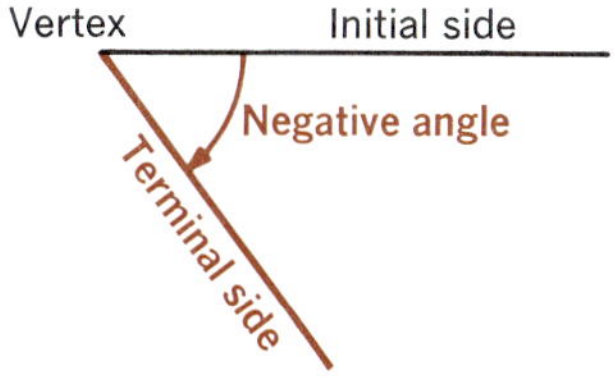

Figure 8-2

Figure 8-3

Since the amount of rotation is arbitrary, the measure of the angle is unlimited. And the measure of the angle is positive or negative depending on whether the rotation is counterclockwise or clockwise.

There are several different units for measuring angles. The definition of an angle suggests one revolution from the position of the initial side, as indicated in Fig. 8-3, as a unit. We shall describe two other units which are used extensively.

If a rotation from the initial side to the terminal side is $\frac{1}{360}$ of a revolution, the angle is said to have a measure of one *degree* (written $1°$). One-sixtieth of a degree is called a *minute* (written $1'$), and one-sixtieth of a minute is called a *second* (written $1''$). These are the units of the *degree system* of angular measure. This system is used in astronomy, surveying, and engineering.

There is another system of angular measure, called the *radian system*, which is preferred in theoretical work, particularly in calculus. The unit of measure in this system is defined as follows:

Definition 8-1 An angle, with its vertex at the center of a circle, which intercepts an arc equal in length to the radius of the circle is said to have a measure of 1 *radian*.*

Angles of 1 radian and $-2\frac{1}{2}$ radians are constructed in Fig. 8-4. The curved arrow points in the direction of rotation in each case.

The ratio of the circumference of a circle to its diameter is denoted by the Greek letter pi (π), whose value is approximately 3.1416. The definition of π leads to the formula $c = 2\pi r$, where c stands for the circumference and r for the radius. Hence, a length

* The part of a circle between any two of its points is called an *arc* of the circle. The two points, of course, separate the circle into two arcs.

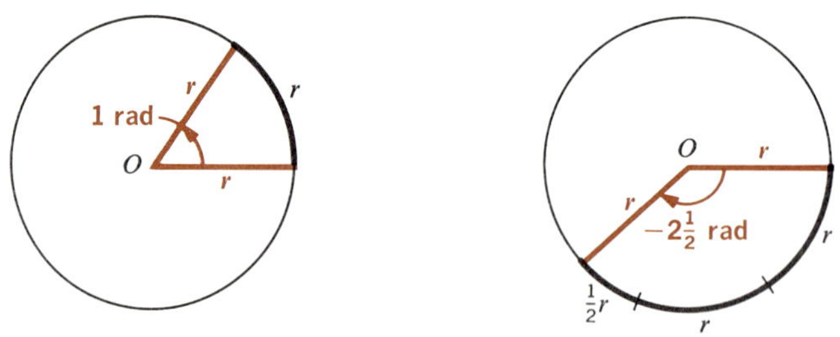

Figure 8-4

of one radius can be measured off along the circumference 2π times without overlapping. This means that the entire circumference subtends a central angle of 2π, or about 6.2832, radians. It should be noticed that 2π radians are obtained regardless of the length of the radius; hence, the radian is a fixed unit of angular measure.

We may find a simple relation among the radius r of a circle, a central angle of measure θ, and the length s of the intercepted arc (Fig. 8-5). Since a central angle of measure 1 radian intercepts an arc equal in length to the radius of the circle, a central angle of measure θ radians intercepts an arc of θ times the radius; that is,

$$s = r\theta$$

This relation among the quantities s, r, and θ may be used to compute the value of any one of them if the values of the other two are known. It is essential in applying the formula that the radius and arc be measured in the same linear units, and that θ be the measure, *in radians,* of the central angle. When the radius has a measure of one unit, we note that the formula $s = r\theta$ becomes simply $s = \theta$. Later, we shall make important use of this.

We can see from our discussion of angles that an angle and its measure are two distinct entities. Notwithstanding this fact, however, mathematicians quite generally use "angle" and "measure of an angle" interchangeably. For example, if θ stands for an angle of measure 60°, we write $\theta = 60°$, when really the measure of the angle is equal to 60°.

Since the entire circumference of a circle subtends a central angle of 2π units in terms of radians and 360 units in terms of degrees,

$$2\pi \text{ radians} = 360°$$

and

$$\pi \text{ radians} = 180°$$

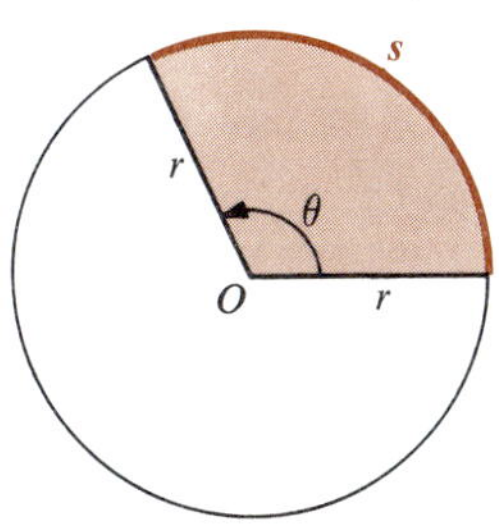

Figure 8-5

By this relation, a radian may be expressed in terms of degrees and a degree in terms of radian measure. Thus,

$$1 \text{ radian} = \frac{180^\circ}{\pi} = 57.296^\circ - = 57^\circ\ 17.7'+$$

and

$$1^\circ = \frac{\pi}{180} \text{ radian} = 0.017453+ \text{ radian}$$

When an angle is expressed in radians, the word "radian" is frequently omitted. Thus $\pi = 180^\circ$ and $\pi/4 = 45^\circ$ are written with the understanding that the radian is the unit of measure in the left members of these equations.

EXAMPLE 1 Express 160° in terms of π radians.

Solution: Since $1^\circ = \pi/180$ radian, we have

$$160^\circ = 160\frac{\pi}{180} = \frac{8\pi}{9}$$

EXAMPLE 2 Change $91^\circ\ 48'$ to radian measure.

Solution: We first express $48'$ as a decimal part of a degree and then use the approximation $1^\circ = 0.01745$ radian. Thus,

$$91^\circ\ 48' = 91.8^\circ = 91.8(0.01745) = 1.602 \text{ radians}$$

EXAMPLE 3 Express $5\pi/6$ in terms of degrees.

Solution: Using the relation $\pi = 180^\circ$, we write

$$\frac{5\pi}{6} = \frac{5}{6}(180^\circ) = 150^\circ$$

EXAMPLE 4 Express 3.2 radians in terms of degrees and minutes.

Solution: From the relation 1 radian $= 57.296^\circ$, we get

$$3.2(57.296^\circ) = 183.347^\circ = 183^\circ\ 21'$$

EXAMPLE 5 Find the central angle subtended by an arc of length 5 if the radius is 2.

Solution: Substituting in the formula $s = r\theta$, we obtain

$$5 = 2\theta \qquad \text{and} \qquad \theta = 2.5 \text{ radians} = 143^\circ\, 14'$$

EXAMPLE 6 A central angle of 38° intercepts an arc of 3 feet. Find the radius of the circle.

Solution: In order to apply the relation $s = r\theta$, we first express 38° in radian measure. Thus,

$$38^\circ = 38(0.01745) = 0.663 \text{ radian}$$

Hence,

$$3 = 0.663r \qquad \text{and} \qquad r = 4.52 \text{ feet}$$

EXERCISE 8-1

Express each of the following angles in terms of π radians.

1. 45°, 60°, 135°, 150°
2. 30°, 90°, 300°, 240°
3. 225°, 315°, −73°, −40°
4. 330°, 20°, −70°, −28°

Express each angle in radian measure, using 1° = 0.01745 radian.

5. 50°, 6° 30′, 80° 36′
6. 46°, 90° 40′, 31° 48′

Change each angle from radian measure to degrees.

7. $\frac{\pi}{6}, \frac{\pi}{4}, \frac{3\pi}{2}, -\frac{\pi}{10}$
8. $\frac{\pi}{3}, \frac{\pi}{9}, \frac{\pi}{15}, -\frac{\pi}{30}$
9. $\frac{3\pi}{4}, \frac{5\pi}{6}, \frac{\pi}{8}, -\frac{\pi}{15}$
10. $\frac{7\pi}{4}, \frac{11\pi}{12}, \frac{7\pi}{24}, -\frac{8\pi}{15}$

Change each angle from radian measure to degrees, expressing a fraction of a degree in minutes. First, use 1 radian = 57.296°.

11. 2, −2.1, 0.6, 0.26
12. 3.2, −0.08, 1.52, 3
13. Find the length of the arc of a circle of radius 32 inches which subtends a central angle of **(a)** 0.8 radian; **(b)** 0.25π; **(c)** 75°.
14. An arc of 16 inches subtends central angles of 1.6 radians, 0.4π, and 20°, respectively, in each of three circles. Find the radii of the circles.

15. Find the central angle in radians which intercepts an arc of 6 feet if the radius of the circle is (**a**) 12 feet; (**b**) 60 inches; (**c**) 2 yards.

16. Kiev, Russia, and Cairo, Egypt have approximately the same longitude. The latitude of Kiev is 50° N and of Cairo 30° N. Find the distance between these cities, taking 4000 miles as the radius of the earth.

17. Quito, Ecuador, and Singapore are located on the earth's equator (almost), and their longitudes differ by 178°. Figuring the radius of the earth as 4000 miles, find the distance between the cities.

18. The pendulum of a clock swings through an angle of 12°. Find the distance which the tip travels in one swing if the length of the pendulum is 40 inches.

19. Find the angle through which a pendulum swings if its length is 30 inches and the tip describes an arc of (**a**) 4 inches; (**b**) 9 inches; (**c**) 13 inches.

20. The minute hand of a clock is 3 inches long. How far does the tip of the hand travel in (**a**) 5 minutes; (**b**) 18 minutes; (**c**) 27 minutes?

21. The moon subtends an angle of about 31.1′ at the center of the earth. Taking the distance from the earth to the moon as 239,000 miles, find the diameter of the moon.

8-3 ANGLE IN STANDARD POSITION

An angle which has its vertex at the origin and its initial side along the positive x axis is said to be in *standard position.* The location of the terminal side is determined by the magnitude and sign of

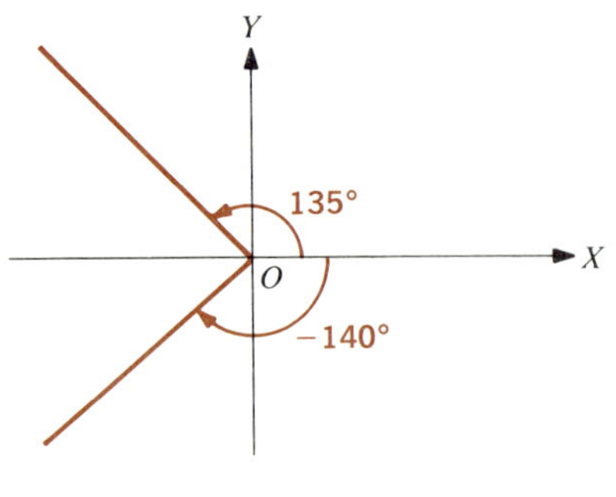

Figure 8-6

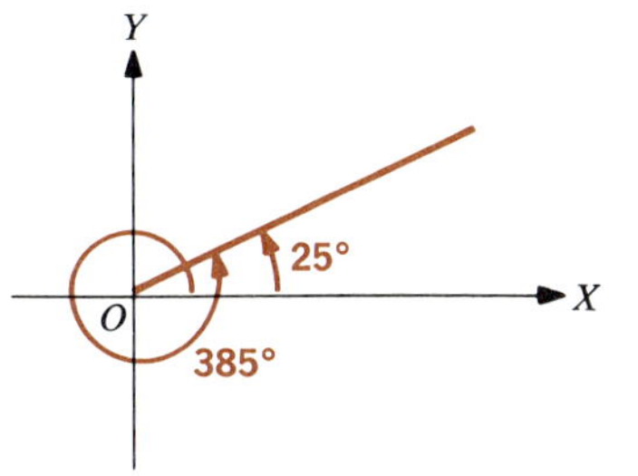

Figure 8-7

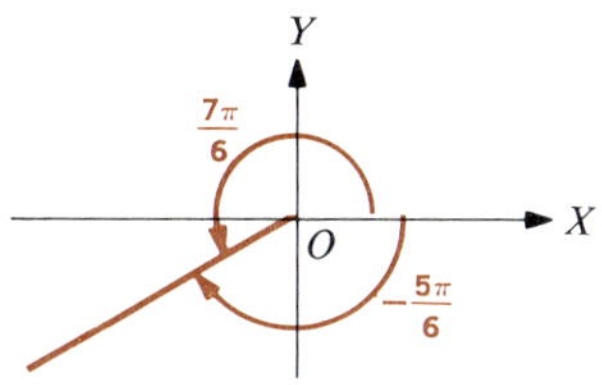

Figure 8-8

the angle. The angle is usually spoken of as being *in the quadrant* which contains the terminal side. For example, we say that 135° is in quadrant II, and −140° is in quadrant III. The angles, of course, really *terminate* in those quadrants (Fig. 8-6).

Angles in standard position and having the same terminal sides are said to be *coterminal.* Two coterminal angles are not necessarily equal. Thus, drawn in standard position, the pair of angles 25° and 385° are coterminal (Fig. 8-7) and the pair $7\pi/6$ and $-5\pi/6$ are coterminal (Fig. 8-8). An angle coterminal with a given angle may be obtained by adding any integral multiple of 360° or 2π. Hence, an angle has an unlimited number of coterminal angles.

EXERCISE 8-2

Construct each of the following angles in standard position and draw a curved arrow to indicate the direction of rotation.

1. 60°, −48°, 210°, −135°

2. 120°, −190°, 320°, −180°

3. 315°, −100°, 440°, −360°

4. 870°, −500°, 270°, −270°

Let each of the following angles be in standard position and name the quadrant in which the terminal side lies.

5. $93°$, $321°$, $-68°$, $-340°$

6. $152°$, $560°$, $-230°$, $-300°$

7. $\frac{5\pi}{3}$, $\frac{3\pi}{4}$, $-\frac{11\pi}{6}$, $-\frac{7\pi}{3}$

8. $\frac{3\pi}{5}$, $\frac{19\pi}{4}$, $-\frac{7\pi}{6}$, $-\frac{13\pi}{3}$

Find two other angles of smallest magnitude which are coterminal with each of the following angles. Assume that the angles are in standard position.

9. $40°$ **10.** $-170°$ **11.** $230°$

12. $-230°$ **13.** $325°$ **14.** $-350°$

15. $\frac{5\pi}{2}$ **16.** 3π **17.** $\frac{7\pi}{2}$

18. $\frac{7\pi}{4}$

Find two other angles of smallest magnitude, one positive and one negative, which are coterminal with each of the following angles. Assume that the angles are in standard position.

19. $230°$ **20.** $-230°$ **21.** $325°$

22. $-350°$ **23.** $390°$ **24.** $580°$

25. $-600°$ **26.** $\frac{7\pi}{3}$ **27.** $\frac{17\pi}{6}$

28. $\frac{10\pi}{3}$

8-4 DEFINITIONS OF THE FUNCTIONS

In Definition 5-3 we specified that the domain and range of a function consist of numbers. In a broader sense, however, the domain or range, or both, may be other objects than numbers. The trigonometric functions, which we are now ready to define, are illustrations of functions whose domains are sets of angles.

Let θ be an angle in standard position (Fig. 8-9) and let $P(x, y)$ be any point, other than the origin, on the terminal side. Then the

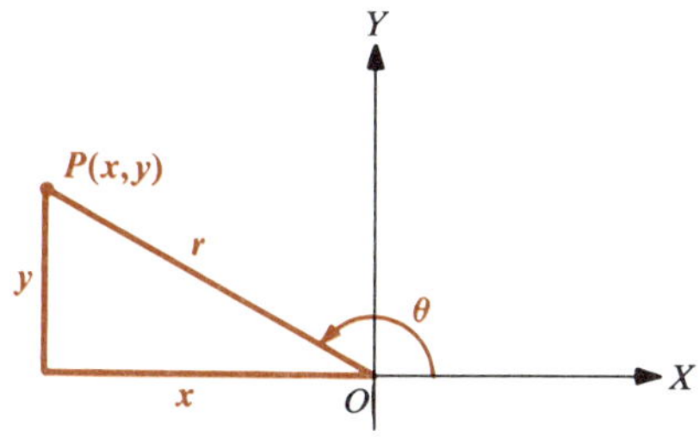

Figure 8-9

abscissa x, the ordinate y, and the distance r have six possible ratios called *trigonometric ratios.* Using these ratios, we define the trigonometric functions by the following equations:

$$\text{sine } \theta = \frac{\text{ordinate}}{\text{distance}} = \frac{y}{r}, \qquad \text{cosecant } \theta = \frac{\text{distance}}{\text{ordinate}} = \frac{r}{y}$$

$$\text{cosine } \theta = \frac{\text{abscissa}}{\text{distance}} = \frac{x}{r}, \qquad \text{secant } \theta = \frac{\text{distance}}{\text{abscissa}} = \frac{r}{x}$$

$$\text{tangent } \theta = \frac{\text{ordinate}}{\text{abscissa}} = \frac{y}{x}, \qquad \text{cotangent } \theta = \frac{\text{abscissa}}{\text{ordinate}} = \frac{x}{y}$$

Each of these equations, as the ensuing discussions will reveal, determines a set of pairs of elements which conform to the idea of a function (Definition 5-3).

The abbreviations $\sin\theta$ for sine θ, $\cos\theta$ for cosine θ, $\tan\theta$ for tangent θ, $\cot\theta$ for cotangent θ, $\sec\theta$ for secant θ, and $\csc\theta$ for cosecant θ are customarily used.

An angle in standard position with its terminal side along a coordinate axis is called a *quadrantal angle.* One of the coordinates of points on the terminal side of a quadrantal angle is equal to zero, and consequently not all of the six trigonometric ratios exist. If the terminal side is along the y axis, the x coordinate is equal to zero. This means that the ratios y/x and r/x do not exist and, for such an angle, the tangent and secant functions are not defined. Similarly, the ratios x/y and r/y do not exist at points on the x axis. For this case the cotangent and cosecant functions are not defined.

8-5 IMMEDIATE CONSEQUENCES OF THE DEFINITIONS

The values of the six ratios appearing in the definitions of the trigonometric functions do not depend on the position of the point P on the terminal side. To show that this is true, let P' be

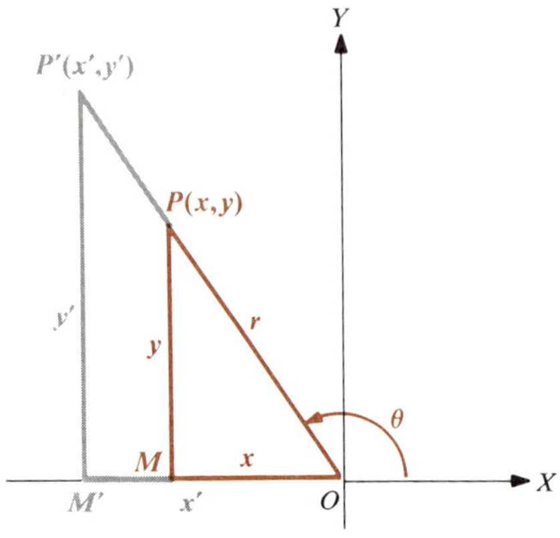

Figure 8-10

another point on the terminal side with coordinates (x', y') and distance r' (Fig. 8-10). The right triangles *OMP* and *OM′P′*, called *reference triangles* for θ, are similar. That is, the corresponding sides are proportional. Hence we have the equal ratio

$$\frac{y}{r} = \frac{y'}{r'}$$

which shows that the same value is obtained for $\sin\theta$ by using either P or P'. Similarly, the value of each of the other functions is independent of the point selected on the terminal side.

We see now that the value of each trigonometric ratio depends solely on the position of the terminal side of the angle. Further, corresponding to an angle θ, each defined trigonometric ratio is equal to a unique real number. Because of this uniqueness, we conclude that each of the six equations of the previous section gives rise to a set of ordered pairs of elements which constitute a function. The domain of the sine and cosine functions is the set of all angles, but certain quadrantal angles are excluded from the domain of each of the other functions. As examples, the sine function comprises the set

$$\left\{(\theta, \sin\theta) \mid \sin\theta = \frac{y}{r}\right\}$$

and the tangent function the set

$$\left\{(\theta, \tan\theta) \mid \tan\theta = \frac{y}{x} \text{ and } x \neq 0\right\}$$

Note that the condition $x \neq 0$ excludes all angles with the terminal side along the y axis.

As we have observed, the value of each trigonometric ratio depends solely on the position of the terminal side of the angle. But the position of the terminal side is determined by the measure of the angle. Because of this fact, we shall equate a trigonometric function of an angle to the function of the measure of the angle. For example, if θ is an angle of measure 60°, we write

$$\sin \theta = \sin 60°$$

Since the value of a trigonometric function of an angle in standard position is determined solely by the position of the terminal side, we see that the function has the same value for any two coterminal angles. This is illustrated by the equation

$$\sin 15° = \sin (15° + 360°)$$

and, more generally, by

$$\sin \theta = \sin (\theta + n \cdot 360°)$$

where n is any integer.

Recalling that the reciprocal of any nonzero number a is $1/a$, we discover that the six trigonometric ratios form three pairs of reciprocals, namely, y/r and r/y, x/r and r/x, y/x and x/y. Since the product of each pair of reciprocals is unity, we have the equations

$$\sin \theta \csc \theta = 1, \qquad \cos \theta \sec \theta = 1, \qquad \tan \theta \cot \theta = 1$$

These equations may be reduced to the equivalent forms

$$\sin \theta = \frac{1}{\csc \theta}, \qquad \cos \theta = \frac{1}{\sec \theta}, \qquad \tan \theta = \frac{1}{\cot \theta}$$

$$\csc \theta = \frac{1}{\sin \theta}, \qquad \sec \theta = \frac{1}{\cos \theta}, \qquad \cot \theta = \frac{1}{\tan \theta}$$

Each of these equations is true for all angles except the quadrantal angles previously mentioned, and therefore is an identity (Sec. 6-1).

The distance r appearing in the definitions of the trigonometric functions is always positive, and consequently the signs of the values of the functions depend on the signs of x and y. Accordingly, $\sin \theta$ and $\csc \theta$, taking the sign of y, are positive if θ is in quadrant I or quadrant II and negative if θ is in quadrant III or quadrant IV.

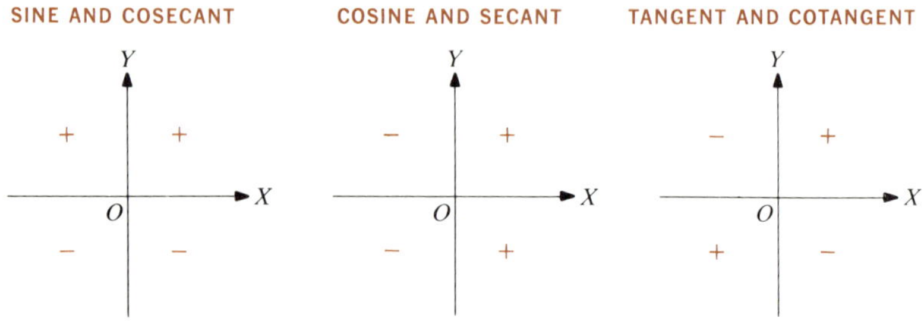

Figure 8-11

The pair $\cos\theta$ and $\sec\theta$ have the same sign as x and are positive if θ is in quadrant I or quadrant IV and negative if θ is in quadrant II or quadrant III. The ratios y/x and x/y are positive if x and y have like signs and negative if x and y have unlike signs. Hence, $\tan\theta$ and $\cot\theta$ are positive if θ is in quadrant I or quadrant III and negative if θ is in quadrant II or quadrant IV.

The signs of the values of the trigonometric functions for the various quadrants are indicated in Fig. 8-11. However, one should be able to visualize the terminal side of an angle in any of the four quadrants and, from the signs of x and y, give immediately the signs of the values of the functions.

EXAMPLE 1 Write the values of the functions of an angle whose terminal side passes through $(-4, -3)$.

Solution: We substitute $x = -4$ and $y = -3$ in the relation $r^2 = x^2 + y^2$ and find $r = 5$. Then we apply the definitions of the functions to the angle (Fig. 8-12) and obtain

$$\sin\theta = \frac{y}{r} = \frac{-3}{5} = -\frac{3}{5} \qquad \csc\theta = \frac{r}{y} = \frac{5}{-3} = -\frac{5}{3}$$

$$\cos\theta = \frac{x}{r} = \frac{-4}{5} = -\frac{4}{5} \qquad \sec\theta = \frac{r}{x} = \frac{5}{-4} = -\frac{5}{4}$$

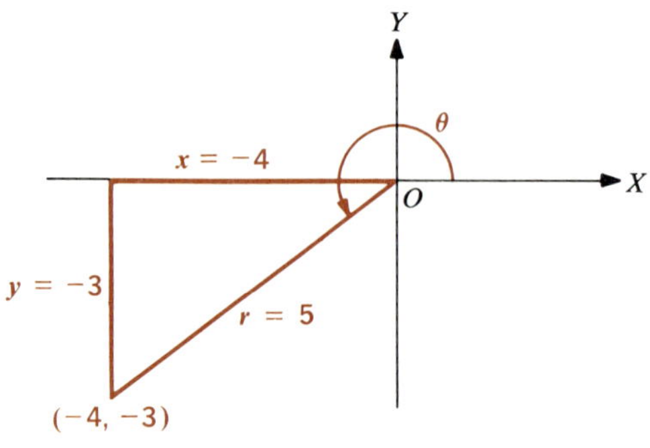

Figure 8-12

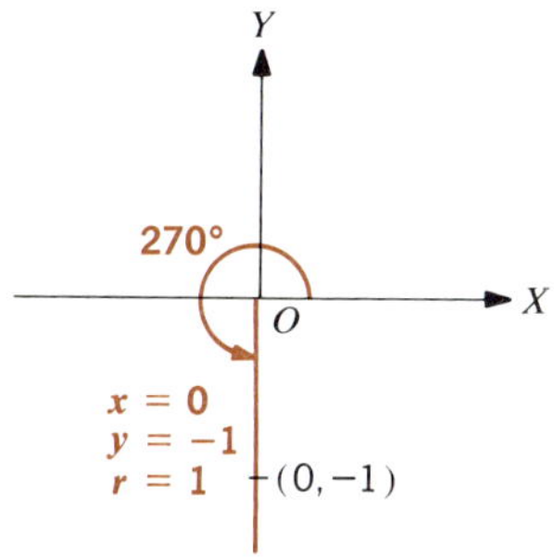

Figure 8-13

$$\tan\theta = \frac{y}{x} = \frac{-3}{-4} = \frac{3}{4} \qquad \cot\theta = \frac{x}{y} = \frac{-4}{-3} = \frac{4}{3}$$

EXAMPLE 2 Find the values of the functions of 270°.

Solution: The terminal side of 270° is along the negative y axis, and for convenience we select the point $(0, -1)$ on this side (Fig. 8-13). The distance r for this point is 1, and applying the definitions of the trigonometric functions, we obtain

$$\sin 270° = \frac{-1}{1} = -1 \qquad \cot 270° = \frac{0}{-1} = 0$$

$$\cos 270° = \frac{0}{1} = 0 \qquad \csc 270° = \frac{1}{-1} = -1$$

Notice that we have omitted tan 270° and sec 270°. The ratios y/x and r/x, corresponding to the tangent and secant functions, have zero for the denominator. But division by zero is forbidden, and therefore 270° has no tangent and no secant. We remark that the values of the four defined functions of 270° are the same for all angles coterminal with 270°. Thus, for example,

$$\sin 270° = \sin(270° + n \cdot 360°)$$

where n is any integer.

We show in tabular form the functions of 0°, 90°, 180°, and 270°. The values for each angle should be verified by the use of a figure.

	sin	cos	tan	cot	sec	csc
0°	0	1	0	None	1	None
90°	1	0	None	0	None	1
180°	0	−1	0	None	−1	None
270°	−1	0	None	0	None	−1

As this table shows, the values of two of the functions of a quadrantal angle are 0, the values of two are 1 and -1, and two of the functions do not exist.

EXAMPLE 3 Given $\csc \theta = -2$ and θ in quadrant III, find the values of the other functions.

Solution: Here, $r/y = -2$, and y must be negative. If we select the point on the terminal side for which $r = 2$, then y must have the value -1. To find x, we substitute these values for r and y in the relation $x^2 + y^2 = r^2$ and find

$$x^2 + (-1)^2 = 2^2, \qquad x^2 = 3, \qquad x = \pm\sqrt{3}$$

In order to obtain a third-quadrant angle, as specified, we choose $x = -\sqrt{3}$ and draw the terminal side through $(-\sqrt{3}, -1)$. Then from Fig. 8-14, we write

$$\sin \theta = -\tfrac{1}{2} \qquad \csc \theta = -2$$

$$\cos \theta = -\frac{\sqrt{3}}{2} \qquad \sec \theta = -\frac{2}{\sqrt{3}}$$

$$\tan \theta = \frac{1}{\sqrt{3}} \qquad \cot \theta = \sqrt{3}$$

EXAMPLE 4 If $\sin \theta = 0$, find the values of the other functions of θ.

Solution: We have $y/r = 0$, and hence $y = 0$. This means that the terminal side must lie on the positive x axis or the negative x axis. The angle θ, therefore, is equal to 0° or 180° (or any angle coterminal with one of these). The values of the other defined functions are

$$\cos 0° = 1 \qquad \cos 180° = -1$$

$$\tan 0° = 0 \qquad \tan 180° = 0$$

$$\sec 0° = 1 \qquad \sec 180° = -1$$

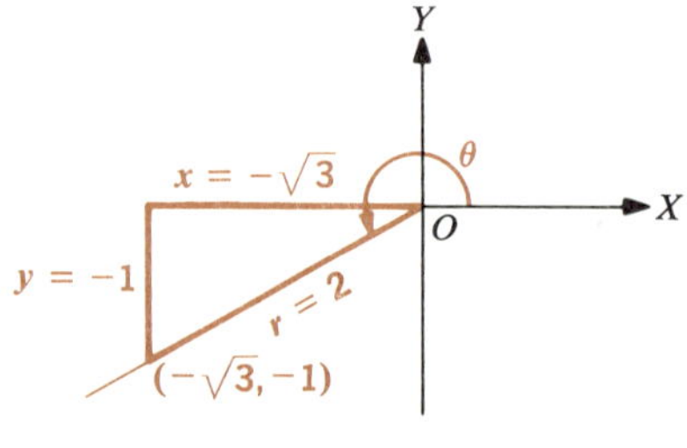

Figure 8-14

EXERCISE 8-3

The terminal side of an angle θ in standard position passes through the given point in each of the following problems. Draw the reference triangle, find r, draw a curved arrow from the initial side to the terminal side, and give the values of the six trigonometric functions of θ.

1. $(4, 3)$ **2.** $(3, 4)$ **3.** $(-5, 12)$

4. $(-12, 5)$ **5.** $(-\sqrt{3}, -1)$ **6.** $(-7, -24)$

7. $(2, -\sqrt{5})$ **8.** $(8, -15)$ **9.** $(-\sqrt{7}, -\sqrt{2})$

10. $(1, -\sqrt{3})$ **11.** $(-\sqrt{5}, \sqrt{11})$ **12.** $(\sqrt{3}, \sqrt{6})$

Name the quadrant or quadrants in which the terminal side of an angle in standard position may be and satisfy the given condition or conditions.

13. The sine is negative. **14.** The cosine is positive.

15. The tangent is positive. **16.** The cotangent is negative.

17. $\sin\theta > 0, \quad \cos\theta < 0$ **18.** $\cos\theta > 0, \quad \tan\theta < 0$

19. $\cos\theta < 0, \quad \cot\theta > 0$ **20.** $\sin\theta < 0, \quad \cos\theta < 0$

Give the signs of the values of the trigonometric functions of each of the following angles.

21. $118°$ **22.** $340°$ **23.** $-35°$ **24.** $-167°$

25. $-128°$ **26.** $\frac{4\pi}{5}$ **27.** $-\frac{5\pi}{3}$ **28.** $-\frac{\pi}{7}$

29. $\frac{7\pi}{6}$ **30.** $\frac{11\pi}{5}$

Construct an angle of measure between $0°$ and $360°$ which satisfies the conditions in each problem 31 through 40 and write the values of the other trigonometric functions. (Q is an abbreviation for quadrant.)

31. $\cos\theta = \frac{5}{13}, \quad \theta$ in Q I **32.** $\sin\theta = -\frac{3}{5}, \quad \theta$ in Q IV

33. $\tan\theta = \frac{8}{15}, \quad \theta$ in Q III **34.** $\cot\theta = -\frac{24}{7}, \quad \theta$ in Q II

35. $\csc\theta = \sqrt{17}, \quad \sec\theta < 0$ **36.** $\sec\theta = \sqrt{10}, \quad \csc\theta > 0$

37. $\sin\theta = 1$ **38.** $\cos\theta = -1$

39. $\sec\theta = 1$ **40.** $\csc\theta = -1$

Write the values of the defined functions of each of the following angles.

41. 3π **42.** $-\frac{7\pi}{2}$ **43.** 10π **44.** $\frac{3\pi}{2}$

Construct two angles in standard position which have the given functional value and different terminal sides, and write the values of the other trigonometric functions of each angle.

45. $\sin\theta = \frac{9}{41}$	**46.** $\cos\theta = \frac{24}{25}$	**47.** $\cot\theta = -\frac{7}{24}$
48. $\csc\theta = \frac{29}{21}$	**49.** $\tan\theta = -4$	**50.** $\sec\theta = 2$
51. $\cos\theta = -\frac{2}{3}$	**52.** $\sin\theta = -\frac{1}{2}$	**53.** $\cot\theta = \frac{5}{3}$
54. $\cot\theta = 0$	**55.** $\cos\theta = 0$	**56.** $\tan\theta = 0$

8-6 TRIGONOMETRIC FUNCTIONS OF NUMBERS

We have defined the trigonometric functions so that the domain of each function is a set of angles and the range a set of real numbers. We shall now discuss a modification of the definitions so that the domains as well as the ranges are sets of real numbers.

The length s of the arc of a circle intercepted by a central angle of θ radians is equal to the product of θ and the radius (Sec. 8-2); that is, $s = r\theta$. If the radius is chosen as the unit of length, this equation reduces to $s = \theta$. Hence, the number of radians in the central angle is equal to the number of units of length of the intercepted arc. This equality suggests that we might seek an interpretation of the trigonometric functions in which the domains are real numbers instead of angles (or measures of angles). Inasmuch as angles have positive, zero, and negative measures, it seems desirable that the domains of the new type of functions should be similarly inclusive. To achieve this end, we refer to Fig. 8-15 in which $r = 1$ and $\theta = s$. Now suppose we let s be a distance directed along the circle measured from $A(1, 0)$ with the positive direction counterclockwise, and the negative direction clockwise. Then, for example, we may interpret $\sin s$ as the sine of a directed distance and take the value of $\sin s$ to be the same as the value of $\sin$ (s radians). With this understanding, the sine function consists of a set of number pairs of the form $(s, \sin s)$ in which the domain is the set of all real numbers. So we may write

$$\sin s = \sin (s \text{ radians}) = \sin \left(\frac{180s}{\pi}\right)^\circ$$

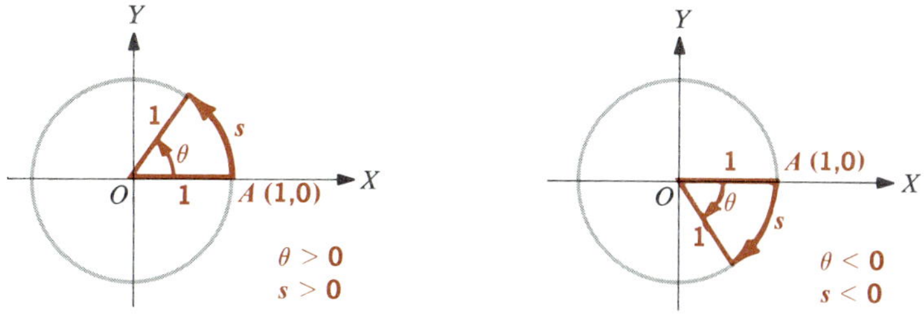

Figure 8-15

The preceding discussion motivates us to make the following definition:

Definition 8-2 If θ is an angle of s radians, then any trigonometric function of θ is equal to the same-named function of s.

On the basis of this definition and the fact that we equate a trigonometric function of an angle to the same function of the measure of the angle, we have

$$\sin\theta = \sin(s \text{ radians}) = \sin s$$

$$\cos\theta = \cos(s \text{ radians}) = \cos s$$

$$\tan\theta = \tan(s \text{ radians}) = \tan s$$

and so on, for the remaining functions.

These new functions, being introduced by use of the unit circle, are sometimes called *circular trigonometric functions* or just *circular functions.*

Clearly, the trigonometric functions and the circular functions differ only in their domains. For example, the ordered pair $(\theta, \sin\theta)$ corresponds to the ordered pair $(s, \sin s)$, where s is the number of radians in the angle θ and $\sin\theta = \sin s$. As a consequence of this close tie between the two types of functions, a property of one of the types has its counterpart in the other type. We shall be studying formulas and equations involving trigonometric function values in which the domain may be regarded either as a set of angles or a set of real numbers.

As the student progresses, he will discover that trigonometric functions of numbers are used freely and effectively in advanced mathematics, especially in calculus. This does not mean, however, that the trigonometric functions of angles are to be supplanted by the trigonometric functions of numbers. In some instances, it is advantageous to use numbers as the domain of a function, and in others, the problem is better handled with angles as the domain.

9

TABLES OF TRIGONOMETRIC FUNCTIONS AND SOLUTIONS OF RIGHT TRIANGLES

9-1 OPERATIONS WITH APPROXIMATE NUMBERS

The results of physical measurements are only approximately correct. The degree of accuracy in a measurement depends, among other things, on the quality of the measuring instrument and the aptness of the observer. The weight of an adult is usually expressed as an integer. Thus, we might give a man's weight as 174 pounds if he weighs between 173.5 and 174.5 pounds. We then consider 174 as an *approximate number.*

The exact area of the surface of the earth is not known. It has been computed to the nearest million to be 197,000,000 square miles. This means the exact area is between 196,500,000 and 197,500,000 square miles. The first three digits in 197,000,000 are correct (the 7 may be in error not to exceed $\frac{1}{2}$ in its place); the zeros are used to fill in the places of unknown digits up to the position of the decimal point. It is common practice to use only zeros for this purpose. Zeros are also used to locate the decimal point to the left of a set of digits. Suppose, for example, the length of an object is measured to be 0.0041 meter. The zeros in this number serve only to fix the position of the decimal point; they do not indicate accuracy of measurement. In each of the two preceding measurements the zero digits and the nonzero digits have different purposes, and we distinguish between them in accordance with the following definition.

Definition 9-1 All the digits of an approximate number except zeros which serve only to fix the position of the decimal point are called *significant digits* or *significant figures.*

The following numbers have their significant digits underscored:

$$0.00\underline{251} \qquad \underline{308} \qquad \underline{2.50} \qquad \underline{500.0} \qquad \underline{23.00}$$

The zero in 2.50 stands for a known digit and is significant. If the digit for that place were unknown, the number would be written as 2.5. The approximate number 2.50 stands for a value between 2.495 and 2.505. Likewise, the zeros in the last two numbers are significant.

The significant digits in decimal notation are apparent from the number itself except in one case. An integer ending with one or more zeros does not reveal the nature of the zero or zeros. In the so-called scientific notation, however, all significant digits are evident. In this notation a number is expressed as a product. The first factor has the significant digits with a decimal point just after the first digit, and the second factor is an integral power of 10. Employing scientific notation, for example, we can indicate the nature of the zeros in 800. To indicate that neither zero, one zero, or both zeros are significant, we express the number, respectively, as

$$8 \times 10^2, \qquad 8.0 \times 10^2, \qquad 8.00 \times 10^2$$

In each form, the significant digits appear in the first factor; and the product of the factors is 800.

A result obtained from calculations in which approximate numbers occur may have digits which are not significant. Such digits need to be discarded. The discarding of extra digits is called *rounding off.* In each case the new number is an approximation to the original number.

EXAMPLES We round off 2.718 to three digits and have 2.72. Here we add 1 to the third digit of the given number in order to obtain the best three-figure approximation. For two-figure and one-figure approximations, we round off 2.718, respectively, to 2.7 and 3.

In rounding off 2.715 to three digits, we observe that 2.71 and 2.72 are equally valid. In all such cases, we shall arbitrarily *round off so that the last digit is an even number.* Accordingly, we choose 2.72.

If any discarded digits are to the left of the decimal point, their places are filled with zeros. Thus, when rounded off to two significant figures, 756.2 becomes 760, with the zero not significant, or, in scientific notation, 7.6×10^2.

We state here two rules for rounding off which are frequently employed and which are usually satisfactory:

Rule 1 To find the sum (or difference) of two approximate numbers, write one number under the other and add (or subtract) in the usual way. Then round off the result so that the last digit retained is in the column farthest to the right in which both given numbers have significant digits.

Rule 2 To find the product (or quotient) of two approximate numbers, multiply (or divide) in the usual way. Then round off the result to the least number of significant digits found in either of the given numbers.

It is sometimes advantageous to do certain rounding off before multiplying or dividing numbers. If the significant digits in one number exceed those in the other number, we suggest that the first number be rounded off so that its excess in significant digits is only one. For example, to find the product of 3.1729 and 2.16, we could replace 3.1729 by 3.173 and then multiply.

EXAMPLE 1 Find the sum and difference of the numbers 84.037 and 2.72.

Solution:

$$\begin{array}{r} 84.037 \\ 2.72 \\ \hline 86.757 \end{array} \quad \text{Sum} \qquad \begin{array}{r} 84.037 \\ 2.72 \\ \hline 81.317 \end{array} \quad \text{Difference}$$

The column farthest to the right in which both given numbers have significant digits corresponds to the fourth digit in each of these results. We round off accordingly and have the approximate values 86.76 and 81.32.

EXAMPLE 2 Find the product of 0.0241 and 0.52.

Solution: The first number has three significant figures and the second two significant figures. Hence, their product should show two significant figures. Multiplying, we get

$$0.0241 \times 0.52 = 0.012532$$

Rounding off this value to two significant figures, we have either 0.013 or 1.3×10^{-2}, as the final result.

EXERCISE 9-1

Consider the following numbers as approximate measurements and give in each case the values between which the correct measurement lies.

1. 605 feet
2. 40.2 pounds
3. 66.0 feet
4. 0.73 inches
5. 8.170 centimeters
6. 63.40 pounds

Round off each of the following numbers to (a) three significant digits, (b) two significant digits, (c) one significant digit.

7. 8555
8. 1745
9. 4070
10. 0.3052
11. 0.9995
12. 72.50
13. 279,252
14. 0.002385
15. Express 6000 in scientific notation, indicating that it has (*a*) three significant digits, (*b*) two significant digits, (*c*) one significant digit.

Perform the indicated operations and express each result with the proper number of significant figures.

16. $5.8 + 3.05$
17. $7.5 + 2.07$
18. $7.56 - 9.80$
19. 2.6×3.3
20. 172×3.0
21. 16.0×22.0
22. $(7.341)(3.60)$
23. $(0.43)(240)$
24. $(813)(0.54)$
25. $78.2 \div 6.7$
26. $6.51 \div 0.13$
27. $(7.00)(4.0)$

9-2 TABLES OF TRIGONOMETRIC FUNCTIONS

The values of the trigonometric functions of angles are required in many problems and investigations. Although the exact values for certain special angles can be found, the exact values for an arbitrary angle cannot be determined. By methods of calculus, however, the values of the functions for any angle can be computed to any desired degree of accuracy. Table 2 in the Appendix gives the values of the functions of angles, at 10′ intervals, from 0° to 90°. This table is called *a table of natural trigonometric functions* to distinguish it from a table of the logarithms of the values of the functions, to be discussed later. The values of the functions, with a few exceptions, are unending decimals, and hence the tabular values are approximations. The table is called *a four-place table,* since the values of the functions are rounded off to four significant figures.

Angles from 0° to 45° are in the first column of the pages, and the next six columns contain the functional values, with a particular function indicated at the top of each column. The functions of each angle are to the right along the line of the angle. Angles from 45° to 90° are in the last column of the pages. The values of the functions of an angle in this rightmost column are to the left along the line of the angle, with the functions indicated at the bottom of the page.

The table may be used to find a function of a given angle or, inversely, to find the angle corresponding to a given functional value. We shall explain the use of the table with some illustrative examples.

EXAMPLE 1 Find the sine, cosine, and tangent of 41° 20′.

Solution: We locate 41° 20′ in the first column of Table 2. Then by looking across the table in line with the angle and reading from the columns headed by Sin, Cos, and Tan, we find

$$\sin 41^\circ 20' = 0.6604$$

$$\cos 41^\circ 20' = 0.7509$$

$$\tan 41^\circ 20' = 0.8796$$

EXAMPLE 2 Find the sine, cosine, and cotangent of 73° 50′.

Solution: We locate 73° 50′ in the last column of Table 2, and read the functional values which are in line with the angle and in the columns with Sin, Cos, and Cot at the bottom. Thus, we obtain

$$\sin 73^\circ\ 50' = 0.9605$$

$$\cos 73^\circ\ 50' = 0.2784$$

$$\cot 73^\circ\ 50' = 0.2899$$

EXAMPLE 3 Given $\sin \theta = 0.8526$, find θ.

Solution: We look for the entry .8526 in a column headed by Sin or in a column with Sin at its foot. To locate the entry readily, we observe that from 0° to 90° the sine increases from 0 to 1. The number .8526 appears in Table 2 in the column with Sin at the bottom, and the angle must be read at the right. We find $\theta = 58^\circ\ 30'$.

EXERCISE 9-2

Find the value of each of the following functions.

1. $\sin 34^\circ\ 50'$ **2.** $\cot 34^\circ\ 00'$ **3.** $\tan 46^\circ\ 30'$

4. $\sec 31^\circ\ 40'$ **5.** $\sin 47^\circ\ 00'$ **6.** $\cos 70^\circ\ 40'$

7. $\cot 65^\circ\ 20'$ **8.** $\sec 45^\circ\ 30'$ **9.** $\csc 2^\circ\ 10'$

10. $\cot 4^\circ\ 00'$ **11.** $\cos 86^\circ\ 00'$ **12.** $\sin 30^\circ\ 10'$

Find the angle corresponding to each of the given functions.

13. $\sin \theta = 0.1248$ **14.** $\cos \theta = 0.9971$

15. $\cot \theta = 7.770$ **16.** $\sec \theta = 1.147$

17. $\cos \theta = 0.8843$ **18.** $\sin \theta = 0.5398$

19. $\tan \theta = 0.6661$ **20.** $\csc \theta = 2.154$

21. $\sin \theta = 0.8124$ **22.** $\cos \theta = 0.5544$

23. $\cot \theta = 0.4592$ **24.** $\sec \theta = 2.650$

25. $\tan \theta = 1.171$ **26.** $\cot \theta = 0.8195$

9-3 INTERPOLATION

In the preceding section, we learned how to find the functions of an angle which appears in the table and also how to find the angle corresponding to a functional entry. When the given angle is not listed in the table, we can find approximate values of its functions by a process called *interpolation.* According to this process, a function of an angle is proportional to the angle for small changes in the angle. Although this assumption is not entirely true, improved accuracy may be obtained by interpolation. Interpolation may also be used to find an angle whose given functional value is not listed. We shall explain the procedure with some examples.

EXAMPLE 1 Find the value of sin 57° 14′.

Solution: The given angle, not listed in Table 2, has a value between the consecutive entries 57° 10′ and 57° 20′. Since 57° 14′ is $\frac{4}{10}$ of the way from 57° 10′ to 57° 20′, we assume that sin 57° 14′ is also $\frac{4}{10}$ of the way from sin 57° 10′ to sin 57° 20′. Reading directly from Table 2, we find $\sin 57°\ 10' = 0.8403$ and $\sin 57°\ 20' = 0.8418$. From 8403 to 8418 (neglecting decimal points for the moment) is 15 units, and $\frac{4}{10} \times 15 = 6$. Adding 6 to 8403, we have 8409; hence $\sin 57°\ 14' = 0.8409$.

The data and computations may be put in the following tabular form:

$$10'\left[\begin{array}{l} 4'\left[\begin{array}{l}\sin 57°\ 10' = 0.8403 \\ \sin 57°\ 14' = \qquad ?\end{array}\right. \\ \quad \sin 57°\ 20' = 0.8418 \end{array}\right]15 \qquad \begin{array}{c} \frac{4}{10} \times 15 = 6 \\ 8403 + 6 = 8409 \\ \sin 57°\ 14' = 0.8409 \end{array}$$

Note: The difference between two consecutive entries of the table is called the *tabular difference.* In the preceding problem the tabular difference of sin 57° 10′ and sin 57° 20′ is 0.0015. The part of the tabular difference used in the interpolation process is called the *correction.* In computing the correction, we may omit the decimal points until we write the final result. Usually it is necessary to round off in finding the correction. Had the preceding example required sin 57° 15′, we would have had, for the correction, $\frac{5}{10} \times 15 = 7.5$. We would round off this number to 8 in accordance with our agreement in Sec. 9-1. In no case should more digits be kept in the interpolated result than are found in the table.

EXAMPLE 2 Find the value of cos 37° 27′.

Solution: As we can see from the table, the values of the cosine decrease as the angle increases from 0° to 90°. Hence, cos 37° 27′ is less than cos 37° 20′ and greater than cos 37° 30′. Accordingly, the correction in this case is to be subtracted from the value of cos 37° 20′. We find $\cos 37°\ 20' = 0.7951$ and $\cos 37°\ 30' =$

0.7934. The tabular difference is 7951 − 7934 = 17 (neglecting decimal points). Since the angle 37° 27′ is $\frac{7}{10}$ of the way from 37° 20′ to 37° 30′, we take $\frac{7}{10}$ of 17 (equals 12, rounded off) as the correction. Subtracting this correction, we have 7951 − 12 = 7939; hence, cos 37° 27′ = 0.7939.

The quantities involved in the preceding discussion are as follows:

$$10'\left[7'\left[\begin{array}{l}\cos 37^\circ\ 20' = 0.7951\\ \cos 37^\circ\ 27' = \qquad ?\end{array}\right. \begin{array}{l}\\ \\ \cos 37^\circ\ 30' = 0.7934\end{array}\right]17 \qquad \begin{array}{r}\frac{7}{10}\times 17 = 12 \text{ (rounded off)}\\ 7951 - 12 = 7939\\ \cos 37^\circ\ 27' = 0.7939\end{array}$$

EXAMPLE 3 Find θ if cot θ = 0.5782.

Solution: We look for .5782 in a column with Cot at the top or Cot at the bottom. Our number is not in any of these columns, but its value is between two consecutive entries with Cot at the bottom of Table 2. Since the name of the function is at the bottom, we read the corresponding angles at the right. The data and computations may be arranged in tabular form as follows:

$$10'\left[\begin{array}{r}\left.\begin{array}{r}\cot 59^\circ\ 50' = 0.5812\\ ? = 0.5782\end{array}\right]30\\ \cot 60^\circ\ 00' = 0.5774\quad\end{array}\right]38$$

$$\tfrac{30}{38}\times 10' = 8' \text{ (nearest integer)}$$

$$59^\circ\ 50' + 8' = 59^\circ\ 58'$$

$$\text{Hence } \theta = 59^\circ\ 58'$$

Note: It is advisable when first interpolating to write the quantities in detail, similar to that of the preceding problems. When the procedure is fully understood, the computations in most cases can be done mentally, and only the final result needs to be written. Also, it may be observed from the table that the sine, tangent, and secant increase as the angle increases from 0° to 90° and the cosine, cotangent, and cosecant decrease.

EXERCISE 9-3

Interpolate to find the value of each trigonometric function.

1. sin 20° 25′
2. sin 36° 23′
3. tan 36° 45′
4. tan 11° 11′
5. sin 63° 47′
6. tan 70° 6′

7. cos 13° 46′	**8.** cos 30° 25′	**9.** cot 23° 33′
10. cot 21° 19′	**11.** cos 52° 18′	**12.** cos 84° 18′
13. sec 15° 41′	**14.** sec 46° 37′	**15.** csc 44° 36′
16. csc 50° 58′	**17.** sin 88° 51′	**18.** cos 5° 46′
19. cot 65° 9′	**20.** sec 71° 5′	**21.** csc 16° 19′

Interpolate to find each angle θ to the nearest minute.

22. $\sin\theta = 0.6470$	**23.** $\sin\theta = 0.5260$
24. $\sin\theta = 0.7483$	**25.** $\sin\theta = 0.9021$
26. $\tan\theta = 0.2367$	**27.** $\tan\theta = 0.7609$
28. $\cos\theta = 0.9845$	**29.** $\cos\theta = 0.8734$
30. $\cos\theta = 0.6785$	**31.** $\cos\theta = 0.3942$
32. $\cot\theta = 4.445$	**33.** $\cot\theta = 1.244$
34. $\sec\theta = 1.067$	**35.** $\sec\theta = 1.428$
36. $\csc\theta = 3.926$	**37.** $\csc\theta = 1.568$
38. $\sin\theta = 0.0931$	**39.** $\cos\theta = 0.9991$
40. $\tan\theta = 10.58$	**41.** $\cot\theta = 0.0912$

9-4 REDUCTION TO POSITIVE ACUTE ANGLES*

Having used a table to find the values of trigonometric functions of angles in the interval 0° to 90°, we next consider the problem of finding the functions of angles outside this interval. Any angle greater than 360° or any negative angle in standard position is coterminal with a positive angle not exceeding 360°. Hence, the functions of any angle are equal to the functions of some angle in the interval 0° to 360°. We shall next show that the functions of angles in the interval 0° to 360°, and therefore of all angles may be expressed in terms of functions of angles in the narrower interval 0° to 90°. This means that trigonometric functions for acute angles will serve also for angles not restricted to that interval. We shall not consider quadrantal angles in this discussion, however,

* An angle whose measure is between 0° and 90° or between 0° and −90° is called an *acute angle*.

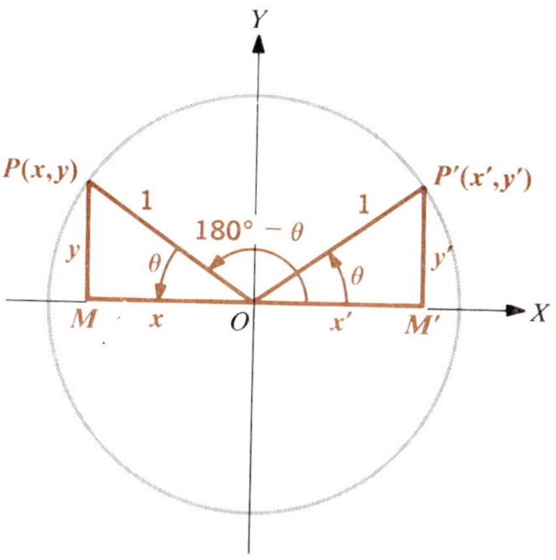

Figure 9-1

since we have already obtained their functions (Sec. 8-5). We begin by introducing the idea of a reference angle.

Definition 9-2 The positive acute angle which the terminal side of an angle in standard position makes with the x axis is called the *reference angle* of the given angle.

In Fig. 9-1, we observe that θ is the reference angle of $180° - \theta$ and that, by definition,

$$\sin(180° - \theta) = y \qquad \cos(180° - \theta) = x$$
$$\sin\theta = y' \qquad \cos\theta = x'$$

Since coordinates of points are directed distances, we see that $x = -x'$ and $y = y'$. And, recalling that $\tan\theta = \sin\theta/\cos\theta$, we may write

$$\sin(180° - \theta) = \sin\theta$$
$$\cos(180° - \theta) = -\cos\theta$$
$$\tan(180° - \theta) = -\tan\theta$$

Referring to Fig. 9-2, where $x = -x'$ and $y = -y'$, we have

$$\sin(180° + \theta) = -\sin\theta$$
$$\cos(180° + \theta) = -\cos\theta$$
$$\tan(180° + \theta) = \tan\theta$$

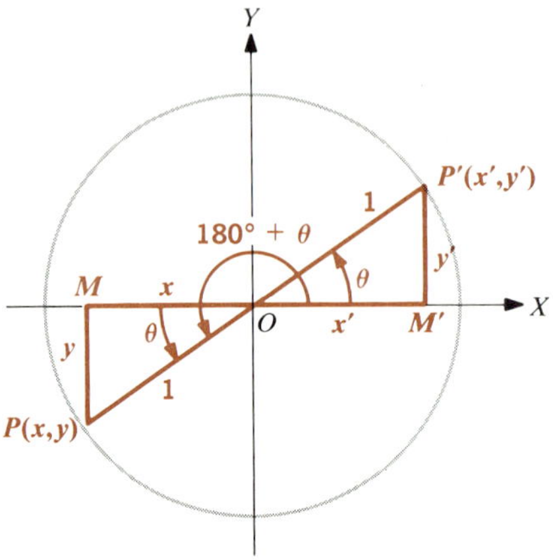

Figure 9-2

Finally, in Fig. 9-3, $x = x'$ and $y = -y'$, and therefore

$$\sin(360° - \theta) = -\sin\theta$$
$$\cos(360° - \theta) = \cos\theta$$
$$\tan(360° - \theta) = -\tan\theta$$

The preceding formulas are called *reduction formulas*. We note that similar reduction formulas could be written for the cotangent, secant, and cosecant functions. We see then that the functions of the second-, third-, and fourth-quadrant angles are expressible as functions of angles between 0° and 90°. It may also be observed that each function is equal to plus or minus the same function of the reference angle. Hence, we state the following theorem:

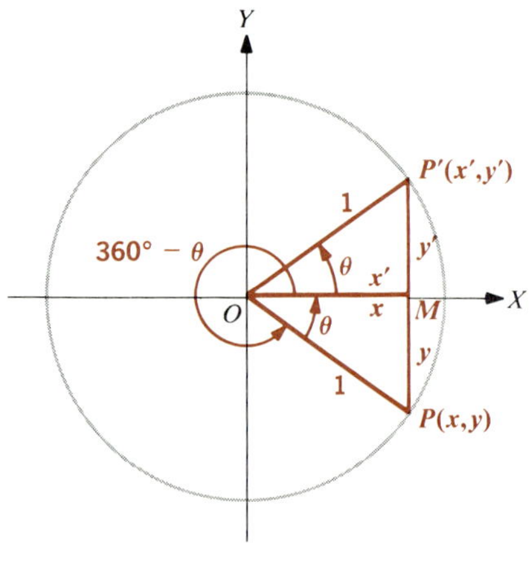

Figure 9-3

Theorem 9-1 A trigonometric function of an angle in any quadrant is equal to plus or minus the same function of the reference angle.

We list the steps for expressing a function of an angle as a function of the reference angle:

1. Determine from the quadrant of the given angle if the function is positive or negative.
2. Compute the reference angle.
3. Use the same function of the reference angle preceded by the proper sign.

Although we have used degree measure in the reduction formulas, we could, of course, convert the angles to radian measure. Thus, for example,

$$\sin(180^\circ + 30^\circ) = \sin(\pi + \tfrac{1}{6}\pi) = -\sin\tfrac{1}{6}\pi$$

EXAMPLE Express $\sin 148^\circ$ and $\tan 148^\circ$ as functions of the reference angle.

Solution: We notice that the terminal side of 148° is in the second quadrant which makes the sine positive and the tangent negative. Since $148^\circ = 180^\circ - 32^\circ$, the reference angle is 32°. Hence,

$$\sin 148^\circ = \sin 32^\circ \qquad \text{and} \qquad \tan 148^\circ = -\tan 32^\circ$$

The sign and the reference angle should be verified in each of the following illustrations:

$$\cos 220^\circ = \cos(180^\circ + 40^\circ) = -\cos 40^\circ$$

$$\cot 290^\circ = \cot(360^\circ - 70^\circ) = -\cot 70^\circ$$

$$\sin\frac{5\pi}{3} = \sin\left(2\pi - \frac{\pi}{3}\right) = -\sin\frac{\pi}{3}$$

$$\tan(-50^\circ) = \tan(360^\circ - 50^\circ) = -\tan 50^\circ$$

EXERCISE 9-4

Draw each of the following angles in standard position and find the reference angle.

1. 153° **2.** 257° **3.** 325° **4.** 465°

5. $\frac{3\pi}{4}$ 6. $\frac{5\pi}{6}$ 7. $\frac{4\pi}{3}$ 8. $\frac{8\pi}{5}$

9. $-318°$ 10. $-245°$ 11. $-140°$ 12. $-55°$

Give the value of each of the following quantities, using Table 2.

13. sin 141°
14. cos 217°
15. tan (−58°)
16. cot 290°
17. sec 223°
18. sin 91° 10′
19. cos 100° 30′
20. sin (−168°)
21. csc (−128°)
22. tan 785°
23. cos 1110°
24. sec (−760°)

9-5 THE RIGHT TRIANGLE

The right triangle is a figure of special importance; it occurs in many ordinary experiences and has numerous applications in engineering, physics, and other sciences. Hence, the right triangle is given special attention in trigonometry courses. We shall begin our study by learning how to find the values of the trigonometric functions of the acute angles from the lengths of the sides of the triangle.

Figure 9-4 shows a right triangle with the acute angle A in standard position. The coordinates of the vertex of angle B are $x = b$, $y = a$, and the distance $r = c$. We observe that a is the length of the side opposite A, b is the length of the side adjacent to A, and c is the length of the hypotenuse. Then using these terms and applying the definitions of the trigonometric functions (Sec. 8-4), we have

$$\sin A = \frac{\text{side opposite}}{\text{hypotenuse}} \qquad \csc A = \frac{\text{hypotenuse}}{\text{side opposite}}$$

$$\cos A = \frac{\text{side adjacent}}{\text{hypotenuse}} \qquad \sec A = \frac{\text{hypotenuse}}{\text{side adjacent}}$$

$$\tan A = \frac{\text{side opposite}}{\text{side adjacent}} \qquad \cot A = \frac{\text{side adjacent}}{\text{side opposite}}$$

The functions of A depend only on the lengths of the sides of the triangle; they are independent of a coordinate system. Henceforth, then, we may obtain the values of the functions of either acute angle of a right triangle by picking out the opposite side and the adjacent side and applying the above results. It is essential to remember, however, that this method applies only to the acute angles of a *right* triangle.

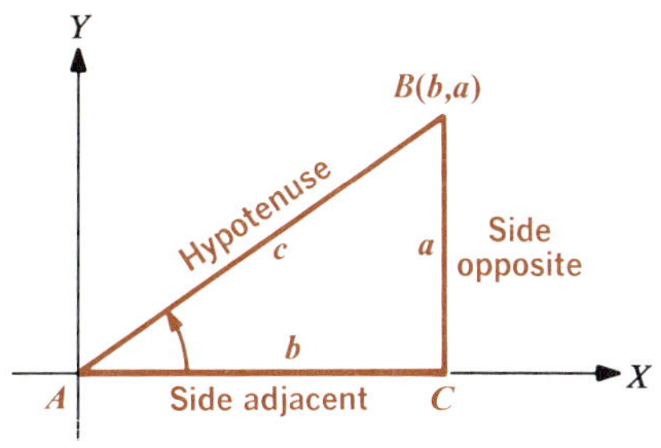

Figure 9-4

In dealing with right triangles, we shall usually let A and B stand for the acute angles and C for the right angle. Also, we shall let a, b, and c stand for the lengths of the sides opposite A, B, and C, respectively.

If the sum of two positive angles is 90°, the angles are said to be *complementary*. It follows from this definition that the acute angles of a right triangle are complementary. We shall use this fact to see how the functions of complementary angles are related. Referring to Fig. 9-5, we observe that side a is opposite angle A and adajcent to angle B, and that side b is opposite B and adjacent to A. Hence, we may write

$$\sin A = \frac{a}{c} = \cos B \qquad \csc A = \frac{c}{a} = \sec B$$

$$\cos A = \frac{b}{c} = \sin B \qquad \sec A = \frac{c}{b} = \csc B$$

$$\tan A = \frac{a}{b} = \cot B \qquad \cot A = \frac{b}{a} = \tan B$$

In these equations the functions appear in the pairs *sine* and *cosine*, *tangent* and *cotangent*, *secant* and *cosecant*. The functions in each of these pairs are called *cofunctions*. For example, the cosine is the cofunction of the sine, and the sine is the cofunction of the cosine. The equations show, then, that a function of A is equal to its cofunction of B. Since any two complementary angles

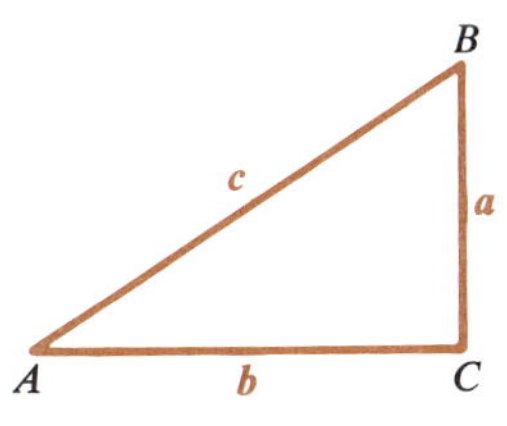

Figure 9-5

may be the acute angles of a right triangle, we have established the following theorem:

Theorem 9-2 Any trigonometric function of a positive acute angle is equal to the corresponding cofunction of the complementary angle.

Conversely, if a function of a positive acute angle is equal to the cofunction of another positive acute angle, then the angles are complementary. The student may prove this statement.

EXAMPLE 1 Write each of the quantities sin 40°, cot 70°, sec 60° as a function of the complementary angle.

Solution: We subtract the angle in each case from 90° to obtain its complement, then we replace each angle by its complement and each function by its cofunction. Hence,

$$\sin 40^\circ = \cos 50^\circ$$

$$\cot 70^\circ = \tan 20^\circ$$

$$\sec 60^\circ = \csc 30^\circ$$

EXAMPLE 2 If $\tan 3\theta = \cot(\theta - 6^\circ)$ and if 3θ and $(\theta - 6^\circ)$ are positive acute angles, then find θ.

Solution: Since the tangent and cotangent are cofunctions, the angles are complementary. Therefore, we have

$$3\theta + (\theta - 6^\circ) = 90^\circ$$

$$4\theta = 96^\circ$$

$$\theta = 24^\circ$$

EXERCISE 9-5

1. Write the values of the trigonometric functions of angles A and B in each triangle of Fig. 9-6.

Draw a right triangle and write the values of the functions of A and B in each of the following problems.

2. $a = 12, \quad b = 5$
3. $a = 24, \quad c = 25$
4. $a = 4, \quad b = 7$
5. $a = 35, \quad c = 37$
6. $a = 2, \quad c = 5$
7. $a = 6, \quad b = 7$

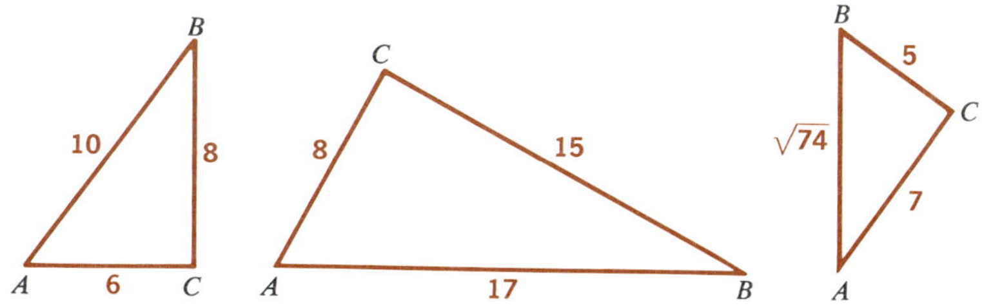

Figure 9-6

Express each of the following quantities as a function of the complementary angle.

8. $\sin 65°$ **9.** $\cos 35°$ **10.** $\tan 61°$

11. $\cot 53°$ **12.** $\sec \dfrac{\pi}{4}$ **13.** $\csc \dfrac{3\pi}{10}$

14. $\cos \dfrac{\pi}{3}$ **15.** $\sin \dfrac{2\pi}{5}$

Find θ if functions of positive acute angles are involved in each equation.

16. $\cot \theta = \tan 65°$ **17.** $\sin \theta = \cos \theta$

18. $\cos 2\theta = \sin \theta$ **19.** $\tan (\theta + 36°) = \cot \theta$

20. $\sin \left(\theta + \dfrac{\pi}{4}\right) = \cos \left(\theta - \dfrac{\pi}{8}\right)$

21. $\csc \left(\theta + \dfrac{\pi}{6}\right) = \sec \left(\theta + \dfrac{\pi}{3}\right)$

22. $\cot \left(2\theta + \dfrac{\pi}{9}\right) = \tan 3\theta$

23. $\sec \left(\dfrac{\pi}{3} + \theta\right) = \csc \left(2\theta - \dfrac{\pi}{4}\right)$

24. If A and B are positive acute angles and $\sin A = \cos B$, prove that $A + B = 90°$.

9-6 FUNCTIONS OF 30°, 45°, AND 60°

There are certain special angles whose functional values can be determined geometrically. Among these are the angles of 30°, 45°, and 60°.

The functions of 45° may be read directly from a right triangle

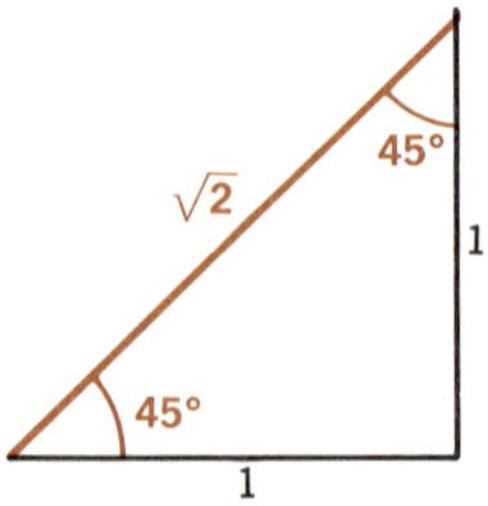

Figure 9-7

whose acute angles are each equal to 45°. For convenience, we let the length of the equal sides be 1, and therefore the hypotenuse is equal to $\sqrt{2}$. Then, from Fig. 9-7, we write

$$\sin 45° = \frac{1}{\sqrt{2}}, \qquad \cos 45° = \frac{1}{\sqrt{2}}, \qquad \tan 45° = 1$$

$$\csc 45° = \sqrt{2}, \qquad \sec 45° = \sqrt{2}, \qquad \cot 45° = 1$$

To obtain the functions of 30° and 60°, we use one of the right triangles formed by bisecting an angle of an equilateral triangle. It is evident that the right triangle thus constructed (Fig. 9-8) has acute angles of 30° and 60° and also that the hypotenuse is double the side opposite the 30° angle. If we select the convenient lengths 2 and 1 for the hypotenuse and the shorter side, then the length of the remaining side is $\sqrt{3}$. From the triangle, we have

$$\sin 30° = \frac{1}{2}, \qquad \cos 30° = \frac{\sqrt{3}}{2}, \qquad \tan 30° = \frac{1}{\sqrt{3}}$$

$$\csc 30° = 2, \qquad \sec 30° = \frac{2}{\sqrt{3}}, \qquad \cot 30° = \sqrt{3}$$

$$\sin 60° = \frac{\sqrt{3}}{2}, \qquad \cos 60° = \frac{1}{2}, \qquad \tan 60° = \sqrt{3}$$

$$\csc 60° = \frac{2}{\sqrt{3}}, \qquad \sec 60° = 2, \qquad \cot 60° = \frac{1}{\sqrt{3}}$$

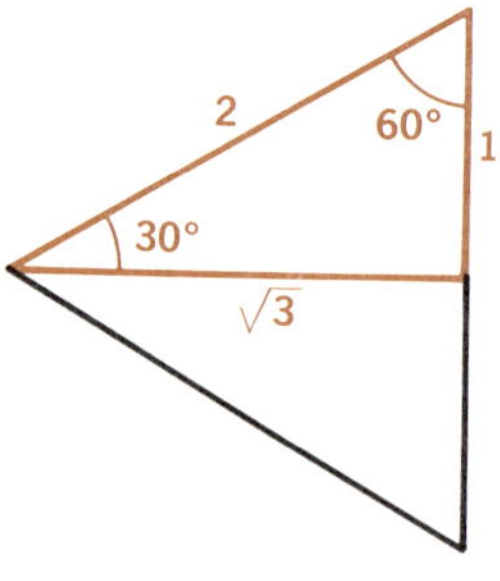

Figure 9-8

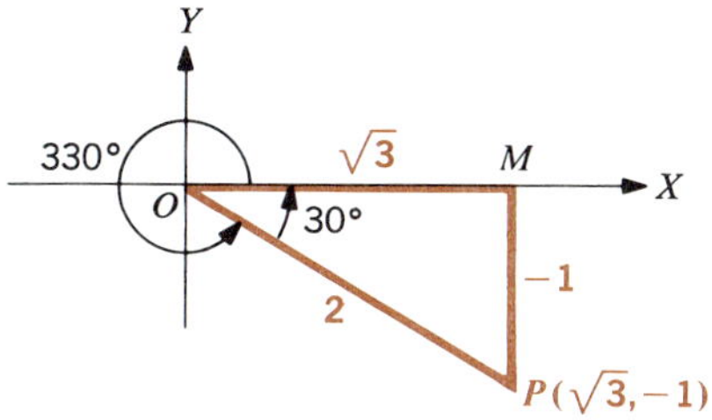

Figure 9-9

It is advantageous to be able to readily write the values of the functions of 30°, 45°, and 60° since these values are used so frequently. A better plan than memorizing the values, however, is to visualize the right triangles of Figs. 9-7 and 9-8 and then read the values from the mental picture.

The methods of evaluating the functions of 30°, 45°, and 60° will also serve to yield the functions of angles in any quadrant if the terminal side makes one of these angles with the positive or negative x axis. This is done by using the appropriate reduction formulas (Sec. 9-4) or, alternatively, from a figure, as illustrated in the following example:

EXAMPLE Construct the angle 330° in standard position and find the sine, cosine, and tangent of the angle.

Solution: The terminal side of 330° is in the fourth quadrant and makes an angle of 30° with the positive x axis (Fig. 9-9). We select a point P on the terminal side 2 units from the origin and draw MP, thus forming the reference triangle OMP. The coordinates of P are $(\sqrt{3}, -1)$. Then, applying the definitions of the trigonometric functions, we obtain

$$\sin 330° = -\tfrac{1}{2}, \qquad \cos 330° = \frac{\sqrt{3}}{2}, \qquad \tan 330° = -\frac{1}{\sqrt{3}}$$

EXERCISE 9-6

Substitute the values of the trigonometric functions and show that each equation 1 through 12 is true.

1. $\sin^2 45° + \cos^2 45° = 1$† **2.** $2 \sin 30° \cos 30° = \sin 60°$

3. $1 + \tan^2 30° = \sec^2 30°$ **4.** $1 + \cot^2 30° = \csc^2 30°$

† The notation $\sin^2 \theta$, read "sine square theta," means $\sin \theta \cdot \sin \theta$ or $(\sin \theta)^2$, and similarly for the other functions.

5. $2\cos^2 30° = 1 + \cos 60°$

6. $\cos^2 30° - \sin^2 30° = \cos 60°$

7. $1 - 2\sin^2 30° = \cos 60°$

8. $1 - \cos^2 60° = \sin^2 60°$

9. $\dfrac{1 - \tan^2 30°}{2\tan 30°} = \dfrac{1}{\tan 60°}$

10. $\dfrac{1 + \tan 60° \tan 30°}{\tan 60° - \tan 30°} = \dfrac{1}{\tan 30°}$

11. $\dfrac{2\cot 30°}{\cot^2 30° - 1} = \dfrac{1}{\cot 60°}$

12. $\dfrac{\sin 60°}{1 + \cos 60°} = \dfrac{1}{\cot 30°}$

Evaluate each of the following expressions.

13. $\dfrac{\cot 30° - \cot 45°}{1 + \cot 30° \cot 45°}$

14. $\dfrac{\cos 60° - \sin 60°}{\cos 30° \sin 30°}$

15. $\sin 60° \cos 30° + \cos 60° \sin 30°$

16. $\cos 30° \cos 60° + \sin 30° \sin 60°$

17. $\sin 45° \cos 30° - \cos 45° \sin 30°$

Draw each of the following angles in standard position, complete a reference triangle, and find the values of its sine, cosine, and tangent.

18. $150°$
19. $-120°$
20. $315°$
21. $-330°$
22. $300°$
23. $-150°$
24. $-330°$
25. $135°$
26. $\dfrac{11\pi}{6}$
27. $\dfrac{5\pi}{4}$
28. $-\dfrac{5\pi}{3}$
29. $-\dfrac{4\pi}{3}$

9-7 SOLUTION OF RIGHT TRIANGLES

The three sides and the three angles of a triangle are called its *parts*. The process of finding the unknown parts from the given parts is called *solving the triangle*. We shall now consider the problem of solving right triangles. It is possible to find the remaining parts of a right triangle if, in addition to the right angle, one side and any other part (side or angle) are known. This is true because a right triangle is determined by three of its parts at least

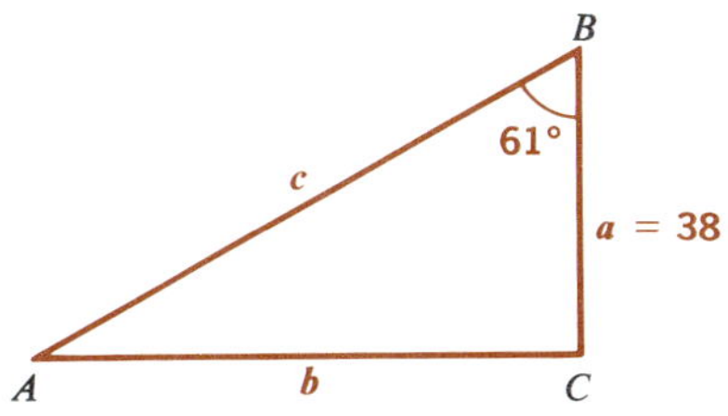

Figure 9-10

one of which is a side. The key principle in solving a right triangle is the fact that a function of an acute angle yields an equation involving three quantities, namely, two sides and the value of the function formed by the ratio of the sides.

EXAMPLE 1 A right triangle has $a = 38$ and $B = 61°$. Find A, b, and c.

Solution: We first draw a right triangle approximately to scale and mark the known parts (Fig. 9-10). The acute angles are complementary, and we have at once

$$A = 90° - 61° = 29°$$

To find b, we write, from the definition of tangent,

$$\frac{b}{38} = \tan 61°$$

$$b = 38 \tan 61°$$

Hence, using Table 2, we find

$$\begin{aligned} b &= 38(1.80) \\ &= 68.4 \\ &= 68 \qquad \text{(rounded off)} \end{aligned}$$

Secant B may conveniently be used in finding c. Thus,

$$\frac{c}{38} = \sec 61°$$

$$\begin{aligned} c &= 38 \sec 61° \\ &= 38(2.06) \\ &= 78 \qquad \text{(rounded off)} \end{aligned}$$

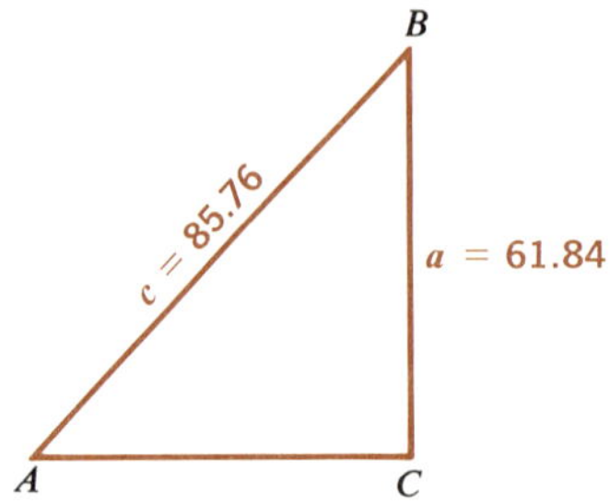

Figure 9-11

From the given parts, we have obtained $A = 29°$, $b = 68$, and $c = 78$. We rounded off the sides to two significant figures because the given side has two significant figures.

EXAMPLE 2 Solve the right triangle which has $a = 61.84$ and $c = 85.76$.

Solution: From the right triangle (Fig. 9-11), we write

$$\sin A = \frac{61.84}{85.76} = 0.7211$$

Using Table 2 and interpolating, we find

$$A = 46°\ 9' \qquad \text{and} \qquad B = 90° - 46°\ 9' = 43°\ 51'$$

From the definition of cosine, we now obtain

$$\frac{b}{85.76} = \cos 46°\ 9'$$

$$b = 85.76 \cos 46°\ 9' = 85.76(0.6928)$$

$$= 59.41 \qquad \text{(rounded off)}$$

The results are $A = 46°\ 9'$, $B = 43°\ 51'$, and $b = 59.41$.

Although the preceding examples show how to solve right triangles, we list the following directions:

1. Use the given parts and construct a right triangle approximately to scale. Mark the known parts in the figure and indicate the unknown parts with letters. The figure, exhibiting roughly the values of the parts to be computed, may serve as a check against large errors.

2. For each unknown part choose a formula which will yield that part. It is better, where possible, to find a part from given parts rather than from computed parts. This plan tends to give more accurate results and avoids carrying forward an erroneous computation.

3. Arrange the work in a neat, orderly way.

9-8 ACCURACY OF RESULTS

In solving triangles, we shall regard the given parts as approximate quantities, just as we did in the illustrative examples. This means that the accuracy of the computed parts depends on the accuracy of the given parts. Accordingly, a computed side should be rounded off to as many significant figures as contained in the given side or sides (Sec. 9-1). Since angles are also involved, we need to know what accuracy in angles is comparable with a given accuracy in the sides. The corresponding accuracies indicated in the table, and which we shall use, are approximately correct.

Sides	Angles
Two significant figures	Nearest degree
Three significant figures	Nearest multiple of 10′
Four significant figures	Nearest minute

A computed angle, by this plan, should be expressed to the nearest degree, nearest multiple of 10′, or nearest minute according as the sides have two, three, or four significant figures. It is a good plan, however, when the sides have fewer significant figures than the table, to carry along in the computing process one step greater accuracy than that of the given data and finally to round off properly the values of the computed parts. Suppose, for example, that the product 53 sin 38° yields a side of a triangle. Since the factor 53 has two significant figures and the table gives sin 38° to four significant figures, we would round off the value of sin 38° to three significant figures before multiplying. This preliminary rounding off makes the work a little shorter and usually, though not always, gives the same final result as would be found otherwise.

We shall assume that the given parts of triangles which are to be solved, in all the problems of this book, are of corresponding accuracies. Thus, in problem 2 of the exercise below, the zero in $B = 40°$ is significant since the given side $a = 26$ has two significant figures. Similarly, in problem 13 the zero in $A = 32°\ 30'$ is not significant because the given side has only three significant figures. This plan of interpreting final zeros holds for the computed parts as given in the answers.

As a further remark concerning the answers, we remind the student that two different methods of computing a part of a triangle will sometimes yield slightly different results.

EXERCISE 9-7

Find the remaining parts of each right triangle. Give final results properly rounded off.

1. $a = 43, \quad A = 37°$

2. $b = 26, \quad B = 40°$

3. $c = 6.4, \quad B = 29°$

4. $c = 28, \quad A = 23°$

5. $b = 45, \quad A = 51°$

6. $a = 18, \quad B = 44°$

7. $b = 70, \quad a = 88$

8. $a = 4.0, \quad b = 5.8$

9. $a = 6.1, \quad c = 9.3$

10. $a = 0.61, \quad b = 0.47$

11. $a = 48, \quad b = 84$

12. $b = 3.3, \quad c = 4.4$

13. $a = 253, \quad A = 32° \, 30'$

14. $b = 312, \quad B = 57° \, 30'$

15. $b = 0.622, \quad A = 51° \, 40'$

16. $a = 23.1, \quad A = 42° \, 40'$

17. $a = 3.52, \quad b = 4.50$

18. $a = 0.352, \quad b = 0.450$

19. $a = 42.0, \quad c = 54.0$

20. $b = 1.84, \quad c = 2.60$

Solve the following right triangles, making the necessary interpolations.

21. $c = 6000, \quad A = 62° \, 30'$

22. $c = 3000, \quad B = 54° \, 13'$

23. $a = 4200, \quad b = 5400$

24. $a = 2.752, \quad c = 5.000$

25. $a = 0.7553, \quad B = 60° \, 40'$

26. $b = 3541, \quad A = 56° \, 17'$

27. $a = 84.32, \quad b = 68.71$

28. $a = 5.320, \quad b = 6.431$

29. The length of the base of an isosceles triangle is 89.0 inches, and the angle opposite the base is 77° 00′. Find the other parts of the triangle.

30. Find the altitude of an equilateral triangle each of whose sides is 306 feet long.

31. Find the length of a side of an equilateral triangle which is inscribed in a circle of radius 11.4.

32. Find the perimeter of a regular pentagon (five-sided polygon with equal sides and angles) which is inscribed in a circle with a radius of 25 inches.

33. A chord of length 12 inches is drawn in a circle of radius 18 inches. Find (*a*) the central angle subtended by the chord; (*b*) the distance from the center of the circle to the midpoint of the chord.

34. Find the perimeter of a regular octagon which is (*a*) inscribed in a circle of radius 42 inches; (*b*) circumscribed about the circle.

9-9 APPLICATIONS OF RIGHT TRIANGLES

In certain problems of a practical nature, new terms arise. At this point we define some terms pertaining to angles.

Suppose E stands for the position of an observer's eye and O for a point on an object which he is viewing (Fig. 9-12). The line from E to O is called the *line of sight.* Let EH be a horizontal line in the vertical plane containing EO. Then, if O is above E, the angle HEO is called the *angle of elevation.* If O is lower than E, the angle HEO is called the *angle of depression.*

In plane surveying and marine navigation, the direction from one point to another is indicated in a special way. Let NS (Fig. 9-13) stand for a horizontal line in the north-south direction. Let O be a point on NS and A a point in the horizontal plane through NS. The acute angle which OA makes with NS is called the *bearing* of the point A from O. The angle is also the bearing of the line OA. The bearing of A may be precisely indicated by giving the size of the angle and specifying whether it is measured from ON or OS and whether to the east or west. This information is expressed in abbreviated form. Thus, in the figure, OA is in a direction 35° east of north, and the bearing is written N 35° E. The bearings of some other points are also indicated.

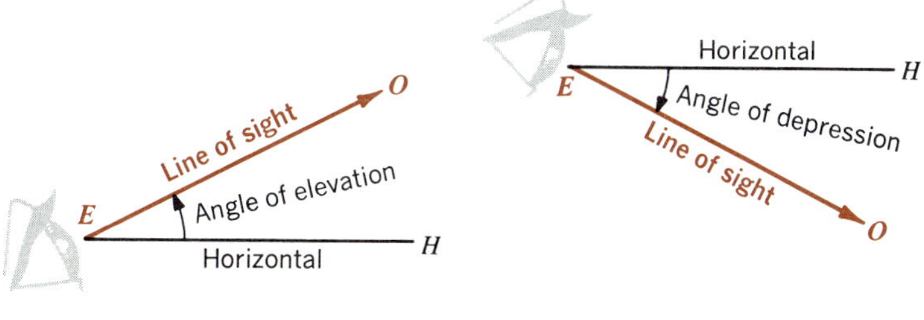

Figure 9-12

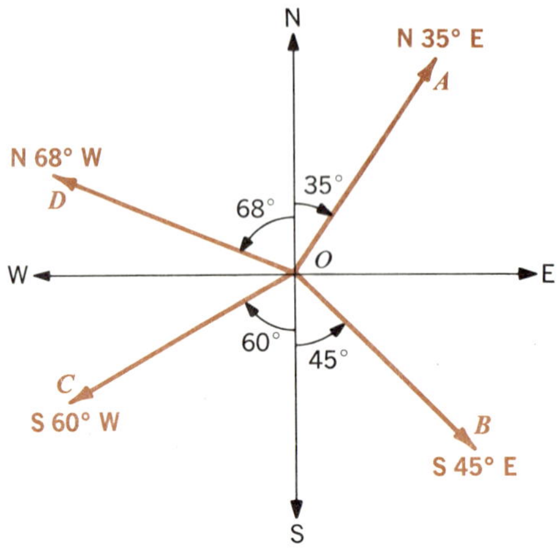

Figure 9-13

EXAMPLE 1 From a point 124 feet from the foot of a tower and on the same level, the angle of elevation of the top of the tower is 36° 20′ and the angle of elevation of the top of a water tank supported by the tower is 40° 50′. Find (*a*) the height of the tower; (*b*) the depth of the tank.

Solution a: To find the height of the tower, we use the right triangle ABC (Fig. 9-14), where CB, or h, denotes the height of the tower. Thus, we have

$$\frac{h}{124} = \tan 36^\circ\ 20'$$

$$h = 124 \tan 36^\circ\ 20'$$

$$= 124(0.7355)$$

$$= 91.20$$

Solution b: Referring to the right triangle ADC, where $x = CD$ denotes the distance from the ground to the top of the tank, we obtain

$$x = 124 \tan 40^\circ\ 50' = 124(0.8642) = 107.2$$

Then $BD = x - h = 107.2 - 91.20 = 16.0$. We now round off to three significant figures and express the height of the tower as 91.2 feet and the depth of the tank as 16.0 feet.

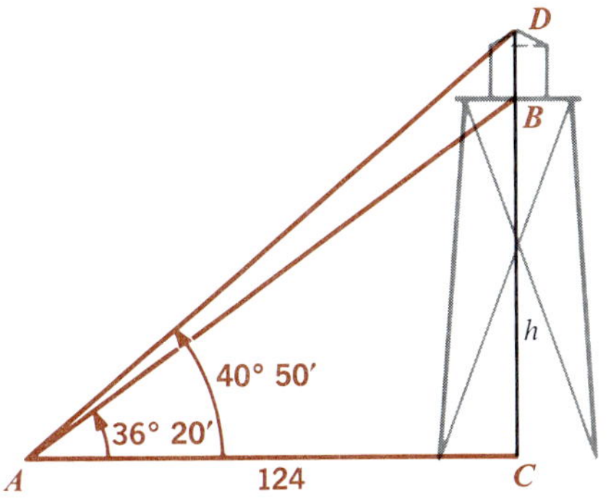

Figure 9-14

EXAMPLE 2 Two points, A and B, are 526 feet apart on a level stretch of road leading to a hill. The angle of elevation of the hilltop from A is $26° 30'$, and the angle of elevation from B is $36° 40'$. Find how high the hill extends above AB produced.

Solution: The given data are marked in Fig. 9-15. We let h stand for the height of the hill and x for the distance BC. From the right triangles ADC and BDC, we write

$$\cot 26° 30' = \frac{526 + x}{h}$$

and

$$\cot 36° 40' = \frac{x}{h}$$

These equations, when cleared of fractions, become

$$h \cot 26° 30' = 526 + x \qquad \text{and} \qquad h \cot 36° 40' = x$$

We observe that each equation contains two unknowns and that the two right triangles cannot be solved separately. Hence, we

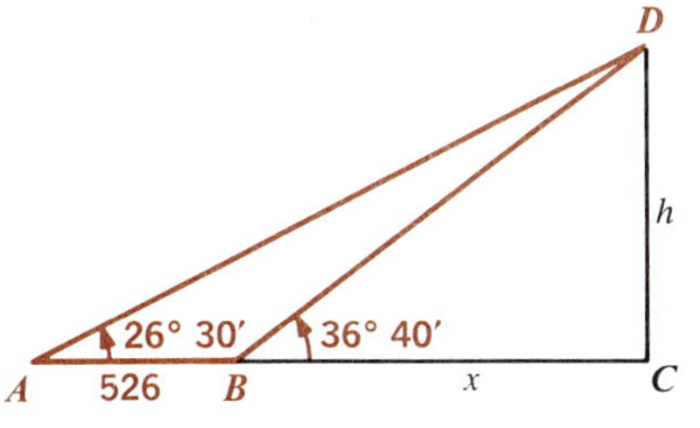

Figure 9-15

combine the equations to eliminate x. Thus we get

$$h \cot 26^\circ\ 30' = 526 + h \cot 36^\circ\ 40'$$

$$h \cot 26^\circ\ 30' - h \cot 36^\circ\ 40' = 526$$

$$h = \frac{526}{\cot 26^\circ\ 30' - \cot 36^\circ\ 40'}$$

$$= \frac{526}{2.006 - 1.343} = \frac{526}{0.663} = 793 \text{ feet}$$

EXERCISE 9-8

1. A ladder 18 feet long is leaning against a house. If the foot of the ladder is 9.6 feet from the house, find the angle of elevation of the ladder and the height it reaches on the house.
2. From a point 224 feet from the base of a smokestack, the angle of elevation of the top is 30° 40′. Find the height of the smokestack.
3. From a tower 64 feet high, the angle of depression of a car is 26°. How far is the car from the tower?
4. From a cliff 140 feet above the shore line, the angle of depression of a ship is 21° 30′. Find the distance from the ship to a point on the shore directly below the observer.
5. The shadow of a vertical pole is 26.4 feet long when the angle of elevation of the sun is 41° 00′. Find the height of the pole.
6. From a point on a lighthouse 104 feet above the water, the angle of depression of a boat is 9° 40′. How far is the boat from the lighthouse?
7. What is the angle of elevation of the sun when a vertical pole 19 feet high casts a shadow 12 feet long?
8. A ship sails S 21° W a distance of 34 miles. How far south and how far west is the ship from its original position?
9. In measuring the length of the east-west boundary of a piece of land, a surveyor comes to a marshy place. He diverts his path from a point P on the boundary line directly north to Q and from Q returns to R on the line. If the bearing of R from Q is S 60° 00′ E and the distance PQ is 312 feet, find the distance PR.

10. Two ships leave the same port at the same time. One goes 24 miles per hour in the direction N 38° E, and the other travels 32 miles per hour in the direction S 52° E. Find the distance between the ships at the end of 2 hours.

11. Town B is 26 miles south and 36 miles west of town A. Find the bearing and distance of B from A.

12. A smokestack is 150 feet from a building. From a window of the building, the angle of elevation of the top of the smokestack is 33° and the angle of depression of the base is 22°. Compute the height of the window and the height of the smokestack.

13. From a point on a level with the base of a building and 410 feet away, the angle of elevation of the bottom of a flagpole on the building is 21° 20′ and the angle of elevation of the top of the pole is 24° 00′. Find the length of the pole.

14. From a point on level ground 260 feet from a church, the angle of elevation of the top of a spire on the building is 27° and the angle of elevation of the bottom of the spire is 25°. What is the length of the spire?

15. Two points A and B are 45 feet apart on a level stretch of road leading to a hill. The angle of elevation of the hilltop from A is 21° and from B is 31°. Find how high the hill extends above the road level.

16. Generalize problem 15 by taking the distance from A to B as a feet and the angles of elevation from A and B, respectively, as θ and ϕ; that is, express the height of the hill in terms of a and functions of θ and ϕ.

17. A balloon is in the vertical plane of two consecutive milestones of a road. The angles of depression of the stones as viewed from the balloon are 16° and 18°. Find the height of the balloon if (a) the stones are on opposite sides of the balloon; (b) the stones are on the same side of the balloon.

18. A balloon is in the vertical plane of two stones a feet apart. The angles of depression of the stones are θ and ϕ. Find the height of the balloon if (a) the stones are on opposite sides of the balloon; (b) the stones are on the same side of the balloon.

19. From a tower 128 feet high, the angles of depression of two rocks which are in a horizontal line through the base of the tower are 17° 00′ and 13° 20′. Find the distance between the rocks if they are on (a) opposite sides of the tower; (b) the same side of the tower.

20. Find the distance between the rocks in problem 19 if one rock is directly south, and the other directly east, of the tower, the other data unchanged.

21. A ship sailing in the direction S 42° W passed point A directly east of a lighthouse and 440 feet farther on was directly south of the lighthouse. If the angle of elevation of the top of the lighthouse from A was 19°, find the angle of elevation when the ship was closest to the lighthouse.

9-10 VECTORS

Many physical quantities possess the properties of magnitude and direction. A quantity of this kind is called a *vector quantity.* A force, for example, is characterized by its magnitude and direction of action. The force would not be completely specified if either the magnitude or direction were unknown. Velocity, acceleration, and displacement are other examples of vector quantities.

A vector quantity is customarily represented by a directed line segment called a *vector.** The length of the vector represents the magnitude of the quantity, and the vector points in the assigned direction. Thus, a vector representing the velocity of a body points in the direction of motion and has a length (in a convenient unit) equal to the speed of the body. We shall denote a vector by a symbol such as **OA** if O is the initial point and A is the terminal point.

The sum of two vectors **OA** and **OB** is defined as the diagonal vector **OC** of the parallelogram determined by **OA** and **OB** (Fig. 9-16). The vectors **OA** and **OB** are called the *components,* and **OC** is called the *resultant.* This method of adding vectors is used in physics, where it is shown, for example, that two forces applied at a point of a body have exactly the same effect as a single force equal to their resultant.

Noticing Fig. 9-16, we discover that the resultant **OC** may be constructed without drawing the complete parallelogram; only the triangle OAC is necessary. Hence, if the second vector is drawn from the terminal point of the first, then the sum vector extends from the initial point of the first vector to the terminal point of the second vector. The triangle thus formed is called a *vector triangle.*

It is sometimes desirable to find two vectors whose sum is equal to a given vector. The given vector is then said to be *resolved* into two components. The components may be along any two direc-

* Two points of a line and the part between the points form a *line segment.* If one direction on a line segment is chosen as positive and the opposite direction as negative, then the line segment is called a *directed line segment.* The positive direction is usually indicated by an arrowhead.

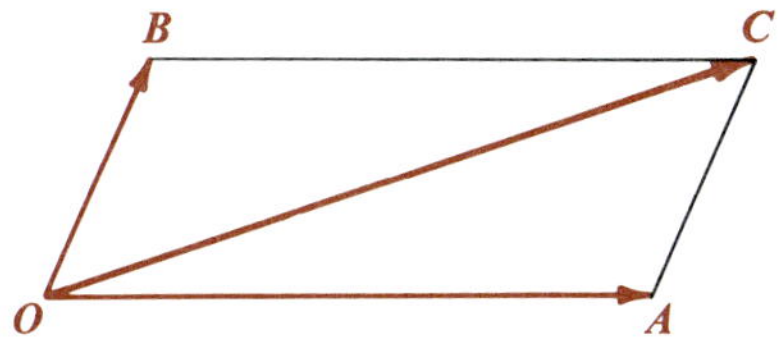

Figure 9-16

tions in a plane containing the given vector. A graphical construction of the two components can be obtained by drawing a parallelogram which has the given vector as a diagonal, one pair of opposite sides along one of the chosen directions, and the other pair along the other direction.

We shall show how to resolve the vector **OA** (Fig. 9-17) into a pair of perpendicular components. Let OX and OY be the axes in a coordinate plane containing **OA.** By drawing perpendiculars from A to OX and OY the projections of **OA** on the axes are determined. We regard these projections as the vectors **OB** and **OC.** Their sum is obviously equal to the vector **OA.** Thus, the vector **OA** has been resolved into two perpendicular components. If x, y, and r stand for the lengths of **OB, OC,** and **OA,** respectively, and if θ is the angle which **OA** makes with the positive x axis, then

$$x = r\cos\theta \qquad \text{and} \qquad y = r\sin\theta$$

EXAMPLE 1 A force of 185 pounds acts perpendicularly to a force of 275 pounds. Find the magnitude of the resultant and the angle which it makes with each of the given forces.

Solution: Let ABC be the vector triangle (Fig. 9-18), where c stands for the magnitude of the resultant. Then

$$\begin{aligned} \tan A &= \tfrac{185}{275} = 0.6727 \\ A &= 33^\circ\,56',\ B = 56^\circ\,4' \\ c &= 185\csc 33^\circ\,56' \\ &= 185(1.791) = 331.3 \end{aligned}$$

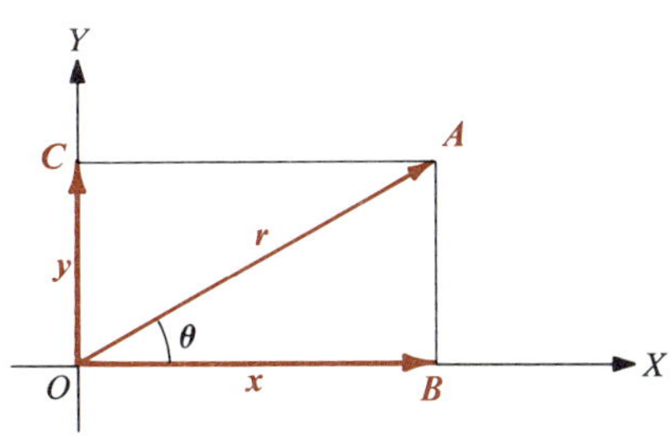

Figure 9-17

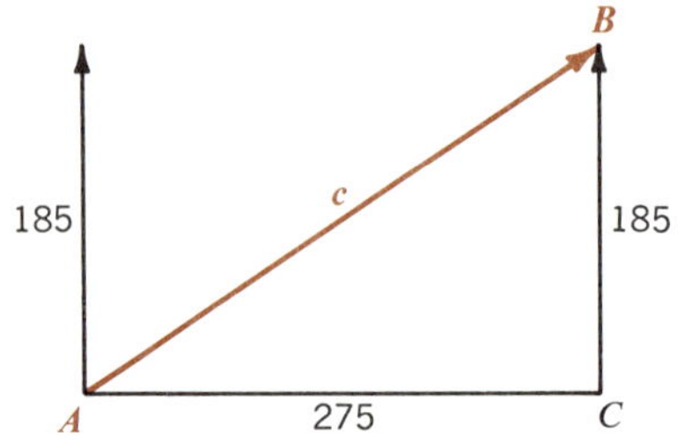

Figure 9-18

The results, rounded off, are $A = 34°\ 00'$, $B = 56°\ 00'$, and $c = 331$ pounds.

EXAMPLE 2 A body weighing 324 pounds is on a smooth inclined plane which makes an angle of 28° 30′ with the horizontal. What force parallel to the plane will just keep the body from slipping, and with what force will the body push against the plane?

Solution: The 324-pound weight acts vertically downward and is represented in Fig. 9-19 by **OC**. This force may be resolved into two components, **OB** parallel to the plane and **OA** perpendicular to the plane. The component **OB** tends to slide the body down the plane; the component **OA** is the force against the plane. In the vector triangle *BCO*, the angle $C = 28°\ 30'$. Hence, we have

$$\frac{c}{324} = \sin 28°\ 30' \qquad\qquad \frac{o}{324} = \cos 28°\ 30'$$

$$\begin{aligned} c &= 324 \sin 28°\ 30' & o &= 324 \cos 28°\ 30' \\ &= 324(0.4772) & &= 324(0.8788) \\ &= 155 \text{ pounds} & &= 285 \text{ pounds} \end{aligned}$$

The weight of the body has a component of 155 pounds down the plane; hence a force of 155 pounds oppositely directed will just keep the body from slipping. The body pushes against the plane with a force of 285 pounds.

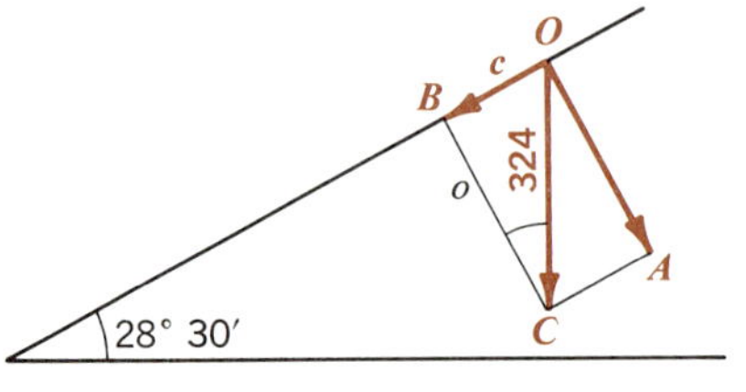

Figure 9-19

Figure 9-20

Vectors also have applications in aviation. We let OB (Fig. 9-20) represent the speed of an airplane relative to the air and the direction in which the airplane points, or heads. If the air were not in motion, the speed and heading would bring the airplane from O to B in one unit of time. If, however, **BC** represents the speed and direction of the wind, the airplane would not travel from O to B, but along the line from O to C, reaching C in one unit of time. Hence the vector **OC**, the sum of the vectors **OB** and **BC**, represents the path and speed of the airplane relative to the ground. The magnitude of **OC** is called the *groundspeed*, and the magnitude of **OB** is called the *airspeed*.

In aviation the bearing of a point, or direction of a course of flight, is measured in a clockwise direction from the north. Thus, in Fig. 9-21, the direction of OA is 55°, that of OB is 220°, and that of OC is 315°.

EXAMPLE 3 A pilot, whose plane has an airspeed of 180 miles per hour, wishes to travel in a due east direction. If the wind is blowing north at 44 miles per hour, find the direction in which the pilot should head the plane and the resulting groundspeed of the plane.

Solution: We let the vector **OB** (Fig. 9-22) represent the airspeed and direction of heading, **BC** the speed and direction of the wind, **OC** the groundspeed and path of travel, and θ the angle BOC. Then, we have

$$\sin\theta = \tfrac{44}{180} = 0.2444$$

$$\theta = 14°\ 10'$$

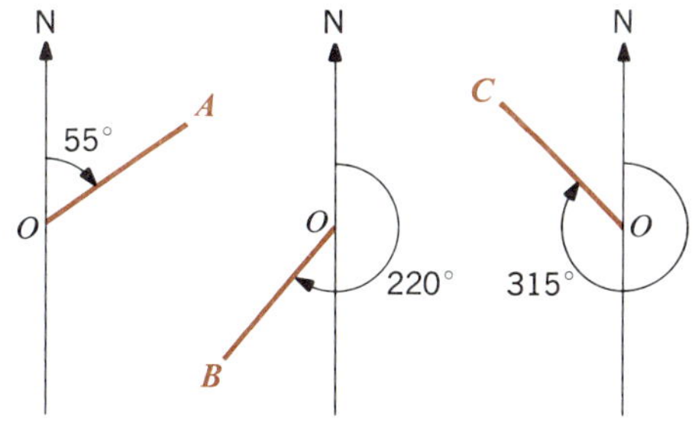

Figure 9-21

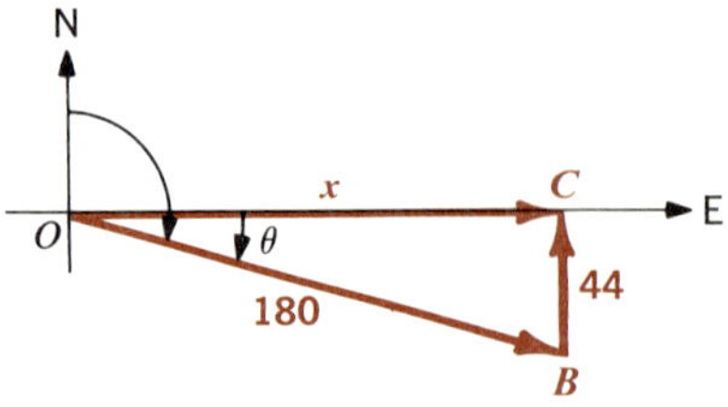

Figure 9-22

Letting x stand for the length of **OC**, we get

$$x = 44 \cot 14^\circ\ 10'$$
$$= 44(3.96) = 152.24$$

Rounding off, we find the heading to be 104° and the groundspeed 150 miles per hour.

EXERCISE 9-9

1. A force of 160 pounds and another of 200 pounds act at right angles to each other. Find the magnitude of the resultant and the angle which it makes with each force if the last digit of each given force is not significant.

2. A force of 210 pounds acts in a due-east direction and another force of 330 pounds acts due south. Consider the zeros of the forces as not significant and find the direction and magnitude of the resultant.

3. An airplane is headed due east with an airspeed of 260 miles per hour. A wind is blowing due north at 39.2 miles per hour. Find the groundspeed and the direction which the plane takes.

4. A boat is going south at the rate of 20.4 miles per hour. A man walks east across the deck at 3.10 miles per hour. Find the magnitude and direction of the man's velocity relative to the surface of the water.

5. A force of 325 pounds makes an angle of 30° 20′ with the vertical. Resolve the force into its horizontal and vertical components.

6. A bullet is fired from a gun with a muzzle velocity of 2100 feet per second. The gun makes an angle of 44° 40′ with the horizontal. Find the magnitudes of the horizontal and verti-

cal components of the velocity of the bullet as it leaves the gun.

7. A 360-pound block is on a smooth plane inclined at 30° to the horizontal. What force parallel to the plane will just keep the block from sliding down the plane, and with what force will the block push against the plane?

8. A car, standing on pavement which makes an angle of 15° with the horizontal, is kept from rolling down by a force of 430 pounds parallel to the pavement. Find the weight of the car and the force it exerts against the plane.

9. A man can row a boat 4.1 miles per hour in still water. He wishes to cross a stream which flows 2.4 miles per hour. In what direction should he head in order that his path will be perpendicular to the stream? Find the actual speed in crossing.

10. The airspeed of an airplane is 240 miles per hour. A 52-mile-per-hour wind is blowing from the west. In what direction should the plane head in order to travel due north?

11. A pilot, flying in a wind which is blowing 38.4 miles per hour due south, finds he moves due east when the plane is headed in the direction 72° 00′. Find the airspeed and the groundspeed of the airplane.

12. An airplane heads in the direction 74° with an airspeed of 180 miles per hour. A wind blowing due south causes the airplane to travel due east. Find the groundspeed of the airplane and the speed of the wind.

13. A man rolls a 460-pound barrel up a smooth inclined plane. If he can push with a force of 165 pounds, find the greatest angle which the plane may make with the horizontal.

14. A force of 195 pounds is just sufficient to keep a weight of 350 pounds from sliding down a smooth inclined plane. Find the angle of inclination of the plane.

15. Two children are carrying an object. One exerts a force of 42 pounds at an angle of 42° with the vertical and the other a force of 62 pounds at an angle of 27° with the vertical. Find the weight of the object. *Hint:* The vertical components of the forces must balance the weight of the object.

16. A crate is supported by two ropes fastened to it. One rope makes an angle of 21° with the vertical and has a tension of

270 pounds, and the other rope is in a horizontal position. Find the weight of the crate and the tension in the horizontal rope (see problem 15).

17. Two men are carrying a box. One man exerts a force of 170 pounds at an angle of $22°$ with the vertical, and the other man exerts a force at an angle of $33°$ with the vertical. Find the force which the other man exerts and the weight of the box. *Hint:* The horizontal components of the forces must balance each other.

18. An airplane is flying at 240 miles per hour in the direction $38° 20'$. Find the component of its velocity in the direction $84° 50'$.

10

TRIGONOMETRIC IDENTITIES

10-1 THE FUNDAMENTAL IDENTITIES

In this chapter, we shall consider equations which involve trigonometric functions of angles. Just as there are conditional algebraic equations and algebraic identities (Sec. 6-1), there are trigonometric equations of these two types. Conditional trigonometric equations and trigonometric identities are of vast importance. Many problems and investigations require solutions of trigonometric equations; and, aside from their employment in purely theoretical considerations, trigonometric identities are invaluable in a great diversity of problems, particularly in physics and the engineering sciences. We shall first study identities, and in Chap. 19 we shall take up conditional trigonometric equations.

Here, we list eight relations of the trigonometric functions. Each relation or equation is true for all angles for which the members of the equation are defined and therefore is an identity. These identities follow readily from the definitions of the trigonometric functions and are called *fundamental identities.* The identities are numbered and will be referred to by the numbers assigned here.

In our discussion, we shall speak of the domain of θ as being a set of angles. The domain, however, may just as well be chosen as a set of real numbers (Sec. 8-6).

$$\sin\theta = \frac{1}{\csc\theta} \qquad \text{or} \qquad \csc\theta = \frac{1}{\sin\theta} \tag{1}$$

$$\cos\theta = \frac{1}{\sec\theta} \qquad \text{or} \qquad \sec\theta = \frac{1}{\cos\theta} \tag{2}$$

$$\tan\theta = \frac{1}{\cot\theta} \qquad \text{or} \qquad \cot\theta = \frac{1}{\tan\theta} \tag{3}$$

$$\tan\theta = \frac{\sin\theta}{\cos\theta} \tag{4}$$

$$\cot\theta = \frac{\cos\theta}{\sin\theta} \tag{5}$$

$$\sin^2\theta + \cos^2\theta = 1 \tag{6}$$

$$1 + \tan^2\theta = \sec^2\theta \tag{7}$$

$$1 + \cot^2\theta = \csc^2\theta \tag{8}$$

We have already obtained identities (1), (2), and (3) in Sec. 8-5. For convenience in deriving the remaining identities, we refer to Fig. 10-1, in which the angle θ is drawn in standard position, and the unit circle has its center at the origin (O). The terminal side of θ therefore cuts the circle at the point P, so that $r = 1$. Then, by definition,

$$\sin\theta = \frac{y}{1} = y \qquad \text{and} \qquad \cos\theta = \frac{x}{1} = x$$

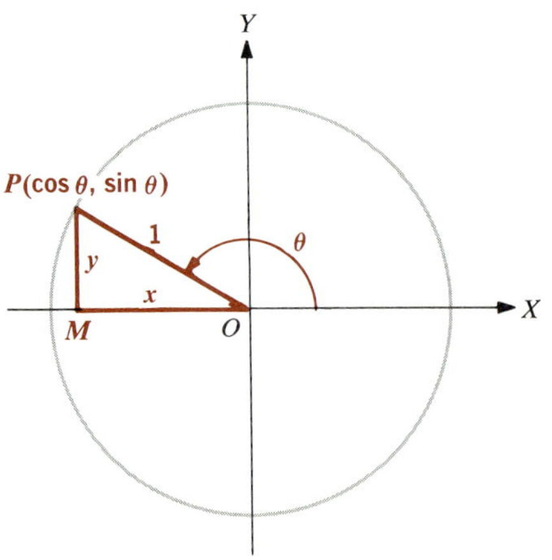

Figure 10-1

Hence, the coordinates of P are $(\cos\theta, \sin\theta)$. From the definitions of $\tan\theta$ and $\cot\theta$, we can now write

$$\tan\theta = \frac{y}{x} = \frac{\sin\theta}{\cos\theta} \qquad \text{and} \qquad \cot\theta = \frac{x}{y} = \frac{\cos\theta}{\sin\theta}$$

Applying the Pythagorean theorem to the right triangle OMP in the figure, we have the equation $y^2 + x^2 = 1$, and since $y = \sin\theta$ and $x = \cos\theta$, we also have the identity

$$\sin^2\theta + \cos^2\theta = 1$$

Next, we divide each member of this equation by $\cos^2\theta$, and get

$$\frac{\sin^2\theta}{\cos^2\theta} + 1 = \frac{1}{\cos^2\theta}$$

Identity (7) now follows because $\tan^2\theta = \sin^2\theta/\cos^2\theta$ and $1/\cos^2\theta = \sec^2\theta$. Finally, we may establish identity (8) by dividing each member of (6) by $\sin^2\theta$.

Identity (6) is true for all angles—positive, negative, and zero. The remaining fundamental identities do not hold for certain quadrantal angles. As we have already observed (Sec. 8-4), the tangent and secant functions are not defined if the terminal side of θ is along the y axis, and the cotangent and cosecant functions are not defined if the terminal side is along the x axis. This means that identities (2), (3), (4), and (7) are not valid if, in radian measure and with n any integer, $\theta = \frac{1}{2}\pi + n\pi$; and (1), (3), (5), and (8) are not valid if $\theta = n\pi$.

Because of their great importance, *these eight identities should be memorized.* Indeed, the student should be so familiar with them that he will be able to recognize them even when they are in slightly different forms. In particular, an identity should be recognizable when it is solved for any function occurring in it. For example, the identity

$$\sin^2\theta + \cos^2\theta = 1$$

when solved for $\sin\theta$, becomes

$$\sin\theta = \pm\sqrt{1 - \cos^2\theta}$$

The double sign is placed before the radical since $\sin\theta$ is positive for some angles and negative for others. We recall that $\sin\theta$ is positive for angles with the terminal side above the x axis and negative for angles with the terminal side below the x axis.

EXAMPLE 1 Given $\sin\theta = -\frac{1}{3}$ and θ in quadrant III, find the other functions of θ by use of the fundamental identities.

Solution: It may be observed that if the sine, cosine, and tangent of an angle are known, the other three functions can be had immediately from the reciprocal relations. To find $\cos\theta$, we use identity (6), and obtain

$$\cos^2\theta = 1 - \sin^2\theta = 1 - (-\tfrac{1}{3})^2 = 1 - \tfrac{1}{9} = \tfrac{8}{9}$$

and

$$\cos\theta = -\frac{2\sqrt{2}}{3}$$

We choose the negative square root because the cosine of a third-quadrant angle is negative. Next, from identity (4),

$$\tan\theta = \frac{\sin\theta}{\cos\theta} = \left(-\frac{1}{3}\right) \div \left(-\frac{2\sqrt{2}}{3}\right) = \frac{1}{2\sqrt{2}}$$

Taking the reciprocals of the sine, cosine, and tangent of θ, we write

$$\sin\theta = -\tfrac{1}{3} \qquad \csc\theta = -3$$

$$\cos\theta = -\frac{2\sqrt{2}}{3} \qquad \sec\theta = -\frac{3}{2\sqrt{2}}$$

$$\tan\theta = \frac{1}{2\sqrt{2}} \qquad \cot\theta = 2\sqrt{2}$$

EXAMPLE 2 Express each of the other trigonometric functions in terms of $\tan\theta$.

Solution: From identity (3), we have immediately

$$\cot\theta = \frac{1}{\tan\theta}$$

Solving $1 + \tan^2\theta = \sec^2\theta$ for $\sec\theta$, we obtain

$$\sec\theta = \pm\sqrt{1 + \tan^2\theta} \qquad \text{and} \qquad \cos\theta = \pm\frac{1}{\sqrt{1 + \tan^2\theta}}$$

Substituting $1/\tan\theta$ for $\cot\theta$ in $\csc^2\theta = 1 + \cot^2\theta$ gives

$$\csc^2\theta = 1 + \frac{1}{\tan^2\theta} = \frac{\tan^2\theta + 1}{\tan^2\theta}$$

Hence,

$$\csc\theta = \pm\frac{\sqrt{1+\tan^2\theta}}{\tan\theta} \qquad \text{and} \qquad \sin\theta = \pm\frac{\tan\theta}{\sqrt{1+\tan^2\theta}}$$

EXERCISE 10-1

1. Solve $\tan\theta = \sin\theta/\cos\theta$ for $\sin\theta$. Solve for $\cos\theta$.
2. Solve $\cot\theta = \cos\theta/\sin\theta$ for $\cos\theta$. Solve for $\sin\theta$.
3. Solve $\sin^2\theta + \cos^2\theta = 1$ for $\sin\theta$. Solve for $\cos\theta$.
4. Solve $1 + \tan^2\theta = \sec^2\theta$ for $\sec\theta$. Solve for $\tan\theta$.
5. Solve $1 + \cot^2\theta = \csc^2\theta$ for $\csc\theta$. Solve for $\cot\theta$.

Tell in which quadrants θ may be in order that each of the following equations will be true.

6. $\cos\theta = \sqrt{1-\sin^2\theta}$
7. $\sin\theta = \sqrt{1-\cos^2\theta}$
8. $\sec\theta = -\sqrt{1+\tan^2\theta}$
9. $\cot\theta = -\sqrt{\csc^2 - 1}$

By use of the fundamental identities, find the values of the other five functions in each problem 10 through 17.

10. $\cos\theta = \frac{4}{5}$, θ in Q I
11. $\sin\theta = \frac{5}{13}$, θ in Q IV
12. $\cos\theta = -\frac{3}{4}$, θ in Q II
13. $\tan\theta = \frac{5}{12}$, θ in Q III
14. $\csc\theta = -\frac{13}{12}$, θ in Q III
15. $\sec\theta = \dfrac{3}{\sqrt{5}}$, θ in Q I
16. $\cos\theta = -\dfrac{\sqrt{11}}{6}$, θ in Q II
17. $\csc\theta = -\dfrac{3}{\sqrt{7}}$
18. Express each of the other functions in terms of $\sin\theta$.
19. Express each of the other functions in terms of $\cos\theta$.
20. Express each of the other functions in terms of $\cot\theta$.

10-2 EQUIVALENT TRIGONOMETRIC EXPRESSIONS

Frequently, it is desirable to change a trigonometric expression into an equivalent expression which is simpler, or into an equivalent expression which contains only a specified function. The original expression and the expression to which it is reduced must be identical; that is, the expressions are to be equal for all angles for which each expression is defined. The form of a trigonometric expression may be changed by substitutions from the fundamental identities. Also algebraic operations, such as combining fractions, making indicated multiplications and divisions, and factoring, will reduce an expression to another form. Usually, a combination of algebraic operations and substitutions from the fundamental identities is needed to simplify a given expression. We shall illustrate this with some examples.

EXAMPLE 1 Reduce the expression $\sin\theta\,(\csc\theta - \sin\theta)$ to $\cos^2\theta$.

Solution: We replace $\csc\theta$ by $1/\sin\theta$ and then perform the indicated multiplication. Thus, we have

$$\begin{aligned}\sin\theta\,(\csc\theta - \sin\theta) &= \sin\theta\left(\frac{1}{\sin\theta} - \sin\theta\right)\\ &= 1 - \sin^2\theta\\ &= \cos^2\theta \qquad \text{By identity (6)}\end{aligned}$$

EXAMPLE 2 Reduce the expression $\dfrac{1 + \cot^2\theta}{1 + \tan^2\theta}$ to $\cot^2\theta$.

Solution:

$$\begin{aligned}\frac{1 + \cot^2\theta}{1 + \tan^2\theta} &= \frac{\csc^2\theta}{\sec^2\theta} \qquad \text{By identities (7) and (8)}\\ &= \frac{1/\sin^2\theta}{1/\cos^2\theta} \qquad \text{By identities (1) and (2)}\\ &= \frac{1}{\sin^2\theta}\cdot\frac{\cos^2\theta}{1} = \frac{\cos^2\theta}{\sin^2\theta}\\ &= \cot^2\theta \qquad \text{By identity (5)}\end{aligned}$$

EXAMPLE 3 Reduce $\dfrac{\tan^2\theta}{\sec\theta + 1}$ to $\sec\theta - 1$.

Solution: Since the result is in terms of the secant, we make a substitution for $\tan^2 \theta$. Thus, we obtain

$$\frac{\tan^2 \theta}{\sec \theta + 1} = \frac{\sec^2 \theta - 1}{\sec \theta + 1} \quad \text{By identity (7)}$$

$$= \frac{(\sec \theta + 1)(\sec \theta - 1)}{\sec \theta + 1} \quad \text{By factoring}$$

$$= \sec \theta - 1$$

EXERCISE 10-2

Reduce the first expression to the second in each problem.

1. $\cos \theta \csc \theta;\ \cot \theta$
2. $\sin \theta \sec \theta;\ \tan \theta$
3. $\tan \theta \cos \theta;\ \sin \theta$
4. $\cot \theta \sin \theta;\ \cos \theta$
5. $\cos \theta \csc \theta \tan \theta;\ 1$
6. $\sec \theta \cot \theta \sin \theta;\ 1$
7. $\dfrac{\csc \theta}{\sec \theta};\ \cot \theta$
8. $\dfrac{\tan \theta}{\sin \theta};\ \sec \theta$
9. $\dfrac{\cos \theta}{\cot \theta};\ \sin \theta$
10. $\dfrac{\sin \theta}{\csc \theta};\ \sin^2 \theta$
11. $\cot^2 \theta\,(1 + \tan^2 \theta);\ \csc^2 \theta$
12. $\tan^2 \theta\,(\csc^2 \theta - 1);\ 1$
13. $\sin^2 \theta\,(\csc^2 \theta - 1);\ \cos^2 \theta$
14. $\sec^2 \theta\,(\sin^2 \theta - 1);\ -1$
15. $\tan \theta\,(\tan \theta + \cot \theta);\ \sec^2 \theta$
16. $\sin \theta\,(\csc \theta - \sin \theta);\ \cos^2 \theta$
17. $\dfrac{\sin^2 \theta}{1 - \cos \theta};\ \dfrac{\sec \theta + 1}{\sec \theta}$
18. $\dfrac{\cos^2 \theta}{1 + \sin \theta};\ 1 - \sin \theta$
19. $\dfrac{\sec^2 \theta - 1}{\sec \theta - 1};\ \sec \theta + 1$
20. $\dfrac{1 - \sin^2 \theta}{1 + \cot^2 \theta};\ \sin^2 \theta \cos^2 \theta$
21. $\dfrac{\cot^2 \theta}{\csc \theta + 1};\ \csc \theta - 1$
22. $\dfrac{1 - \cos^2 \theta}{1 + \tan^2 \theta};\ \sin^2 \theta \cos^2 \theta$

10-3 PROVING IDENTITIES

In the preceding section, we transformed trigonometric expressions into other equivalent expressions. The algebraic operations and the substitutions altered the form but not the value of the expression.

This means that in each case the original expression and final expression, as well as the intermediate expressions, are identically equal. Any two of the expressions could be written as the members of an equation.

We shall next consider the problem of proving identities, that is, of proving one member of an equation is identically equal to the other member. A trigonometric identity is really a theorem, and the process of establishing the theorem consists in making logical deductions from known facts, just as proofs are made in geometry. The tools which we have for this purpose are the fundamental identities and algebraic operations. If by these means we reduce one member of a trigonometric equation to the form of the other, then the identity (or theorem) is established.

It is not possible to give specific rules for proving identities. However, we make the following suggestions:

1. It is usually better to change the more complicated member of the equation into the form of the other member.
2. If one member contains only one function of an angle, it might be of advantage to express the functions in the other member in terms of the single function.
3. Usually it is preferable to avoid substitutions from the fundamental identities which introduce radicals.
4. Frequently it is helpful to combine fractions, perform indicated multiplications, or make divisions. In other cases, factoring is useful, and sometimes a fraction needs to be written as the sum of two or more fractions.
5. While often not the most direct way, many reductions can be made satisfactorily by first changing all the functions to sines and cosines.
6. Sometimes the proof can be facilitated by multiplying the numerator and denominator of a fraction by the same factor.
7. In all steps of the process, the member which is not being altered should be kept in view. This member is the goal, and reductions leading to it must be discovered.

EXAMPLE 1 Prove that the equation

$$\frac{\cos^2 \theta + \tan^2 \theta - 1}{\sin^2 \theta} = \tan^2 \theta$$

is an identity.

Solution: We shall work with the left member since it is more complicated than the right member. First, expressing the left member as the sum of two fractions, we have

$$\begin{aligned}
\frac{\cos^2\theta + \tan^2\theta - 1}{\sin^2\theta} &= \frac{\cos^2\theta - 1}{\sin^2\theta} + \frac{\tan^2\theta}{\sin^2\theta} \\
&= \frac{-\sin^2\theta}{\sin^2\theta} + \frac{\sin^2\theta}{\cos^2\theta}\cdot\frac{1}{\sin^2\theta} && \text{By identities (6) and (4)} \\
&= -1 + \sec^2\theta && \text{By identity (2)} \\
&= \tan^2\theta && \text{By identity (7)}
\end{aligned}$$

We have transformed one member of the equation to the form of the other; hence, the given equation is an identity.

EXAMPLE 2 Prove the identity

$$\frac{1}{1-\cos\theta} - \frac{1}{1+\cos\theta} = 2\cot\theta\csc\theta$$

Solution: Combining the fractions of the left member, we get

$$\begin{aligned}
\frac{1}{1-\cos\theta} - \frac{1}{1+\cos\theta} &= \frac{(1+\cos\theta)-(1-\cos\theta)}{(1-\cos\theta)(1+\cos\theta)} \\
&= \frac{2\cos\theta}{1-\cos^2\theta} \\
&= \frac{2\cos\theta}{\sin^2\theta} && \text{By identity (6)} \\
&= \frac{2\cos\theta}{\sin\theta}\cdot\frac{1}{\sin\theta} \\
&= 2\cot\theta\csc\theta && \text{By identities (1) and (5)}
\end{aligned}$$

EXAMPLE 3 Prove the identity

$$\frac{\cot\theta}{1+\tan\theta} = \frac{\cot\theta - 1}{2-\sec^2\theta}$$

Solution: We seek to express the left member as a fraction with the denominator like that of the right member. The first step is to multiply the numerator and denominator of the left member by

$1 - \tan\theta$. Thus, we get

$$\frac{\cot\theta}{1+\tan\theta} = \frac{\cot\theta\,(1-\tan\theta)}{1-\tan^2\theta}$$

$$= \frac{\cot\theta - \cot\theta\tan\theta}{1-(\sec^2\theta - 1)} \qquad \text{By identity (7)}$$

$$= \frac{\cot\theta - 1}{2-\sec^2\theta} \qquad \text{By identity (3)}$$

EXAMPLE 4 Prove the identity

$$\frac{\tan^2\theta}{\tan^2\theta - 1} + \frac{\csc^2\theta}{\sec^2\theta - \csc^2\theta} = \frac{1}{\sin^2\theta - \cos^2\theta}$$

Solution: The right member has only sine and cosine functions. This suggests the changing of all the functions in the left member to sines and cosines. Thus, applying identities (4), (1), and (2), we get

$$\frac{\tan^2\theta}{\tan^2\theta - 1} + \frac{\csc^2\theta}{\sec^2\theta - \csc^2\theta} = \frac{\sin^2\theta/\cos^2\theta}{(\sin^2\theta/\cos^2\theta) - 1} + \frac{1/\sin^2\theta}{(1/\cos^2\theta) - (1/\sin^2\theta)}$$

The complex fractions may be simplified by multiplying the numerator and denominator of the first by $\cos^2\theta$ and the second by $\sin^2\theta\cos^2\theta$. Thus, the above equation becomes

$$\frac{\tan^2\theta}{\tan^2\theta - 1} + \frac{\csc^2\theta}{\sec^2\theta - \csc^2\theta} = \frac{\sin^2\theta}{\sin^2\theta - \cos^2\theta} + \frac{\cos^2\theta}{\sin^2\theta - \cos^2\theta}$$

$$= \frac{\sin^2\theta + \cos^2\theta}{\sin^2\theta - \cos^2\theta}$$

$$= \frac{1}{\sin^2\theta - \cos^2\theta} \qquad \text{By identity (6)}$$

EXERCISE 10-3

Prove that the following equations are identities.

1. $(\sec\theta - 1)(\sec\theta + 1) = \tan^2\theta$
2. $(1 - \cos\theta)(1 + \cos\theta) = \sin^2\theta$

3. $\cot\theta\,(\cot\theta + \tan\theta) = \csc^2\theta$

4. $\csc^2\theta\,(\sin^2\theta - 1) = -\cot^2\theta$

5. $\cos\theta\,(\cos\theta - \sec\theta) = -\sin^2\theta$

6. $\cos^2\theta - \sin^2\theta = 1 - 2\sin^2\theta$

7. $\cos^2\theta - \sin^2\theta = 2\cos^2\theta - 1$

8. $(\cos\theta - \sin\theta)^2 = 1 - 2\sin\theta\cos\theta$

9. $(\tan\theta + \cos\theta)^2 = \csc^2\theta + \sec^2\theta$

10. $\dfrac{\cos\theta + \sin\theta}{\sin\theta} = 1 + \cot\theta$

11. $\dfrac{\cos\theta + \tan\theta}{\sin\theta} = \cot\theta + \sec\theta$

12. $\dfrac{\sec\theta + \sin\theta}{\cos\theta} = \sec^2\theta + \tan\theta$

13. $\dfrac{\cos^2\theta - \sin^2\theta}{\sin\theta\cos\theta} = \cot\theta - \tan\theta$

14. $\dfrac{\cos\theta}{\sec\theta} + \dfrac{\sin\theta}{\csc\theta} = 1$

15. $\dfrac{\csc\theta}{\sec\theta} + \dfrac{\cos\theta}{\sin\theta} = 2\cot\theta$

16. $(1 - \cos\theta)(1 + \sec\theta) = \sec\theta - \cos\theta$

17. $\cos\theta + \tan\theta\sin\theta = \sec\theta$

18. $\sec^2\theta + \csc^2\theta = \sec^2\theta\csc^2\theta$

19. $\tan^2\theta - \sin^2\theta = \tan^2\theta\sin^2\theta$

20. $\cot^2\theta - \cos^2\theta = \cot^2\theta\cos^2\theta$

21. $\csc\theta - \cot\theta\cos\theta = \sin\theta$

22. $\sec\theta - \tan\theta\sin\theta = \cos\theta$

23. $\csc\theta - \sin\theta = \cos\theta\cot\theta$

24. $\tan\theta + \cot\theta = \sec\theta\csc\theta$

25. $(\cot\theta - \csc\theta)(\cos\theta + 1) = -\sin\theta$

26. $(\sec\theta + \csc\theta)(\cos\theta - \sin\theta) = \cot\theta - \tan\theta$

27. $(\cos\theta - \cot\theta)(\sec\theta + \tan\theta) = \sin\theta - \csc\theta$

28. $(\sin\theta - \tan\theta)(\csc\theta + \cot\theta) = \cos\theta - \sec\theta$

29. $\cot^2\theta + \cos^2\theta + \sin^2\theta = \csc^2\theta$

30. $\sin^2\theta + \tan^2\theta + \cos^2\theta = \sec^2\theta$

31. $\dfrac{\cot^2\theta - \cos^2\theta}{\cos^2\theta} = \cot^2\theta$

32. $\dfrac{\tan^2\theta - \sin^2\theta}{\sin^2\theta} = \tan^2\theta$

33. $\dfrac{\cot\theta - \tan\theta}{\sin\theta\cos\theta} = \csc^2\theta - \sec^2\theta$

34. $\dfrac{\sin^2\theta + \cos^2\theta}{\sin\theta\cos\theta} = \cot\theta + \tan\theta$

35. $\dfrac{1 + \tan^2\theta}{1 + \cot^2\theta} = \tan^2\theta$

36. $\dfrac{\sin\theta}{\csc\theta} + \dfrac{\cos\theta}{\sec\theta} = \sin\theta\csc\theta$

37. $\sec^2\theta + \cot^2\theta = \csc^2\theta + \tan^2\theta$

38. $(\sin\theta + \cos\theta)(\tan\theta + \cot\theta) = \sec\theta + \csc\theta$

39. $\dfrac{\sin^2\theta + \cot^2\theta - 1}{\cos^2\theta} = \cot^2\theta$

40. $\dfrac{\sec^2\theta - \sin^2\theta}{\tan^2\theta} = \csc^2\theta - \cos^2\theta$

41. $\dfrac{\csc^2\theta - \cos^2\theta}{\cot^2\theta} = \sec^2\theta - \sin^2\theta$

42. $\dfrac{1 - \sin^2\theta}{1 + \cot^2\theta} = \sin^2\theta\cos^2\theta$

43. $\dfrac{\sec^2\theta}{\sec^2\theta - 1} = \csc^2\theta$

44. $\dfrac{\cot^2\theta + \sec^2\theta + 1}{\cot^2\theta} = \sec^4\theta$

45. $\dfrac{1 + \csc^2\theta + \tan^2\theta}{\sec^2\theta} = \csc^2\theta$

46. $\dfrac{\cos^2\theta - \sin^2\theta}{\cos\theta + \sin\theta} = \cos\theta - \sin\theta$

47. $\dfrac{\sin^4\theta - \cos^4\theta}{\sin^2\theta - \cos^2\theta} = 1$

48. $\dfrac{\tan^3 \theta - 1}{\tan \theta - 1} = \sec^2 \theta + \tan \theta$

49. $\dfrac{1 + \cot^3 \theta}{1 + \cot \theta} = \csc^2 \theta - \cot \theta$

50. $\dfrac{1}{1 - \sin \theta} + \dfrac{1}{1 + \sin \theta} = 2 \sec^2 \theta$

51. $\dfrac{1}{\tan \theta - \sec \theta} + \dfrac{1}{\tan \theta + \sec \theta} = -2 \tan \theta$

52. $\dfrac{1}{\csc \theta - \cot \theta} - \dfrac{1}{\csc \theta + \cot \theta} = 2 \cot \theta$

53. $\dfrac{1 + \cos \theta}{1 - \cos \theta} - \dfrac{1 - \cos \theta}{1 + \cos \theta} = 4 \cot \theta \csc \theta$

54. $\dfrac{\tan \theta}{1 + \cos \theta} + \dfrac{\sin \theta}{1 - \cos \theta} = \cos \theta + \sec \theta \csc \theta$

55. $\dfrac{1 + \sec \theta}{1 - \sec \theta} = \dfrac{\cos \theta + 1}{\cos \theta - 1}$

56. $\dfrac{\cot \theta - 2}{\cot \theta + 2} = \dfrac{1 - 2 \tan \theta}{1 + 2 \tan \theta}$

57. $\dfrac{\cos \theta - \sin \theta}{\cos \theta + \sin \theta} = \dfrac{\cot \theta - 1}{\cot \theta + 1}$

58. $\dfrac{\csc \theta - \sec \theta}{\csc \theta + \sec \theta} = \dfrac{\cos \theta - \sin \theta}{\cos \theta + \sin \theta}$

59. $\dfrac{\cos \theta - 3 \sin \theta}{\cos \theta + 2 \sin \theta} = \dfrac{\cot \theta - 3}{\cot \theta + 2}$

60. $\dfrac{\sec \theta - \cos \theta}{\sec \theta + \cos \theta} = \dfrac{\sin^2 \theta}{1 + \cos^2 \theta}$

10-4 COSINE OF THE DIFFERENCE AND SUM OF TWO ANGLES

The eight fundamental identities (Sec. 10-1) involve functions of a single angle. In the remainder of this chapter, we shall consider identities involving functions of more than one angle. We start by proving that the cosine of the difference and sum of two angles are given by the formulas

$$\cos(A - B) = \cos A \cos B + \sin A \sin B \tag{9}$$

$$\cos(A + B) = \cos A \cos B - \sin A \sin B \tag{10}$$

The angles A and B (Fig. 10-2) are constructed in standard position, so that angle P_1OP_2 is equal to $B - A$. We construct angle P_0OP_3 the same size as $B - A$ but measured clockwise. Hence, angle $P_0OP_3 = A - B$. It is then evident that chords P_1P_2 and P_0P_3 have the same length. Recalling that the terminal side of an angle cuts the unit circle at a point whose coordinates are the cosine and sine of the angle (Fig. 10-1), we see that the terminal sides of A, B, and $A - B$ cut the unit circle at the points whose coordinates are, respectively,

$$P_1(\cos A, \sin A)$$

$$P_2(\cos B, \sin B)$$

$$P_3[\cos(A - B), \sin(A - B)]$$

Since the chords P_1P_2 and P_0P_3 have the same length, we may apply the distance formula (Sec. 5-4), and write

$$\begin{aligned}
|P_1P_2|^2 &= (\cos B - \cos A)^2 + (\sin B - \sin A)^2 \\
&= 2 - 2(\cos A \cos B + \sin A \sin B) \\
|P_0P_3|^2 &= [\cos(A - B) - 1]^2 + \sin^2(A - B) \\
&= \cos^2(A - B) - 2\cos(A - B) + 1 + \sin^2(A - B) \\
&= 2 - 2\cos(A - B)
\end{aligned}$$

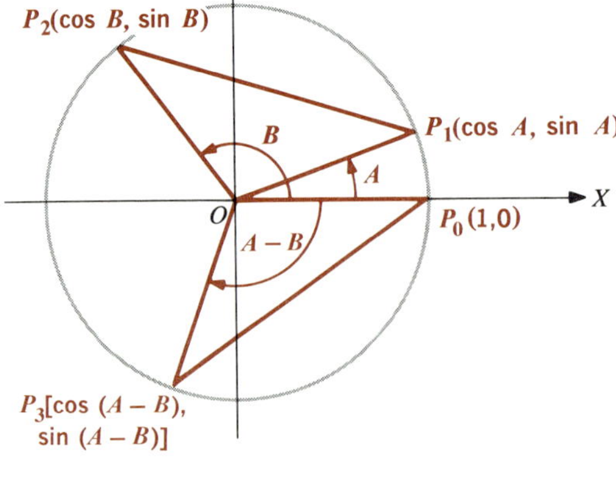

Figure 10-2

We now equate the quantities which are equal to the squares of the length of P_1P_2 and P_0P_3 and have

$$\cos (A - B) = \cos A \cos B + \sin A \sin B \tag{9}$$

The method of deriving this formula will apply regardless of the measures of A and B. Hence, the formula holds where A and B are any angles whatever, positive, zero, or negative. We state the formula in words:

The cosine of the difference of two angles is equal to the cosine of the first times the cosine of the second plus the sine of the first times the sine of the second.

Before establishing the formula for the cosine of the sum of two angles, we use formula (9) to develop some additional identities. Replacing A by $\pi/2$ and recalling that $\cos \pi/2 = 0$ and $\sin \pi/2 = 1$, we have

$$\cos \left(\frac{\pi}{2} - B\right) = \cos \frac{\pi}{2} \cos B + \sin \frac{\pi}{2} \sin B$$

and therefore

$$\cos \left(\frac{\pi}{2} - B\right) = \sin B \tag{a}$$

Formula (a) holds for all values of B, and in particular when B is replaced by $\pi/2 - B$. Thus

$$\cos \left[\frac{\pi}{2} - \left(\frac{\pi}{2} - B\right)\right] = \sin \left(\frac{\pi}{2} - B\right)$$

Since $\pi/2 - (\pi/2 - B) = B$, we conclude that

$$\sin \left(\frac{\pi}{2} - B\right) = \cos B \tag{b}$$

If we let $A = 0$ in formula (9) and use $\cos 0 = 1$ and $\sin 0 = 0$, we get

$$\cos (0 - B) = \cos 0 \cos B + \sin 0 \sin B$$

$$\cos (-B) = \cos B \tag{c}$$

In formula (a), we replace B by $-B$ and find that

$$\sin(-B) = \cos\left(\frac{\pi}{2} + B\right)$$

$$= \cos\left[B - \left(-\frac{\pi}{2}\right)\right]$$

$$= \cos B \cos\left(-\frac{\pi}{2}\right) + \sin B \sin\left(-\frac{\pi}{2}\right)$$

Hence,

$$\sin(-B) = -\sin B \tag{d}$$

Formulas (c) and (d) now yield

$$\tan(-B) = -\tan B \tag{e}$$

We may now use formulas (c), (d), and (9) in finding the cosine of the sum of two angles. Thus,

$$\cos(A + B) = \cos[A - (-B)]$$

$$= \cos A \cos(-B) + \sin A \sin(-B)$$

Hence,

$$\cos(A + B) = \cos A \cos B - \sin A \sin B \tag{10}$$

EXAMPLE 1 Find the exact value of $\cos(\pi/12)$.

Solution: Observing that $\pi/12 = \pi/3 - \pi/4$, we apply formula (9) and get

$$\cos\frac{\pi}{12} = \cos\left(\frac{\pi}{3} - \frac{\pi}{4}\right)$$

$$= \cos\frac{\pi}{3}\cos\frac{\pi}{4} + \sin\frac{\pi}{3}\sin\frac{\pi}{4}$$

$$= \frac{1}{2}\cdot\frac{\sqrt{2}}{2} + \frac{\sqrt{3}}{2}\cdot\frac{\sqrt{2}}{2} = \frac{\sqrt{2} + \sqrt{6}}{4}$$

EXAMPLE 2 If $\sin A = \frac{3}{5}$ and $\cos B = \frac{5}{13}$ and if A is in quadrant II and B is in quadrant IV, find $\cos(A - B)$ and $\cos(A + B)$.

Solution: From the quadrants of A and B, $\cos A$ is negative and $\sin B$ is negative. Hence,

$$\cos A = -\sqrt{1 - \sin^2 A} = -\sqrt{1 - \tfrac{9}{25}} = -\tfrac{4}{5}$$
$$\sin B = -\sqrt{1 - \cos^2 B} = -\sqrt{1 - \tfrac{25}{169}} = -\tfrac{12}{13}$$

Applying formulas (9) and (10), we obtain

$$\cos (A - B) = \cos A \cos B + \sin A \sin B$$
$$= -\frac{4}{5}\cdot\frac{5}{13} - \frac{3}{5}\cdot\frac{12}{13} = -\frac{56}{65}$$

and

$$\cos (A + B) = \cos A \cos B - \sin A \sin B$$
$$= -\frac{4}{5}\cdot\frac{5}{13} + \frac{3}{5}\cdot\frac{12}{13} = \frac{16}{65}$$

EXERCISE 10-4

Use formulas (9) and (10) and find, without tables, the cosine of each of the following angles.

1. $-15°$ **2.** $75°$ **3.** $105°$ **4.** $-75°$

5. $195°$ **6.** $165°$ **7.** $255°$ **8.** $285°$

9. Show that $\cos (60° + 45°) \neq \cos 60° + \cos 45°$.

10. Show that $\cos (45° + 30°) \neq \cos 45° + \cos 30°$.

11. Show that $\cos 180° \neq 2 \cos 90°$

12. Show that $\cos (180° - 30°) \neq \cos 180° - \cos 30°$.

By means of formulas (9) and (10) express each of the following quantities as a function of θ.

13. $\cos (90° + \theta)$ **14.** $\cos (\pi + \theta)$ **15.** $\cos (\pi - \theta)$

16. $\cos (270° - \theta)$ **17.** $\cos (3\pi - \theta)$ **18.** $\cos (2\pi + \theta)$

Verify the following equations and identities.

19. $\cos 130° \cos 40° + \sin 130° \sin 40° = 0$

20. $\cos 35° \cos 25° - \sin 35° \sin 25° = \frac{1}{2}$

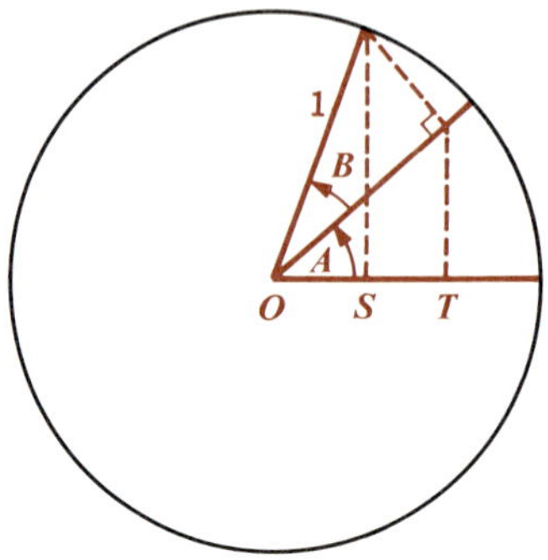

Figure 10-3

21. $\cos 118° \cos 62° - \sin 118° \sin 62° = -1$

22. $\cos(40° + \theta)\cos(10° + \theta) + \sin(40° + \theta)\sin(10° + \theta) = \frac{\sqrt{3}}{2}$

23. $\cos 4A \cos A - \sin 4A \sin A = \cos 5A$

24. $\cos 7A \cos 5A + \sin 7A \sin 5A = \cos 2A$

25. $\cos(30° - A) - \cos(30° + A) = \sin A$

26. $\cos(60° + A) + \cos(60° - A) = \cos A$

In the following problems find the values of $\cos(A - B)$ *and* $\cos(A + B)$.

27. A and B in Q I, $\cos A = \frac{12}{13}$, $\sin B = \frac{3}{5}$

28. A and B in Q II, $\sin A = \frac{3}{5}$, $\cos B = -\frac{7}{25}$

29. A in Q III, B in Q IV, $\tan A = \frac{3}{4}$, $\tan B = -\frac{15}{8}$

30. A and B in Q II, $\sin A = \frac{3}{5}$, $\sin B = \frac{8}{17}$

31. Use Fig. 10-3 and show that $OS = \cos(A + B)$, $OT = \cos A \cos B$, $ST = \sin A \sin B$, and consequently $\cos(A + B) = \cos A \cos B - \sin A \sin B$.

32. Using identities (c), (d), and (e) above, prove the following theorem:

Theorem: Changing the sign of an angle does not change the value of its cosine and secant. Changing the sign of an angle reverses the sign of the sine and cosecant, and the tangent and cotangent.

Establish this theorem also by referring to Fig. 10-4. Notice that the theorem is true for any value of θ and $-\theta$ except for the undefined functions of a quadrantal angle.

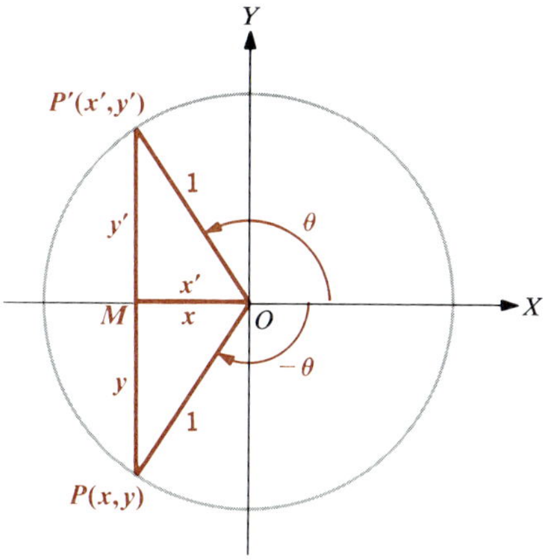

Figure 10-4

10-5 SINE AND TANGENT OF THE SUM AND DIFFERENCE OF TWO ANGLES

We now apply formulas (a) and (b) of the previous section to derive the formulas

$$\sin(A + B) = \sin A \cos B + \cos A \sin B \tag{11}$$

$$\sin(A - B) = \sin A \cos B - \cos A \sin B \tag{12}$$

Replacing B by $(A + B)$ in (a) and reversing the members of the equation gives

$$\sin(A + B) = \cos\left[\frac{\pi}{2} - (A + B)\right] = \cos\left[\left(\frac{\pi}{2} - A\right) - B\right]$$

$$= \cos\left(\frac{\pi}{2} - A\right)\cos B + \sin\left(\frac{\pi}{2} - A\right)\sin B$$

Then, using (a) and (b) to change the form of the right member of this equation, we have

$$\sin(A + B) = \sin A \cos B + \cos A \sin B$$

To obtain the sine of the difference of two angles, we replace B by $-B$ in formula (11). This gives

$$\sin[A + (-B)] = \sin A \cos(-B) + \cos A \sin(-B)$$

Formulas (c) and (d) of Sec. 10-4 permit us to write

$$\sin(A - B) = \sin A \cos B - \cos A \sin B$$

Formulas (9) through (12) enable us to obtain the formulas

$$\tan(A + B) = \frac{\tan A + \tan B}{1 - \tan A \tan B} \tag{13}$$

$$\tan(A - B) = \frac{\tan A - \tan B}{1 + \tan A \tan B} \tag{14}$$

Since $\tan\theta = \sin\theta/\cos\theta$, we have

$$\tan(A + B) = \frac{\sin(A + B)}{\cos(A + B)} = \frac{\sin A \cos B + \cos A \sin B}{\cos A \cos B - \sin A \sin B}$$

If we divide each term of the numerator and each term of the denominator of the right member of this equation by $\cos A \cos B$ and simplify, we obtain formula (13). Formula (14) can be derived in a similar way.

EXAMPLE Find tan 15° by using tangents of 60° and 45°.

Solution: Since $15° = 60° - 45°$, by formula (14), we have

$$\begin{aligned} \tan 15° &= \tan(60° - 45°) \\ &= \frac{\tan 60° - \tan 45°}{1 + \tan 60° \tan 45°} \\ &= \frac{\sqrt{3} - 1}{1 + \sqrt{3}} = 2 - \sqrt{3} \end{aligned}$$

EXERCISE 10-5

By means of formulas (11) through (14) find, without tables, the values of the sine and tangent of each of the following angles.

1. $-15°$ **2.** $75°$ **3.** $105°$ **4.** $-75°$

5. $195°$ **6.** $165°$ **7.** $255°$ **8.** $285°$

9. Show that $\sin(45° + 30°) \neq \sin 45° + \sin 30°$.

10. Show that $\tan(60° + 45°) \neq \tan 60° + \tan 45°$.

11. Show that $\tan 120° \neq 2 \tan 60°$.

12. Show that $\sin(30° - 20°) \neq \sin 30° - \sin 20°$.

13. Show that the reduction formulas of Sec. 9-4 involving θ are valid for θ any angle whatever.

Verify the following equations and identities.

14. $\sin 160^\circ \cos 20^\circ + \cos 160^\circ \sin 20^\circ = 0$

15. $\sin 35^\circ \cos 5^\circ - \cos 35^\circ \sin 5^\circ = \frac{1}{2}$

16. $\sin 220^\circ \cos 130^\circ - \cos 220^\circ \sin 130^\circ = 1$

17. $\sin 3\theta \cos \theta + \cos 3\theta \sin \theta = \sin 4\theta$

18. $\sin (45^\circ + \theta) - \sin (45^\circ - \theta) = \sqrt{2} \sin \theta$

19. $\dfrac{\tan 46^\circ + \tan 14^\circ}{1 - \tan 46^\circ \tan 14^\circ} = \sqrt{3}$

20. $\dfrac{\tan 68^\circ - \tan 38^\circ}{1 + \tan 68^\circ \tan 38^\circ} = \dfrac{1}{\sqrt{3}}$

21. $\dfrac{\tan 4\theta - \tan 3\theta}{1 + \tan 4\theta \tan 3\theta} = \tan \theta$

22. $\dfrac{\tan 5A + \tan 2A}{1 - \tan 5A \tan 2A} = \tan 7A$

23. $\dfrac{\sin (x + y)}{\sin (x - y)} = \dfrac{\tan x + \tan y}{\tan x - \tan y}$

24. $\dfrac{\sin (x + y)}{\cos (x - y)} = \dfrac{\tan x + \tan y}{1 + \tan x \tan y}$

25. $\dfrac{\cos (x - y)}{\cos (x + y)} = \dfrac{1 + \tan x \tan y}{1 - \tan x \tan y}$

26. $\dfrac{\tan (45^\circ - x)}{\tan (45^\circ + x)} = \dfrac{(1 - \tan x)^2}{(1 + \tan x)^2}$

27. $\sin (A + B + C) = \sin A \cos B \cos C + \cos A \sin B \cos C + \cos A \cos B \sin C - \sin A \sin B \sin C$

28. $\cos (A + B + C) = \cos A \cos B \cos C - \sin A \sin B \cos C - \sin A \cos B \sin C - \cos A \sin B \sin C$

29. $\dfrac{\sin (\theta + h) - \sin \theta}{h} = \cos \theta \dfrac{\sin h}{h} - \sin \theta \dfrac{1 - \cos h}{h}$

30. $\dfrac{\cos (\theta + h) - \cos \theta}{h} = -\sin \theta \dfrac{\sin h}{h} - \cos \theta \dfrac{1 - \cos h}{h}$

31. $\dfrac{\tan (\theta + h) - \tan \theta}{h} = \sec^2 \theta \dfrac{\tan h}{h(1 - \tan \theta \tan h)}$

In problems 32 through 35 find the values of the sine and tangent of $(A + B)$ and of $(A - B)$. Give the quadrant in which $(A + B)$ terminates and the quadrant in which $(A - B)$ terminates.

32. A and B in Q I, $\sin A = \frac{3}{5}$, $\sin B = \frac{5}{13}$

33. A in Q II, B in Q I, $\sin A = \frac{4}{5}$, $\sin B = \frac{7}{25}$

34. A in Q IV, B in Q III, $\tan A = -\frac{3}{4}$, $\tan B = \frac{15}{8}$

35. A in Q III, B in Q II, $\sin A = -\frac{4}{5}$, $\tan B = -1$

36. Using formula (13) and the identity $\cot \theta = 1/\tan \theta$, show that

$$\cot (A + B) = \frac{\cot A \cot B - 1}{\cot B + \cot A}$$

37. Use $\sin (A - B)$ and $\cos (A - B)$ and derive the formula

$$\cot (A - B) = \frac{\cot A \cot B + 1}{\cot B - \cot A}$$

38. Locate line segments in Fig. 10-3, Sec. 10-4, which are equal to $\sin (A + B)$, $\sin A \cos B$, and $\cos A \sin B$; then show that $\sin (A + B) = \sin A \cos B + \cos A \sin B$.

10-6 FUNCTIONS OF TWICE AN ANGLE

Formulas for the functions of twice an angle may be had readily from the functions of the sum of two angles. Setting $B = A$ in formulas (11), (10), and (13), we obtain the following results.

$$\sin (A + A) = \sin A \cos A + \cos A \sin A$$

$$\cos (A + A) = \cos A \cos A - \sin A \sin A$$

$$\tan (A + A) = \frac{\tan A + \tan A}{1 - \tan A \tan A}$$

Hence,

$$\sin 2A = 2 \sin A \cos A \tag{15}$$

$$\begin{aligned} \cos 2A &= \cos^2 A - \sin^2 A \\ &= 1 - 2 \sin^2 A \qquad (16) \\ &= 2 \cos^2 A - 1 \end{aligned}$$

$$\tan 2A = \frac{2 \tan A}{1 - \tan^2 A} \tag{17}$$

Relations (15) to (17) are sometimes called the *double-angle formulas.* The student should state these formulas in words. It is essential to notice that the formulas are identities and are true if the angle appearing in the left member is twice the angle of the right member. Thus, as illustrations, we write

$$\sin 20° = 2 \sin 10° \cos 10°$$

$$\cos 4x = \cos^2 2x - \sin^2 2x$$

$$\tan \theta = \frac{2 \tan \frac{1}{2}\theta}{1 - \tan^2 \frac{1}{2}\theta}$$

EXAMPLE 1 Find the functions of $2x$ if x is in quadrant II and $\sin x = \frac{3}{5}$.

Solution: First, we notice that $\cos x = -\frac{4}{5}$ and $\tan x = -\frac{3}{4}$. Then from formulas (15) to (17), we get

$$\sin 2x = 2 \sin x \cos x = 2(\tfrac{3}{5})(-\tfrac{4}{5}) = -\tfrac{24}{25}$$

$$\cos 2x = \cos^2 x - \sin^2 x = (-\tfrac{4}{5})^2 - (\tfrac{3}{5})^2 = \tfrac{7}{25}$$

$$\tan 2x = \frac{2 \tan x}{1 - \tan^2 x} = \frac{2(-\frac{3}{4})}{1 - (-\frac{3}{4})^2} = -\frac{24}{7}$$

The other three functions may be obtained as the reciprocals of the three computed here.

EXAMPLE 2 Express $\sin 3\theta$ in terms of $\sin \theta$.

Solution: First, we write $3\theta = 2\theta + \theta$; then from formulas (11), (15), and (16), we obtain

$$\begin{aligned}
\sin (2\theta + \theta) &= \sin 2\theta \cos \theta + \cos 2\theta \sin \theta \\
&= (2 \sin \theta \cos \theta) \cos \theta + (1 - 2 \sin^2 \theta) \sin \theta \\
&= 2 \sin \theta \cos^2 \theta + \sin \theta - 2 \sin^3 \theta \\
&= 2 \sin \theta (1 - \sin^2 \theta) + \sin \theta - 2 \sin^3 \theta \\
&= 3 \sin \theta - 4 \sin^3 \theta
\end{aligned}$$

Hence,

$$\sin 3\theta = 3 \sin \theta - 4 \sin^3 \theta$$

EXAMPLE 2 Prove the identity

$$\frac{\sin 2x}{1 - \cos 2x} = \cot x$$

Solution: From (15) and the second form of (16), we have

$$\frac{\sin 2x}{1 - \cos 2x} = \frac{2 \sin x \cos x}{1 - (1 - 2 \sin^2 x)}$$

$$= \frac{2 \sin x \cos x}{2 \sin^2 x} = \frac{\cos x}{\sin x} = \cot x$$

EXERCISE 10-6

1. Verify formulas (15) and (17) when (*a*) $A = 30°$; (*b*) $A = 45°$; (*c*) $A = 60°$.
2. Find the values of sin 90° and cos 90° from functions of 45°.
3. Find the values of the sine, cosine, and tangent of 120° from functions of 60°.

Express each of the following expressions as a function of twice the given angle.

4. $2 \sin 16° \cos 16°$
5. $2 \sin 55° \cos 55°$
6. $\cos^2 52° - \sin^2 52°$
7. $2 \cos^2 20° - 1$
8. $1 - 2 \sin^2 38°$
9. $\sin 70° \cos 70°$
10. $\cos^2 \frac{1}{2}x - \frac{1}{2}$
11. $\frac{1}{2} - \sin^2 5x$
12. $\dfrac{2 \tan 30°}{1 - \tan^2 30°}$
13. $\dfrac{1 - \tan^2 21°}{2 \tan 21°}$
14. $\dfrac{1}{1 - 2 \sin^2 3x}$
15. $\dfrac{10 \tan 4x}{1 - \tan^2 4x}$

In problems 16 through 19 find the values of the functions of 2A and the quadrant in which the angle 2A terminates.

16. A in Q I, $\cos A = \frac{12}{13}$

17. A in Q II, $\tan A = -\frac{3}{4}$

18. A in Q III, $\sin A = -\frac{8}{17}$

19. A in Q IV, $\cos A = \frac{1}{3}$

Prove the following identities.

20. $\sec 2\theta = \dfrac{1}{2\cos^2\theta - 1}$

21. $\sec 2\theta = \dfrac{\sec^2\theta}{2 - \sec^2\theta}$

22. $\csc 2\theta = \dfrac{\sec\theta \csc\theta}{2}$

23. $\cot 2\theta = \dfrac{\cot^2\theta - 1}{2\cot\theta}$

24. $\dfrac{\sin 2x}{1 + \cos 2x} = \tan x$

25. $\cot 4x = \dfrac{1 - \tan^2 2x}{2\tan 2x}$

26. $4\sin^2\theta\cos^2\theta = 1 - \cos^2 2\theta$

27. $\cos 4x = 1 - 8\sin^2 x\cos^2 x$

28. $\cos 3x = 4\cos^3 x - 3\cos x$

29. $\cos 2x + 2\sin^2 x = 1$

30. $\sin 4x = 4\sin x\cos x\,(2\cos^2 x - 1)$

31. $\tan(\theta - 45°) + \tan(\theta + 45°) = 2\tan 2\theta$

32. $\dfrac{1 + \sin 2x - \cos 2x}{1 + \sin 2x + \cos 2x} = \tan x$

33. $\tan 3\theta = \dfrac{\tan\theta(3 - \tan^2\theta)}{1 - 3\tan^2\theta}$

10-7 FUNCTIONS OF HALF AN ANGLE

Frequently, it is desirable to obtain the functions of half an angle from the known functions of the whole angle. For this purpose the second and third forms of formula (16), Sec. 10-6, may be conveniently used. To emphasize that the angle in the right member is half the angle appearing in the left member, we replace A by $\frac{1}{2}A$ and $2A$ by A and write the identities

$$\cos A = 1 - 2\sin^2 \tfrac{1}{2}A \qquad \text{and} \qquad \cos A = 2\cos^2 \tfrac{1}{2}A - 1$$

We solve the first of the equations for $\sin \frac{1}{2}A$ and the second for $\cos \frac{1}{2}A$. Thus, we get

$$2\sin^2 \tfrac{1}{2}A = 1 - \cos A$$

$$\sin^2 \tfrac{1}{2}A = \frac{1 - \cos A}{2}$$

$$\sin \tfrac{1}{2}A = \pm\sqrt{\frac{1 - \cos A}{2}} \tag{18}$$

And similarly,

$$\cos \tfrac{1}{2}A = \pm\sqrt{\frac{1 + \cos A}{2}} \tag{19}$$

We next divide the members of identity (18) by the corresponding members of identity (19) and obtain

$$\tan \tfrac{1}{2}A = \pm\sqrt{\frac{1 - \cos A}{1 + \cos A}} \tag{20}$$

The double signs are placed before the radicals of formulas (18) to (20) because $\frac{1}{2}A$ may terminate in any of the four quadrants. The proper sign for each function must be chosen when the quadrant of $\frac{1}{2}A$ is determined. Thus, if $\frac{1}{2}A$ is in the first or second quadrant, $\sin \frac{1}{2}A$ is positive; but if $\frac{1}{2}A$ is a third- or fourth-quadrant angle, then $\sin \frac{1}{2}A$ is negative. For example, if $180° \leq A \leq 360°$ so that $90° \leq \frac{1}{2}A \leq 180°$, then

$$\sin \tfrac{1}{2}A = \sqrt{\frac{1 - \cos A}{2}} \qquad \text{and} \qquad \cos \tfrac{1}{2}A = -\sqrt{\frac{1 + \cos A}{2}}$$

Two rational expressions for $\tan \frac{1}{2}A$ may be derived as follows:

$$\tan \tfrac{1}{2}A = \frac{\sin \frac{1}{2}A}{\cos \frac{1}{2}A} = \frac{\sin \frac{1}{2}A}{\cos \frac{1}{2}A} \cdot \frac{2\sin \frac{1}{2}A}{2\sin \frac{1}{2}A} = \frac{2\sin^2 \frac{1}{2}A}{2\sin \frac{1}{2}A \cos \frac{1}{2}A}$$

Applying formulas (15) and (18), we have

$$\tan \tfrac{1}{2}A = \frac{1 - \cos A}{\sin A} \tag{21}$$

Similarly,

$$\tan \tfrac{1}{2}A = \frac{\sin \frac{1}{2}A}{\cos \frac{1}{2}A} = \frac{2\sin \frac{1}{2}A \cos \frac{1}{2}A}{2\cos^2 \frac{1}{2}A}$$

Hence, employing (15) and (19), we get

$$\tan \tfrac{1}{2}A = \frac{\sin A}{1 + \cos A} \tag{22}$$

EXAMPLE 1 Find sin 105° and tan 105°.

Solution: If we use formula (18) with $\frac{1}{2}A = 105°$ and $A = 210°$, we have

$$\sin 105° = +\sqrt{\frac{1 - \cos 210°}{2}}$$

The positive sign is chosen because the sine of a second-quadrant angle is positive. Since $\cos 210° = -\frac{1}{2}\sqrt{3}$, we obtain

$$\sin 105° = \sqrt{\frac{1 + \frac{1}{2}\sqrt{3}}{2}} = \sqrt{\frac{2 + \sqrt{3}}{4}} = \frac{\sqrt{2 + \sqrt{3}}}{2}$$

To calculate tan 105°, we may use formula (20), (21), or (22). Substituting in (21), we get

$$\tan 105° = \frac{1 - \cos 210°}{\sin 210°} = \frac{1 + \frac{1}{2}\sqrt{3}}{-\frac{1}{2}} = -(2 + \sqrt{3})$$

EXAMPLE 2 If θ is an angle in the third quadrant and $\cos \theta = -\frac{5}{9}$ (Fig. 10-5), find the values of the sine, cosine, and tangent of $\frac{1}{2}\theta$.

Solution: Since $\frac{1}{2}\theta$ is between 90° and 135°, $\sin \frac{1}{2}\theta$ is positive and the cosine and tangent of $\frac{1}{2}\theta$ are negative. Then substituting

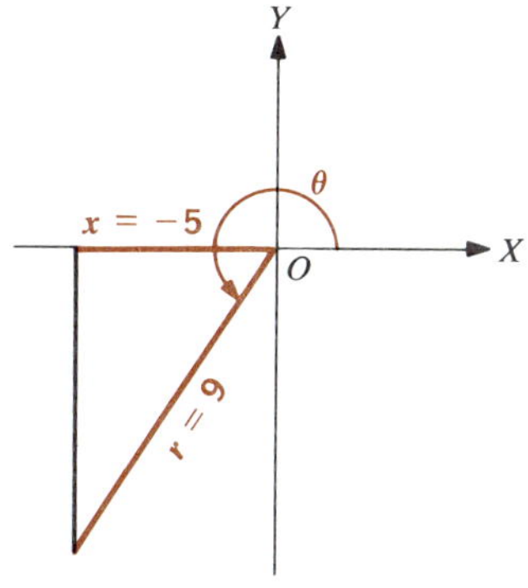

Figure 10-5

in formulas (18) to (20), we obtain

$$\sin \tfrac{1}{2}\theta = \sqrt{\frac{1 + \frac{5}{9}}{2}} = \sqrt{\frac{14}{18}} = \frac{\sqrt{7}}{3}$$

$$\cos \tfrac{1}{2}\theta = -\sqrt{\frac{1 - \frac{5}{9}}{2}} = -\sqrt{\frac{4}{18}} = -\frac{\sqrt{2}}{3}$$

$$\tan \tfrac{1}{2}\theta = -\sqrt{\frac{1 + \frac{5}{9}}{1 - \frac{5}{9}}} = -\sqrt{\frac{14}{4}} = -\frac{\sqrt{14}}{2}$$

EXAMPLE 3 Prove the identity

$$\cos^4 \theta = \tfrac{1}{8}(3 + 4 \cos 2\theta + \cos 4\theta)$$

Solution: The square of both members of formula (19) yields $\cos^2 \frac{1}{2}A = \frac{1}{2}(1 + \cos A)$. We use this identity twice, first, with $A = 2\theta$, and then with $A = 4\theta$, to obtain

$$\begin{aligned}\cos^4 \theta &= (\cos^2 \theta)^2 = [\tfrac{1}{2}(1 + \cos 2\theta)]^2 \\ &= \tfrac{1}{4}(1 + 2 \cos 2\theta + \cos^2 2\theta) \\ &= \tfrac{1}{4}[1 + 2 \cos 2\theta + \tfrac{1}{2}(1 + \cos 4\theta)] \\ &= \tfrac{1}{8}(3 + 4 \cos 2\theta + \cos 4\theta)\end{aligned}$$

EXERCISE 10-7

Use the half-angle formulas to find the exact values of the sine, cosine, and tangent of each of the following angles.

1. $30°$ **2.** $15°$ **3.** $75°$ **4.** $67\frac{1}{2}°$

5. $22\frac{1}{2}°$ **6.** $112\frac{1}{2}°$ **7.** $105°$ **8.** $157\frac{1}{2}°$

Express each of the following as a function of half the given angle.

9. $\sqrt{\dfrac{1 + \cos 42°}{2}}$ **10.** $\sqrt{\dfrac{1 - \cos 32°}{2}}$

11. $\sqrt{\dfrac{1 - \cos 160°}{1 + \cos 160°}}$ **12.** $\sqrt{\dfrac{1 + \cos 80°}{1 - \cos 80°}}$

13. $\dfrac{1 - \cos 44°}{\sin 44°}$ **14.** $\dfrac{\sin 55°}{1 + \cos 55°}$

15. $1 + \cos 70°$ **16.** $1 - \cos 230°$

Express each of the following in terms of the cosine of twice the given angle.

17. $\cos^2 \frac{1}{2}x$

18. $\sin^2 \frac{1}{2}x$

19. $\tan^2 \frac{1}{2}x$

20. $\sin^2 6x$

21. $\cos^2 5x$

22. $\tan^2 4x$

In problems 23 through 28 find the sine, cosine, and tangent of $\frac{1}{2}\theta$ if θ is between $0°$ and $360°$.

23. $\tan \theta = -\frac{3}{4}$, θ in Q II

24. $\sin \theta = -\frac{7}{25}$, θ in Q IV

25. $\cos \theta = -\frac{1}{3}$, θ in Q III

26. $\sec \theta = \frac{9}{5}$, θ in Q I

27. $\tan \theta = 3$, θ in Q I

28. $\cos \theta = -\frac{1}{4}$, θ in Q II

Prove the following identities.

29. $\cot \frac{1}{2}\theta = \dfrac{1 + \cos \theta}{\sin \theta}$

30. $\cot \frac{1}{2}\theta = \dfrac{\sin \theta}{1 - \cos \theta}$

31. $\sec^2 \frac{1}{2}\theta = \dfrac{2}{1 + \cos \theta}$

32. $\csc^2 \frac{1}{2}\theta = \dfrac{2}{1 - \cos \theta}$

33. $\sin^2 \frac{1}{2}\theta = \dfrac{\sec \theta - 1}{2 \sec \theta}$

34. $\cot^2 \frac{1}{2}\theta = \dfrac{1 + \cos \theta}{1 - \cos \theta}$

35. $\tan \frac{1}{2}A \sin A = \dfrac{\tan A - \sin A}{\sin A \sec A}$

36. $\dfrac{1 - \tan^2 \frac{1}{2}A}{\cos A} = \dfrac{2(1 - \cos A)}{\sin^2 A}$

37. $8 \sin^2 \frac{1}{2}x \cos^2 \frac{1}{2}x = 1 - \cos 2x$

38. $\sin^4 \theta = \frac{1}{8}(3 - 4 \cos 2\theta + \cos 4\theta)$

39. $\sin^6 \theta = \frac{1}{16}(5 - 7 \cos 2\theta + 3 \cos 4\theta - \cos 2\theta \cos 4\theta)$

40. $\sin^4 \theta \cos^2 \theta = \frac{1}{16}(1 - \cos 4\theta - 2 \sin^2 2\theta \cos 2\theta)$

41. Multiply the numerator and denominator of the radicand in formula (20) by $1 - \cos A$ and simplify to obtain formula (21).

42. Obtain formula (22) directly from formula (20).

10-8 THE PRODUCT AND SUM FORMULAS

We shall next derive formulas which express the product of two functions as an algebraic sum of two functions and, conversely, the sum in terms of products. To make these derivations we rewrite formulas (9) to (12):

$$\sin A \cos B + \cos A \sin B = \sin (A + B)$$

$$\sin A \cos B - \cos A \sin B = \sin (A - B)$$

$$\cos A \cos B - \sin A \sin B = \cos (A + B)$$

$$\cos A \cos B + \sin A \sin B = \cos (A - B)$$

By forming the sum and difference of the members of the first two of these equations and the sum and difference of the last two, we get, respectively,

$$2 \sin A \cos B = \sin (A + B) + \sin (A - B) \tag{23}$$

$$2 \cos A \sin B = \sin (A + B) - \sin (A - B) \tag{24}$$

$$2 \cos A \cos B = \cos (A + B) + \cos (A - B) \tag{25}$$

$$2 \sin A \sin B = -\cos (A + B) + \cos (A - B) \tag{26}$$

These identities are called the *product formulas.* They express the product of a sine and a cosine, the product of two cosines, and the product of two sines as a sum or difference of sines and cosines. In short, the formulas express the product of sines and cosines as algebraic sums of sines and cosines. There are situations, especially in the calculus, in which the formulas may be used advantageously.

We can convert the relations (23) to (26) into other useful forms by letting

$$A + B = x \qquad \text{and} \qquad A - B = y$$

or, when solved for A and B,

$$A = \tfrac{1}{2}(x + y) \qquad \text{and} \qquad B = \tfrac{1}{2}(x - y)$$

Making these substitutions for $A + B$, $A - B$, A, and B, we obtain

$$\sin x + \sin y = 2 \sin \tfrac{1}{2}(x + y) \cos \tfrac{1}{2}(x - y) \tag{27}$$

$$\sin x - \sin y = 2 \cos \tfrac{1}{2}(x + y) \sin \tfrac{1}{2}(x - y) \tag{28}$$

$$\cos x + \cos y = 2 \cos \tfrac{1}{2}(x + y) \cos \tfrac{1}{2}(x - y) \tag{29}$$

$$\cos x - \cos y = -2 \sin \tfrac{1}{2}(x + y) \sin \tfrac{1}{2}(x - y) \tag{30}$$

These identities are called the *sum formulas.* They express algebraic sums of functions in terms of products.

EXAMPLE 1 Express the product $\sin 28^\circ \cos 20^\circ$ as the sum of two functions.

Solution: From formula (23), we have

$$2 \sin 28^\circ \cos 20^\circ = \sin (28^\circ + 20^\circ) + \sin (28^\circ - 20^\circ)$$
$$\sin 28^\circ \cos 20^\circ = \tfrac{1}{2}(\sin 48^\circ + \sin 8^\circ)$$

EXAMPLE 2 Express the algebraic sum $\sin 5x - \sin 3x$ as a product.

Solution: Using formula (28), we get at once

$$\sin 5x - \sin 3x = 2 \cos 4x \sin x$$

EXAMPLE 3 Prove the identity

$$\frac{\sin 5x + \sin x}{\cos 5x - \cos x} = -\cot 2x$$

Solution: Applying formula (27) to the numerator and formula (30) to the denominator, we obtain

$$\frac{\sin 5x + \sin x}{\cos 5x - \cos x} = \frac{2 \sin 3x \cos 2x}{-2 \sin 3x \sin 2x} = -\cot 2x$$

EXERCISE 10-8

Express each of the following products as an algebraic sum.

1. $2 \sin 45^\circ \cos 15^\circ$ **2.** $2 \sin 30^\circ \cos 20^\circ$

3. $\cos 42^\circ \cos 24^\circ$ **4.** $\sin 38^\circ \sin 25^\circ$

5. $2 \sin 3\theta \cos \theta$ **6.** $2 \cos 6\theta \cos 2\theta$

7. $\sin 40^\circ \cos 48^\circ$ **8.** $\sin 19^\circ \cos 42^\circ$

9. $\cos 3\theta \cos 4\theta$ **10.** $\sin \theta \sin 5\theta$

Express each of the following algebraic sums as a product.

11. $\sin 32^\circ + \sin 16^\circ$ **12.** $\sin 44^\circ - \sin 10^\circ$

13. $\cos 25^\circ + \cos 5^\circ$ **14.** $\cos 55^\circ - \cos 45^\circ$

15. $\cos 4x - \cos 2x$

16. $\cos 3x + \cos 5x$

17. $\sin 9x - \sin 5x$

18. $\sin 3x + \sin 6x$

19. $\cos 25° - \cos 45°$

20. $\sin 32° - \sin 42°$

Use one of the formulas (23) to (30) and find the value of each of the following expressions.

21. $\sin 75° \cos 15°$

22. $2 \sin 45° \sin 15°$

23. $\cos 52\frac{1}{2}° \cos 7\frac{1}{2}°$

24. $\cos 75° + \cos 15°$

25. $\sin 165° - \sin 75°$

26. $\cos 165° - \cos 105°$

Prove the following identities.

27. $\dfrac{\cos 3x + \cos x}{\sin 3x + \sin x} = \cot 2x$

28. $\dfrac{\cos 5x - \cos 7x}{\sin 7x - \sin 5x} = \tan 6x$

29. $\dfrac{\sin 5x + \sin 3x}{\sin 5x - \sin 3x} = \dfrac{\cot x}{\cot 4x}$

30. $\dfrac{\cos 2x - \cos 4x}{\cos 2x + \cos 4x} = \dfrac{\tan x}{\cot 3x}$

31. $\dfrac{\sin 4x + \sin 3x + \sin 2x}{\cos 4x + \cos 3x + \cos 2x} = \tan 3x$

32. $\dfrac{\cos x - \cos 2x + \cos 3x}{\sin x - \sin 2x + \sin 3x} = \cot 2x$

33. $\sin 5\theta - \sin 3\theta + 2 \sin \theta \cos 2\theta = 2 \sin 2\theta \cos 3\theta$

34. $\cos 7\theta + \cos 5\theta + \cos 3\theta + \cos \theta = 4 \cos \theta \cos 2\theta \cos 4\theta$

10-9 SUMMARY OF FORMULAS

For convenient reference we list here the identities (9) to (30) of this chapter.

Functions of the sum and difference of two angles:

$$\cos (A - B) = \cos A \cos B + \sin A \sin B \tag{9}$$

$$\cos (A + B) = \cos A \cos B - \sin A \sin B \tag{10}$$

$$\sin (A + B) = \sin A \cos B + \cos A \sin B \tag{11}$$

$$\sin (A - B) = \sin A \cos B - \cos A \sin B \tag{12}$$

$$\tan (A + B) = \frac{\tan A + \tan B}{1 - \tan A \tan B} \tag{13}$$

$$\tan (A - B) = \frac{\tan A - \tan B}{1 + \tan A \tan B} \tag{14}$$

The double-angle formulas:

$$\sin 2A = 2 \sin A \cos A \tag{15}$$

$$\begin{aligned} \cos 2A &= \cos^2 A - \sin^2 A \\ &= 1 - 2 \sin^2 A \\ &= 2 \cos^2 A - 1 \end{aligned} \tag{16}$$

$$\tan 2A = \frac{2 \tan A}{1 - \tan^2 A} \tag{17}$$

The half-angle formulas:

$$\sin \tfrac{1}{2}A = \pm \sqrt{\frac{1 - \cos A}{2}} \tag{18}$$

$$\cos \tfrac{1}{2}A = \pm \sqrt{\frac{1 + \cos A}{2}} \tag{19}$$

$$\tan \tfrac{1}{2}A = \pm \sqrt{\frac{1 - \cos A}{1 + \cos A}} \tag{20}$$

$$\tan \tfrac{1}{2}A = \frac{1 - \cos A}{\sin A} \tag{21}$$

$$\tan \tfrac{1}{2}A = \frac{\sin A}{1 + \cos A} \tag{22}$$

The product formulas:

$$2 \sin A \cos B = \sin (A + B) + \sin (A - B) \tag{23}$$

$$2 \cos A \sin B = \sin (A + B) - \sin (A - B) \tag{24}$$

$$2 \cos A \cos B = \cos (A + B) + \cos (A - B) \tag{25}$$

$$2 \sin A \sin B = -\cos (A + B) + \cos (A - B) \tag{26}$$

The sum formulas:

$$\sin A + \sin B = 2 \sin \tfrac{1}{2}(A + B) \cos \tfrac{1}{2}(A - B) \tag{27}$$

$$\sin A - \sin B = 2 \cos \tfrac{1}{2}(A + B) \sin \tfrac{1}{2}(A - B) \tag{28}$$

$$\cos A + \cos B = 2 \cos \tfrac{1}{2}(A + B) \cos \tfrac{1}{2}(A - B) \tag{29}$$

$$\cos A - \cos B = -2 \sin \tfrac{1}{2}(A + B) \sin \tfrac{1}{2}(A - B) \tag{30}$$

EXERCISE 10-9

Prove the following identities.

1. $\sec x \sec y \sin (x - y) = \tan x - \tan y$

2. $2 \csc 2x \sin (x + y) = \cos y \sec x + \sin y \csc x$

3. $\sin (x + y) \sin (x - y) = \cos^2 y - \cos^2 x$

4. $\cos (x + y) \cos (x - y) = \cos^2 x - \sin^2 y$

5. $(\sin \theta - \cos \theta)^2 = 1 - \sin 2\theta$

6. $(\sec \theta + \csc \theta)^2 = \sec^2 \theta + \csc^2 \theta + 4 \csc 2\theta$

7. $\tan \theta + \tan \frac{1}{2}\theta = \csc \theta - 2 \cot 2\theta$

8. $(1 - \tan^2 \theta) \cot^2 \theta = \cos 2\theta \csc^2 \theta$

9. $2 \tan \theta \cot 2\theta + \tan \frac{1}{2}\theta = 1 + \csc \theta - \cot \theta - \tan^2 \theta$

10. $\tan \frac{1}{2}\theta + 2 \sin^2 \theta + \cos 2\theta = 1 - \cot \theta + \csc \theta$

11. $\dfrac{\sin 3x}{\sin 2x} - \dfrac{\cos 3x}{\cos 2x} = \dfrac{2 \sin x}{\sin 4x}$

12. $\dfrac{\sin 4x}{\cos 2x} + \dfrac{\cos 4x}{\sin 2x} = \csc 2x$

13. $\dfrac{\cos 2x}{\sin 3x} - \dfrac{\sin 2x}{\cos 3x} = \dfrac{2 \cos 5x}{\sin 6x}$

14. $\dfrac{\sin 3x}{\cos 2x} - \dfrac{\sin 2x}{\cos x} = \dfrac{\tan 2x}{2 \cos x}$

15. $\dfrac{\cos 2x}{\cos x} - \dfrac{\cos 4x}{\cos 3x} = \dfrac{2 \sin^2 x}{\cos 3x}$

16. $\dfrac{\sin 3x}{\sin x} - \dfrac{\sin 4x}{\sin 2x} = 1$

17. $\cos^4 \theta - \sin^4 \theta = \dfrac{\sin 4\theta}{2 \sin 2\theta}$

18. $\dfrac{\sin^2 2x}{2 \cos^2 x} = 1 - \cos 2x$

19. $\sin 2x = \dfrac{2 \cot x}{1 + \cot^2 x}$

20. $\cos 2x = \dfrac{\cot^2 x - 1}{\cot^2 x + 1}$

21. $\dfrac{1 - \cos x}{1 + \cos \frac{1}{2}x} = 2(1 - \cos \frac{1}{2}x)$

22. $\dfrac{1 + \cos 2x}{1 - \sin x} = 2(1 + \sin x)$

23. $\dfrac{1 + \cos^2 \frac{1}{2}x}{1 + \sin^2 \frac{1}{2}x} = \dfrac{3 \sec x + 1}{3 \sec x - 1}$

24. $\dfrac{\cos \frac{1}{2}x + \sin \frac{1}{2}x}{\cos \frac{1}{2}x - \sin \frac{1}{2}x} = \dfrac{1 + \sin x}{\cos x}$

25. $\sin x + \tan x = \dfrac{\sin x \tan x}{\tan \frac{1}{2}x}$

26. $\csc 2x = \dfrac{1 + \tan^2 x}{2 \tan x}$

27. $\dfrac{\sin^3 \theta - \cos^3 \theta}{\sin \theta - \cos \theta} = \dfrac{2 + \sin 2\theta}{2}$

28. $\dfrac{1 + \tan^2 \frac{1}{2}\theta}{(1 + \tan \frac{1}{2}\theta)^2} = \dfrac{1}{1 + \sin \theta}$

29. $\dfrac{\tan^2 \frac{1}{2}\theta}{\cos^2 \theta} = \dfrac{\tan^2 \theta}{4 \cos^4 \frac{1}{2}\theta}$

30. $\dfrac{4 \cos^3 \theta - \cos 3\theta}{4 \sin^3 \theta + \sin 3\theta} = \cot \theta$

31. $\dfrac{\cos 2\theta - \sin 2\theta}{\sin \theta \cos \theta} = \cot \theta - \tan \theta - 2$

32. $\tan (x + y) + \tan (x - y) = \dfrac{2 \tan x \sec^2 y}{1 - \tan^2 x \tan^2 y}$

33. $\dfrac{4}{\cos \theta - \cos 3\theta} - \dfrac{4}{\sin \theta + \sin 3\theta} = \sec \theta \csc \theta(\csc \theta - \sec \theta)$

34. $\dfrac{2}{\cos 6\theta - \cos 2\theta} + \dfrac{2}{\sin 6\theta - \sin 2\theta} = \csc 2\theta(\sec 4\theta - \csc 4\theta)$

35. $\sin 3\theta \cos 2\theta + \cos 6\theta \sin \theta = \sin 4\theta \cos 3\theta$

36. $(\cos x - \cos 5x)(\cos x + \cos 5x) = \sin 4x \sin 6x$

37. $\sin^2 3x - \sin^2 x = \sin 2x \sin 4x$

38. $\cos^2 3x - \cos^2 x = -\sin 2x \sin 4x$

39. $4 \sin x \sin y \sin (x - y) = \sin 2y - \sin 2x - \sin (2y - 2x)$

11

COMPLEX NUMBERS

11-1 THE IMAGINARY UNIT

In our treatment of fractional exponents (Sec. 7-3), we excluded even roots of negative numbers. But even roots, especially square roots, of negative numbers are of great importance in mathematics. Their use has contributed to the development of a large body of mathematics, much of which has vital applications in the engineering and physical sciences.

Many equations cannot be solved and many problems cannot be investigated in the real number system. The equation $x^2 + 1 = 0$, for example, has a solution if and only if $x^2 = -1$. But the square of any real number is not negative, and consequently the equation has no real solution. In order to solve this equation and certain more complicated equations which arise in later chapters, we need to enlarge our number system to include a new set of numbers.

As a first step in extending the real number system, we introduce the number $\sqrt{-1}$, called the *imaginary unit*, whose square is defined to be -1. The imaginary unit is customarily denoted by the letter i; that is, $i = \sqrt{-1}$ and $i^2 = -1$. Also, we agree to express the product of i and a real number b by bi and the sum of a real number a and bi by $a + bi$.

Now that we have introduced a square root of -1 into our system of numbers, let us consider the square roots of any negative number $-p$. Since $i^2 = -1$, we may write

$$(i\sqrt{p})^2 = i^2p = -p \qquad \text{and} \qquad (-i\sqrt{p})^2 = i^2p = -p$$

Hence, the square roots of $-p$ are $i\sqrt{p}$ and $-i\sqrt{p}$. To distinguish between these roots, we let

$$\sqrt{-p} = i\sqrt{p} \qquad \text{and} \qquad -\sqrt{-p} = -i\sqrt{p}$$

As examples, we have

$$\sqrt{-5} = i\sqrt{5}$$
$$\sqrt{-9} = i\sqrt{9} = 3i$$
$$-\sqrt{-4} = -i\sqrt{4} = -2i$$

Using $i^2 = -1$, we see that

$$i^3 = i^2 \cdot i = -i$$
$$i^4 = i^2 \cdot i^2 = (-1)(-1) = 1$$

Since $i^4 = 1$, it follows that i with an exponent which is an integral multiple of 4 is also equal to 1. Consequently, integral powers of i yield only the numbers i, -1, $-i$, and 1. Thus,

$$i^{13} = i^{12} \cdot i = i$$
$$i^{-6} = i^{-6} \cdot i^8 = i^2 = -1$$

11-2 COMPLEX NUMBERS

Having enlarged the number system by introducing the imaginary unit i and numbers of the form bi, b a real number, we now find it useful to make a further enlargement.

Definition 11-1 A number of the form $a + bi$, with a and b real constants, is called a *complex number*. The number a is called the *real component* and b is called the *imaginary component*.

Depending on the values of a and b, we have the following special cases of a complex number.

Definition 11-2

$a + bi$ is called an *imaginary number* if $b \neq 0$.

$a + bi$ is called a *pure imaginary number* if $a = 0$ and $b \neq 0$.

$a + bi$ is called a *real number* if $b = 0$.

EXAMPLES

$2i$, $3 - 5i$, and $-4 + i$ are imaginary numbers.

$2i$, $-6i$, and $i\sqrt{5}$ are pure imaginary numbers.

7, -2.5, and 0 are real numbers.

The set of numbers $x + yi$, where $x \in R$ and $y \in R$, comprises the *complex number system.* Any real number x and the complex number $x + 0i$ are considered to be the same; and if $y \in R$, the complex number $0 + yi$ and the pure imaginary number yi are also considered to be the same. Hence, as examples, we write

$$2 + 0i = 2 \qquad \text{and} \qquad 0 - 5i = -5i$$

We conclude, then, that the set of real numbers and the set of pure imaginary numbers are proper subsets of the system of complex numbers.

Definition 11-3

Two complex numbers are equal if and only if the real components are equal and the imaginary components are equal; that is,

$$a + bi = c + di \qquad \text{if and only if} \qquad a = c \text{ and } b = d$$

EXAMPLE If $2x - 5 + (4x - 1)i = y + 3 + (3x - y)i$, find the values of x and y.

Solution: The real components of the members of the equation must be equal and the imaginary components must be equal. Hence, we have the system of equations

$$2x - 5 = y + 3$$
$$4x - 1 = 3x - y$$

The solution of this system is $x = 3$, $y = -2$.

11-3 FUNDAMENTAL OPERATIONS ON COMPLEX NUMBERS

In this section, we define the operations of addition, multiplication, subtraction, and division of complex numbers. The operations are designed so that complex numbers obey the field axioms (Sec. 2-2).

Definition 11-4 The sum and product of two complex numbers are given, respectively, by the following equations:

Sum: $(a + bi) + (c + di) = (a + c) + (b + d)i$

Product: $(a + bi)(c + di) = (ac - bd) + (bc + ad)i$

We note that the sum and product are obtained by carrying out the operations as though i were a real number and, for the product, replacing i^2 by -1.

From the definition of addition, we see that the sum of $a + bi$ and $-a - bi$ is equal to zero. Accordingly, $-a - bi$ is called the *additive inverse of* $a + bi$. Using the idea of the additive inverse of a number, we define subtraction by the equation

$$(a + bi) - (c + di) = (a + bi) + (-c - di)$$

Then, applying the definition of addition to the right member of this equation, we have

$$(a + bi) - (c + di) = (a - c) + (b - d)i$$

The multiplicative inverse of a complex number $c + di$ is defined to be the number $x + yi$ such that

$$(c + di)(x + yi) = 1$$

or, multiplying (Definition 11-4),

$$(cx - dy) + (dx + cy)i = 1$$

Equating the real components and the imaginary components of the equal complex numbers in this equation (Definition 11-3), we have the system of equations

$$cx - dy = 1$$

$$dx + cy = 0$$

The solution of this system in the unknowns x and y yields

$$\frac{c}{c^2 + d^2} - \frac{d}{c^2 + d^2}i \qquad \text{or} \qquad \frac{c - di}{c^2 + d^2}$$

as the multiplicative inverse of $c + di$.

We now define the quotient of two complex numbers as the product of the dividend and the multiplicative inverse of the divisor, and have

$$\begin{aligned}\frac{a + bi}{c + di} &= (a + bi)\left(\frac{c - di}{c^2 + d^2}\right) \\ &= \frac{(ac + bd) + (bc - ad)i}{c^2 + d^2}\end{aligned}$$

Summarizing, we express the difference and quotient of two complex numbers as follows:

Difference: $(a + bi) - (c + di) = (a - c) + (b - d)i$

Quotient: $\dfrac{a + bi}{c + di} = \dfrac{(ac + bd) + (bc - ad)i}{c^2 + d^2}$

We remark that the quotient may be obtained by multiplying the dividend and divisor by $c - di$. Thus,

$$\frac{a + bi}{c + di} = \frac{(a + bi)(c - di)}{(c + di)(c - di)} = \frac{(ac + bd) + (bc - ad)i}{c^2 + d^2}$$

The numbers $c + di$ and $c - di$ are said to be *conjugate complex numbers.* Each is the conjugate of the other. Thus, $4 + 3i$, $4 - 3i$ and $5i$, $-5i$ are pairs of conjugate complex numbers.

The closure, commutative, associative, and distributive laws of real numbers (Sec. 2-2) are valid for complex numbers. We shall omit the proofs of these properties except for the distributive law.

Theorem 11-1 The distributive law holds for complex numbers; that is,

$$(a + bi)[(c + di) + (e + fi)] = (a + bi)(c + di) + (a + bi)(e + fi)$$

Proof: By employing Definition 11-4 and assuming that the associative law is valid for complex numbers, we reduce the left member of this equation to the form of the right member. All steps should be followed:

$$
\begin{aligned}
(a + bi)[(c + di) + (e + fi)]& \\
&= (a + bi)[(c + e) + (d + f)i] \\
&= [a(c + e) - b(d + f)] + [b(c + e) + a(d + f)]i \\
&= [(ac - bd) + (ae - bf)] + [(bc + ad)i + (be + af)i] \\
&= [(ac - bd) + (bc + ad)i] + [(ae - bf) + (be + af)i] \\
&= (a + bi)(c + di) + (a + bi)(e + fi)
\end{aligned}
$$

EXAMPLE 1 Find the product of $\sqrt{-5}$ and $\sqrt{-7}$.

Solution: We first express each number as a real number times i and then multiply. Thus,

$$\sqrt{-5} \cdot \sqrt{-7} = i\sqrt{5} \cdot i\sqrt{7} = i^2\sqrt{35} = -\sqrt{35}$$

This example illustrates that the law for multiplying two radicals of like order (Sec. 7-4) is not applicable for imaginary numbers. Although $\sqrt{5} \cdot \sqrt{7} = \sqrt{35}$, the product $\sqrt{-5} \cdot \sqrt{-7} \neq \sqrt{35}$. An error is less likely to be made in a product of this kind if each factor is expressed as a real number times i before multiplying.

EXAMPLE 2 Find the quotient of $2 - \sqrt{-9}$ divided by $4 + \sqrt{-25}$.

Solution: We express these numbers as $2 - 3i$ and $4 + 5i$. Then, remembering that $i^2 = -1$, we get

$$
\begin{aligned}
\frac{2 - 3i}{4 + 5i} &= \frac{2 - 3i}{4 + 5i} \cdot \frac{4 - 5i}{4 - 5i} \\
&= \frac{8 + 15i^2 - 12i - 10i}{16 - 25i^2} \\
&= \frac{-7 - 22i}{41}
\end{aligned}
$$

In practice, the method used in this example for finding the quotient of two complex numbers is more convenient than using the quotient formula.

EXERCISE 11-1

Simplify the following expressions.

1. $2i^8 - 3i^3 + 4i^6$
2. $3i^{11} - 5i^7 + 6i^3$
3. $i^{19} - 2i^{14} - 5i^{23}$
4. $10i^{10} + 9i^{11} - 8i^{12}$
5. $4i^7 - 9i^{11} + 5i^{15}$
6. $6i^3 + 2i^{16} - 6i^{19}$

Solve each of the following equations for x and y.

7. $4 + ix = y + 3i$
8. $x - iy = 2 + 5i$
9. $2y - 3 + (4x + 8)i = 0$
10. $3x + (y - x)i = 6$
11. $x + 2y + 3i = 3 + (2y - x)i$
12. $7x + (x - 3y)i = 3y + 9i$

Perform the indicated operations, leaving each result in the form $a + bi$.

13. $(4 + 3i) + (7 + 5i)$
14. $(2 - i) + (-5 + 3i)$
15. $(1 - 2i) - (6 + 4i)$
16. $(9 + 3i) - (7 - 4i)$
17. $(6 - 6i) - (-4 - 3i)$
18. $(-1 + 7i) - (-8 - 3i)$
19. $(3 - i) + (-2 + 2i) + (4 + i)$
20. $(8 - 4i) - (9 + 3i) + (8 - i)$
21. $5\sqrt{-9} + 2\sqrt{-16} - 5\sqrt{-25} + 4\sqrt{-36}$
22. $(5 + \sqrt{-81}) + (6 - \sqrt{-49}) + (1 + \sqrt{-1})$
23. $2\sqrt{-8} - 5\sqrt{-2} + 3\sqrt{-18} + \sqrt{-32}$
24. $(1 - \sqrt{-12}) + (2 + \sqrt{-27}) + (3 - \sqrt{-48})$
25. $(2 + i)(5 + 4i)$
26. $(5 + 2i)(5 + 3i)$
27. $(4 - 6i)(4 + 6i)$
28. $(7 + 3i)(7 - 3i)$
29. $(2 - 8i)(1 - i)$
30. $(5 + 2i)(2 + 5i)$
31. $(1 + i)(2 - i)(3 + 2i)$
32. $(4 - 2i)(2 + 4i)(-1 - 3i)$
33. $\dfrac{5 + i}{i}$
34. $\dfrac{6 - 12i}{3i}$
35. $\dfrac{4}{4 - 3i}$

36. $\dfrac{7i}{4+i}$ 37. $\dfrac{2-i}{6+i}$ 38. $\dfrac{3+i}{3-2i}$

39. $\dfrac{17-6i}{2-3i}$ 40. $\dfrac{2+3i}{2-3i}$ 41. $\dfrac{9-7i}{7+9i}$

42. Prove that $x + yi = 0$ if and only if $x = 0$ and $y = 0$. *Hint:* Use Definition 11-3.

43. Prove that the sum and product of two conjugate complex numbers are both real.

44. Prove that the conjugate of the sum of two complex numbers is equal to the sum of their conjugates.

45. If the product of two complex numbers is zero, prove that at least one of the factors is zero.

46. Prove that addition and multiplication of complex numbers is commutative.

11-4 GRAPHICAL REPRESENTATION OF COMPLEX NUMBERS

Complex numbers are usually represented by points in the plane of a rectangular coordinate system. The number $a + bi$ is represented by the point whose coordinates are (a, b). Hence, any complex number determines a point in the plane. Conversely, any point of the plane determines a complex number. The points on the axes correspond to special complex numbers. Points on the x axis represent real numbers because the ordinates of the points are zero. Similarly, the points on the y axis have abscissas equal to zero and therefore correspond to pure imaginary numbers. Thus, the point $(a, 0)$ corresponds to the real number $a = a + 0i$, and the point $(0, b)$ corresponds to the pure imaginary number $bi = 0 + bi$.

In Fig. 11-1, the correspondence between some complex numbers and points is indicated.

When a rectangular coordinate plane is used to represent complex numbers, the x axis is called the *real axis*, the y axis the *pure imaginary axis*, and the plane the *complex plane*.

The sum of two complex numbers may be found graphically by constructing a parallelogram. The two points representing the given numbers are opposite vertices, and the origin is a third vertex. The fourth vertex of the parallelogram thus determined represents the sum. This method of adding may be verified by

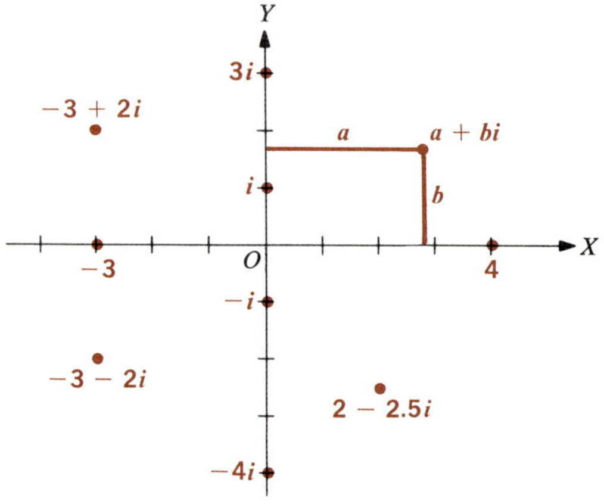

Figure 11-1

observing that the opposite sides of the quadrilateral (Fig. 11-2) are parallel. Furthermore, since the coordinates of the vertices are (0, 0), (a, b), (c, d), and $(a + c, b + d)$, the distance formula (Sec. 5-4) reveals that $\sqrt{a^2 + b^2}$ is the length of each of one pair of opposite sides and $\sqrt{c^2 + d^2}$ is the length of each of the other sides.

The parallelogram method of adding two complex numbers is exactly the method of adding two vectors (Sec. 9-10). This suggests that a complex number represented by a point may also be represented by a vector drawn from the origin to the point. Often complex numbers are so represented. Figure 11-3 shows the vectors for the numbers $8 + 2i$ and $-4 + 3i$ and the vector for their sum.

11-5 TRIGONOMETRIC FORM OF A COMPLEX NUMBER

A complex number expressed as $a + bi$ is said to be in *algebraic, or rectangular, form.* We next show how to express a complex number in another useful form.

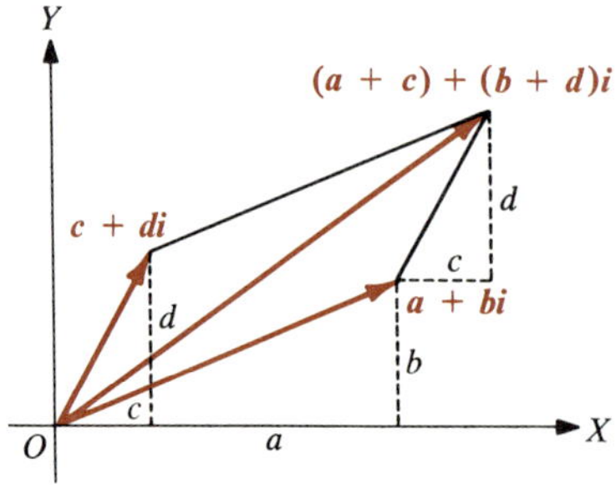

Figure 11-2

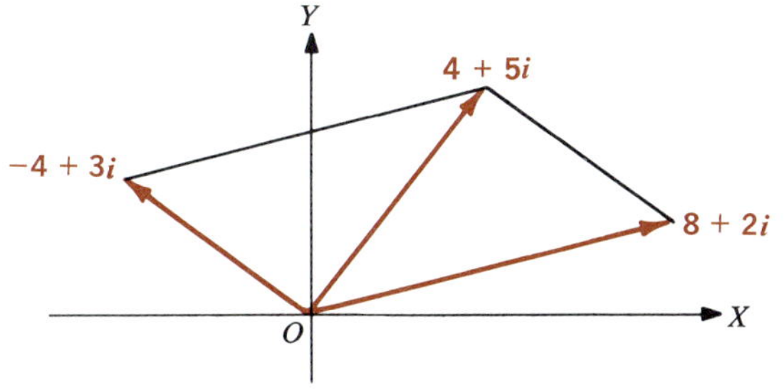

Figure 11-3

The point P (Fig. 11-4) corresponding to the number $a + bi$ has the coordinates (a, b). The distance from the origin to P is denoted by r, and the angle from the positive real axis to the distance segment is denoted by θ. The distance r (chosen positive when P is any point other than the origin) is called the *absolute value*, or *modulus*. The angle θ is called the *amplitude*, or *argument*.

Referring to the diagram, we have the relations

$$r = \sqrt{a^2 + b^2} \qquad \tan\theta = \frac{b}{a}$$

and

$$a = r\cos\theta \qquad b = r\sin\theta$$

Hence, it follows that

$$a + bi = r(\cos\theta + i\sin\theta)$$

The expression $r(\cos\theta + i\sin\theta)$ is called the *trigonometric*, or *polar*, *form* of a complex number. The number r is never negative since it is the modulus. The factor $(\cos\theta + i\sin\theta)$ reveals the direction of the point $P(a + bi)$ from the origin. This is true because θ must satisfy the equations $r\cos\theta = a$ and $r\sin\theta = b$, with r positive. The preceding formulas, then, are sufficient for changing a number from algebraic form to trigonometric form and vice versa.

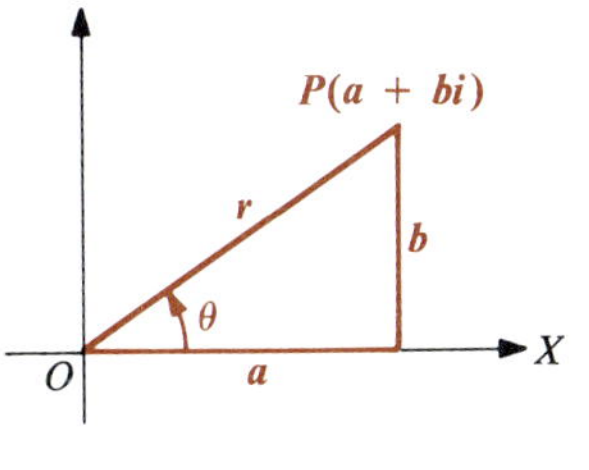

Figure 11-4

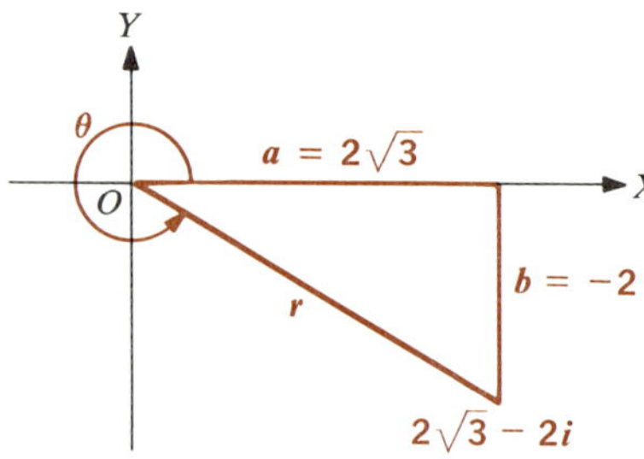

Figure 11-5

EXAMPLE 1 Plot the number $2\sqrt{3} - 2i$ and change to the trigonometric form.

Solution: The point $(2\sqrt{3}, -2)$ represents the number (Fig. 11-5). Letting $a = 2\sqrt{3}$ and $b = -2$ in the preceding formulas, we obtain

$$r = \sqrt{12 + 4} = 4, \qquad \cos\theta = \frac{\sqrt{3}}{2}, \qquad \sin\theta = -\frac{1}{2}$$

We choose the smallest positive value of θ which satisfies the last two equations and have

$$2\sqrt{3} - 2i = 4(\cos 330^\circ + i \sin 330^\circ)$$

EXAMPLE 2 Plot the number $5(\cos 120^\circ + i \sin 120^\circ)$ and change to algebraic form.

Solution: The point corresponding to this number is located by drawing the angle 120° in standard position and measuring 5 units off along the terminal side. From Fig. 11-6, we have

$$a = 5\cos 120^\circ = 5(-\tfrac{1}{2}) = -\tfrac{5}{2}$$

$$b = 5\sin 120^\circ = 5\left(\frac{\sqrt{3}}{2}\right) = \frac{5\sqrt{3}}{2}$$

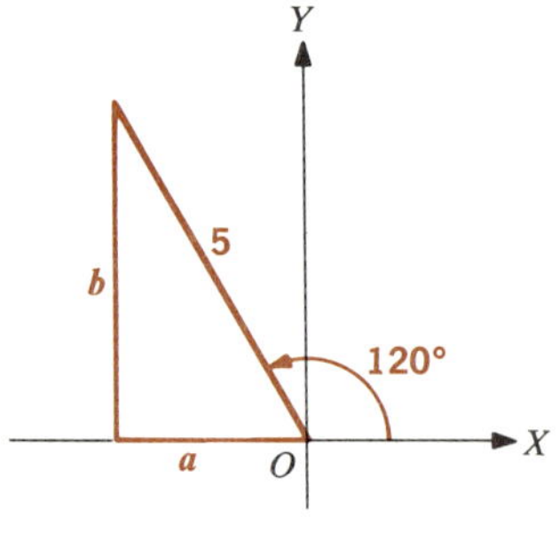

Figure 11-6

and therefore

$$5(\cos 120° + i \sin 120°) = -\frac{5}{2} + \frac{5\sqrt{3}}{2}i$$

EXAMPLE 3 Express the number $4 + 3i$ in trigonometric form.

Solution: Since $a = 4$ and $b = 3$, we have

$$r = \sqrt{16 + 9} = 5 \qquad \cos\theta = \tfrac{4}{5} \qquad \sin\theta = \tfrac{3}{5}$$

Referring to Table 2, we find $\theta = 36° \, 52'$, and hence

$$4 + 3i = 5(\cos 36° \, 52' + i \sin 36° \, 52')$$

EXERCISE 11-2

Write the conjugate of each complex number. Plot the number and its conjugate.

1. $2 + 3i$ **2.** $6 - 5i$ **3.** $0 + 2i$

4. $-4 + 0i$ **5.** $-2 + 6i$ **6.** $7 - 4i$

7. $-4 - 3i$ **8.** $-3i + 4$

Plot each complex number and write the corresponding trigonometric form.

9. $-1 - i$ **10.** $1 + i$ **11.** $4 + 0i$

12. $\sqrt{3} + i$ **13.** $2\sqrt{3} - 2i$ **14.** $1 - i\sqrt{3}$

15. $0 - 3i$ **16.** $-4 + 0i$ **17.** $0 + 2i$

18. $\sqrt{2} + i\sqrt{2}$ **19.** $6 - 6i\sqrt{3}$ **20.** $-\sqrt{15} + i\sqrt{5}$

21. $-\sqrt{7} - i\sqrt{21}$ **22.** $3 + 4i$ **23.** $2 + i$

Change each number to algebraic form.

24. $4(\cos 60° + i \sin 60°)$ **25.** $2(\cos 30° + i \sin 30°)$

26. $\cos 45° + i \sin 45°$ **27.** $2(\cos 135° + i \sin 135°)$

28. $8(\cos 150° + i \sin 150°)$ **29.** $3(\cos 90° + i \sin 90°)$

30. $5(\cos 180° + i \sin 180°)$ **31.** $9(\cos 270° + i \sin 270°)$

32. $\cos 78° + i \sin 78°$ **33.** $10(\cos 228° + i \sin 228°)$

Plot each pair of numbers and their sum. Draw the parallelogram determined by the three points and the origin.

34. $-2 + 3i, \quad 3 + i$ **35.** $4 + 2i, \quad 2 + 5i$

36. $3 - 2i, \quad -2 - 4i$ **37.** $1 - 4i, \quad 1 + 2i$

11-6 MULTIPLICATION AND DIVISION OF NUMBERS IN POLAR FORM

The product and quotient of two complex numbers may be written immediately if the numbers are in polar form.

Theorem 11-2 The absolute value of the product of two complex numbers is equal to the product of their absolute values. The amplitude of the product of two complex numbers is equal to the sum of their amplitudes.

Proof: Let the two complex numbers be denoted by

$$z_1 = r_1(\cos\theta_1 + i\sin\theta_1) \qquad \text{and} \qquad z_2 = r_2(\cos\theta_2 + i\sin\theta_2)$$

Then, multiplying in the usual way, we obtain

$$\begin{aligned} z_1z_2 &= r_1(\cos\theta_1 + i\sin\theta_1)\cdot r_2(\cos\theta_2 + i\sin\theta_2) \\ &= r_1r_2[(\cos\theta_1\cos\theta_2 + i^2\sin\theta_1\sin\theta_2) \\ &\qquad + i(\sin\theta_1\cos\theta_2 + \sin\theta_2\cos\theta_1)] \end{aligned}$$

Replacing i^2 by -1 and applying identities (10) and (11) of Sec. 10-9 for the cosine and the sine of the sum of two angles, we get

$$\begin{aligned} &r_1(\cos\theta_1 + i\sin\theta_1)\cdot r_2(\cos\theta_2 + i\sin\theta_2) \\ &\qquad = r_1r_2[\cos(\theta_1 + \theta_2) + i\sin(\theta_1 + \theta_2)] \end{aligned}$$

This law of multiplication may be applied repeatedly to give the product of three or more complex numbers. For the three numbers

$$\begin{aligned} z_1 &= r_1(\cos\theta_1 + i\sin\theta_1) \\ z_2 &= r_2(\cos\theta_2 + i\sin\theta_2) \\ z_3 &= r_3(\cos\theta_3 + i\sin\theta_3) \end{aligned}$$

we have

$$z_1z_2 = r_1r_2[\cos(\theta_1 + \theta_2) + i \sin(\theta_1 + \theta_2)]$$

and then

$$z_1z_2z_3 = r_1r_2r_3[\cos(\theta_1 + \theta_2 + \theta_3) + i \sin(\theta_1 + \theta_2 + \theta_3)]$$

The product of n complex numbers is given by

$$z_1z_2 \cdots z_n = r_1r_2 \cdots r_n[\cos(\theta_1 + \theta_2 + \cdots + \theta_n) + i \sin(\theta_1 + \theta_2 + \cdots + \theta_n)]$$

We suggest that the student prove the associative property of complex numbers for multiplication; that is, prove that $(z_1z_2)z_3 = z_1(z_2z_3)$.

Theorem 11-3 The absolute value of the quotient of two complex numbers is the quotient of their absolute values. The amplitude is the amplitude of the dividend minus the amplitude of the divisor.

Proof: To find the quotient of z_1 divided by z_2, we multiply the dividend and divisor by the conjugate of the divisor. Thus, we have

$$\frac{z_1}{z_2} = \frac{r_1(\cos\theta_1 + i \sin\theta_1)(\cos\theta_2 - i \sin\theta_2)}{r_2(\cos\theta_2 + i \sin\theta_2)(\cos\theta_2 - i \sin\theta_2)}$$

The product of the factors in the numerator is

$$r_1[(\cos\theta_1 \cos\theta_2 + \sin\theta_1 \sin\theta_2) + i(\sin\theta_1 \cos\theta_2 - \cos\theta_1 \sin\theta_2)]$$

and the product of the factors of the denominator is

$$r_2(\cos^2\theta_2 - i^2 \sin^2\theta_2) = r_2$$

We now apply identities (9) and (12) for the cosine and sine of the difference of two angles and express the numerator in the form $r_1[\cos(\theta_1 - \theta_2) + i \sin(\theta_1 - \theta_2)]$. Then, we obtain the formula

$$\frac{r_1(\cos\theta_1 + i \sin\theta_1)}{r_2(\cos\theta_2 + i \sin\theta_2)} = \frac{r_1}{r_2}[\cos(\theta_1 - \theta_2) + i \sin(\theta_1 - \theta_2)]$$

The two preceding formulas may be applied graphically to give the product and quotient of two complex numbers. To find the product of $r_1(\cos\theta_1 + i\sin\theta_1)$ and $r_2(\cos\theta_2 + i\sin\theta_2)$, draw the angle $(\theta_1 + \theta_2)$ in standard position and measure off a distance on the terminal side equal to r_1r_2. To find the quotient of $r_1(\cos\theta_1 + i\sin\theta_1)$ divided by $r_2(\cos\theta_2 + i\sin\theta_2)$, draw the angle $(\theta_1 - \theta_2)$ in standard position and measure off a distance along the terminal side equal to r_1/r_2.

EXAMPLE 1 Express the numbers $1 + i\sqrt{3}$ and $3 - 3i$ in polar forms and find their product.

Solution: We have for the polar forms

$$1 + i\sqrt{3} = 2(\cos 60° + i\sin 60°)$$

and

$$3 - 3i = 3\sqrt{2}(\cos 315° + i\sin 315°)$$

Hence,

$$\begin{aligned}(1 + i\sqrt{3})(3 - 3i) &= 6\sqrt{2}(\cos 375° + i\sin 375°)\\ &= 6\sqrt{2}(\cos 15° + i\sin 15°)\end{aligned}$$

EXAMPLE 2 Find the quotient of $12(\cos 30° + i\sin 30°)$ divided by $3(\cos 70° + i\sin 70°)$.

Solution: Applying the law for division, we obtain

$$\begin{aligned}\frac{12(\cos 30° + i\sin 30°)}{3(\cos 70° + i\sin 70°)} &= 4[\cos(-40°) + i\sin(-40°)]\\ &= 4(\cos 320° + i\sin 320°)\end{aligned}$$

EXERCISE 11-3

Perform the indicated operations, leaving each result in the form $a + bi$.

1. $2(\cos 30° + i\sin 30°) \cdot 3(\cos 60° + i\sin 60°)$
2. $5(\cos 120° + i\sin 120°) \cdot 4(\cos 60° + i\sin 60°)$

3. $3(\cos 50° + i \sin 50°) \cdot 6(\cos 70° + i \sin 70°)$

4. $7(\cos 130° + i \sin 130°) \cdot 2(\cos 95° + i \sin 95°)$

5. $9(\cos 175° + i \sin 175°) \cdot 3(\cos 275° + i \sin 275°)$

6. $\dfrac{\cos 130° + i \sin 130°}{\cos 40° + i \sin 40°}$

7. $\dfrac{4(\cos 266° + i \sin 266°)}{2(\cos 86° + i \sin 86°)}$

8. $\dfrac{9(\cos 313° + i \sin 313°)}{3(\cos 268° + i \sin 268°)}$

9. $\dfrac{21(\cos 33° + i \sin 33°)}{3(\cos 93° + i \sin 93°)}$

Express each number in trigonometric form and then perform the indicated operations. Leave the results in trigonometric form.

10. $(1 - i)(1 + i\sqrt{3})$

11. $(\sqrt{3} + i)(1 + i)$

12. $(3 - 3i)(4 - 4i)$

13. $4i(5 + 5i)$

14. $3i(1 - i\sqrt{3})(\sqrt{3} + i)$

15. $(2 + 2i)(-\sqrt{3} + i)(-1 + i\sqrt{3})$

16. $\dfrac{3}{2 - 2i}$

17. $\dfrac{2\sqrt{3} - 2i}{1 - i\sqrt{3}}$

18. $\dfrac{4}{3\sqrt{3} - 3i}$

19. $\dfrac{1 + i}{1 + i\sqrt{3}}$

Perform graphically the indicated operations.

20. $2(\cos 45° + i \sin 45°) \cdot 3(\cos 45° + i \sin 45°)$

21. $4(\cos 30° + i \sin 30°) \cdot (\cos 90° + i \sin 90°)$

22. $5(\cos 90° + i \sin 90°) \cdot 2(\cos 60° + i \sin 60°)$

23. $\dfrac{4(\cos 225° + i \sin 225°)}{2(\cos 90° + i \sin 90°)}$

24. $\dfrac{6(\cos 270° + i \sin 270°)}{3(\cos 135° + i \sin 135°)}$

11-7 POWERS AND ROOTS OF NUMBERS

In this section, we shall be interested in finding positive integral powers of complex numbers and roots of complex numbers. The following theorem opens the way for the performance of these operations.

Theorem 11-4 If n is any real number, the nth power of the complex number $r(\cos\theta + i\sin\theta)$ is given by the formula

$$[r(\cos\theta + i\sin\theta)]^n = r^n(\cos n\theta + i\sin n\theta) \tag{1}$$

We shall not prove this theorem. We point out, however, that the theorem may be established at once for the case in which n is a positive integer. This is accomplished simply by letting each of the factors in the formula for the product of n complex numbers be equal to $r(\cos\theta + i\sin\theta)$. When n is a positive integer, the preceding theorem is called De Moivre's theorem in honor of the French mathematician Abraham De Moivre.

Formula (1) yields readily a positive integral power of a complex number. To find roots of numbers, however, we convert the formula to a more useful form.

The number $s(\cos\alpha + i\sin\alpha)$ is, by definition, an nth root of $r(\cos\theta + i\sin\theta)$ if

$$[s(\cos\alpha + i\sin\alpha)]^n = r(\cos\theta + i\sin\theta)$$

We apply De Moivre's formula (1) to the left member of this equation and, in the right member, replace θ by $\theta + k\cdot 360°$, where k is any integer. This replacement is justified by the fact that the values of the trigonometric functions of θ and $\theta + k\cdot 360°$ are the same (Sec. 8-5). Thus, we have

$$s^n(\cos n\alpha + i\sin n\alpha) = r[\cos(\theta + k\cdot 360°) + i\sin(\theta + k\cdot 360°)]$$

We seek values of s and α so that this equation will be satisfied. Consequently, we equate the moduli and amplitudes of the members of the equation. This gives

$$s^n = r \qquad \text{and} \qquad n\alpha = \theta + k\cdot 360°$$

hence, solving for s and α, we have

$$s = \sqrt[n]{r} \qquad \text{and} \qquad \alpha = \frac{\theta + k\cdot 360°}{n}$$

Hence,

$$s(\cos\alpha + i\sin\alpha) = \sqrt[n]{r}\left(\cos\frac{\theta + k\cdot 360°}{n} + i\sin\frac{\theta + k\cdot 360°}{n}\right)$$

The right member of this equation may be made to take n distinct values by giving k the values 0, 1, 2, . . . , $n - 1$. Any other integral value of k yields a repetition and not a new root. Thus, there are n different nth roots of any nonzero number. And it may be observed that the amplitudes are such that graphically the roots are equally spaced about the circumference of a circle of radius $\sqrt[n]{r}$. Hence, we have the following theorem.

Theorem 11-5 A nonzero number $r(\cos\theta + i\sin\theta)$ has n nth roots which are given by the formula

$$\sqrt[n]{r}\left(\cos\frac{\theta + k\cdot 360^\circ}{n} + i\sin\frac{\theta + k\cdot 360^\circ}{n}\right) \tag{2}$$

where $k = 0, 1, 2, \ldots, n - 1$.

EXAMPLE 1 Find the fourth power of $1 - i\sqrt{3}$.

Solution: First changing to polar form, we obtain

$$1 - i\sqrt{3} = 2(\cos 300^\circ + i\sin 300^\circ)$$

Applying formula (1) with $n = 4$, we get

$$\begin{aligned}(1 - i\sqrt{3})^4 &= [2(\cos 300^\circ + i\sin 300^\circ)]^4\\ &= 2^4(\cos 1200^\circ + i\sin 1200^\circ)\\ &= 16(\cos 120^\circ + i\sin 120^\circ)\\ &= -8 + 8i\sqrt{3}\end{aligned}$$

EXAMPLE 2 Find the cube roots of -8.

Solution: We first express -8 in polar form. Thus,

$$-8 = 8[\cos(180^\circ + k\cdot 360^\circ) + i\sin(180^\circ + k\cdot 360^\circ)]$$

We now apply formula (2) for roots of a number and have

$$\begin{aligned}&\sqrt[3]{8}\left(\cos\frac{180^\circ + k\cdot 360^\circ}{3} + i\sin\frac{180^\circ + k\cdot 360^\circ}{3}\right)\\ &\quad = 2[\cos(60^\circ + k\cdot 120^\circ) + i\sin(60^\circ + k\cdot 120^\circ)]\end{aligned}$$

Assigning k the values 0, 1, and 2, in succession, we find the three cube roots of -8 to be

$$2(\cos 60^\circ + i \sin 60^\circ) = 1 + i\sqrt{3}$$
$$2(\cos 180^\circ + i \sin 180^\circ) = -2$$
$$2(\cos 300^\circ + i \sin 300^\circ) = 1 - i\sqrt{3}$$

The numbers $1 + i\sqrt{3}$, -2, and $1 - i\sqrt{3}$ are the rectangular forms of the cube roots of -8. These are all the cube roots of -8; any other integral value of k gives a repetition of one of these roots. In particular, if the values 3, 4, and 5 are assigned to k, the roots are repeated in the order just mentioned.

EXAMPLE 3 Find the five fifth roots of $-16 + 16i\sqrt{3}$.

Solution: The polar form of $-16 + 16i\sqrt{3}$ is

$$32[\cos(120^\circ + k \cdot 360^\circ) + i \sin(120^\circ + k \cdot 360^\circ)]$$

From formula (2), we write

$$\sqrt[5]{32}\left(\cos\frac{120^\circ + k \cdot 360^\circ}{5} + i \sin\frac{120^\circ + k \cdot 360^\circ}{5}\right)$$
$$= 2[\cos(24^\circ + k \cdot 72^\circ) + i \sin(24^\circ + k \cdot 72^\circ)]$$

The desired roots are obtainable by assigning k the values 0, 1, 2, 3, 4, in succession, in the last expression. Thus, the five fifth roots of $-16 + 16i\sqrt{3}$ are

$$2(\cos 24^\circ + i \sin 24^\circ)$$
$$2(\cos 96^\circ + i \sin 96^\circ)$$
$$2(\cos 168^\circ + i \sin 168^\circ)$$
$$2(\cos 240^\circ + i \sin 240^\circ)$$
$$2(\cos 312^\circ + i \sin 312^\circ)$$

The points representing these roots are equally spaced about the circle of radius 2 in Fig. 11-7.

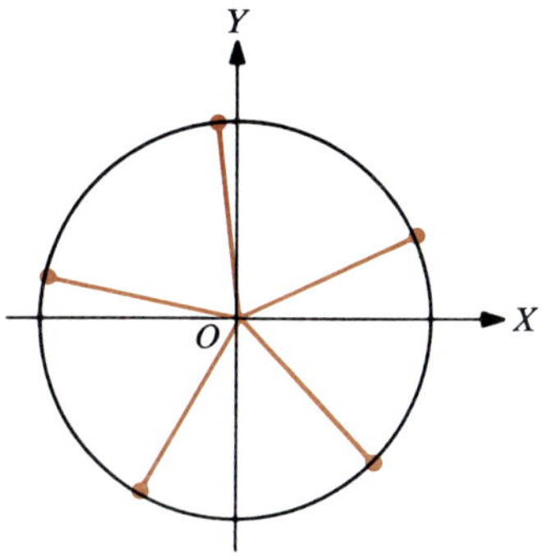

Figure 11-7

EXERCISE 11-4

Find the indicated powers in the following problems. Express each result in trigonometric form.

1. $[2(\cos 35^\circ + i \sin 35^\circ)]^2$ **2.** $[3(\cos 20^\circ + i \sin 20^\circ)]^3$

3. $[4(\cos 140^\circ + i \sin 140^\circ)]^3$ **4.** $[5(\cos 206^\circ + i \sin 206^\circ)]^4$

5. $(\cos 65^\circ + i \sin 65^\circ)^9$ **6.** $(\cos 200^\circ + i \sin 200^\circ)^8$

7. $(1 + i)^4$ **8.** $(1 - i)^6$ **9.** $(-1 + i\sqrt{3})^5$

10. $(\sqrt{3} - i)^6$ **11.** $(-2\sqrt{3} - 2i)^5$ **12.** $(2 + 2i)^7$

Find the indicated roots of the given numbers. Leave each result in trigonometric form.

13. The square roots of $4(\cos 42^\circ + i \sin 42^\circ)$

14. The square roots of $16(\cos 128^\circ + i \sin 128^\circ)$

15. The cube roots of $64(\cos 147^\circ + i \sin 147^\circ)$

16. The fourth roots of $81(\cos 144^\circ + i \sin 144^\circ)$

17. The fourth roots of $16(\cos 84^\circ + i \sin 84^\circ)$

18. The fifth roots of $\cos 200^\circ + i \sin 200^\circ$

19. The cube roots of $1 + i$

20. The cube roots of i

21. The fourth roots of $-8i$

22. The fourth roots of 32

23. The fifth roots of $-i$

24. The fifth roots of $-1 - i$

Find the indicated powers and roots in the following problems, giving each result in algebraic form.

25. $[2(\cos 25^\circ + i \sin 25^\circ)]^6$

26. $[3(\cos 36^\circ + i \sin 36^\circ)]^5$

27. $(\cos 162^\circ + i \sin 162^\circ)^5$

28. $(\cos 125^\circ + i \sin 125^\circ)^6$

29. $(\sqrt{2} + i\sqrt{2})^{10}$

30. $\left(\frac{\sqrt{3}}{2} + \frac{i}{2}\right)^{20}$

31. The cube roots of $-4\sqrt{3} - 4i$

32. The cube roots of $13.5(1 + i\sqrt{3})$

33. The cube roots of $3 - 2i$

34. The fourth roots of $1 + i\sqrt{3}$

Solve the following equations.

35. $x^4 + 1 = 0$

36. $x^6 - 1 = 0$

37. $x^4 - i = 0$

12

QUADRATIC EQUATIONS

12-1 SOLUTION OF QUADRATIC EQUATIONS BY FACTORING

An equation which can be expressed in the form

$$ax^2 + bx + c = 0 \tag{1}$$

where a, b, and c are constants with $a \neq 0$, is called a *quadratic equation in x*.

An important operation on a quadratic equation is that of finding the *solution set*, or *roots*, of the equation. If the first-degree term is missing, the roots can be readily obtained. For example, the roots of $x^2 - 16 = 0$ are $x = \pm 4$, and the roots of $x^2 + 9 = 0$ are $x = \pm\sqrt{-9} = \pm 3i$.

If the left member of an equation of the form (1) can be separated into two linear factors, the roots can be had at once from the factors. The key principle in this method is the fact that the product of two factors is equal to zero if either of the factors is equal to zero (Theorem 2-7, Sec. 2-3). Hence the two linear equations obtained by setting each factor equal to zero yield the roots.

EXAMPLE Solve the equation $8x^2 - 2x - 3 = 0$.

Solution: Factoring the left member gives

$$(2x + 1)(4x - 3) = 0$$

Equating each factor to zero, we have the linear equations

$$2x + 1 = 0 \qquad \text{and} \qquad 4x - 3 = 0$$

whose solutions are, respectively, $x = -\frac{1}{2}$ and $x = \frac{3}{4}$. We check by substituting each of these values, in turn, for x in the given equation. Thus, we find

$$8(-\tfrac{1}{2})^2 - 2(-\tfrac{1}{2}) - 3 = 2 + 1 - 3 = 0$$
$$8(\tfrac{3}{4})^2 - 2(\tfrac{3}{4}) - 3 = \tfrac{9}{2} - \tfrac{3}{2} - 3 = 0$$

Each value satisfies the quadratic equation, and therefore the solution set is $\{-\frac{1}{2}, \frac{3}{4}\}$.

EXERCISE 12-1

Solve the following quadratic equations for x.

1. $x^2 - 5 = 0$
2. $x^2 - a^2 = 0$
3. $x^2 + 16 = 0$
4. $9x^2 + 4 = 0$
5. $a^2x^2 + b^2 = 0$
6. $25x^2 + 1 = 0$

Solve each of the following equations for x by the factoring method. Check your solutions.

7. $x^2 + x - 2 = 0$
8. $x^2 - x - 6 = 0$
9. $x^2 - 9x + 8 = 0$
10. $2x^2 + x = 10$
11. $2x^2 - 3x + 1 = 0$
12. $7x^2 + 11x = 0$
13. $81x^2 - 4 = 0$
14. $25x^2 - 9 = 0$
15. $49x^2 - 16 = 0$
16. $6x^2 - 11x = -3$
17. $2x^2 + 5x - 3 = 0$
18. $6x^2 + 5x + 1 = 0$
19. $18x^2 + 29x + 3 = 0$
20. $8x^2 + 14x + 5 = 0$
21. $12x^2 - 5x - 2 = 0$
22. $x^2 - (a + 1)x + a = 0$
23. $x^2 + ax - bx - ab = 0$
24. $a^2x^2 - abx - 2b^2 = 0$
25. $ax^2 + abx - x - b = 0$

12-2 SOLUTION BY FORMULA

The roots of a quadratic equation are best obtained by the factoring method if the factors can be found readily. In some cases, however, there may be no real linear factors, and in other cases the factors, though real, may not be discovered quickly. So we need a practicable procedure which is applicable to any quadratic equation. If by some means we solve the equation $ax^2 + bx + c = 0$, the roots (in terms of the literal coefficients a, b, and c) can be used for solving other quadratic equations. In this connection, we have the following theorem.

Theorem 12-1 The roots x_1 and x_2 of the quadratic equation

$$ax^2 + bx + c = 0 \qquad a \neq 0$$

are

$$x_1 = \frac{-b + \sqrt{b^2 - 4ac}}{2a} \qquad \text{and} \qquad x_2 = \frac{-b - \sqrt{b^2 - 4ac}}{2a}$$

or, expressed in set form, the solution is

$$\left\{\frac{-b + \sqrt{b^2 - 4ac}}{2a}, \frac{-b - \sqrt{b^2 - 4ac}}{2a}\right\}$$

Proof:

$ax^2 + bx + c = 0$	Given equation
$ax^2 + bx = -c$	Adding $-c$ to both sides
$x^2 + \frac{b}{a}x = -\frac{c}{a}$	Dividing both sides by a
$x^2 + \frac{b}{a}x + \left(\frac{b}{2a}\right)^2 = \left(\frac{b}{2a}\right)^2 - \frac{c}{a}$	Adding $\left(\frac{b}{2a}\right)^2$ to both sides
$\left(x + \frac{b}{2a}\right)^2 = \frac{b^2 - 4ac}{4a^2}$	Simplifying
$x + \frac{b}{2a} = \frac{\pm\sqrt{b^2 - 4ac}}{2a}$	Taking square roots
$x = \frac{-b \pm \sqrt{b^2 - 4ac}}{2a}$	Adding $-\frac{b}{2a}$ to both sides

This last equation, yielding the roots as stated in the theorem, is called the *quadratic formula.*

The process by which we derived the quadratic formula is called *completing the square;* that is, we manipulated the equation to make the left member a perfect square.

EXAMPLE 1 Solve the equation $3x^2 - 5x - 2 = 0$ by completing the square.

Solution:

$3x^2 - 5x - 2 = 0$	Given equation
$3x^2 - 5x = 2$	Adding 2 to both sides
$x^2 - \frac{5}{3}x = \frac{2}{3}$	Dividing both sides by 3
$x^2 - \frac{5}{3}x + \frac{25}{36} = \frac{2}{3} + \frac{25}{36}$	Adding $(\frac{5}{6})^2$ to both sides
$(x - \frac{5}{6})^2 = \frac{49}{36}$	Simplifying
$x - \frac{5}{6} = \pm\frac{7}{6}$	Taking square roots
$x = \frac{5}{6} \pm \frac{7}{6}$	Adding $\frac{5}{6}$ to both sides
$x = 2$ and	$x = -\frac{1}{3}$

Hence, the solution set is $\{2, -\frac{1}{3}\}$.

EXAMPLE 2 Use the quadratic formula to solve the equation

$$9x^2 - 12x + 8 = 0$$

Solution: Comparing the coefficients of this equation with the corresponding coefficients of the equation $ax^2 + bx + c = 0$, we see that $a = 9$, $b = -12$, and $c = 8$. Then substituting these values for a, b, and c in the quadratic formula, we get

$$x = \frac{12 \pm \sqrt{(-12)^2 - 4(9)(8)}}{2(9)} = \frac{12 \pm \sqrt{-144}}{18}$$

$$= \frac{12 \pm 12i}{18} = \frac{2 \pm 2i}{3}$$

The solution set is $\left\{\frac{2 + 2i}{3}, \frac{2 - 2i}{3}\right\}$.

Check: We show that these values of x reduce the left member of the given equation to zero, and therefore are the roots.

$$9\left(\frac{2 \pm 2i}{3}\right)^2 - 12\left(\frac{2 \pm 2i}{3}\right) + 8 = 4 \pm 8i - 4 - (8 \pm 8i) + 8 = 0$$

EXERCISE 12-2

Solve the following equations by completing the square and check each result.

1. $x^2 + 4x + 3 = 0$

2. $x^2 - 6x + 8 = 0$

3. $x^2 + x - 20 = 0$

4. $x^2 - 7x + 12 = 0$

5. $x^2 + x - 30 = 0$

6. $2x^2 - x - 3 = 0$

7. $2x^2 = 5x - 2$

8. $6x^2 + 1 = 5x$

9. $x^2 - 2x + 1 = 0$

10. $x^2 + 4x + 1 = 0$

11. $2x^2 + 4x + 1 = 0$

12. $9x^2 + 11 = 24x$

13. $x^2 + 6x + 13 = 0$

14. $x^2 - 8x + 25 = 0$

15. $9x^2 - 12x + 10 = 0$

16. $2x^2 - 10x + 9 = 0$

Solve by the quadratic formula, and check each result.

17. $x^2 + 6x + 8 = 0$

18. $x^2 - 8x + 15 = 0$

19. $2x^2 - 3x - 2 = 0$

20. $3x^2 + x - 2 = 0$

21. $3x^2 + 2x - 5 = 0$

22. $20x^2 + 9x + 1 = 0$

23. $2x^2 + 5x + 1 = 0$

24. $3x^2 + 10x + 4 = 0$

25. $4x^2 - 9x + 3 = 0$

26. $3x^2 + 11x + 7 = 0$

27. $8x^2 - 10x + 1 = 0$

28. $11x^2 + 20x + 8 = 0$

29. $9x^2 - 10x + 2 = 0$

30. $2x^2 + 14x + 11 = 0$

31. $2x^2 + 2x + 5 = 0$

32. $2x^2 - 10x + 17 = 0$

33. $5x^2 + 12x + 10 = 0$

34. $3x^2 + 8x + 6 = 0$

35. $5x^2 - 16x + 14 = 0$

36. $11x^2 + 12x + 4 = 0$

37. $13x^2 - 16x + 6 = 0$

38. $7x^2 - 9x + 3 = 0$

39. $5x^2 - 7x + 3 = 0$

40. $7x^2 - 13x + 7 = 0$

41. $mx^2 - (1 + m)x + 1 = 0$

42. $(m + n)^2x^2 + (m + n)x - 2 = 0$

12-3 WORD PROBLEMS

There are many kinds of word problems which can be solved by means of quadratic equations. The procedure for setting up the necessary equation in such a problem is much the same as that in Sec. 6-4. Frequently, only one root of the equation has meaning in the problem at hand. A root not satisfying the conditions of the problem must be rejected. It is necessary, therefore, to examine the roots to determine whether one or both may be used.

EXAMPLE 1 A rectangular pasture is to be fenced along four sides and divided into three parts by two fences parallel to two of the sides. Find the dimensions of the pasture if the total amount of fencing is 400 yards and the area is (*a*) 4800 square yards, (*b*) 5000 square yards.

Solution: Referring to Fig. 12-1, where the width is x yards and the length is $200 - 2x$ yards, we have

$$(200 - 2x)x = 4800$$
$$x^2 - 100x + 2400 = 0$$
$$(x - 40)(x - 60) = 0$$
$$x = 40 \text{ or } 60$$

If the width is 40 yards, the length is 120 yards; if the width is 60 yards, the length is 80 yards. Each of these solutions satisfies the conditions of the problem. The student should show that part (*b*) has only one solution.

EXAMPLE 2 A car which travels 20 miles per hour faster than a truck goes 360 miles in 3 hours less time than required by the truck to travel the same distance. Find the rate of each vehicle in miles per hour.

Solution: If x miles per hour is the rate of the truck, then $x + 20$ miles per hour is the rate of the car, $360/x$ hours is the time for the truck, and $360/(x + 20)$ hours is the time for the car. Since

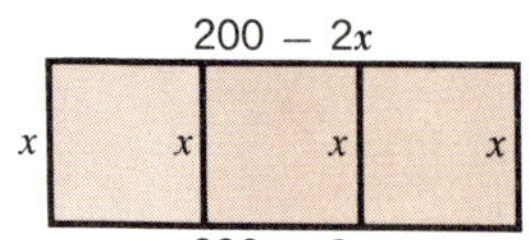

Figure 12-1

the difference of the times is 3 hours, we write the equation

$$\frac{360}{x} - \frac{360}{x + 20} = 3$$

This equation, by clearing of fractions and simplifying, can be reduced to the form

$$(x - 40)(x + 60) = 0$$

whose roots are $x = 40$ and $x = -60$. Rejecting the negative value, we find the truck's rate to be 40 miles per hour and the car's rate to be 60 miles per hour.

EXERCISE 12-3

1. Find two numbers whose difference is 6 and whose product is 55.

2. Two numbers differ by 1; their cubes differ by 127. Find the numbers.

3. The product of two consecutive even integers is 168. Find the integers.

4. The sum of a number and its reciprocal is $\frac{37}{6}$. Find the number.

5. A gardener wishes to enclose a rectangular piece of ground with 200 yards of fencing. A straight riverbank runs along one side, and no fence is needed there. Find the dimensions of the rectangle if (*a*) the area is 4800 square yards, (*b*) 5000 square yards.

6. The four margins of a page of a book are all the same width. If the area of the margins is 38 square inches and the dimensions of the page (printed area plus the margins) are 9 inches by 12 inches, find the width of the margins.

7. A page, with length 3 inches more than its width, has 63 square inches of printed area. The margins at the top and bottom are each $1\frac{1}{2}$ inches wide and the side margins are each 1 inch wide. Find the dimensions of the page.

8. A car travels 15 miles per hour faster than a truck and goes a distance of 360 miles in 2 hours less time than required by

the truck to travel the same distance. Find the speed of each vehicle.

9. A jet plane flying 500 miles per hour faster than a propeller plane travels 700 miles in 2.5 hours less time than is required for the propeller plane to travel the same distance. Find the speeds of the planes.

10. A man can do a job in 9 hours less time than is required by a boy. Working together, they can do the job in 20 hours. Find how long it would take each alone to do the job.

11. A rectangular pasture is to be fenced along four sides and divided into two parts by a fence parallel to one of the sides. Find the dimensions of the pasture if the total amount of fencing is 600 yards and the area of the pasture is (*a*) 11,250 square yards, (*b*) 15,000 square yards.

12. Two boys are to mow a rectangular grass area 90 feet by 120 feet. One of the boys starts on the outside and mows around until he has mowed half the grass area. How wide a strip does he mow from each side?

13. A railroad and highway intersect at right angles. A train is 10 miles from the intersection at the same time a car is 6 miles from the intersection. If each is traveling at the rate of 0.5 mile per minute, at what time will they be 4 miles apart?

Do problems 14 through 18, using the following formulas.

1. *If an object is thrown vertically downward with a velocity v_0, the distance s which the object will fall in t seconds is given by the formula*

$$s = v_0t + \tfrac{1}{2}gt^2$$

where s is in feet, v_0 in feet per second, and g is 32 (approximately).

2. *If the object is projected vertically upward, the formula is*

$$s = v_0t - \tfrac{1}{2}gt^2$$

14. If a ball is dropped ($v_0 = 0$) from a tower, find how long it takes to reach the ground if the height of the tower is (*a*) 256 feet; (*b*) 144 feet; (*c*) 64 feet.

15. From a balloon a marble is thrown downward with a velocity of 64 feet per second. Find (*a*) how far the ball falls in 3 seconds; (*b*) how long it takes to fall 512 feet.

16. A ball is thrown vertically upward with a starting velocity of 80 feet per second. In how many seconds will the ball be (*a*) 100 feet high; (*b*) 96 feet high? Explain the two answers for part (*b*).

17. A bullet is shot vertically upward at a balloon with a velocity of 1100 feet per second. If the balloon is 9400 feet high, find the time required for the bullet to reach the balloon.

18. If a ball is thrown upward with a velocity of 48 feet per second, find when the ball will be 32 feet above the starting point. Explain the two answers.

12-4 EQUATIONS IN QUADRATIC FORM

An equation is often met which is not quadratic but can be reduced to a quadratic by the substitution of a new unknown.

EXAMPLE 1 Solve the equation $y^4 + 3y^2 - 4 = 0$

Solution: We see that this fourth-degree equation when written as

$$(y^2)^2 + 3y^2 - 4 = 0$$

is quadratic in y^2. The substitution $y^2 = x$ gives the quadratic equation.

$$x^2 + 3x - 4 = 0$$

whose roots are $x = 1$ and $x = -4$. Then, replacing x by y^2, we have

$$y^2 = 1 \qquad \text{and} \qquad y^2 = -4$$

$$y = \pm 1 \qquad \text{and} \qquad y = \pm 2i$$

The solution set is therefore $\{-1, 1, -2i, 2i\}$. The student may check each solution.

Alternatively, we could find y^2 directly by using the quadratic formula or by factoring. Thus, by factoring, we find that

$$(y^2 - 1)(y^2 + 4) = 0$$

The four roots are obtainable by equating each factor to zero.

EXAMPLE 2 Solve the equation $x - 2x^{1/2} - 8 = 0$.

Solution: This equation is quadratic in $x^{1/2}$, and applying the quadratic formula, we obtain

$$x^{1/2} = \frac{2 \pm \sqrt{4 + 32}}{2} = \frac{2 \pm 6}{2}$$

$$x^{1/2} = 4 \qquad \text{and} \qquad x^{1/2} = -2$$

We recall that $x^{1/2} = +\sqrt{x}$, not $-\sqrt{x}$, and therefore $x^{1/2} = -2$ has no solution. The equation $x^{1/2} = 4$ yields $x = 16$, which is the only root of the given equation.

EXAMPLE 3 Solve the equation

$$\frac{x^2 - 4}{x} - \frac{4x}{x^2 - 4} - 3 = 0$$

Solution: Substituting $y = (x^2 - 4)/x$, we get

$$y - \frac{4}{y} - 3 = 0$$

$$y^2 - 3y - 4 = 0$$

The roots of the last equation are $y = -1$ and $y = 4$. Hence, we form the equations

$$\frac{x^2 - 4}{x} = 4 \qquad \text{and} \qquad \frac{x^2 - 4}{x} = -1$$

These equations yield the quadratic equations

$$x^2 - 4x - 4 = 0 \qquad \text{and} \qquad x^2 + x - 4 = 0$$

whose solutions are, respectively,

$$x = 2 \pm 2\sqrt{2} \qquad \text{and} \qquad x = \frac{-1 \pm \sqrt{17}}{2}$$

Expressing the solutions in set form, we have

$$\left\{2 + 2\sqrt{2}, 2 - 2\sqrt{2}, \frac{-1 + \sqrt{17}}{2}, \frac{-1 - \sqrt{17}}{2}\right\}$$

EXERCISE 12-4

Solve the following equations.

1. $x^4 - 13x^2 + 36 = 0$
2. $x^4 + x^2 - 2 = 0$
3. $4x^4 - 9x^2 + 2 = 0$
4. $x^4 + 7x^2 + 12 = 0$
5. $3x^4 + 5x^2 - 12 = 0$
6. $2x^4 + 7x^2 - 4 = 0$
7. $x + 2x^{1/2} - 8 = 0$
8. $x + 3x^{1/2} - 10 = 0$
9. $x^{2/3} - 7x^{1/3} + 12 = 0$
10. $x^{4/3} - 10x^{2/3} + 9 = 0$
11. $2x^{-2} + x^{-1} - 1 = 0$
12. $x^{-2} - x^{-1} - 12 = 0$
13. $4x^{-2} - x^{-1} - 5 = 0$
14. $3x^{-2} + 4x^{-1} - 4 = 0$
15. $(x^2 - 5x)^2 + (x^2 - 5x) - 12 = 0$
16. $(2x^2 + x)^2 - 4(2x^2 + x) + 3 = 0$
17. $(x^2 - 3x) - 3(x^2 - 3x)^{1/2} + 2 = 0$
18. $(x^2 - x) - 8(x^2 - x)^{1/2} + 12 = 0$
19. $\left(\frac{2}{x - 1}\right)^2 - 7\left(\frac{2}{x - 1}\right) - 30 = 0$
20. $\left(\frac{x - 1}{x}\right)^2 + \frac{x - 1}{x} - 2 = 0$
21. $5\left(\frac{x + 1}{x - 1}\right)^2 - 3\left(\frac{x + 1}{x - 1}\right) - 2 = 0$
22. $\frac{x + 1}{x - 3} - \left(\frac{x + 1}{x - 3}\right)^{1/2} - 2 = 0$
23. $\frac{2x - 3}{x} + \frac{7x}{2x - 3} - 8 = 0$
24. $4x^2 - 3x + \frac{5}{4x^2 - 3x} + 6 = 0$
25. $\frac{x^2 - 3}{x - 1} - 2\left(\frac{x - 1}{x^2 - 3}\right) + 1 = 0$
26. $\frac{x + 1}{x - 3} + \left(\frac{x + 1}{x - 3}\right)^{1/2} - 2 = 0$

12-5 EQUATIONS CONTAINING RADICALS

An equation in which the unknown appears under a radical sign is called a *radical,* or an *irrational,* equation. We shall consider the problem of solving equations which contain second-order radicals. In this process, we shall use the operation of squaring the members of equations. A new equation thus obtained has all the roots of the original equation. This is true because the squares of two equal quantities are equal. The new equation, however, may have roots other than those of the original equation. Because of this possibility, a check must always be made in this type of problem by substituting all supposed roots in the original equation.

We list the steps for solving second-order radical equations:

1. Arrange the members of the equation so that a single radical constitutes one member.
2. Eliminate this radical by squaring both members of the equation.
3. Repeat steps 1 and 2, if necessary, until an equation free of radicals is obtained.
4. Solve the resulting equation and check all of its roots in the given equation. Reject any extraneous roots.

EXAMPLE 1 Solve the equation $\sqrt{3x-5}+\sqrt{x-2}-3=0$.

Solution:

$\sqrt{3x-5}+\sqrt{x-2}-3=0$	Given equation
$\sqrt{3x-5}=3-\sqrt{x-2}$	Adding $3-\sqrt{x-2}$ to both sides
$3x-5=9-6\sqrt{x-2}+x-2$	Squaring both sides
$6\sqrt{x-2}=12-2x$	Isolating the radical
$3\sqrt{x-2}=6-x$	Dividing both sides by 2
$9(x-2)=36-12x+x^2$	Squaring both sides
$x^2-21x+54=0$	Simplifying
$(x-3)(x-18)=0$	Factoring

The roots of the last equation are $x = 3$ and $x = 18$. We check these values by substituting 3 and 18, in turn, in the given equation.

$$\sqrt{9-5}+\sqrt{3-2}-3=2+1-3=0$$

$$\sqrt{54-5}+\sqrt{18-2}-3=7+4-3\neq 0$$

Although $x = 18$ is a root of the resulting quadratic equation, it is not a root of the given equation. Hence, 3 is the only root of the radical equation, and therefore the solution set is $\{3\}$.

EXAMPLE 2 Solve the equation

$$\sqrt{7x-5}-\sqrt{x+2}-\sqrt{2x-3}=0$$

Solution: This equation has three radicals but the solution process is exactly the same as that of the previous example. All steps should be followed:

$$\sqrt{7x-5}=\sqrt{x+2}+\sqrt{2x-3}$$

$$7x-5=x+2+2\sqrt{(x+2)(2x-3)}+2x-3$$

$$4x-4=2\sqrt{2x^2+x-6}$$

$$2x-2=\sqrt{2x^2+x-6}$$

$$4x^2-8x+4=2x^2+x-6$$

$$2x^2-9x+10=0$$

$$(x-2)(2x-5)=0$$

The roots of the final equation are $x = 2$ and $x = \frac{5}{2}$. These values are also roots of the given equation. We leave the verifications to the student.

EXERCISE 12-5

Solve the following radical equations. Check your results.

1. $\sqrt{x^2-5}=2$

2. $\sqrt{x^2-7}=3$

3. $\sqrt{4x-3}=x$

4. $\sqrt{2x-5}=x-4$

5. $\sqrt{1-x}=x+5$

6. $\sqrt{2+x}=x+2$

7. $\sqrt{3x+1}=-2$

8. $\sqrt{x+1}+\sqrt{2x-3}=0$

9. $\sqrt{2x-4}=\sqrt{x+5}$

10. $\sqrt{3x+7}=\sqrt{19-x}$

11. $\sqrt{5x-1}+\sqrt{3x+3}=0$

12. $\sqrt{2x+2}=\sqrt{3x-5}$

13. $\sqrt{2x + 11} - \sqrt{x + 2} = 2$

14. $1 + \sqrt{x + 9} = \sqrt{2x + 11}$

15. $\sqrt{3x - 5} + \sqrt{x + 2} = 3$

16. $\sqrt{13 - 4x} + \sqrt{5 - 4x} = 4$

17. $\sqrt{6x + 6} = \sqrt{4x + 5} + 1$

18. $\sqrt{x - 4} + 1 = \sqrt{2x - 7}$

19. $\sqrt{3x + 4} + \sqrt{x} = \sqrt{6x + 12}$

20. $\sqrt{6x + 6} = \sqrt{x} + \sqrt{x + 6}$

21. $2\sqrt{2x} - \sqrt{3x + 3} = \sqrt{x - 1}$

22. $3\sqrt{x - 5} + \sqrt{x - 9} = 2\sqrt{x}$

23. $\sqrt{2x - 3} + \sqrt{x + 2} = \sqrt{7x - 5}$

24. $\sqrt{2x - 7} + \sqrt{3x + 1} = \sqrt{7x + 8}$

25. $\sqrt{3x + 3} + \sqrt{2x + 3} = \sqrt{10x + 11}$

26. $\sqrt{7 - 2x} - \sqrt{3 - x} = \sqrt{4 - x}$

27. $1 + \sqrt{x^2 + x - 1} = \sqrt{x^2 + x + 2}$

28. $\sqrt{x^2 - 4x - 1} = \sqrt{x^2 - x - 4}$

12-6 NATURE OF THE ROOTS

The roots of the quadratic equation $ax^2 + bx + c = 0$ are given by the formulas

$$x_1 = \frac{-b + \sqrt{b^2 - 4ac}}{2a} \qquad \text{and} \qquad x_2 = \frac{-b - \sqrt{b^2 - 4ac}}{2a}$$

In deriving these formulas, we placed no restriction on the coefficients a, b, and c except $a \neq 0$. With this single exception, the formulas hold where each coefficient is any real number or any imaginary number. We now point out certain properties of the roots when the coefficients are real numbers and also when they are rational numbers. The properties depend on the radicand $b^2 - 4ac$, which is called the *discriminant:*

1. If $b^2 - 4ac = 0$, the roots are real and equal.
2. If $b^2 - 4ac > 0$, the roots are real and unequal.
3. If $b^2 - 4ac < 0$, the roots are imaginary and unequal.

If we further restrict the coefficients a, b, and c to be rational numbers and the discriminant to be positive, the following statements are true:

4. If $b^2 - 4ac$ is a perfect square, the roots are rational.
5. If $b^2 - 4ac$ is not a perfect square, the roots are irrational.

These conclusions can readily be verified. If the discriminant is zero, the roots are both equal to $-b/2a$. Clearly, the roots are unequal when the discriminant is not zero. Furthermore, the roots are imaginary if and only if the discriminant is negative. If a, b, and c are rational, then $b^2 - 4ac$ is rational and, if positive, statements 4 and 5 follow at once.

EXAMPLE 1 Without solving the equation, determine the nature of the roots of $3x^2 - 6x + 2 = 0$.

Solution: Computing, we find $b^2 - 4ac = 36 - 4(3)(2) = 12$. The discriminant is positive and not a perfect square; hence, the roots are real, irrational, and unequal.

EXAMPLE 2 Compute the value of the discriminant and determine the nature of the roots of the equation $\sqrt{3}x^2 + 2x + \sqrt{2} = 0$.

Solution: $b^2 - 4ac = 2^2 - 4(\sqrt{3})(\sqrt{2}) = 4 - 4\sqrt{6} < 0$. The discriminant is negative, and therefore the roots are imaginary and unequal.

12-7 SUM AND PRODUCT OF THE ROOTS

We can quite easily find the sum and product of the roots of the equation $ax^2 + bx + c = 0$ in terms of the coefficients a, b, and c. Letting x_1 and x_2 denote the roots, we have

$$x_1 = \frac{-b + \sqrt{b^2 - 4ac}}{2a} \qquad \text{and} \qquad x_2 = \frac{-b - \sqrt{b^2 - 4ac}}{2a}$$

and the sum

$$x_1 + x_2 = -\frac{b}{a}$$

For the product, we have either

$$x_1x_2 = \frac{-b + \sqrt{b^2 - 4ac}}{2a} \cdot \frac{-b - \sqrt{b^2 - 4ac}}{2a}$$

$$= \frac{b^2 - (b^2 - 4ac)}{4a^2} = \frac{4ac}{4a^2}$$

or

$$x_1x_2 = \frac{c}{a}$$

These results enable us to find the sum and product of the roots of a quadratic equation, at a glance. The formulas may also be used as a rapid check after solving a quadratic equation. The test consists simply of seeing whether or not the sum and product of the supposed roots satisfy the conditions $-b/a$ and c/a.

We note that the equation $x^2 + bx + c = 0$ has $a = 1$, consequently, the sum of the roots is $-b$, and the product c. This information may be used to form a quadratic equation whose roots are two given numbers.

EXAMPLE 1 Form a quadratic equation whose roots are $(3 + \sqrt{2})/5$ and $(3 - \sqrt{2})/5$.

Solution: For the sum and product of the roots, we get

$$\frac{3 + \sqrt{2}}{5} + \frac{3 - \sqrt{2}}{5} = \frac{6}{5}$$

$$\frac{3 + \sqrt{2}}{5} \cdot \frac{3 - \sqrt{2}}{5} = \frac{9 - 2}{25} = \frac{7}{25}$$

From these results, we write the equation as either

$$x^2 - \tfrac{6}{5}x + \tfrac{7}{25} = 0 \qquad \text{or} \qquad 25x^2 - 30x + 7 = 0$$

A quadratic equation with two given roots may also be formed by reversing the operation of solving an equation by the factoring method.

EXAMPLE 2 Write a quadratic equation whose roots are $-\frac{2}{3}$ and 4.

Solution: We obtain the required equation by setting the product of $x + \frac{2}{3}$ and $x - 4$ equal to zero. To avoid fractions, however, we may replace $x + \frac{2}{3}$ by $3x + 2$. Thus, we have either

$$(3x + 2)(x - 4) = 0 \qquad \text{or} \qquad 3x^2 - 10x - 8 = 0$$

It is immediately clear that, here, the first equation and, therefore, the second equation both have $-\frac{2}{3}$ and 4 as roots.

EXERCISE 12-6

Without solving, determine the nature of the roots of each equation. Find also the sum and product.

1. $x^2 - 6x + 5 = 0$

2. $x^2 - 2x + 6 = 0$

3. $2x^2 + x - 7 = 0$

4. $2x^2 - 5x + 5 = 0$

5. $4x^2 + 12x + 9 = 0$

6. $3x^2 - 7x - 1 = 0$

7. $5x^2 + 11 = 0$

8. $10x^2 + 3x = 0$

9. $7x^2 - 4 = 0$

10. $\sqrt{2}x^2 - 2x - 1 = 0$

11. $\sqrt{3}x^2 + 4x + \sqrt{2} = 0$

12. $x^2 - \sqrt{5}x - \sqrt{2} = 0$

13. $\sqrt{2}x^2 + \sqrt{5}x - 3\sqrt{2} = 0$

Find a quadratic equation with integral coefficients whose roots are the given numbers in each problem.

14. $1, \quad -2$

15. $-4, \quad -3$

16. $6, \quad 2$

17. $\frac{4}{3}, \quad -\frac{2}{3}$

18. $-\frac{5}{2}, \quad -\frac{2}{5}$

19. $-\frac{4}{5}, \quad \frac{3}{2}$

20. $\sqrt{3}, \quad -\sqrt{3}$

21. $5i, \quad -5i$

22. $3 - 2i, \quad 3 + 2i$

23. $6 \pm \sqrt{6}$

24. $\sqrt{5} \pm \sqrt{2}$

25. $\sqrt{7} \pm i\sqrt{3}$

Determine the value of k in each problem so that the given condition is satisfied.

26. $2x^2 - 3x + k + 6 = 0$, one root is 0

27. $4x^2 + 5x + k^2 - 4 = 0$, one root is 0

28. $5x^2 - (k + 2)x + 7k - 6 = 0$, one root is 2

29. $3kx^2 + 5x - 4k - 6 = 0$, one root is -1

30. $kx^2 + (3k - 4)x - 6 = 0$, one root is 3

31. $kx^2 - 6x + 32 = 0$, sum of roots is -12

32. $4x^2 - kx + 7 = 0$, sum of roots is 5

33. $3x^2 - 41x + k = 0$, product of roots is -6

34. $kx^2 + 11x - 3 = 0$, product of roots is $\frac{1}{2}$

35. $2x^2 - kx + k = 0$, the roots are equal

36. $kx^2 + kx + 5 = 0$, the roots are equal

37. $3x^2 + (k - 4)x - 7 = 0$, one root is the negative of the other

38. $x^2 + (k^2 - 3k)x + 1 = 0$, one root is the negative of the other

39. $x^2 - 8x + k = 0$, the difference of the roots is 6

12-8 THE GRAPH OF A QUADRATIC FUNCTION

The function defined by the equation

$$y = ax^2 + bx + c \qquad a \neq 0$$

is called a *quadratic function*. The graph of the function is called a *parabola*. (We assume that a, b, and c are real numbers.) If the graph has a point on the x axis, the abscissa of the point is the value of x which makes y equal to zero. This value of x is a zero of the function (Sec. 5-3), and is a root of the equation $y = ax^2 + bx + c = 0$. Hence, the real roots, if any, of a quadratic equation may be obtained (at least approximately) from a graph. In finding the zeros of a quadratic function, we shall make use of a theorem which is proved in calculus.

Theorem 12-2 If $f(x)$ denotes a polynomial with real coefficients and if a and b are two real numbers such that $f(a)$ and $f(b)$ have opposite signs, then the equation $f(x) = 0$ has at least one real root between a and b.

Interpreted geometrically, this theorem means that the graph of $y = f(x)$ crosses the x axis at least once between any two of its points which are on opposite sides of the x axis.

EXAMPLE Construct the graph of the function defined by the equation.

$$y = 2 - 2x - x^2$$

Solution: Proceeding as in Sec. 5-3, we prepare a table of corresponding values of x and y, and then draw a smooth curve through the points thus determined (Fig. 12-2).

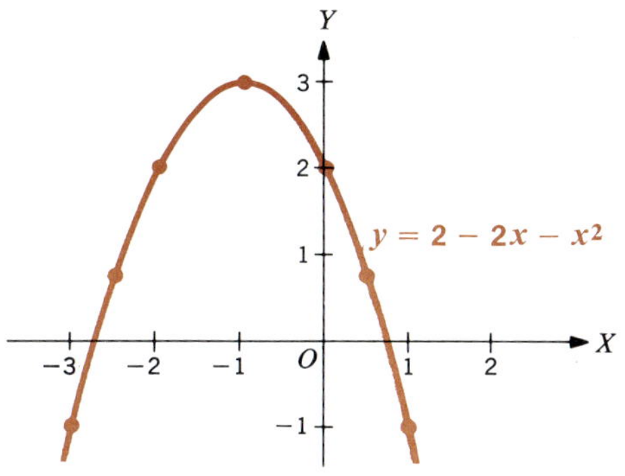

Figure 12-2

x	-3	-2.5	-2	-1	0	0.5	1
y	-1	0.75	2	3	2	0.75	-1

The graph is a parabola which opens downward and crosses the x axis at two points. We estimate the abscissas of these points to be $x = -2.7$ and $x = 0.7$. These values are approximations to the zeros of the function and the roots of the equation $2 - 2x - x^2 = 0$.

We may observe that the graph of $y = ax^2 + bx + c$ does not touch the x axis when $b^2 - 4ac < 0$. The negative discriminant means the function has no real zero and the corresponding quadratic equation has no real root. On the other hand, the graph crosses the x axis at two points if $b^2 - 4ac > 0$, and touches the x axis at one point only if $b^2 - 4ac = 0$. Hence, the discriminant reveals if the graph of a quadratic function crosses the x axis at two points, touches the x axis at one point only, or has no point on the x axis. These three possibilities are illustrated by the graphs (Fig. 12-3) of the functions defined by $y = x^2 - 2x - 1$, $y = x^2 -$

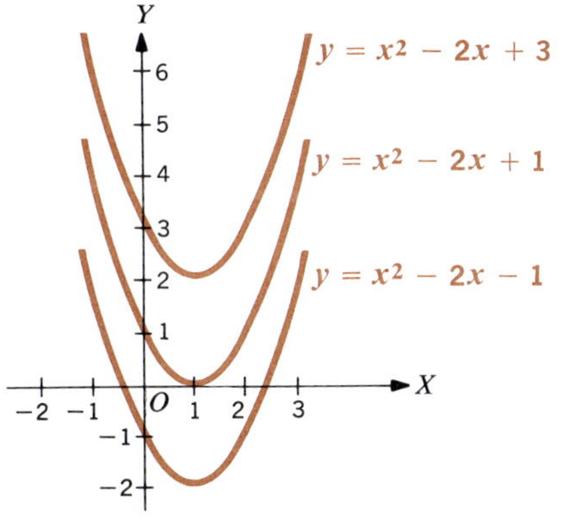

Figure 12-3

$2x + 1$, and $y = x^2 - 2x + 3$, whose discriminants are positive, zero, and negative, respectively.

12-9 MAXIMUM AND MINIMUM VALUES

The parabola in Fig. 12-2 opens downward and the parabolas in Fig. 12-3 open upward. The direction in which the graph of the equation $y = ax^2 + bx + c$ opens depends on the sign of the coefficient of x^2, as we shall prove.

Theorem 12-3 The graph of the function defined by the equation

$$y = ax^2 + bx + c$$

opens upward from a lowest point if a is positive and opens downward from a highest point if a is negative.

Proof: By steps easy to follow, we have

$$\begin{aligned} y &= ax^2 + bx + c \\ &= a\left(x^2 + \frac{b}{a}x + \frac{c}{a}\right) \\ &= a\left(x^2 + \frac{b}{a}x + \frac{b^2}{4a^2} + \frac{c}{a} - \frac{b^2}{4a^2}\right) \\ &= a\left(x + \frac{b}{2a}\right)^2 + \frac{4ac - b^2}{4a} \end{aligned}$$

The quantity $(x + b/2a)^2$ is equal to zero when $x = -b/2a$ and is positive for all other values of x. Consequently, this value of x yields the least value of y when a is positive and the greatest value when a is negative. Hence, the point with coordinates

$$\left(-\frac{b}{2a}, \frac{4ac - b^2}{4a}\right)$$

is the lowest point or *minimum point* of the graph if $a > 0$ and the highest point or *maximum point* if $a < 0$. This point is called the *vertex* of the parabola. The ordinate of the vertex is the minimum value or maximum value of the function depending on the direction in which the graph opens.

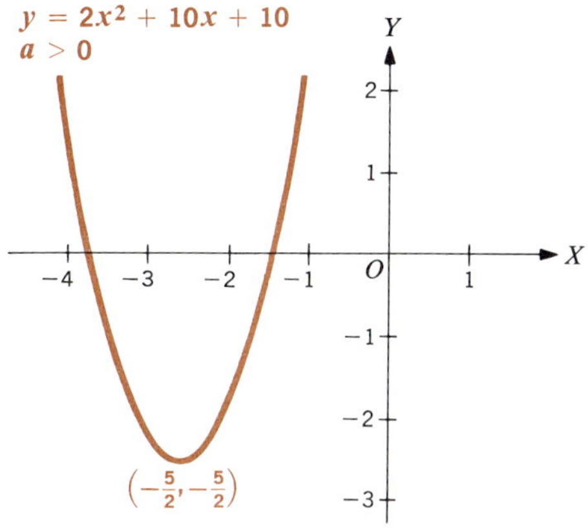

Figure 12-4

EXAMPLE 1 Find the coordinates of the minimum point of the graph of the equation $y = 2x^2 + 10x + 10$.

Solution: We change the form of the right member by adding and subtracting $\frac{25}{2}$. Thus,

$$\begin{aligned} y &= 2(x^2 + 5x) + 10 \\ &= 2(x^2 + 5x + \tfrac{25}{4}) + 10 - \tfrac{25}{2} \\ &= 2(x + \tfrac{5}{2})^2 - \tfrac{5}{2} \end{aligned}$$

Clearly, y takes its minimum value $-\frac{5}{2}$ when $x = -\frac{5}{2}$; any other value of x would require the addition of a positive number to $-\frac{5}{2}$. Hence, the coordinates of the minimum point of the graph (Fig. 12-4) are $(-\frac{5}{2}, -\frac{5}{2})$.

EXAMPLE 2 Find the maximum value of the function defined by the equation $y = -3x^2 + 12x + 5$.

Solution: By subtracting and adding 12 in the right member, we have

$$\begin{aligned} y &= -3(x^2 - 4x + 4) + 5 + 12 \\ &= -3(x - 2)^2 + 17 \end{aligned}$$

Since the expression $-3(x - 2)^2$ is never positive, the function has its greatest value when the expression is equal to zero. Hence, we set $x = 2$ and obtain $y = 17$, which is the maximum value of the function. The coordinates of the maximum point of the graph (Fig. 12-5) are (2, 17).

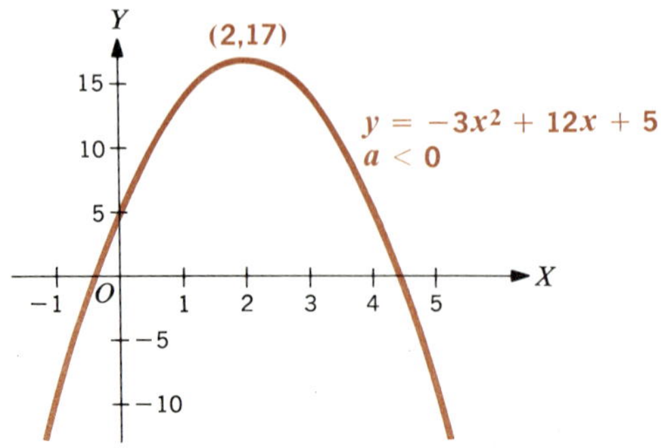

Figure 12-5

EXAMPLE 3 Find the area of the largest rectangular plot which can be fenced on three of its sides by 80 yards of fencing.

Solution: If x yards is the width of the rectangle, then $80 - 2x$ yards is the length. Letting y square yards stand for the area, we have

$$\begin{aligned} y &= 80x - 2x^2 \\ &= -2(x^2 - 40x + 400) + 800 \\ &= -2(x - 20)^2 + 800 \end{aligned}$$

The maximum area, obtained when $x = 20$, is 800 square yards. The graph (Fig. 12-6) clearly shows that the domain and range of the function arising in this problem are, respectively,

$$\{x|0 < x < 40\} \qquad \text{and} \qquad \{y|0 < y \leq 800\}$$

To construct the graph of the equation of a parabola, it is usually best to locate the vertex, as we have done in the preceding examples. Then two or three additional points are sufficient for plotting a reasonably accurate graph.

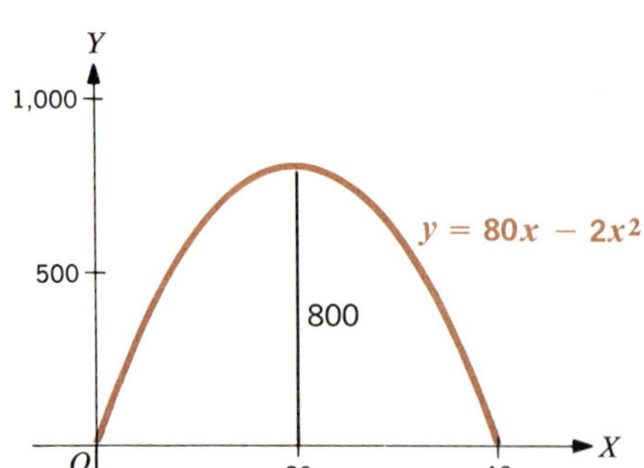

Figure 12-6

EXERCISE 12-7

Find the coordinates of the maximum or minimum point of the graph of each of the following equations. From the graph give (or estimate) the real zeros, if any, of the function determined by the equation.

1. $y = x^2 + 4x + 5$

2. $y = x^2 - 8x + 9$

3. $y = 3 + 8x - x^2$

4. $y = 2 - 6x - x^2$

5. $y = 2x^2 - 3x + 1$

6. $y = 4x^2 + 4x + 3$

7. $y = -2x^2 + 2x - 3$

8. $y = 10 - 6x - x^2$

9. $y = -6x^2 + 4x + 2$

10. $y = 3x^2 - 3x - 5$

11. What are the dimensions of the largest rectangular field which can be enclosed with 240 rods of fencing?

12. Separate 28 into two parts so that the product of the parts is a maximum.

13. A rectangular pasture is to be fenced along four sides and divided into three equal parts by two fences parallel to two of the sides. Find the greatest possible area of the pasture if the total length of the fencing is 400 rods.

14. Find two numbers whose sum is 50 and the sum of their squares is a minimum.

15. If 40 children attend a skating party, the price of each ticket will be \$1. If the number of children exceeds 40, the price of a ticket for each child will be reduced by 2 cents for each child in excess of 40. Under this plan, what attendance will yield the greatest receipts?

12-10 THE GRAPH OF A QUADRATIC EQUATION IN x AND y

An equation of the form

$$Ax^2 + Bxy + Cy^2 + Dx + Ey + F = 0 \qquad (1)$$

where A, B, C, D, E, and F are constants with A, B, and C not all zero, is called a *quadratic equation* in the variables x and y. The graph of an equation of this kind is called a *conic section,* so-named because an identical curve can be obtained by the intersection of a plane and a right circular cone. To construct the graph of an equation of the form (1) in which all terms are present is a tedious task. We shall, however, consider certain special cases of the equation whose graphs can be drawn with little difficulty.

In Sec. 12-8, we constructed graphs, called *parabolas*, of equations of the form $y = ax^2 + bx + c$. An equation of this form may be obtained from Eq. (1) by assigning to B and C the value zero and solving for y. Similarly, the equation

$$x = ay^2 + by + c$$

is a special case of Eq. (1). The graph of an equation of this form is a parabola which opens from a leftmost point (vertex) to the right if $a > 0$, and from a rightmost point to the left if $a < 0$.

We shall next consider other special quadratic equations in x and y whose graphs, like the parabola, are conic sections with distinguishing names. We list here the steps involved in the graphing process:

1. Solve the equation for y in terms of x.
2. Assign values to x and compute the corresponding values of y.
3. Plot the points thus determined and draw a smooth curve through them.

If it is easier to solve for x in terms of y, interchange x and y in these directions.

Definition 12-1 The graph of an equation of the form

$$ax^2 + by^2 = c$$

with $a, b, c > 0$, and $a \neq b$ is called an *ellipse.*

We require that $a \neq b$ in this definition because the condition $a = b$ would yield an equation whose graph is a circle (Sec. 5-4).

EXAMPLE 1 Draw the graph of the equation $9x^2 + 16y^2 = 144$.

Solution: Solving for y in terms of x and for x in terms of y, we get

$$y = \pm\tfrac{3}{4}\sqrt{16 - x^2} \qquad \text{and} \qquad x = \pm\tfrac{4}{3}\sqrt{9 - y^2}$$

From the first equation, we see that the radicand is negative when $x^2 > 16$. This means, for the purpose of graphing, that x may take only the values from -4 to 4, inclusive, and the second equation

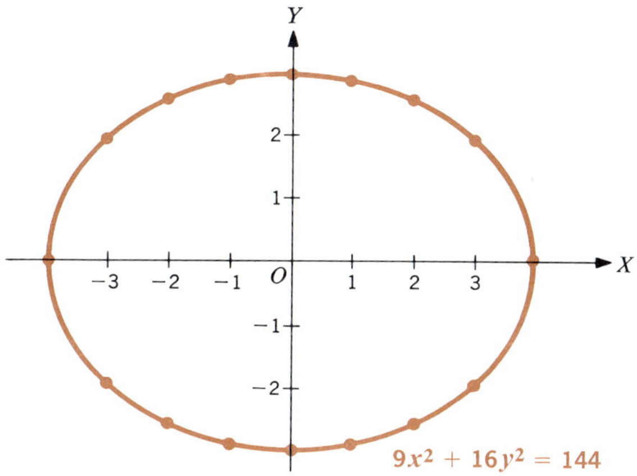

Figure 12-7

reveals that y is restricted to the values -3 to 3, inclusive. Selecting the first equation, we prepare a table of corresponding values of x and y. The graph drawn through the points thus determined is shown in Fig. 12-7.

x	-4	-3	-2	-1	0	1	2	3	4
y	0	± 2.0	± 2.6	± 2.9	± 3	± 2.9	± 2.6	± 2.0	0

Definition 12-2 The graph of an equation of the form

$$ax^2 - by^2 = c \qquad \text{or} \qquad ay^2 - bx^2 = c$$

where a, b, and c are positive constants is called a *hyperbola.*

FXAMPLE 2 Construct the graph of the equation $25x^2 - 36y^2 = 100$.

Solution: We observe that y is imaginary for values of x between -2 and 2 but is real for all other values of x. Hence, the graph consists of two *branches,* one starting at (2, 0) and extending to the right and the other extending to the left from $(-2, 0)$. The graph, drawn by the use of the pairs of values in the table, is in Fig. 12-8.

x	-7	-5	-3	-2	2	3	5	7
y	± 5.6	± 3.8	± 1.9	0	0	± 1.9	± 3.8	± 5.6

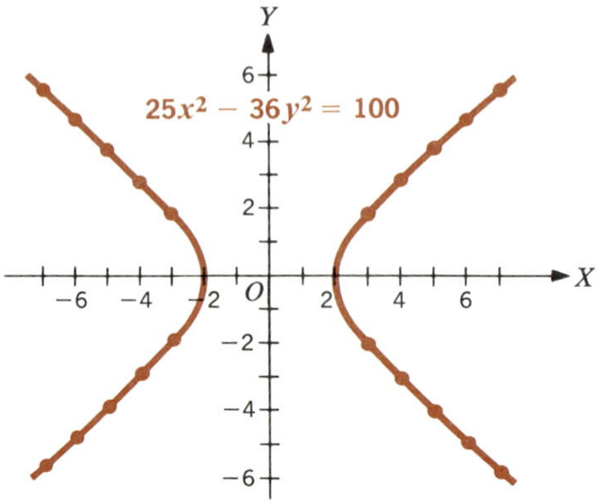

Figure 12-8

The graph of $ay^2 - bx^2 = c$, where a, b, and c are positive constants, has one branch opening upward and the other branch opening downward. This is illustrated (Sec. 12-11) in Figs. 12-11 and 12-12.

EXAMPLE 3 Draw the graph of the equation $xy = 3$.

Solution: In this equation x may have any value except zero. For $x = 0$, there is no value for y which makes the product xy equal to zero. Likewise, y may be given any value except zero. The graph is a hyperbola with one branch in the first quadrant and the other in the third quadrant, as drawn in Fig. 12-9.

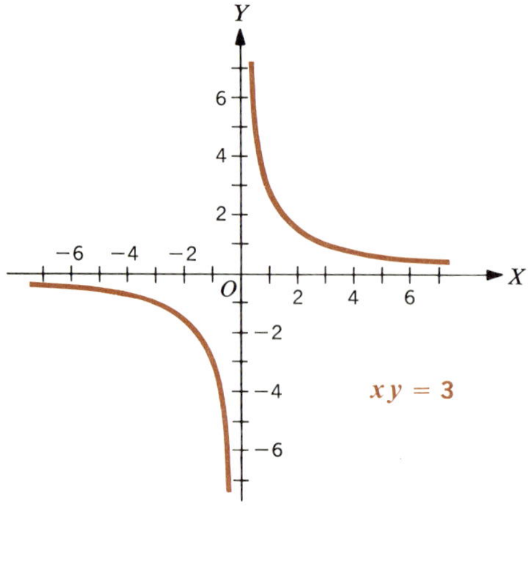

Figure 12-9

12-11 SOLUTION OF A QUADRATIC SYSTEM BY GRAPHING

In Sec. 6-5 we solved systems of linear equations by graphing each equation of a system and reading the coordinates of the point of intersection. The same procedure may be applied to second-degree equations. However, the graphs are not restricted to straight lines, and there may be more than one point of intersection. Only real solutions are obtained by this process and these, as a rule, are approximations.

EXAMPLE 1 Graphically, solve the system of equations

$$3x - 2y + 4 = 0$$
$$y^2 - 4x - 12 = 0$$

Solution: The graphs of these equations, one a straight line and the other a parabola, are shown in Fig. 12-10. There are two intersection points, whose coordinates we estimate to be $(-2.4, -1.6)$ and $(1.5, 4.2)$. These value pairs are good approximations of the solutions which may be verified by substituting in the given equations.

EXAMPLE 2 Graphically, solve the system of equations

$$4y^2 + 9x^2 = 36$$
$$9y^2 - 4x^2 = 36$$

Solution: The graph of the first equation is an ellipse, and the graph of the second equation is a hyperbola. We estimate the coordinates of the intersection points (Fig. 12-11) to be

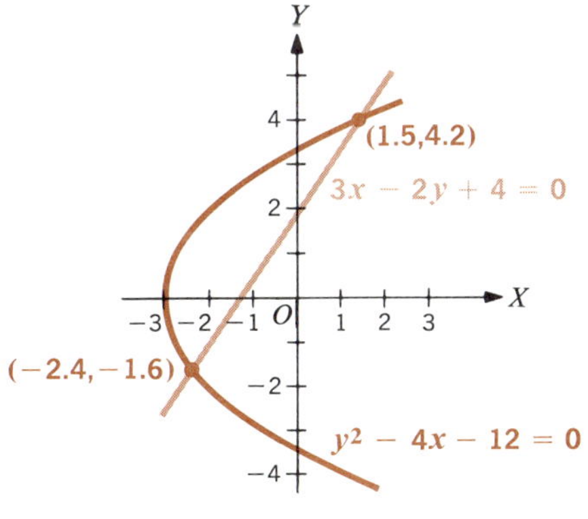

Figure 12-10

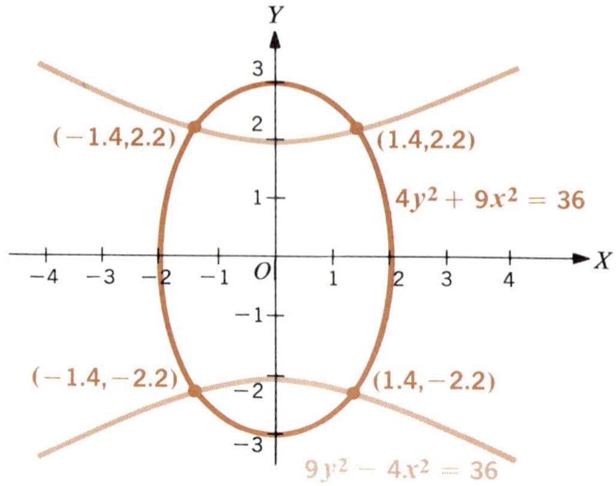

Figure 12-11

$(-1.4, -2.2)$, $(-1.4, 2.2)$, $(1.4, -2.2)$, and $(1.4, 2.2)$. Each of these number pairs is a close approximation to a solution of the system.

EXERCISE 12-8

Solve the following systems of equations graphically. Estimate the results to one decimal place. (See Table 1 in the Appendix to find square roots of numbers.)

1. $y^2 - 3x = 0$, $x - y = 1$
2. $x^2 - 4y = 0$, $x - y = 0$
3. $x^2 + y^2 = 25$, $y - x = 1$
4. $xy - 12 = 0$, $x - 2y = 2$
5. $x^2 + y^2 = 8$, $x^2 + 4y^2 = 16$
6. $2x^2 + y^2 = 9$, $x - y = 1$
7. $x^2 - y^2 = 8$, $x^2 + 2y^2 = 14$
8. $x^2 + y^2 = 10$, $xy - 4 = 0$
9. $4y^2 - x^2 = 16$, $2xy - 15 = 0$
10. $2x^2 + y^2 = 18$, $x^2 + 2y^2 = 33$
11. $x^2 + y^2 = 16$, $x^2 = 2y + 8$
12. $2x^2 + 3y^2 = 20$, $y = x^2 - 2$

12-12 SOLUTION BY ALGEBRAIC METHODS

We consider next the problem of solving a system of quadratic equations by algebraic methods.

Any two independent quadratic equations in two variables can be solved by algebraic methods. An algebraic method yields the exact values of all solutions, including both real and imaginary solutions. However, a general algebraic method leads to a fourth-degree equation in one of the variables. Although an equation of this kind can be solved to obtain its roots exactly, the process is not simple. Consequently, we shall consider certain special quadratic systems which can be solved by processes which we have already discussed.

If both equations of a system are of the form

$$Ax^2 + Cy^2 + F = 0$$

we may use one of the methods for solving two linear equations in two variables (Sec. 6-6). If we treat the equations as linear in x^2 and y^2, values for the squares of the variables can be found. Then the variables x and y are obtained by taking square roots.

EXAMPLE 1 Solve the system of equations

$$4y^2 + 3x^2 = 24$$
$$3y^2 - 2x^2 = 35$$

Solution: We multiply the members of the first equation by 2 and those of the second equation by 3 to get

$$\begin{aligned} 8y^2 + 6x^2 &= 48 \\ 9y^2 - 6x^2 &= 105 \\ \hline 17y^2 &= 153 \\ y &= \pm 3 \end{aligned}$$

Then, substituting for y in the first of the given equations, we have

$$36 + 3x^2 = 24$$
$$x = \pm 2i$$

Pairing the values for x and y, we have the solution set

$$\{(-2i, 3), (2i, 3), (-2i, -3), (2i, -3)\}$$

These solutions are not real; and, consequently, the graphs of the given equations (Fig. 12-12) do not intersect. Each value pair, however, satisfies the given equations.

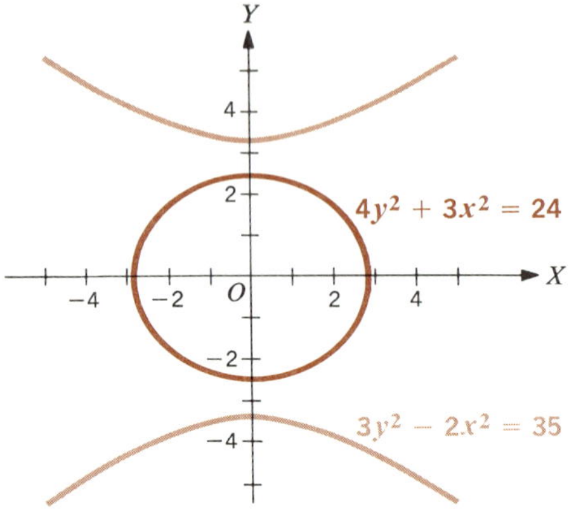

Figure 12-12

If a system has both a linear and a quadratic equation, the solutions can readily be obtained. The process consists in solving the linear equation for one of its unknowns in terms of the other and then substituting in the quadratic equation. This step yields a quadratic equation in one unknown. We illustrate the procedure.

EXAMPLE 2 Solve the system of equations

$$3x^2 - 2xy + 4y^2 = 3$$
$$3x - 2y + 2 = 0$$

Solution: Solving the linear equation for y and substituting the result in the quadratic equation, we find

$$3x^2 - 2x\left(\frac{3x + 2}{2}\right) + 4\left(\frac{3x + 2}{2}\right)^2 = 3$$
$$3x^2 - 3x^2 - 2x + 9x^2 + 12x + 4 = 3$$
$$9x^2 + 10x + 1 = 0$$

The solutions of this equation are $x = -1$ and $x = -\frac{1}{9}$. Substituting each of these values, in turn, in the given linear equation, we obtain $y = -\frac{1}{2}$ and $y = \frac{5}{6}$. Pairing the corresponding values of x and y gives the solution set

$$\{(-1, -\tfrac{1}{2}), (-\tfrac{1}{9}, \tfrac{5}{6})\}$$

If both equations of a system are of the form

$$Ax^2 + Bxy + Cy^2 + F = 0$$

the system may be reduced to two (sometimes only one) systems of equations, each consisting of a linear equation and a quadratic equation. Each of the resulting systems can then be handled as in the example above. The linear equations, as we shall illustrate, come from a new quadratic equation free of a constant term.

EXAMPLE 3 Solve the system of equations

$$\begin{aligned} 2x^2 + 4xy - 3y^2 &= 3 \\ x^2 + xy - 2y^2 &= 4 \end{aligned}$$

Solution: We multiply the members of the first equation by 4, and those of the second equation by 3, and have the equivalent system

$$\begin{aligned} 8x^2 + 16xy - 12y^2 &= 12 \\ 3x^2 + 3xy - 6y^2 &= 12 \end{aligned}$$

Subtracting the members of the second of these equations from the corresponding members of the first equation, the equation we obtain is either

$$5x^2 + 13xy - 6y^2 = 0 \qquad \text{or} \qquad (5x - 2y)(x + 3y) = 0$$

From the last equation, we get the linear equations $5x - 2y = 0$ and $x + 3y = 0$, each of which may be paired with either of the given equations. Thus, pairing with the second equation, we form the systems

$$\begin{cases} x^2 + xy - 2y^2 = 4 \\ \qquad 5x - 2y = 0 \end{cases} \qquad \text{and} \qquad \begin{cases} x^2 + xy - 2y^2 = 4 \\ \qquad x + 3y = 0 \end{cases}$$

The solution sets of these systems are, respectively,

$$\left\{\left(\frac{2i}{3}, \frac{5i}{3}\right), \left(-\frac{2i}{3}, -\frac{5i}{3}\right)\right\} \qquad \text{and} \qquad \{(3, -1), (-3, 1)\}$$

The union of these solution sets

$$\left\{\left(\frac{2i}{3} \frac{5i}{3}\right), \left(-\frac{2i}{3}, -\frac{5i}{3}\right), (3, -1), (-3, 1)\right\}$$

yields the solution set of the given system.

Each of the given equations of this example has the constant F different from zero. If $F = 0$ in either equation of a system of this kind, the left member of that equation can be factored to get two linear equations, provided the factors are distinct. Then the linear equations (or equation) can be paired with the unused equation of the given system.

EXERCISE 12-9

Solve each of the following systems of equations.

1. $3x^2 - 2y^2 = 58$
$x^2 - y^2 = 11$

2. $x^2 - y^2 = 7$
$x^2 + y^2 = 25$

3. $y^2 + 7x^2 = 46$
$4y^2 - 3x^2 = 29$

4. $3x^2 - 2y^2 = 15$
$2x^2 + 3y^2 = 23$

5. $5x^2 - 9y^2 = 31$
$6x^2 - 11y^2 = 38$

6. $x^2 + y^2 = 5$
$5x^2 + 2y^2 = 37$

7. $4y^2 - 5x^2 = 35$
$5y^2 - 6x^2 = 43$

8. $5x^2 - 4y^2 = 13$
$8x^2 + 3y^2 = 2$

9. $3x^2 + 4y^2 = 16$
$4x^2 + 5y^2 = 21$

10. $y^2 + 2x = 7$
$x - 2y = -7$

11. $x^2 - y = 5$
$2x + y = 3$

12. $x^2 + y^2 = 1$
$2x - y = 4$

13. $3x^2 + 2y^2 = 25$
$x + y = 5$

14. $xy + x - y = 13$
$4x - 3y - 7 = 0$

15. $x^2 + y^2 = 20$
$3x - y = 2$

16. $2x^2 + xy + y^2 = 2$
$x^2 - 2xy - 2y^2 = 1$

17. $x^2 - y^2 = 3$
$2x^2 - xy = 6$

18. $x^2 - 2xy - y^2 = 2$
$3x^2 - 5xy - 4y^2 = 8$

19. $4x^2 + 4xy = -16$
$5x^2 + 2xy + y^2 = 16$

20. $4x^2 - 6xy + 3y^2 = 4$
$2x^2 - 3xy + 2y^2 = 4$

21. $2x^2 - 4xy - 3y^2 = 3$
$x^2 - 2xy - 2y^2 = 2$

22. $x^2 - 4xy + 4y^2 = 0$
$3x^2 - 3xy + y^2 = 7$

23. $x^2 - 3xy + 2y^2 = 15$
$2xy + y^2 = 0$

24. $3x^2 - 3xy - 5y^2 = 9$
$4x^2 - 5xy - 6y^2 = 0$

25. $3x^2 + 8xy - 3y^2 = 0$
$5x^2 + 3xy - 2y^2 = 34$

26. $x^2 + 2xy - 5y^2 = -1$
$6x^2 - 13xy + 6y^2 = 0$

27. $2x^2 + 7xy + 6y^2 = 0$
$2x^2 + xy + y^2 = 16$

28. Show that the substitutions $x = u + v$ and $y = u - v$ reduce the system

$$x^2 + B_1xy + y^2 + D_1x + D_1y + F_1 = 0$$
$$x^2 + B_2xy + y^2 + D_2x + D_2y + F_2 = 0$$

to

$$(2 - B_1)v^2 + (2 + B_1)u^2 + 2D_1u + F_1 = 0$$
$$(2 - B_2)v^2 + (2 + B_2)u^2 + 2D_2u + F_2 = 0$$

Observe now that the new system can be solved by first eliminating v^2 to obtain a quadratic equation in u. And then x and y can easily be found, thus giving the solutions of the original system.

Use the information of problem 28 and solve the following systems.

29. $x^2 + y^2 - 10 = 0$
$xy - x - y - 7 = 0$

30. $x^2 + y^2 + x + y - 4 = 0$
$2xy + 3x + 3y + 4 = 0$

31. $x^2 + y^2 + 10x + 10y = -40$
$xy + 5x + 5y + 22 = 0$

32. $x^2 + y^2 + xy - 3x - 3y = 18$
$x^2 + y^2 - 2x - 2y = 15$

13

INEQUALITIES

13-1 THE ORDER AXIOMS

In Chap. 2, we introduced a set of undefined elements R, called *real numbers,* which possess the properties embodied in the six field axioms (Sec. 2-2). The real numbers, satisfying the field axioms, are said to form a field. We now ascribe additional properties on the set R by requiring the set to satisfy the so-called *order axioms.* The order symbols are made and described as follows:

$<$ means "is less than"

$>$ means "is greater than"

Axiom 7 *The trichotomy axiom:* If a and $b \in R$, then one and only one of the following relations holds:

$$a < b \qquad a = b \qquad a > b$$

Axiom 8 *The transitive axiom:* If a, b, and $c \in R$ such that $a > b$ and $b > c$, then $a > c$.

Axiom 9 *The addition axiom:* If a, b, and $c \in R$ such that $a > b$, then $a + c > b + c$.

Axiom 10 *The multiplicative axiom:* If a, b, and $c \in R$ such that $a > b$ and $c > 0$, then $ac > bc$.

Definition 13-1 A field which satisfies the order axioms is called an *ordered* field.

Definition 13-2 If a and $b \in R$, then $a < b$ if and only if $b > a$.

Definition 13-3 A real number a is positive if $a > 0$ and negative if $a < 0$.

Definition 13-4 A statement that one quantity is greater than or less than another quantity is called an *inequality*.

If we set $b = 0$ in Axiom 7, it follows that $a < 0$, $a = 0$, or $a > 0$. From this fact and Definition 13-3, we conclude that 0 is neither positive nor negative.

Sometimes it is convenient to combine an inequality with an equality. Thus,

$a \leq b$ means "a is less than or equal to b"
$a \geq b$ means "a is greater than or equal to b"

For example, $a \leq 0$ means a is not positive, and $a \geq 0$ means a is not negative. Clearly, if the two statements hold simultaneously, then $a = 0$.

There is a great variety of useful theorems involving inequalities. We shall prove certain theorems to be used later and state other theorems which the student may prove. Where no reason is given for a step in the proof of a theorem, the student should give the reason.

Theorem 13-1 If $a > 0$ and $b > 0$, then $a + b > 0$.

Proof

$a > 0$	Given
$a + b > 0 + b = b$	Axiom 9
$b > 0$	Given
$a + b > 0$	Axiom 8

Theorem 13-2 If $a > 0$ and $b > 0$, then $ab > 0$.

These theorems tell us that the sum, or the product, of two positive numbers is positive. In Secs. 3-2 and 3-3, we assumed these closure properties for addition and multiplication of positive numbers. Then we proceeded to prove that the product (or quotient) of two numbers of like signs is positive and the product (or quotient) of two numbers of unlike signs is negative.

Theorem 13-3 If a and $b \in R$, then $a > b$ if and only if $-a < -b$.

Proof: We first prove that if $a > b$, then $-a < -b$.

$$a > b \qquad \text{Given}$$

$$a + [(-a) + (-b)] > b + [(-a) + (-b)] \qquad \text{Axiom 9}$$

$$-b > -a$$

$$-a < -b \qquad \text{Definition 13-2}$$

Next, we must show that if $-a < -b$, then $a > b$. The steps in the first part of the proof are reversible. Hence, the proof would consist in starting with $-a < -b$ and rewriting the inequalities back to $a > b$.

Corollary If $a > 0$, then $-a < 0$; and if $-a < 0$, then $a > 0$.

Theorem 13-4 Let a, b, and $c \in R$ with $a > b$ and $c < 0$. It then follows that $ac < bc$.

Proof: Since $c < 0$, we conclude from the preceding corollary that $-c > 0$. Then we have

$$a > b \qquad \text{Given}$$

$$-ac > -bc \qquad \text{Axiom 10}$$

$$ac < bc \qquad \text{Theorem 13-3}$$

Axiom 10 tells us that both members of an inequality may be multiplied by a positive number if the *sense* (direction of pointing of the inequality symbol) of the inequality is unchanged. And Theorem 13-4 tells us that the sense must be changed if the multiplier is a negative number.

Theorem 13-5 If a, b, c, and $d \in R$ with $a > b$ and $c > d$, then $a + c > b + d$.

Proof

$a > b$ and $c > d$ — Given

$a + c > b + c$ and $b + c > b + d$ — Axiom 9

$a + c > b + d$ — Axiom 8

Theorem 13-6 If $a > b$ and $c > d$, with a, b, c, and $d > 0$, then $ac > bd$.

Theorem 13-7 If $a > 0$, $b > 0$, and $a > b$, then $1/a < 1/b$.

Proof

$$a > b \qquad \text{Given}$$

$$a\left(\frac{1}{ab}\right) > b\left(\frac{1}{ab}\right) \qquad \text{Axiom 10}$$

$$\frac{1}{b} > \frac{1}{a}$$

$$\frac{1}{a} < \frac{1}{b}$$

EXERCISE 13-1

1. If $a \in R$, tell why the inequality $a > a$ is impossible.
2. If $b \in R$, prove that $b < 0$ if and only if $-b > 0$. Prove also that $b > 0$ if and only if $-b < 0$. [*Hint:* Use Theorem 13-4.]
3. If a, b, c, and $d \in R$ with $b > 0$ and $d > 0$, prove that

$$\frac{a}{b} > \frac{c}{d} \qquad \text{if and only if} \qquad ad > bc$$

4. Verify Theorems 13-1 to 13-7 by using properly chosen integers.
5. Show that $a > b$ if and only if $a - b > 0$.
6. If $a \neq 0$, prove that $a^2 > 0$.

7. If $a > b$ and $c > 0$, show that $a/c > b/c$.

8. If $a > b$ and $c < 0$, show that $a/c < b/c$.

13-2 SOLUTION OF INEQUALITIES

In the remainder of this chapter, we shall consider inequalities in which one or both members contain a variable (or variables). In Sec. 6-1, we defined two special types of equations—conditional equations and identities. Analogously, there are, in general, two special types of inequalities, which we now define.

Definition 13-5 An inequality which is satisfied by some, but not all, of the permissible values of the variables involved is called a *conditional inequality.**

Definition 13-6 An inequality which is satisfied by all the permissible values of the variables involved is called an *absolute inequality.*

Thus, $x + 3 < 4$ is a conditional inequality; it is true only if $x < 1$. But $x^2 + 1 > 0$, valid for all real values of x, is an absolute inequality.

To solve an inequality means to find all the values of the variable or variables which satisfy the inequality. The axioms and theorems of the preceding section may be applied in solving inequalities. In particular, we may use the following properties of inequalities:

1. The sense of an inequality is not changed when the same number is added to, or subtracted from, both members.
2. The sense of an inequality is not changed when both members are multiplied by, or divided by, the same positive number.
3. The sense of an inequality is reversed when both members are multiplied by, or divided by, the same negative number.

Each of these operations changes an inequality into an equivalent inequality (i.e. an inequality with the same solution set). And the solution process consists in obtaining a series of equivalent inequalities which lead to a final inequality whose solution is evident.

* The permissible values include all the values for which both members of an inequality are defined.

EXAMPLE 1 Solve the inequality $2 - 3x \leq 2x + 12$.

Solution: Subtracting $2x + 2$ from both sides, we obtain

$$-5x \leq 10$$

Then dividing by -5 and reversing the sense of the inequality gives

$$x \geq -2$$

This shows that any solution of the given inequality is a solution of the last inequality. We could now start with $x \geq -2$, reverse the operations, and work back to the given inequality, which would establish the equivalence of the given inequality and the final inequality. We conclude, then, that the solution of the given inequality, in set notation, is

$$\{x|x \geq -2\}$$

EXAMPLE 2 Solve the inequality $2x^2 - x - 15 < 0$.

Solution: The first step is to factor the left member. Thus, we have

$$(2x + 5)(x - 3) < 0 \qquad \text{or} \qquad 2(x + \tfrac{5}{2})(x - 3) < 0$$

In the factored form, we see that the zeros of $2x^2 - x - 15$ are $x = -\frac{5}{2}$ and $x = 3$. For all other values of x the left side is either positive or negative. Hence, we seek the values of x which make one of the factors positive and the other negative. The factor $x + \frac{5}{2}$ is negative when $x < -\frac{5}{2}$ and positive when $x > -\frac{5}{2}$. This is pictured on the number axis (Fig. 13-1) by the minus signs to the left of $-\frac{5}{2}$ and the plus signs to the right. Similarly, the factor $x - 3$ is negative to the left of 3 and positive to the right. From the diagram, it is easy to see that the factors have opposite signs for all values of x between $-\frac{5}{2}$ and 3. The solution of the given inequality may be expressed by either

$$\{x| -\tfrac{5}{2} < x < 3\} \qquad \text{or} \qquad \{-\tfrac{5}{2} < x < 3\}$$

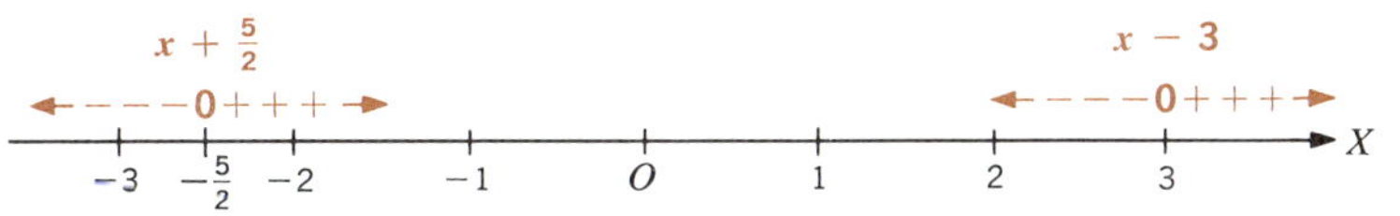

Figure 13-1

By referring to Fig. 13-1, we see that the factors of $2x^2 - x - 15$ are both negative when $x < -\frac{5}{2}$ and both are positive when $x > 3$. Hence, the solution of the inequality

$$2x^2 - x - 15 > 0$$

is $\{x|x < -\frac{5}{2} \text{ or } x > 3\}$.

EXERCISE 13-2

Solve the following inequalities.

1. $4x > 12$

2. $2x - 4 < 6$

3. $5x - 1 > 3x + 7$

4. $6x + 3 < x - 9$

5. $x + 6 < 4 - 3x$

6. $5x - 7 < 3x + 2$

7. $x^2 - 4 < 0$

8. $x^2 - 4 \geq 0$

9. $x^2 - 25 \leq 0$

10. $x(x - 2) < 0$

11. $x^2 - 4x + 3 < 0$

12. $x^2 - 3x - 10 < 0$

13. $2x^2 - x - 10 < 0$

14. $3x^2 - 4x - 4 \leq 0$

15. $2x^2 - 9x + 7 < 0$

16. $\dfrac{x - 1}{x} < 0$

17. $\dfrac{x}{x - 1} > 0$

18. $\dfrac{x - 4}{x + 4} < 0$

19. $\dfrac{x - 4}{x + 4} > 0$

20. $\dfrac{x - 1}{x} < 1$

21. $\dfrac{x - 1}{x} > 1$

13-3 ABSOLUTE INEQUALITIES

In this section, we shall deal with absolute inequalities. To establish an inequality of this type, we need to show that it is true for all permissible values of the variables which appear in the inequality. The examples illustrate the process.

EXAMPLE 1 Show that the inequality $2x^2 - 4x + 3 > 0$ is an absolute inequality.

Solution: We start with the given inequality and reduce it to a form which clearly holds for all real values of x. Thus,

$$2x^2 - 4x + 3 > 0$$
$$2(x^2 - 2x) + 3 > 0$$
$$2(x^2 - 2x + 1) + 3 - 2 > 0$$
$$2(x - 1)^2 + 1 > 0$$

The term $2(x - 1)^2$ of the last inequality cannot be negative when x has any real value. Accordingly, the inequality is true for all values of x. But the left member of the last inequality and the left member of the given inequality are identical, and therefore the given inequality, holding for all values of x, is an absolute inequality.

EXAMPLE 2 Prove the inequality $a + (1/a) \geq 2$, $a > 0$.

Proof: Since $a > 0$, we multiply both members of the inequality by a and leave the sense unchanged. Hence,

$$a^2 + 1 \geq 2a$$
$$a^2 - 2a + 1 \geq 0$$
$$(a - 1)^2 \geq 0$$

The last inequality is true since $(a - 1)^2$ is never negative. Then to complete the proof we start with the last inequality and work back to the first. Thus, by obvious operations, we have

$$(a - 1)^2 \geq 0$$
$$a^2 - 2a + 1 \geq 0$$
$$a^2 + 1 \geq 2a$$
$$a + \frac{1}{a} \geq 2$$

EXERCISE 13-3

Prove that each of the following inequalities is an absolute inequality.

1. $x^2 + 2x + 2 > 0$

2. $x^2 - 10x + 26 > 0$

3. $x^2 + 1 \geq 2x$

4. $4x^2 + 1 \geq 4x$

5. $2x^2 - 4x + 3 > 0$

6. $9x^2 - 6x + 2 > 0$

7. $16x^2 + 1 \geq 8x$

8. $3x^2 - 6x + 4 > 0$

If a, b, and c stand for unequal positive numbers, prove the following inequalities.

9. $(a + b)^2 > a^2 + b^2$

10. $a^2 + b^2 > 2ab$

11. $\frac{a}{b} + \frac{b}{a} > 2$

12. $\frac{a + b}{2} > \sqrt{ab}$

13. $\frac{a + b}{4a} > \frac{b}{a + b}$

14. $\frac{a^3 + b^3}{a + b} > ab$

15. $a^2 + b^2 + c^2 < (a + b + c)^2$

16. $a^2b + ab^2 < a^3 + b^3$

17. $a^3 + 3ab^2 > b^3 + 3a^2b, \quad a > b$

18. $\frac{a^2}{b} + \frac{b^2}{a} > a + b$

19. If x and y are any real numbers, show that

    a. $|x + y| \leq |x| + |y|$
    b. $|x - y| \geq |x| - |y|$

    Hint: Consider the cases in which x and y are of the same sign and, next, of different signs.

13-4 SYSTEMS OF INEQUALITIES

In this section, we shall solve certain simple systems of inequalities. The solution set of a system of one variable consists of all numbers which satisfy both inequalities. The solution set of a system of two variables consists of the totality of ordered number pairs which satisfy both inequalities.

EXAMPLE 1 Solve the system of inequalities

$$2x - 7 < 5 - x$$

$$11 - 5x < 1$$

Solution: We seek the set of values for x which make each inequality true. The first inequality is true for all values of $x < 4$ and the second for all values of $x > 2$. Hence, the solution of the system is given by the intersection of these two sets, which may be expressed by either

$$\{x|2 < x < 4\} \quad \text{or} \quad \{2 < x < 4\}$$

EXAMPLE 2 Solve the inequality $|2x - 3| < 7$.

Solution: Recalling the definition of the absolute value of a number (Sec. 3-1), we see that the given inequality is equivalent to the system

$$2x - 3 < 7$$
$$2x - 3 > -7$$

For convenience in finding the solution, we first express the system in the equivalent extended form

$$-7 < 2x - 3 < 7$$

Now add 3:

$$-4 < 2x < 10$$

Divide by 2:

$$-2 < x < 5$$

Hence, the solution of the given inequality consists of the set of all numbers between -2 and 5.

EXAMPLE 3 Solve graphically the inequality $2x - 3y - 6 < 0$.

Solution: This is a linear inequality in the variables x and y. Solving for y, we have the equivalent inequality

$$y > \frac{2x - 6}{3}$$

The graph of $y = (2x - 6)/3$ is the line drawn in Fig. 13-2. The coordinates (x_1, y_1) of any point of the line satisfy this equation. But the coordinates of a point (x_1, y) above the line have $y > y_1$, and consequently satisfy the given inequality. Hence, the graph of the solution set of the inequality is the half-plane above the line.

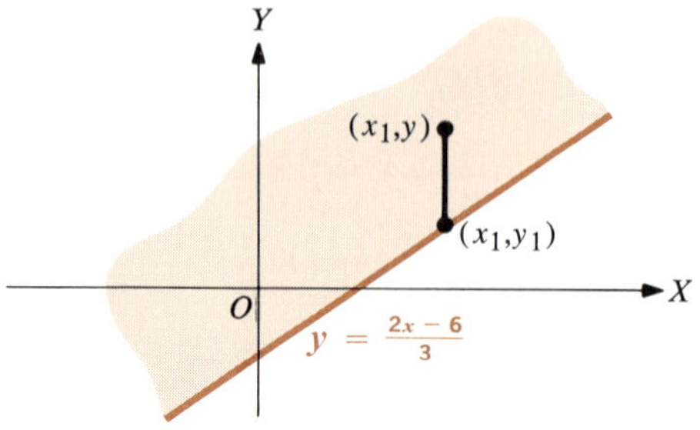

Figure 13-2

In a similar way, we could show that the graph of the solution set of $y < (2x - 6)/3$ consists of all points below the line in Fig. 13-2.

EXAMPLE 4 Solve the system of inequalities

$$x^2 - 6x + 2y - 1 < 0$$
$$3x + 7y - 7 > 0$$

Solution: The graph of the equation $x^2 - 6x + 2y - 1 = 0$ is a parabola and the graph of the equation $3x + 7y - 7 = 0$ is a straight line (Fig. 13-3). By solving each inequality for y, we obtain the equivalent system

$$y < \frac{1 + 6x - x^2}{2}$$
$$y > \frac{7 - 3x}{7}$$

We observe from these inequalities that the coordinates of all points inside, or below, the parabola satisfy the first inequality and

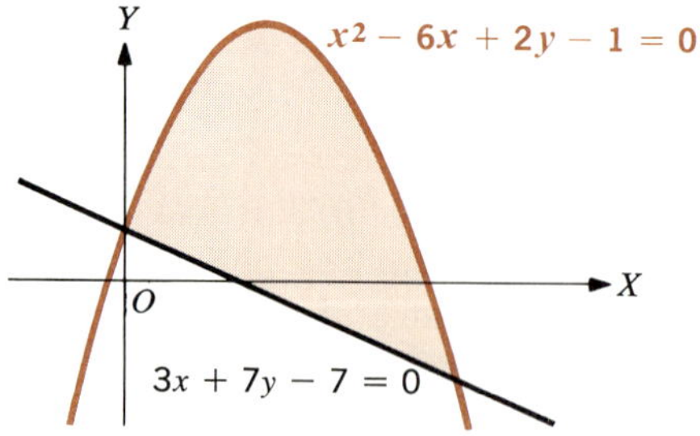

Figure 13-3

that the coordinates of all points above the line satisfy the second inequality. The graph of the solution set of the given inequalities is the shaded area in the figure. We write the solution set as

$$\{(x, y)|x^2 - 6x + 2y - 1 < 0\} \cap \{(x, y)|3x + 7y - 7 > 0\}$$

EXERCISE 13-4

Solve each system of inequalities algebraically.

1. $2x + 2 < x + 5$
$3x + 6 > 2 - x$

2. $8 - 3x > 5x - 10$
$2x - 15 < 5$

3. $10x + 6 > 2x - 2$
$4x - 7 < 3 - x$

4. $7x + 50 > 10 - 3x$
$7x + 4 > 2x - 6$

5. $5x - 6 < x + 2$
$4x - 7 > 11 - 2x$

6. $6x - 2 > 16$
$4x - 7 < 1$

7. $|x - 3| < 2$

8. $|3x - 4| \leq 5$

9. $|2x - 9| < 5$

10. $|3 - 2x| \leq 9$

11. $|3 - 5x| \leq 12$

12. $|7 - x| \leq 6$

Solve the following systems of inequalities graphically.

13. $2x - y - 1 > 0$
$x - 8y + 1 < 0$

14. $x - 2y \leq 4$
$x + y \geq 1$

15. $2x - y > 2$
$2x + y > 2$

16. $y + x < 3$
$y + x > 2$

17. $4x + 3y - 6 > 0$
$4x + 3y + 1 > 0$

18. $x - 3y \leq 0$
$x + 3y \leq 0$

19. $3x + 4y + 14 < 0$
$2x - 5y - 6 > 0$

20. $x + y > 4$
$x + y < 2$

21. $2x - 4y - 5 > 0$
$x - 2y + 3 < 0$

22. $x^2 - 4y < 0$
$y < 4$

23. $x^2 + y^2 < 16$
$y - 3x + 9 < 0$

24. $y^2 - 4x < 0$
$x - y < 0$

25. $x^2 + y^2 < 16$
$x^2 - 4y < 0$

26. $x^2 + y^2 < 16$
$x^2 - y > 0$

14

PROGRESSIONS

14-1 SEQUENCES

A succession of numbers of which one number is designated as the first, another as the second, another as the third, and so on, is called a *sequence*. Each number of the sequence is called a *term*. The set of numbers

$$1, 4, 7, 10, 13, 16$$

is an example of a sequence of six terms.

A sequence which has a fixed number of terms is called a *finite* sequence; otherwise the sequence is an *infinite* sequence. The set of all positive even integers, which we denote by

$$2, 4, 6, 8, \ldots$$

is an infinite sequence.

Since the terms of a sequence have the order first, second, third, and so on, we see that there is a one-to-one correspondence between a set of positive consecutive integers, starting with 1, and the terms of a sequence. Hence, the sequence constitutes the range of a function whose domain is the set of integers. For example, pairing the terms of the sequence 1, 3, 5, 7, 9, 11 with the first six positive integers, we form the function

$$\{(1, 1), (2, 3), (3, 5), (4, 7), (5, 9), (6, 11)\}$$

If we let a_1 stand for the first term of a sequence, a_2 for the second term, a_3 for the third term, and so on, we can denote the sequence by

$$a_1, a_2, a_3, \ldots, a_n, \ldots$$

The term a_n is called a *general term* or the nth *term*. Now suppose that an expression is given in terms of n by which the members of a sequence can be obtained simply by substituting 1, 2, 3, and so on for n. In such a case, we may write as many terms of the sequence as we please. Thus, for example, the equation $a_n = n^2$ defines a function whose values are terms of the sequence

$$1, 4, 9, 16, 25, \ldots, n^2, \ldots$$

As another example, let $a_n = n/(n^2 + 1)$. The terms of the sequence thus determined are

$$\frac{1}{2}, \frac{2}{5}, \frac{3}{10}, \frac{4}{17}, \frac{5}{26}, \ldots, \frac{n}{n^2 + 1}, \ldots$$

The terms of a sequence also may be obtained under the following conditions. Let the first term a_1 be given and in addition a formula by which each term of a sequence, after the first, may be computed from the preceding term. We illustrate by choosing $a_n = 3a_{n-1} + 2$ as the formula and letting $a_1 = 3$. The process consists in substituting 2 for n to find the second term, 3 for n to find the third term, and so on. Then for the first four terms, we have

$$a_1 = 3$$

$$a_2 = 3 \cdot 3 + 2 = 11$$

$$a_3 = 3 \cdot 11 + 2 = 35$$

$$a_4 = 3 \cdot 35 + 2 = 107$$

EXERCISE 14-1

Write the first four terms of the sequence defined by the given a_n. Also find the twentieth terms.

1. $a_n = 2n - 1$ **2.** $a_n = 3n - 2$ **3.** $a_n = (-1)^n 2^n$

4. $a_n = \dfrac{2n}{n+2}$ **5.** $a_n = \dfrac{1}{n(n+1)}$ **6.** $a_n = \dfrac{n}{2n-1}$

7. $a_n = \dfrac{(-1)^n}{n}$ **8.** $a_n = \dfrac{n^2-1}{n^2+1}$ **9.** $a_n = \dfrac{2^n}{n^2+1}$

Write the first four terms determined by the given conditions in each problem.

10. $a_1 = 1, \quad a_n = a_{n-1} + 3$ **11.** $a_1 = 3, \quad a_n = 2a_{n-1} + 3$

12. $a_1 = 4, \quad a_n = \dfrac{a_{n-1}}{2}$ **13.** $a_1 = 5, \quad a_n = \dfrac{1}{a_{n-1}}$

14. $a_1 = 3, \quad a_n = \dfrac{(-1)^n a_{n-1}}{3}$ **15.** $a_1 = -2, \quad a_n = \dfrac{a_{n-1}}{n}$

14-2 ARITHMETIC PROGRESSIONS

A great variety of sequences occurs in mathematics. However, we shall limit our study to three kinds of sequences which, though comparatively simple, are of special importance.

Definition 14-1 A sequence in which each term after the first is formed by adding a fixed number to the preceding term is called an *arithmetic progression*. The fixed number is called the *common difference.*

EXAMPLE 1 The sequence 2, 5, 8, 11, 14 is an arithmetic progression with the common difference 3.

EXAMPLE 2 The sequence 10, 7, 4, 1, -2, -5 is an arithmetic progression with the common difference -3.

If a_1 stands for the first term and d for the common difference of an arithmetic progression, we can write the successive terms by repeated addition of d. Thus for n terms, we have

$$a_1, \quad a_1 + d, \quad a_1 + 2d, \quad a_1 + 3d, \quad \ldots, \quad a_1 + (n-1)d$$

We obtain the last or nth term by observing that the coefficient of d in each term is one less than the corresponding order number of the term. Denoting the nth term by a_n, we have the formula

$$a_n = a_1 + (n - 1)d \tag{1}$$

With a_1 and d given, we may find any desired term of the progression by substituting the proper positive integer for n.

EXAMPLE 1 Find the thirtieth term of the arithmetic progression 2, 6, 10, ...

Solution: Here, $a_1 = 2$, $d = 4$, and $n = 30$. Then using formula (1) for a_n, we find that

$$a_{30} = 2 + (30 - 1)4 = 2 + 116 = 118$$

The first and last terms of an arithmetic progression are called *extremes*, and the terms in between are called *arithmetic means*. If we wish to insert a certain number of arithmetic means between two given numbers, we may use the formula for a_n to find d. Then the terms of the progression can be written by repeated additions of d.

EXAMPLE 2 Insert five arithmetic means between 7 and 25.

Solution: In this case, $a_1 = 7$ and $a_n = 25$. The terms of the progression consist of the two given terms plus the five means, giving $n = 7$. Substituting in formula (1), we find

$$25 = 7 + 6d \qquad \text{and} \qquad d = 3$$

The progression is 7, 10, 13, 16, 19, 22, 25.

We next derive a formula for the sum of the terms of an arithmetic progression. Either we may write the terms by starting with the first term a_1 and adding d repeatedly or we can start with the last term a_n and subtract d repeatedly. Then letting S_n stand for the sum of the terms, we have the two equations

$$S_n = a_1 + (a_1 + d) + (a_1 + 2d) + \cdots + (a_n - d) + a_n$$

$$S_n = a_n + (a_n - d) + (a_n - 2d) + \cdots + (a_1 + d) + a_1$$

Adding the members of these equations, we get

$$2S_n = (a_1 + a_n) + (a_1 + a_n) + (a_1 + a_n) + \cdots + (a_1 + a_n) + (a_1 + a_n)$$

Since the progression has n terms, the sum $(a_1 + a_n)$ appears n times, and therefore

$$S_n = \frac{n}{2}(a_1 + a_n)$$

We obtain another formula for S_n by using formula (1) to replace a_n. Thus,

$$S_n = \frac{n}{2}[a_1 + a_1 + (n - 1)d]$$

We now have three useful formulas which are applicable to arithmetic progressions. The formulas, rewritten for convenient reference, are

$$a_n = a_1 + (n - 1)d \tag{1}$$

$$S_n = \frac{n}{2}(a_1 + a_n) \tag{2}$$

$$S_n = \frac{n}{2}[2a_1 + (n - 1)d] \tag{3}$$

The five quantities a_1, a_n, d, n, and S_n are called the *elements* of an arithmetic progression. If any three of the elements are given, the remaining two may be found by use of the formulas (1) to (3).

EXAMPLE 3 Given the elements $d = 3$, $a_n = 33$, and $S_n = 195$, find a_1 and a_n.

Solution: Substituting the values of the known elements in formulas (1) and (2) gives the system

$$33 = a_1 + (n - 1)3$$

$$195 = \frac{n}{2}(a_1 + 33)$$

which reduces to

$$a_1 + 3n = 36$$

$$n(a_1 + 33) = 390$$

By solving the first equation for a_1 and substituting the result in the second equation, we get the quadratic equation $n^2 - 23n + 130 = 0$. The roots of this equation are $n = 10$, $n = 13$, and the corresponding values of a_1 are 6 and -3. Hence, the two solutions are $n = 10$, $a_1 = 6$ and $n = 13$, $a_1 = -3$.

When a_1 and n are the unknowns, as in this example, and when a_n and n are the unknowns, formulas (1) and (2) give a linear equation and a second-degree equation in the unknowns. This means there may be two solutions. A value of n other than a positive integer, however, must be rejected.

EXERCISE 14-2

Find the last term a_n and the sum S_n of each arithmetic progression.

1. 1, 4, 7, . . . to 8 terms

2. 4, 6, 8, . . . to 20 terms

3. 1.5, 2, 2.5, . . . to 38 terms

4. $-7, -3, 1, \ldots$ to 19 terms

5. $9, 1, -7, \ldots$ to 16 terms

6. $x - 2, 3x + 3, 5x + 8, \ldots$ to 11 terms

7. $2x - y, 5x + y, 8x + 3y, \ldots$ to 13 terms

8. $x, y, 2y - x, \ldots$ to 14 terms

9. Find the sum of the even integers between 19 and 79.

10. Find the sum of odd integers between 2 and 90.

11. Find the sum of the integers between 2 and 103 which are divisible by 3.

12. Find the sum of the integers between 50 and 150 which are divisible by 5.

13. Insert four arithmetic means between 4 and 19.

14. Insert five arithmetic means between -8 and 10.

15. Insert six arithmetic means between -43 and 34.

16. Prove that the sum of the first n even positive integers is $n(n + 1)$.

17. Prove that the sum of the first n odd positive integers is n^2.

18. Find the sum of the first n positive integers which are divisible by 3.

Find the two missing elements in each problem.

19. $a_1 = 36, \quad a_n = 8, \quad d = -2$

20. $a_1 = 6, \quad a_n = 33, \quad n = 10$

21. $d = 5, \quad n = 6, \quad S_n = 51$

22. $a_n = 62, \quad d = 6, \quad n = 18$

23. $a_n = 6, \quad d = -2, \quad S_n = 50$

24. $a_n = 49, \quad n = 14, \quad S_n = 231$

25. $d = -2, \quad a_n = 8, \quad S_n = 44$

26. $d = 4, \quad a_n = 28, \quad S_n = 108$

27. $a_n = -15, \quad d = -2, \quad S_n = -48$

28. $d = -3, \quad n = 10, \quad S_n = -55$

29. A body dropped from a height falls 16 feet during the first second, 48 feet the second second, and 80 feet the third second, and so on. Find how far it falls during the sixth second and how far it falls during the first 6 seconds.

30. An object is dropped from a height of 1024 feet. Find the time required for the object to reach the ground. (See problem 29.)

31. A man deposits \$500 in a savings account and the same amount each year thereafter until he has made 10 deposits. If his money draws 4% per year simple interest, how much will he have in the account just after he makes the tenth deposit?

32. A man buys a car priced at \$3,300. He agrees to pay each month, beginning one month after the purchase date, \$100 on the debt plus an interest charge of 1% of the outstanding debt during the month. Find the amount of interest he will pay in settling the debt.

33. A man borrows \$16,800. He agrees to settle the debt by paying \$280 on the principal at the end of each 3 months plus an interest charge on the indebtedness during the 3-month period. If the interest payments are \$210, \$206.50, \$203, and so on, find the total amount of interest which will be paid in settling the debt.

34. An employer engages a man at a beginning salary of \$6,000 per year. He offers to increase the salary \$400 per year at the end of each year or, alternatively, he will increase the pay \$100 for each six months at the end of each half year. Find the total earnings under each proposition at the end of 4 years.

35. A farmer pays \$11,000 for a tractor which depreciates in value 20% the first year, 17% of the original value the second year, and 14% of the original value the third year, and so on. Find the value of the tractor at the end of 7 years.

36. A man accepts a position for a period of 5 years. He can choose either an increase in pay of \$30 per month at the end of each 6 months or an increase of \$67 per month at the end of each year. Which, and by how much, is the better proposition for the employee?

14-3 GEOMETRIC PROGRESSIONS

As we pointed out in Sec. 14-1, a sequence is a set of functional values whose domain is the set of positive integers or a proper subset of the positive integers, depending on whether the sequence is infinite or finite. We have introduced the arithmetic progressions as a special type of sequences, and now we consider another important special type of sequences.

Definition 14-2 A sequence in which each term after the first is a fixed number times the preceding term is called a *geometric progression.* The fixed number, denoted by r, is called the *common ratio.*

EXAMPLES

2, 4, 8, 16	Common ratio 2
$1, -3, 9, -27$	Common ratio -3

The first term, the common ratio, the number of terms, the nth term, and the sum of n terms constitute the five elements of a geometric progression. We denote the elements, in the order named, by a_1, r, n, a_n, and S_n. The first and last terms of a geometric progression are called the *extremes* and the terms between are called *means.*

We can write the terms of a geometric progression from the elements a_1, r, and n as

$$a_1, a_1r, a_1r^2, a_1r^3, \ldots, a_1r^{n-1}$$

We obtain the last, or the nth term, by observing that the exponent of r in each term is 1 less than the number of the term. Denoting the nth term by a_n, we have the formula

$$a_n = a_1r^{n-1} \tag{1}$$

EXAMPLE 1 Find the sixth term of the geometric progression 9, -6, 4, ...

Solution: In this case we see that the first term multiplied by $-\frac{2}{3}$ gives the second term, and likewise the product of the second term and $-\frac{2}{3}$ gives the third term. Then using $a_1 = 9$, $r = -\frac{2}{3}$, and $n = 6$ in formula (1), we find

$$a_n = 9(-\tfrac{2}{3})^{6-1} = 9(-\tfrac{32}{243}) = -\tfrac{32}{27}$$

EXAMPLE 2 Insert three geometric means between 81 and 16.

Solution: We have $a_1 = 81$, $a_n = 16$, and, counting the two extremes and the three means, $n = 5$. Hence, from formula (1)

$$16 = 81r^4$$
$$r^4 = \tfrac{16}{81}$$
$$r = \tfrac{2}{3} \quad \text{and} \quad -\tfrac{2}{3}$$

Using these possible values of r, we write the two progressions

$$81, 54, 36, 24, 16 \qquad \text{and} \qquad 81, -54, 36, -24, 16$$

To derive a formula for the sum S_n of n terms of a geometric progression, we write the two equations

$$S_n = a_1 + a_1r + a_1r^2 + a_1r^3 + \cdots + a_1r^{n-2} + a_1r^{n-1}$$
$$rS_n = \qquad a_1r + a_1r^2 + a_1r^3 + \cdots + a_1r^{n-2} + a_1r^{n-1} + a_1r^n$$

The second equation is obtained by multiplying each member of the first equation by r and placing the like terms of the right members one under the other. By subtracting the members of the second equation from the corresponding members of the first equation, we get

$$S_n - rS_n = a_1 - a_1r^n$$

$$(1 - r)S_n = a_1 - a_1r^n$$

$$S_n = \frac{a_1(1 - r^n)}{1 - r}$$

This formula gives the sum in terms of a_1, r, and n. The sum in terms of a_1, r, and a_n can be obtained by using formula (1). Thus, $a_n = a_1r^{n-1}$ or $ra_n = a_1r^n$, and therefore

$$S_n = \frac{a_1 - a_1r^n}{1 - r} = \frac{a_1 - ra_n}{1 - r}$$

We now have the following formulas involving the elements of a geometric progression:

$$a_n = a_1r^{n-1} \tag{1}$$

$$S_n = \frac{a_1(1 - r^n)}{1 - r} \tag{2}$$

$$S_n = \frac{a_1 - ra_n}{1 - r} \tag{3}$$

With any three elements known, these formulas suffice to yield the remaining elements. In some cases, there will be more than one solution.

We remark that formulas (2) and (3) are not applicable when $r = 1$. The sum for this trivial case, however, is otherwise easily determined.

EXAMPLE 3 Compute the sum of the first six terms of the geometric progression 2, -6, 18, . . . , using formulas (2) and (3).

Solution: Here $a_1 = 2$, $r = -3$, $n = 6$, and therefore

$$a_6 = 2(-3)^5 = -486$$

By formula (2),

$$S_6 = \frac{2[1 - (-3)^6]}{1 - (-3)} = \frac{2(1 - 729)}{4} = -364$$

and by formula (3),

$$S_6 = \frac{2 + 3(-486)}{1 - (-3)} = \frac{2 - 1458}{4} = -364$$

EXERCISE 14-3

Find the last term a_n and the sum S_n of each geometric progression.

1. 1, 3, 9, . . . to seven terms

2. 3, 6, 12, . . . to eight terms

3. 4, −8, 16, . . . to six terms

4. 1, −3, 9, . . . to six terms

5. 36, 12, 4, . . . to five terms

6. 64, −96, 144, . . . to five terms

7. 243, −81, 27, . . . to seven terms

8. 256, 64, 16, . . . to six terms

9. Insert three geometric means between 162 and 2.

10. Insert six geometric means between 1 and 128.

11. Insert four geometric means between 8 and $-\frac{1}{4}$.

12. Insert four geometric means between $\frac{27}{4}$ and $\frac{8}{9}$.

Find the two missing elements in each geometric progression.

13. $a_1 = 3, \quad r = -2, \quad a_n = 768$

14. $a_1 = 6, \quad r = 3, \quad a_n = 486$

15. $a_n = 81, \quad r = 3, \quad n = 6$

16. $a_1 = -3, \quad r = -4, \quad a_n = 3072$

17. $a_1 = \frac{27}{64}, \quad a_n = \frac{1}{8}, \quad n = 4$

18. $a_1 = \frac{3}{2}, \quad a_n = 48, \quad r = 2$

19. $a_1 = 15, \quad r = \frac{2}{5}, \quad S_n = \frac{117}{5}$

20. $r = -\frac{3}{2}, \quad n = 4, \quad S_n = -\frac{52}{5}$

21. $a_1 = 16, \quad n = 3, \quad S_n = 21$

22. $n = 3, \quad a_n = \frac{2}{3}, \quad S_n = \frac{62}{3}$

23. If at the end of each year the value of a car is two-thirds of its value at the beginning of the year, find the value of a \$3,300 car at the end of 4 years.

24. The population of a city is 30,000. Figuring that the population will increase by 50% every 5 years, find what the population will be in 20 years.

25. The number of bacteria in a certain culture doubles every 3 hours. If there are N bacteria at the start, find the number that there will be in 24 hours.

26. If the population of a state is 2 million and if the rate of increase is 10% each year, what will the population be at the end of 6 years?

27. If a stroke of a vacuum pump removes 10% of the air from a container, how much of the original air remains after 5 strokes? How many strokes are required to remove 50% of the air?

28. A ball is thrown vertically upward to a height of 100 feet. Each time the ball strikes the ground the rebound is one-half of the previous height. Find how far the ball has traveled up and down when it strikes the ground the fifth time.

29. A certain ball when dropped from a height rebounds two-thirds of the distance from which it fell. Find the total distance traveled by the ball from the time it is dropped from a height of 60 feet until it strikes the ground the fifth time.

30. A 10-quart container is filled with water. One quart of water is drained out and replaced with alcohol. After mixing, a quart of the solution is drained out and replaced with alcohol. This process is continued until 5 quarts of alcohol have been put into the container. The solution is then what percent alcohol?

31. A man can have a position at a beginning salary of \$4,000 per year with an increase of \$450 per year each year, or he can begin at the same annual pay and each following year receive 10% more than during the preceding year. Which proposition will yield the better income during the fifth year? Which proposition will yield the better income for the period of 5 years?

32. According to a story, the owner of a horse made the following agreement. For shoeing the horse, which requires 32 nails, he agreed to pay 1 cent for the first nail, 2 cents for the second nail, 4 cents for the third nail, 8 cents for the fourth nail, and so on. Under this plan, what would be the cost for shoeing the horse?

14-4 INFINITE GEOMETRIC PROGRESSIONS

In dividing 1 by 3, we obtain the unending decimal fraction 0.3333 . . . . Each digit of this fraction is a term of a geometric progression. This is evident when we express the values of the digits as

$$0.3,\ 0.03,\ 0.003,\ 0.0003,\ \ldots$$

Denoting the sum of this infinite geometric progression by S, we write

$$S = 0.3 + 0.03 + 0.003 + 0.0003 + \cdots$$

The right member here is called an *infinite geometric progression* or an *infinite geometric series*. The sum of the first n terms of this series, by formula (2) of the preceding section, is

$$S_n = \frac{0.3[1 - (0.1)^n]}{1 - 0.1} = \frac{0.3[1 - (0.1)^n]}{0.9}$$

$$= \tfrac{1}{3}[1 - (0.1)^n] = \tfrac{1}{3} - \tfrac{1}{3}(0.1)^n$$

Clearly, we can make the last term on the right as small as we please by taking n large enough. When $n = 10$, for example, the last term is equal to 1/30,000,000,000, and S_{10}, therefore, differs from $\frac{1}{3}$ by this very small number. Larger values of n would, of course, make S_n still closer to $\frac{1}{3}$. Because of this, we shall say that the "sum" of the infinite geometric series is $\frac{1}{3}$.

Suppose that we now consider the infinite geometric series

$$S = a_1 + a_1r + a_2r^2 + \cdots + a_1r^n + \cdots$$

The formula for the sum of n terms is given by

$$S_n = \frac{a_1[1 - r^n]}{1 - r} = \frac{a_1}{1 - r} - \frac{a_1 \cdot r^n}{1 - r}$$

If $|r| < 1$, or $-1 < r < 1$, the last term of this sum can be made arbitrarily close to zero by taking n sufficiently large. Consequently, S_n can be made arbitrarily close to $a_1/(1 - r)$. We describe this situation, using the symbol ∞ (infinity), by writing

$$\lim_{n \to \infty} S_n = \frac{a_1}{1 - r}$$

The left side of this equation is read "the limit of S_n as n approaches infinity." And we define the right side as the sum of the infinite geometric series. Then, letting S stand for the sum of the series, we have the simple formula

$$S = \frac{a_1}{1 - r} \qquad |r| < 1$$

When $r \leq -1$ and when $r \geq 1$, the absolute values of the terms either increase from term to term or remain the same. For these values of r, S_n has no limit as n approaches infinity.

If a_1 and r are rational numbers and $|r| < 1$, the sum of the corresponding infinite geometric series is a rational number.

EXAMPLE 1 Find the sum of the infinite geometric progression $\frac{3}{2}$, $-1, \frac{2}{3}, -\frac{4}{9}, \ldots$.

Solution: In this progression $a_1 = \frac{3}{2}$ and $r = -\frac{2}{3}$. Hence,

$$S = \frac{\frac{3}{2}}{1 - (-\frac{2}{3})} = \frac{\frac{3}{2}}{1 + \frac{2}{3}} = \frac{9}{10}$$

EXAMPLE 2 Express the repeating decimal of $4.767676 \cdots$ as a rational number.

Solution: The fraction $0.767676 \cdots$ may be written in the form

$$0.76 + 0.0076 + 0.000076 + 0.00000076 + \cdots$$

To find the sum of this series, we note that $a_1 = 0.76$ and $r = 0.01$ and, therefore,

$$S = \frac{0.76}{1 - 0.01} = \frac{0.76}{0.99} = \frac{76}{99}$$

Hence,

$$4.767676 \cdots = 4 + \tfrac{76}{99} = \tfrac{472}{99}$$

EXERCISE 14-4

Find the sum of each infinite geometric series.

1. $\frac{1}{2} + \frac{1}{4} + \frac{1}{8} + \cdots$

2. $\frac{1}{3} + \frac{1}{9} + \frac{1}{27} + \cdots$

3. $5 - \frac{5}{2} + \frac{5}{4} - \cdots$

4. $-6 + 2 - \frac{2}{3} + \cdots$

5. $-5 + 3 - \frac{9}{5} + \cdots$

6. $7 - 3 + \frac{9}{7} - \cdots$

7. $8 + 1.6 + 0.32 + \cdots$

8. $-5 + 2 - 0.8 + \cdots$

Express each of the following repeating decimals in rational form.

9. $0.666\cdots$

10. $0.999\cdots$

11. $0.4343\cdots$

12. $0.1212\cdots$

13. $0.5151\cdots$

14. $0.7474\cdots$

15. $3.132132\cdots$

16. $2.135135\cdots$

17. $4.3535353\cdots$

18. $1.4818181\cdots$

19. $0.142857142857\cdots$

20. $2.7898989\cdots$

21. Compare the sum of the infinite geometric series $1 + \frac{1}{2} + \frac{1}{4} + \cdots$ with the sum of the first 10 terms.

22. Compare the sum of the infinite geometric series $1 - \frac{1}{9} + \frac{1}{81} - \frac{1}{729} + \cdots$ with the sum of the first four terms.

23. A certain ball rebounds four-fifths of the distance from which it has fallen. Find how far the ball will travel before coming to rest if it is dropped from a height of 120 feet.

24. Find the total distance which the tip of a pendulum will travel if the distance of the first swing is 5 inches and the distance of each succeeding swing is 0.97 of the previous distance.

15

PERMUTATIONS, COMBINATIONS, AND PROBABILITY

15-1 FUNDAMENTAL PRINCIPLE

Permutations, combinations, and probability, which we shall define and study in this chapter, have a strong recreational appeal to many people. But aside from this happy interest, they have important practical and theoretical applications. As a first step, we shall examine a principle which is fundamental in the study of these topics.

An illustration will be helpful in understanding the principle. Suppose a governor is to be chosen from three candidates and a lieutenant governor from four other candidates. Let us determine the number of different ways the two positions can be filled. There are three ways of filling the office of governor and with each of these possible ways there are four choices for a lieutenant governor. Hence, in all, there are $3 \cdot 4 = 12$ different ways of making the two choices. In order to check this result we designate the candidates for governor by A, B, and C and the candidates for lieutenant governor by w, x, y, and z. Then the positions can be filled in any of the following 12 ways:

$$\begin{array}{cccc} A\,w & A\,x & A\,y & A\,z \\ B\,w & B\,x & B\,y & B\,z \\ C\,w & C\,x & C\,y & C\,z \end{array}$$

By generalizing the reasoning in this illustration we may make the following statement:

Fundamental Principle If a thing can be done in any of m different ways and then a second thing can be done in any of n different ways, it follows that the total number of different ways in which the first and then the second thing can be done is m times n.

The truth of this principle is easily seen since for each of the m ways of doing the first thing there are n ways of doing the second. It is also equally evident that the principle may be extended to three or more things.

EXAMPLE 1 How many two-digit numbers can be made from the digits 1, 2, 3, 4, 5 if (*a*) no digit is repeated in a number; (*b*) repetitions are allowed?

Solution:

(*a*) There are five choices for the tens digit, and after a choice is made there remain four choices for the units digit. Hence the total possibilities are $5 \cdot 4 = 20$. The same result, of course, would be obtained by first selecting the units digit and then the tens digit.

(*b*) If repetitions are allowed there are five choices for the tens digit and five choices for the units digit, making $5 \cdot 5$, or 25 possible two-digit numbers.

EXAMPLE 2 In how many ways can five people sit in a car, two in the front seat and three in the back seat, if a particular two of the five are to avoid the driver's position?

Solution: Let us designate the five positions by the letters A, B, C, D, E with A indicating the driver's seat. Taking these places in the order named, let us write above each letter the number which tells how many ways it can be filled.

3	4	3	2	1
A	B	C	D	E

We have three choices for a driver, four choices for place B, three choices for place C, two choices for place D, and one choice for place E. Hence, the total number of seating arrangements is $3 \cdot 4 \cdot 3 \cdot 2 \cdot 1 = 72$.

EXERCISE 15-1

1. How many two-digit numbers can be made with the digits 2, 4, 6, 8 if (*a*) repetitions are allowed; (*b*) no digit is to be repeated in a number?
2. How many three-digit numbers can be formed with the digits 1, 2, 3, 4, 5 if (*a*) repetitions are allowed; (*b*) no repetitions are allowed?
3. How many three-digit numbers (zero not the first digit) can be formed with digits 0, 1, 2, 3, 4 if (*a*) repetitions are allowed; (*b*) no repetitions are allowed?
4. How many four-digit numbers, each less than 5000, can be formed with the digits 1, 2, 4, 6, 8 if (*a*) repetitions are allowed; (*b*) no repetitions are allowed?
5. How many committees consisting of 1 freshman, 1 sophomore, and 1 junior can be selected from 40 freshmen, 30 sophomores, and 25 juniors?
6. Five trails lead to a hilltop. In how many ways can one ascend and descend by (*a*) a different trail; (*b*) any trail?
7. Five boys are in a room which has 4 doors. In how many different ways can they leave the room?
8. A building has six outside doors. In how many ways can one enter and leave (*a*) by a different door; (*b*) by any door?
9. How many signals can be made with 5 different kinds of flags placed on a flagpole one under another from the highest? How many signals can be made if only 4 of the flags are used at a time?
10. License plates are to be distinguished by using 2 letters followed by four digits. How many different plates are possible

(*a*) if letters and digits may be repeated; (*b*) if no repetitions are allowed?

11. In how many ways can 3 people be seated in a room where there are 7 seats?

12. In how many ways can 5 books be arranged on a shelf? How many ways if a certain 2 are not to be separated?

13. Four couples are to eat at a round table with the men and women alternating. If the hostess reserves a place for herself, in how many ways can she assign seats to the others?

14. How many seating arrangements are possible in problem 13 if the host is not to sit by the hostess?

15. A club has 21 members. In how many ways can a president and a vice-president be elected from the group if no member is eligible for both positions?

16. In how many ways can 5 men and 4 women be seated in a row of 9 seats (*a*) with the men and women alternating; (*b*) if the women are not to be separated; (*c*) if the women are to be together and the men together?

17. Towns A and B are connected by 3 roads and towns B and C by 4 roads. In how many ways can one drive from A to B to C and return to A via B if (*a*) the same road is not to be traveled twice; (*b*) the same road may be traveled twice?

18. In how many ways can a club of 21 members choose a president, a secretary, and a treasurer if no member may hold more than one of the positions?

19. In how many ways can 9 books, 6 with green covers and 3 with red covers, be arranged on a shelf if (*a*) the red books are not to be separated; (*b*) the green books are not to be separated; (*c*) books of the same color are kept together?

20. How many odd, three-digit numbers (zero not the first digit) can be formed with the digits 0, 1, 2, 3, 4, 5, 6 if (*a*) no digit is repeated in a number; (*b*) repetitions are allowed?

15-2 PERMUTATIONS

The fundamental principle of the preceding section can be applied effectively in studying permutations, which we define in the following way:

Definition 15-1 An arrangement of objects in a definite order is called a *permutation* of the objects.

The letters a, b, c, for example, have the following possible permutations or arrangements:

$$a\ b\ c \qquad a\ c\ b \qquad b\ a\ c \qquad b\ c\ a \qquad c\ a\ b \qquad c\ b\ a$$

There are six permutations. It is easy to determine the number of permutations, however, without writing a list. To do so, we notice that there are three positions to be filled. Hence, there are 3 choices for the first position, then 2 for the second position, and 1 for the third position. According to the fundamental principle, there are $3 \cdot 2 \cdot 1$ different orders in which the letters can be written.

Let us now apply this reasoning to the case of n things, where r of the things (r a part or all) are used at a time. We may consider this problem as one in which selections are to be made from n things to fill r positions. Hence, there are n choices for the first position, $n - 1$ choices for the second position, $n - 2$ choices for the third position, and so on. We observe that a continuation of this process leaves $n - (r - 1) = n - r + 1$ choices for the rth position. We discover, then, that the number of permutations of n things taken r at a time is given by the product of r integers from n to $n - r + 1$, inclusive. Denoting this product by $P(n, r)$, we write the formula

$$P(n, r) = n(n - 1)(n - 2) \cdots (n - r + 1) \tag{1}$$

To find the number of permutations of n things taken all at a time, we simply let $r = n$ in the formula. We then have the product of all the integers from 1 to n inclusive. The symbol $n!$ (read "factorial n") is used as a shorthand notation for this product. According to this definition of $n!$, n a positive integer, we have the illustrations

$$1! = 1, \qquad 2! = 1 \cdot 2, \qquad 3! = 1 \cdot 2 \cdot 3, \qquad 4! = 1 \cdot 2 \cdot 3 \cdot 4$$

Using the factorial notation, we express the formula for the number of permutations of n things taken all at a time by

$$P(n, n) = n! \tag{1A}$$

The factorial notation permits us to express formula (1) in a more compact form. The product in formula (1) has as factors all the integers from n down to $n - r + 1$ inclusive. If we multiply this product by $(n - r)!$, the result is $n!$. Then multiplying and dividing the right member of formula (1) by $(n - r)!$, we obtain the formula

$$P(n, r) = \frac{n!}{(n - r)!} \tag{2}$$

We see that this formula gives the number of permutations of n things taken r at a time if r is any positive integer less than n. However, we would like the formula to serve when $r = n$, that is, to yield $n!$. But for this value of r, the formula becomes $n!/0!$, and would be valid only if $0!$ is interpreted to be equal to 1. This, in fact, is the customary definition of factorial zero. So, adopting this definition, we see that formulas (1) and (2) are applicable if r is any positive integer from 1 to n, inclusive, and we have the following theorem:

Theorem 15-1 The number of permutations of n things taken r at a time is given either by

$$P(n, r) = n(n - 1)(n - 2) \cdots (n - r + 1) \tag{1}$$

or by

$$P(n, r) = \frac{n!}{(n - r)!} \tag{2}$$

EXAMPLE How many permutations can be made from the letters in the word MONDAY if (a) 4 letters are used at a time; (b) all letters are used; (c) all letters are used but the first is a vowel?

Solution:

(a) We want to find the number of permutations of 6 things taken 4 at a time. Using either formula (1) or the fundamental principle, we have

$$P(6, 4) = 6 \cdot 5 \cdot 4 \cdot 3 = 360$$

(b) From formula (1A), or returning to the fundamental principle, we have

$$P(6, 6) = 6 \cdot 5 \cdot 4 \cdot 3 \cdot 2 \cdot 1 = 720$$

(*c*) The first letter in each permutation is to be either "*o*" or "*a*." Therefore, we have two choices for the first letter. But after selecting the first letter, we can select the remaining 5 without restriction, and consequently in 5! ways. The total number of ways of selecting the first letter and then the others is $2 \cdot 5! = 240$.

15-3 PERMUTATIONS OF THINGS NOT ALL DIFFERENT

The formulas of Theorem 15-1 were derived on the assumption that the set of n things, or objects, includes objects that are all different. The formulas do not apply if some of the objects are alike. The word ALASKA, for example, has 6 letters. But we cannot make 6! permutations using the 6 letters at a time because 3 letters are alike. To aid in finding the number of different permutations, let us assign subscripts to the 3 A's. Thus, writing

$$A_1 \quad L \quad A_2 \quad S \quad K \quad A_3$$

we have 6 distinct objects. Having distinguished the A's, we now have 6! possible permutations. Letting L, S, and K remain in fixed positions, we permute the A's in their positions. This can be done in 3! ways as shown.

$$A_1\ L\ A_2\ S\ K\ A_3 \qquad A_1\ L\ A_3\ S\ K\ A_2 \qquad A_2\ L\ A_1\ S\ K\ A_3$$

$$A_2\ L\ A_3\ S\ K\ A_1 \qquad A_3\ L\ A_1\ S\ K\ A_2 \qquad A_3\ L\ A_2\ S\ K\ A_1$$

These 6 arrangements would be indistinguishable if the subscripts were removed. Thus, we can see that any permutation of the 6 letters can be replaced by 3! permutations by making the A's distinct. Hence, the number of permutations P of 6 objects with 3 alike when multiplied by 3! gives the number of permutations of 6 different objects. That is,

$$3!P = 6! \qquad \text{and} \qquad P = \frac{6!}{3!} = 120$$

This reasoning can be applied to the general case of n objects of which n_1 are identical. In this case any arrangement of the n objects could be replaced by $n_1!$ permutations by making the n_1 objects distinct. Hence, we write the following theorem:

Theorem 15-2 The number of permutations P of n things taken all at a time, where n_1 of the objects are alike and the others distinct, is

$$P = \frac{n!}{n_1!}$$

This formula may be extended to the cases in which there are 2, 3, or more groups of like *objects* among the n objects. Thus, we write

$$P = \frac{n!}{n_1!n_2!n_3! \cdots}$$

EXAMPLE 1 How many permutations can be made with the 9 letters in the word TENNESSEE?

Solution: There are 2 S's, 2 N's, and 4 E's; therefore,

$$P = \frac{9!}{2!2!4!} = 3780$$

EXAMPLE 2 A man tosses 1 half dollar, 2 quarters, 3 dimes, and 4 nickels among 10 boys. In how many different ways can the boys profit if each is to get a coin?

Solution: If the coins were all different, there would be 10! ways in which each boy could get a coin. But since there are groups of 1, 2, 3, and 4 like coins, we have

$$P = \frac{10!}{2!3!4!} = 12{,}600$$

In our discussion thus far, we have supposed that a permutation is formed by placing one object after another in a line. Thus, there is a first and last object. If the objects are arranged in a circle, however, there is no first or last object, and the formulas of Theorem 15-1 are not directly applicable. A circular arrangement presents a new aspect because a mere rotation would not change the position of any object relative to the others. Hence, a rotation does not constitute a different permutation.

To find the number of ways of arranging n different objects in a circle, we first select a position for one of the objects. Then the others can be placed in their positions in $(n - 1)!$ different ways. Six people, for example, can form a circle in $5!$ ways. They could, of course, form a line in $6!$ ways.

EXAMPLE 3 In how many ways can four couples be seated at a round table with the men and women alternating?

Solution: We first think of assigning one, say a woman, to a chair. This fixes 3 particular places for the other women and 4 particular places for the men. The women, therefore, can be assigned to their places in $3!$ ways and the men in $4!$ ways. Applying the fundamental principle, we get the result $3!4! = 144$.

EXERCISE 15-2

1. Evaluate (*a*) $P(7, 1)$; (*b*) $P(7, 3)$; (*c*) $P(7, 4)$; (*d*) $P(7, 7)$.
2. Evaluate $P(4, 1) + P(4, 2) + P(4, 3) + P(4, 4)$.
3. Show that (*a*) $P(5, 4) = P(5, 5)$; (*b*) $P(6, 5) = P(6, 6)$; (*c*) $P(n, n - 1) = P(n, n)$.
4. How many ways can the letters of the word TEXAS be arranged if (*a*) all letters are used; (*b*) 2 are used at a time; (*c*) 3 are used at a time?
5. Find the number of permutations, each with 7 letters, which can be made with the letters in (*a*) WYOMING; (*b*) KENTUCKY; (*c*) WASHINGTON.
6. How many three-digit numbers can be made with the digits 1, 2, 3, 4, 5, 6 if no digit is repeated in a number? How many four-digit numbers?
7. Find how many even numbers of three digits each can be made with the digits 1, 2, 3, 4, 6, 7 if no digit is repeated.
8. How many five-digit numbers, each less than 70,000, can be made with the digits 1, 2, 5, 7, 9 if no digit is repeated?
9. Seven songs are to be given in a program. In how many different orders could they be rendered?
10. In how many ways can 6 people be seated in a row of 9 chairs?
11. In how many ways can 8 books be arranged on a shelf if a certain 3 are (*a*) not to be separated; (*b*) not to be separated and left in a particular order?

12. Find the number of possible batting orders for a baseball team of 9 players if the pitcher is to bat last.

13. There are 5 different mathematics books, 4 different physics books, and 2 different history books to be placed on a shelf with the books of each subject kept together. Find the number of ways in which the books can be placed.

14. How many permutations can be made by using all the letters in (*a*) OHIO; (*b*) INDIANA; (*c*) ASAMAYAMA?

15. Find the number of permutations which can be formed by using all the letters in (*a*) ALABAMA; (*b*) ARKANSAS; (*c*) MISSISSIPPI.

16. How many signals can be made by arranging 9 flags in a line if 4 are red, 3 are white, and 2 are blue?

17. There are 5 copies of a mathematics book, 4 copies of a physics book, and 4 copies of a history book to be placed on a shelf. If the covers are of different colors for each kind of book, how many noticeably different ways can the books be placed on the shelf?

18. In how many ways is it possible for 10 customers to buy 5 cans of peaches, 3 cans of pineapple, and 2 cans of tomatoes if each customer gets 1 can?

19. Find the number of ways 2 half dollars, 4 quarters, and 6 dimes can be distributed among 12 boys if each gets a coin?

20. Seven children join hands and form a circle. In how many ways could the circle be formed? In how many ways could they make a line?

21. A line is formed by 5 girls and 5 boys, with the boys and girls alternating. Find the number of ways of making the line. How many different ways could they form a circle in which the boys and girls alternate?

22. Ten people are to sit at a round table. Find the number of seating arrangements if the host and hostess are to sit opposite each other.

23. In how many ways can 8 persons be seated at a round table if a certain 2 are (*a*) to sit next to each other; (*b*) not to sit next to each other?

24. Four couples are to be seated at a round table. Find the number of seating arrangements if each couple is not to be separated.

15-4 COMBINATIONS

A part or all of a set of objects is called a *combination*. A combination, in distinction from a permutation, does not involve the order of selection of the members of the combination or their arrangement. The interest in a committee selected from a group of people is in the individuals rather than any arrangement of the committeemen. When all of a set of objects are taken at a time, there is of course only one combination. But a part of a set of objects obviously can be chosen in more than one way. We shall be interested in learning how to determine the number of combinations in situations of this kind.

The letters a, b, c, d taken 3 at a time have the combinations

$$a\ b\ c \qquad a\ b\ d \qquad a\ c\ d \qquad b\ c\ d$$

Changing the order of one of these combinations does not make a new combination. Thus, $a\ b\ c$ and $b\ c\ a$ are two permutations of the same things, or of a single combination.

We next consider the general case of the number of combinations of n things taken r at a time.

Theorem 15-3 If $C(n, r)$ stands for the number of combinations of n things taken r at a time $(r \leq n)$, then

$$C(n, r) = \frac{n(n-1)(n-2)\cdots(n-r+1)}{r!} \tag{1}$$

or, alternatively,

$$C(n, r) = \frac{n!}{r!(n-r)!} \tag{2}$$

Proof: Now each combination of r things can be permuted in $r!$ ways. But if each of the combinations is thus permuted, then all possible permutations of n things taken r at a time are obtained. Hence, by Theorem 15-1

$$r!C(n, r) = n(n-1)(n-2)\cdots(n-r+1)$$

and

$$C(n, r) = \frac{P(n, r)}{r!} = \frac{n(n-1)(n-2)\cdots(n-r+1)}{r!}$$

or

$$C(n, r) = \frac{n!}{r!(n - r)!}$$

When r things are taken from a group of n things, $n - r$ things are left. Thus we see that for each combination of r things there is a corresponding combination of $n - r$ things. We conclude, therefore, that the number of combinations of n things taken r at a time is equal to the number of combinations of n things taken $n - r$ at a time. This conclusion can be reached also by replacing r by $n - r$ in formula (2) above. This gives

$$C(n, n - r) = \frac{n!}{[n - (n - r)]!(n - r)!} = \frac{n!}{r!(n - r)!}$$

Hence,

$$C(n, r) = C(n, n - r)$$

EXAMPLE 1 In how many ways can a committee be selected from 18 persons if the committee is to have (*a*) 3 members; (*b*) 14 members?

Solution:

(*a*) We wish to find the number of combinations of 18 things taken 3 at a time. Hence, by formula (1),

$$C(18, 3) = \frac{18 \cdot 17 \cdot 16}{3!} = 816$$

(*b*) We may substitute in formula (1) and evaluate $C(18, 14)$, but it is easier to recall that $C(18, 14) = C(18, 4)$ and write

$$C(18, 4) = \frac{18 \cdot 17 \cdot 16 \cdot 15}{1 \cdot 2 \cdot 3 \cdot 4} = 3060$$

EXAMPLE 2 In how many ways can 7 mathematics teachers be employed from 10 men applicants and 7 lady applicants if (*a*) 3 are to be men; (*b*) 3 or 4 are to be men?

Solution:

(*a*) The 3 men can be had in $C(10, 3)$ different ways. The remaining 4 can be had from the lady applicants in $C(7, 4)$ ways.

Hence the total number of ways is

$$C(10, 3) \cdot C(7, 4) = \frac{10 \cdot 9 \cdot 8}{1 \cdot 2 \cdot 3} \cdot \frac{7 \cdot 6 \cdot 5 \cdot 4}{1 \cdot 2 \cdot 3 \cdot 4} = 4200$$

(*b*) To find the number of ways of filling the vacancies with either 3 or 4 men among those employed, we add the number of ways of filling the vacancies with 3 men and 4 ladies to the number of ways of employing 4 men and 3 ladies. That is, we have

$$C(10, 3) \cdot C(7, 4) + C(10, 4) \cdot C(7, 3) = 4200 + 7350 = 11{,}550$$

EXERCISE 15-3

1. Evaluate (*a*) $C(31, 2)$; (*b*) $C(26, 3)$; (*c*) $C(24, 4)$; (*d*) $C(13, 5)$.
2. Taking full advantage of possible cancellations, evaluate $C(16, 4)$ and $C(16, 12)$ by formula (1). Evaluate both by use of formula (2).
3. How many different ways can a tennis team of 4 be chosen from 17 players?
4. Twelve people meet in a room and each shakes hands with all the others. Find the number of handshakes.
5. Nine points, no 3 of which are on a straight line, are marked on a blackboard. How many lines, each through 2 of the points, can be drawn? How many triangles are determined by the points?
6. Find the number of choices which a student has for doing 6 problems from a set of 11 problems.
7. Seven different coins are tossed simultaneously. In how many ways can 3 heads and 4 tails come up?
8. In how many ways can a committee of 5 be chosen from a club of 12 members if (*a*) the president must be on the committee; (*b*) the president is not eligible, (*c*) a certain two members refuse to serve together on the committee?
9. In how many ways can a court of 9 judges make (*a*) a 5 to 4 decision; (*b*) a majority decision?
10. A presiding officer is authorized to appoint a committee of 5

from 4 elders and 8 deacons. How many different committees are possible if 3 elders are to be members?

11. A committee of 7 is to be composed of 4 Republicans and 3 Democrats. Find the number of choices if 7 Republicans and 8 Democrats are available for appointment.

12. A grocery store manager wishes to employ 6 checkers from 4 men applicants and 5 lady applicants. Find the number of ways of employing the 6 if (*a*) 3 are to be men; (*b*) 2 or 3 are to be men.

13. How many different groups of 9 baseball players can be chosen from 3 who play only as pitchers, 2 who play only as catchers, and 14 who play in any position other than pitcher or catcher?

14. Find the number of ways of selecting 8 books from 10 different mathematics books and 7 different physics books if (*a*) 5 are to be mathematics books; (*b*) 3 or 4 are to be physics books.

15. Find the number of ways of selecting 9 balls from 6 red balls, 5 white balls, and 4 blue balls if each selection consists of 3 balls of each color.

16. A committee of 9 is to be chosen from 10 seniors, 8 juniors, and 7 sophomores. Find the number of choices if the committee is to have 4 seniors, 3 juniors, and 2 sophomores.

17. Find how many sums of money, each consisting of 3 or more coins, can be formed from 6 different kinds of coins.

18. In how many ways can a group of 3 or more be chosen from 8 people?

19. Find the number of ways in which a librarian can check out 8 different kinds of books to two students so that each gets 4 books. *Hint:* Find the number of ways of giving 4 books to the first student.

20. In how many ways can 6 different kinds of books be distributed among three children if each child is to have 2 books?

21. In how many ways can 12 people be separated into two groups of 5 and 7?

22. Find the number of ways in which 11 people can be separated into two groups of 5 and 6 if a certain one of the 11 is to be in the smaller group.

15-5 PROBABILITY IN SINGLE EVENTS

Probability is a fascinating subject and has held the interest of mathematicians during the past two or three centuries. Receiving its first impetus from games of chance, the theory of probability has been highly developed, and now has wide and important applications in the fields of insurance, annuities, and the social sciences. Many people who have made no formal or philosophical study of probability have some understanding of its meaning and applications. As is commonly understood, probability has to do with the likelihood of the occurrence of events. We often hear remarks like "the probability of rain is 30%," "he has a fifty–fifty chance of being elected," and "the probability of a knockout in the first round is small." But statements of this kind, though giving some general evaluation, are far from explicit. The conclusions are drawn without full knowledge or significance of the variables involved.

In our study of probability, we shall definitely spell out the conditions which favor the happening of an event and those which oppose the happening. Although the treatment will be quite elementary, the principles introduced are fundamental in the theory of probability. We begin our study by examining outcomes of some simple experiments.

First, consider the experiment of tossing a coin. There are obviously two possible outcomes—the head H will show or the tail T will show. So the set of possible outcomes is $\{H, T\}$. Next, in the fair toss of a die, with faces numbered 1 to 6, the set of possible outcomes is $\{1, 2, 3, 4, 5, 6\}$. The sets of outcomes in these experiments are called the sample space for the toss of a coin and the sample space for the toss of a die. The idea of a sample space, which we now define generally, is fundamental in the theory of probability.

Definition 15-2 The *sample space* U for an experiment is the set of all possible outcomes of the experiment.

EXAMPLE 1 Suppose that an urn contains a white (W) ball, a red (R) ball, and a green (G) ball, and let the experiment be that of drawing one ball from the urn and then another ball. To determine the sample space, we note that the first ball removed will be W, R, or G. If the first ball is W, the next ball will be either R or G. Hence, there are two possible outcomes, if the first ball is W. Similarly, there are two possible outcomes if either of the other colors shows up first. Hence, the sample space for this experiment is $\{WR, WG, RW, RG, GW, GR\}$. Each element of this set represents the order in which the two balls are drawn. Thus, for example, WR and RW are distinct elements.

We shall often be interested in certain combinations of the elements of a sample space. For example, we may wish to focus our attention on the elements 1, 2, 3 of the sample space for the toss of a die. Then this combination of elements, according to the following definition, is an event.

Definition 15-3 Any subset of a sample space is called an *event*.

Thus, the subsets for the experiment of tossing a coin, $U = \{H, T\}$, are

$$\varnothing, \{H\}, \{T\}, \{H, T\}$$

As we see here, four events are associated with a sample space of two elements. The subset $\varnothing$, however, having no element, represents an impossible event. Each of the two subsets with a single element is a possible event in the toss of the coin, and the subset with two elements is an event certain to occur.

If A is any subset of a sample space U, we shall denote the probability of the occurrence of event A in a performance of the experiment by $P(A)$. Then we require that $P(A)$ shall conform to the axioms:

Axiom 15-1

$$0 \leq P(A) \leq 1$$

Axiom 15-2

$$P(U) = 1$$

Axiom 15-3

$$P(A \cup B) = P(A) + P(B)$$

provided that A and B are disjoint subsets of U.

Axiom 15-1 sets the bounds for the measure of the probability of any event so that the least probability is 0 and the greatest is 1. Hence, we assign 0 to the probability of any result which cannot occur and 1 to the probability of an event which is certain to occur. Accordingly, the probability of drawing a red ball from a bag which contains only white balls is 0 and the probability of a white ball is 1. Axiom 15-2 tells us that one of the possible outcomes in a trial of an experiment is certain to occur. Axiom 15-3 states that the probability of the occurrence of either event A or event B for an experiment is equal to the sum of the probabilities of the separate events, provided that A and B have no common element. Axiom 15-3 can be extended to include three or more disjoint subsets. Thus, for three disjoint subsets, the probability of event A or event B or event C is given by the equation

$$\begin{aligned} P(A \cup B \cup C) &= P[(A \cup B) \cup C] \\ &= P(A \cup B) + P(C) \\ &= P(A) + P(B) + P(C) \end{aligned}$$

We are led, intuitively, to think that in some experiments the elements of a sample space all have the same probability. This situation is described by saying the outcomes are all "equally likely." Thus for the experiment of tossing a die, we have

$$P(U) = P(1) + P(2) + P(3) + P(4) + P(5) + P(6) = 1$$

If the 6 probabilities are all equally likely, then the probability that a particular face will be up is $\frac{1}{6}$. Similarly, in the toss of a coin the probabilities are $P(H) = \frac{1}{2}$ and $P(T) = \frac{1}{2}$. And, in general, if a sample space has n elements each with the same probability, then the probability of the occurrence of any one of the elements is $1/n$.

Theorem 15-4 Let a sample space U have n elements with each element possessing the same probability, and let A, consisting of k elements, be a subset of U. Then the probability of the occurrence of some one of the elements of A is equal to k/n.

Proof: Denoting the elements of A by $a_1, a_2, a_3, \ldots a_k$, we have

$$\begin{aligned} P(a_1 \cup a_2 \cup a_3 \cup \cdots \cup a_k) &= P(a_1) + P(a_2) + P(a_3) + \cdots + P(a_k) \\ &= \frac{1}{n} + \frac{1}{n} + \frac{1}{n} + \cdots + \frac{1}{n} \\ &= \frac{k}{n} \end{aligned}$$

This theorem states that the probability of an event is equal to the number of possible ways in which the event can occur divided by the number of possible outcomes of the experiment. We remark that k in the theorem may be any integer from 1 to n inclusive.

EXAMPLE 2 Six pennies are tossed, simultaneously. What is the probability that (*a*) all are heads, (*b*) 3 are heads and 3 are tails?

Solution:

(*a*) There are two ways for a penny to fall, and by the fundamental principle (Sec. 15-1), 6 pennies can fall in $2^6 = 64$ ways. All heads can occur in only 1 way. Therefore, the probability of obtaining all heads is $\frac{1}{64}$.

(*b*) A particular 3 of the pennies can fall heads and the remaining pennies fall tails in only 1 way. But 3 pennies can be chosen in $C(6, 3) = 20$ ways. Hence, the probability of 3 heads and 3 tails is $\frac{5}{16}$.

EXAMPLE 3 The names of 5 boys and 4 girls are written on cards. If 6 of the cards are drawn at random, what is the probability of getting the names of 3 boys and 3 girls?

Solution: There are $C(5, 3)$ ways of drawing the names of 3 boys and $C(4, 3)$ ways of drawing the names of 3 girls. Hence, $C(5, 3) \cdot C(4, 3)$ is the number of ways the event can occur. But the total number of ways of drawing 6 names is $C(9, 6)$. Hence,

$$P(3 \text{ boys and } 3 \text{ girls}) = \frac{C(5, 3) \cdot C(4, 3)}{C(9, 6)} = \frac{40}{84} = \frac{10}{21}$$

Theorem 15-5 If A and A' are complementary subsets of a sample space U, then $P(A) = 1 - P(A')$.

Proof: Since $U = A \cup A'$ and $A \cap A' = \varnothing$, we have

$$P(A) + P(A') = 1$$
$$P(A) = 1 - P(A')$$

We note that $P(A)$ is the probability that event A will occur and $P(A')$ is the probability the event will not occur.

EXAMPLE 4 A pair of dice is tossed. Find the probability that the sum of the numbers coming up is less than 11.

Solution: The sum of the numbers can be from 2 to 12 inclusive. To find the probabilities of getting the sums 2 to 10 inclusive requires several calculations. So we attack this problem by first finding the probabilities of getting 11 or 12. Then for this approach

$$A' = \{(5, 6), (6, 5), (6, 6)\}$$

which means there are two ways for 11 and one way for 12. In this experiment $U = 6 \cdot 6 = 36$, and consequently

$$P(A') = \tfrac{3}{36} = \tfrac{1}{12} \qquad \text{and} \qquad P(A) = 1 - \tfrac{1}{12} = \tfrac{11}{12}$$

EXERCISE 15-4

1. Eight white balls and 2 black balls are in a box. What is the probability that in drawing 1 ball it is (*a*) black, (*b*) white?
2. If a single die is cast, find the probability that it falls (*a*) a 1 or 2, (*b*) a 4, 5, or 6.
3. Two pennies are tossed, simultaneously. Find the probability that (*a*) both fall heads, (*b*) 1 falls heads and the other tails.
4. In a throw of a pair of dice (i.e., 2), find the probability that (*a*) both are aces, (*b*) the sum of the spots is 7, (*c*) just one is an ace. (An ace has 1 spot.)
5. Five pennies are tossed, simultaneously. Find the probability that (*a*) all are heads, (*b*) 4 are heads and 1 is tails, (*c*) 2 are heads and 3 are tails.
6. In an examination, the questions and answers are given but are not paired. A student succeeds in pairing all but 5 questions and 5 answers. These he does wholly by guessing. What is the probability that he pairs (*a*) all of them correctly, (*b*) only 3 correctly?
7. A baby in playing with the wooden digits 1, 2, 3, 4, 5 places them in a row. What is the probability that the number thus formed is (*a*) less than 20,000, (*b*) more than 40,000?
8. At a party, 8 men draw for partners from 8 cards which have the names of their wives. What is the probability that each man will draw the name of his wife?
9. From a bag containing 6 red balls and 5 black balls, 3 balls

are drawn. Find the probability that (*a*) all are red, (*b*) all are black, (*c*) 2 are red and 1 is black.

10. Five coins are tossed, simultaneously. Find the probability that (*a*) at least 1 head is up, (*b*) at least 2 heads are up.

11. A committee of 5 is chosen by lot from a group of 5 seniors and 4 juniors. Find the probability that the committee will have (*a*) 3 seniors and 2 juniors, (*b*) at least 2 seniors, (*c*) at least 1 junior.

12. From a bag containing 6 green balls, 7 black balls, and 8 yellow balls, 3 balls are drawn. Find the probability that they are (*a*) of different colors, (*b*) all of the same color.

13. If 2 cards are drawn from a deck of 52 cards, what is the probability that both are (*a*) queens, (*b*) hearts, (*c*) of the same suit?

14. From a deck of 52 cards, 3 cards are drawn. Find the probability that they are (*a*) hearts, (*b*) of the same suit, (*c*) of different suits.

15. Eight children, 4 boys and 4 girls, draw for places to form a line. Find the probability that (*a*) the boys will be together in the line, (*b*) a particular couple will be together.

16. Ten people, 5 boys and 5 girls, are grouped into pairs as partners. If the pairs are formed by drawing, find the probability that each boy has a girl for his partner.

17. A party of 5 men and 5 women are to eat at a round table. If the hostess reserves a place for herself and the others take places at random, find the probability that the men and women alternate.

18. Eight books, 5 red and 3 green, are taken from a table one after the other at random. Find the probability that (*a*) all the red books are taken first, (*b*) all the books of one color are taken first.

15-6 PROBABILITY IN MULTIPLE EVENTS

In the preceding section, we used Theorem 15-4 to find the probability of an event A. In this section, we shall consider the probability of the occurrence of two or more events.

Theorem 15-6 If A and B are subsets of a sample space U, then

$$P(A \cup B) = P(A) + P(B) - P(A \cap B)$$

Proof: If A and B are disjoint sets, $A \cap B = \varnothing$ and the equation is true by Axiom 15-3, Sec. 15-5. We assume then that A and B contain one or more common elements of U. Under this condition the sum $P(A) + P(B)$ includes the probability of the common elements twice. This sum, therefore, needs to be diminished by $P(A \cap B)$. Hence, the probability that at least one of two events occurs is equal to the sum of the probabilities of the separate events minus the probability of the occurrence of both events.

EXAMPLE 1 A card is drawn from a deck of 52 cards. Find the probability that the card is a queen or a heart.

Solution: Let A be the set of queens and B the set of hearts. Then $A \cap B = \{\text{queen of hearts}\}$, and therefore

$$\begin{aligned} P(A \cup B) &= P(A) + P(B) - P(A \cap B) \\ &= \tfrac{4}{52} + \tfrac{13}{52} - \tfrac{1}{52} = \tfrac{4}{13} \end{aligned}$$

Definition 15-4 If the occurrence of an event A does not affect the occurrence of another event B, then event B is said to be *independent* of event A. If, however, the probability of event B is affected by the prior occurrence of event A, then B is said to *depend* on A.

If an event B depends on event A, the probability of B after the occurrence of A is denoted by $P(B|A)$. If B is independent of A, then

$$P(B|A) = P(B)$$

EXAMPLE 2 Suppose a bag contains 4 green balls and 3 red balls, and that A represents the event of drawing a red ball from the bag and B the event of drawing a green ball after the occurrence of event A. After a red ball is drawn, the bag has 4 green balls and only 2 red balls. Hence, the probability that the second ball drawn is green is

$$P(B|A) = \tfrac{4}{6} = \tfrac{2}{3}$$

If, however, the red ball is replaced before the second drawing, then event B is independent of event A. So in this case B represents the event of drawing a green ball from a bag containing 4 green balls and 3 red balls, and consequently

$$P(B|A) = P(B) = \tfrac{4}{7}$$

Theorem 15-7 If A and B are events in a sample space, then

$$P(A \text{ and } B) = P(A \cap B) = P(A) \cdot P(B|A)$$

Proof: If n_1 is the number of ways A can occur and N_1 is the number of ways A can occur or fail to occur, then $P(A) = n_1/N_1$. Following the occurrence of A, let n_2 be the number of ways B can occur and N_2 the number of ways B can occur or fail to occur. Then $P(B|A) = n_2/N_2$. By the fundamental principle (Sec. 15-1), both events can occur n_1n_2 ways out of a total of N_1N_2 ways of occurring or failing to occur. Therefore,

$$P(A \cap B) = \frac{n_1n_2}{N_1N_2} = \frac{n_1}{N_1} \cdot \frac{n_2}{N_2} = P(A) \cdot P(B|A)$$

We note that if event B is independent of event A, then $P(B|A) = P(B)$, and therefore the formula becomes

$$P(A \cap B) = P(A) \cdot P(B)$$

EXAMPLE 3 A card is drawn from a deck of 52 cards and then a second card is drawn. Find the probability that the first card is a spade and the second a club if (*a*) the first card is replaced before the second card is drawn, (*b*) the first card is not replaced.

Solution:

(*a*) The probability of drawing a spade the first time is $\frac{1}{4}$, and the probability of a club on the second drawing is also $\frac{1}{4}$. Hence, the probability of the occurrence of both events is $\frac{1}{16}$.

(*b*) After the event of drawing a spade, the deck has 51 cards, 13 of which are clubs. Hence, the probability of a spade S and then a club C is

$$P(S \cap C) = P(S) \cdot P(C|S) = \tfrac{1}{4} \cdot \tfrac{13}{51} = \tfrac{13}{204}$$

Theorem 15-7 can be extended to include three or more events. Thus letting $P(B|A)$ stand for the probability of event B after the occurrence of A and $P(C|A \cap B)$ stand for the probability of the occurrence of event C after the occurrence of A and then B, we have

$$P(A \cap B \cap C) = P(A) \cdot P(B|A) \cdot P(C|A \cap B)$$

If the events are independent, the formula for the probability of A, B, and C becomes

$$P(A \cap B \cap C) = P(A) \cdot P(B) \cdot P(C)$$

EXERCISE 15-5

1. Events A and B of a sample space U are

   $$A = \{3, 4, 5\}, \qquad B = \{1, 3, 4, 6\}, \qquad U = \{1, 2, 3, 4, 5, 6\}$$

   Find $P(A \cup B)$ by (*a*) using Theorem 15-6, (*b*) using the fact that $A \cup B = \{1, 3, 4, 5, 6\}$.

2. A ball is drawn from a box containing 4 red balls and 5 green balls. It is then returned to the box and a second ball is drawn. Find the probability that (*a*) both balls are red, (*b*) both are green, (*c*) the first is red and the second green.

3. Repeat problem 2 if the first ball is not returned.

4. A coin is tossed 3 times in succession, find the probability that it falls (*a*) heads each time, (*b*) heads first and tails on the other throws.

5. One card is drawn from each of three decks of 52 cards. What is the probability that (*a*) all are aces, (*b*) all are face cards, (*c*) all are spades?

6. Three cards are drawn from a deck one at a time and are not replaced. What is the probability that (*a*) all are aces, (*b*) all are face cards, (*c*) all are spades?

7. If the probability that a brush salesman will make a sale at one house is $\frac{2}{3}$ and the probability that he will make a sale at a second house is $\frac{1}{2}$, find the probability that he will make (*a*) both sales, (*b*) neither sale, (*c*) a sale at the first house but not at the second house.

8. A store entrance has 4 doors. What is the probability that the next three customers, coming singly, will enter by (*a*) the same door, (*b*) different doors?

9. The names of 9 freshmen, 8 sophomores, and 7 juniors are written on cards and placed in a box. If three cards are drawn from the box, one after the other, find the probability of getting the names of (*a*) 3 freshmen, (*b*) one from each class, (*c*) all from one class.

10. A student is to take 3 tests. If the separate probabilities of his passing the tests are $\frac{4}{5}$, $\frac{3}{4}$, and $\frac{2}{3}$, find the probability that he will pass (*a*) all tests, (*b*) none of the tests, (*c*) exactly two of the tests.

11. To be initiated into an organization a candidate must pass at least two of three tests. If the separate probabilities of his passing the tests are $\frac{2}{3}$, $\frac{1}{4}$, and $\frac{4}{5}$, find the probability that he will win the right to be initiated.

12. Three balls are drawn from a box one at a time and are not replaced. If the box had 5 white balls and 7 green balls, find the probability that the 3 balls are drawn in the order (*a*) white, green, green, (*b*) white, white, green, (*c*) green, green, green.

13. The names of 6 girls and 5 boys are drawn, one at a time, from a box. Find the probability that (*a*) the girls' names are drawn first, (*b*) the names of the girls and boys alternate in the drawing.

14. Three balls are drawn from a box one at a time and are not replaced. If the box had 5 red balls, 6 white balls, and 7 blue balls, find the probability that the 3 balls drawn are (*a*) of the same color, (*b*) of different colors.

15. To determine which will do a task, 3 men match coins on the basis that the odd man is to do the task. Find the probability that the decision will be made on (*a*) the first trial, (*b*) the second trial.

16. Repeat problem 15 if the odd man is relieved of the task and the others then match to determine the issue.

17. An urn contains 5 black balls and 2 green balls and another urn contains 3 black balls and 3 green balls. If 1 ball is to be drawn, what is the probability of drawing (*a*) a black ball, (*b*) a green ball?

18. A, B, and C are to decide on one of them to do a task. A and B first draw lots, and the loser then draws with C. Find each man's probability of having to do the task.

If the probability that a person will receive a sum of A dollars is p, then pA is called the value of his expectation.

19. A man has 3 entries in a horse race. He figures that the separate probabilities for his horses to be first are $\frac{1}{3}$, $\frac{1}{4}$, and $\frac{1}{6}$. On this basis, find the probability that one of his horses wins. If there is a prize of \$500 for the winner, what is the value of the man's expectation?

20. The separate probabilities that marksmen A, B, and C will hit a target are $\frac{2}{3}$, $\frac{3}{4}$, and $\frac{4}{5}$. Shooting in the order A, B, and C, the first to hit the target gets a prize of \$100, but not more than a total of 3 shots are allowed. Find the value of each man's mathematical expectation.

21. A person draws 3 balls in succession from a box containing 5 red balls, 6 yellow balls, and 7 green balls. He is to receive a prize of \$100 if the balls are drawn in the order red, yellow, and green. Find the value of his expectation if (*a*) each ball is replaced before the next is drawn, (*b*) if the balls are not replaced.

22. Find the value of the man's expectation in problem 21 if a second chance is granted in case of failure in the first attempt.

15-7 PROBABILITIES IN REPEATED TRIALS

Let us consider a common type experiment in which the outcome of any trial is independent of the outcome of any other trial. Then let us find the probability that an event A will occur 0, 1, 2, 3, and so on, times in n trials.

We let p be the probability that A will occur in a single trial and q be the probability that A fails to occur. Clearly, the probability that A fails to occur on all of the n trials is q^n. The probability of the occurrence of A on a specified trial and failures on all the other trials is pq^{n-1}. But the occurrence could happen on any of the n trials, and therefore in $C(n, 1)$ ways. So the probability for exactly one occurrence of A is $C(n, 1)pq^{n-1}$. Similarly, the probability that A occurs on r specified trials and fails to occur on the remaining $n - r$ trials is p^rq^{n-r}. Since the r occurrences of A

could come in $C(n, r)$ ways, we conclude that the probability that A occurs exactly r times in n trials is $C(n, r)p^r q^{n-r}$. We state this result as a theorem.

Theorem 15-8 If p is the probability that an event will occur in a single trial of an experiment and q is the probability the event will fail to occur, then the probability that the event will occur exactly r times in n trials is

$$C(n, r)p^r q^{n-r}$$

EXAMPLE In 4 throws of a die, what is the probability of (*a*) exactly 2 aces, (*b*) 2 or more aces?

Solution:

(*a*) The probability of one ace on a single throw is $\frac{1}{6}$, and of failure $\frac{5}{6}$. Hence, the probability of two aces is

$$C(n, r)p^r q^{n-r} = C(4, 2)(\tfrac{1}{6})^2(\tfrac{5}{6})^2 = \tfrac{25}{216}$$

(*b*) The probability of 2, 3, or 4 aces is equal to the sum of the separate probabilities of those events. Hence, we have

$$\begin{aligned} C(4, 2)(\tfrac{1}{6})^2(\tfrac{5}{6})^2 + C(4, 3)(\tfrac{1}{6})^3(\tfrac{5}{6}) + (\tfrac{1}{6})^4 &= 150(\tfrac{1}{6})^4 + 20(\tfrac{1}{6})^4 + (\tfrac{1}{6})^4 \\ &= \frac{171}{6^4} = \frac{171}{1296} \end{aligned}$$

EXERCISE 15-6

1. A coin is tossed 4 times. Find the probability of (*a*) exactly 2 heads, (*b*) exactly 3 heads, (*c*) fewer than 2 heads.
2. Find the probability that in 3 throws of a die there will be (*a*) no six, (*b*) 1 six, (*c*) 2 or 3 sixes.
3. A box contains 4 red balls and 3 white balls. If 4 balls are drawn in succession and each is replaced before the next is taken, find the probability that (*a*) all are white, (*b*) exactly 2 are white, (*c*) 3 or more are red.
4. In 6 tosses of a coin, find the probabilities of exactly 0, 1, 2, 3, 4, 5, and 6 heads. What number of heads is most likely?

5. The probability that a certain ball player will get a hit in one time at bat is 0.3. Find the probability that in 4 times at bat he will (*a*) go hitless, (*b*) make exactly 1 hit, (*c*) make more than 1 hit.

6. If the probability that a rifleman will hit a target in a single trial is 0.9, find the probability that in 4 trials he will make (*a*) no hit, (*b*) exactly 3 hits, (*c*) 4 hits.

7. The probability that team A will win a basketball game from team B is 0.4. In a series of 5 games, what number of games is A most likely to win? Is A more likely to win exactly 1 or exactly 3 games?

16

THE BINOMIAL THEOREM AND MATHEMATICAL INDUCTION

16-1 THE BINOMIAL EXPANSION

In Chap. 3, we learned to write the square and cube of a binomial. We wish now to find out how to write any positive integral power of a binomial as a polynomial.

We may show by multiplying that

$$(a + x)^1 = a + x$$

$$(a + x)^2 = a^2 + 2ax + x^2$$

$$(a + x)^3 = a^3 + 3a^2x + 3ax^2 + x^3$$

$$(a + x)^4 = a^4 + 4a^3x + 6a^2x^2 + 4ax^3 + x^4$$

$$(a + x)^5 = a^5 + 5a^4x + 10a^3x^2 + 10a^2x^3 + 5ax^4 + x^5$$

The right members of these identities are the expanded forms of the positive integral powers of $a + x$ from the first through the fifth. In each of the expansions we may observe the following properties:

1. The number of terms in each expansion is one more than the exponent of the binomial.

2. The first term is a with an exponent the same as the exponent of the binomial, and the exponent decreases by 1 from term to term.

3. The exponent of x in the second term is 1, and it increases by 1 from term to term.

4. The sum of the exponents of a and x in any term is equal to the exponent of the binomial.

5. The coefficients of the terms equidistant from the first term and the last term are equal.

6. The coefficient of the second term is the same as the exponent of the binomial. The coefficient of any term farther on may be computed from the previous term by multiplying that term's coefficient by the exponent of a and dividing by 1 more than the exponent of x.

The student should verify property 6 in the foregoing expansions. To check the expansion of $(a + x)^5$, we observe that the second term is $5a^4x$. Then the coefficient of the next term is $(5 \cdot 4) \div 2 = 10$, which gives $10a^3x^2$ for the third term. A continuation of this simple scheme will yield the expansion.

The six properties which we have stated are true for the expansion of $(a + x)^n$ if n is equal to 1, 2, 3, 4, or 5. It is natural to ask if the properties are true where n is any positive integer. We answer the question by proving the so-called binomial theorem.

Theorem 16-1 If n is any positive integer, then

$$(a + x)^n = a^n + na^{n-1}x + \frac{n(n-1)}{2!}a^{n-2}x^2 + \frac{n(n-1)(n-2)}{3!}a^{n-3}x^3 + \frac{n(n-1)(n-2)(n-3)}{4!}a^{n-4}x^4 + \cdots + x^n \qquad (1)$$

Proof: By definition $(a + x)^n$ means the product of n factors each of which is $a + x$. Thus,

$$(a + x)^n = (a + x)(a + x)(a + x) \cdots (a + x) \qquad n \text{ factors}$$

From the distributive axiom, we know that the expansion consists of the sum of all possible products formed by taking one term from each of the n factors. Thus choosing a from every factor, we have the product a^n. Clearly, there is only one way of forming this product. If we choose a from all but one of the factors and x from the remaining factor, we obtain the product $a^{n-1}x$. But x can come from any of the n factors. Hence, $a^{n-1}x$ occurs $n = C(n, 1)$ times, giving $C(n, 1)a^{n-1}x$ as a term of the expansion. Now, more generally, suppose we choose a from $n - r$ of the factors and x from the remaining r factors. This can be done $C(n, r)$ times, which gives $C(n, r)a^{n-r}x^r$ as a term of the expansion. Using the definition of $C(n, r)$, we may write

$$C(n, r)a^{n-r}x^r = \frac{n(n-1)(n-2)\cdots(n-r+1)}{r!}a^{n-r}x^r \qquad (2)$$

If we give r, successively, the values $1, 2, 3, \ldots, n$, we obtain in order the terms of the expansion following the first term a^n. We note that the expansion thus obtained agrees with the right member of equation (1), and therefore the theorem is proved.

The coefficients in (1) are called the *binomial coefficients*. The binomial coefficients, as we have seen, may be expressed in the notation of combinations. This permits us to write (1) in the more compact form

$$(a + x)^n = a^n + C(n, 1)a^{n-1}x + C(n, 2)a^{n-2}x^2 + \cdots + C(n, r)\cdot a^{n-r}x^r + \cdots + C(n, n-1)ax^{n-1} + C(n, n)x^n \qquad (3)$$

The binomial theorem provides a formula for writing the expansion of any positive integral power of a binomial. Examining formula (1), we see that the binomial coefficients after the second term may be computed in accordance with property 6. And, as we surmised, the coefficients equidistant from the first term and the last term are equal. This conclusion follows from the fact that $C(n, r) = C(n, n - r)$. Hence, the coefficients in the first half of an expansion may be used, in reverse order, as the coefficients in the second half, provided the expansion has an even number of terms. If the expansion has an odd number of terms, the coefficients equidistant from the middle term are equal. This property of the binomial coefficients may be used to advantage in expanding binomials.

Examining the right member of equation (3), we see that x, x^2, x^3, and so on, are the respective factors of the second, third, fourth, and so on, terms of the expansion. Accordingly, the rth term of the expansion is given by $C(n, r - 1)a^{n-r+1}x^{r-1}$. Hence, we may write

$$r\text{th term} = \frac{n(n-1)(n-2)\cdots(n-r+2)}{(r-1)!}a^{n-r+1}x^{r-1} \qquad (4)$$

We now have sufficient information for finding any particular term of a binomial expansion without finding the previous terms. The first and second terms, of course, may be written at once, and formula (4) will yield readily any other chosen term of the expansion by assigning to r the appropriate value.

EXAMPLE 1 Expand $(2 + x)^6$ and simplify each term.

Solution: Applying formula (1) with $a = 2$ and $n = 6$, we get

$$(2+x)^6 = 2^6 + 6(2)^5x + \frac{6\cdot 5}{2!}(2)^4x^2 + \frac{6\cdot 5\cdot 4}{3!}(2)^3x^3$$
$$+ \frac{6\cdot 5\cdot 4\cdot 3}{4!}(2)^2x^4 + \frac{6\cdot 5\cdot 4\cdot 3\cdot 2}{5!}(2)x^5 + x^6$$

We now simplify the coefficients, and thus have

$$(2+x)^6 = 64 + 192x + 240x^2 + 160x^3 + 60x^4 + 12x^5 + x^6$$

In this example we expressed the binomial coefficient of each term independently of any other term. The expansion may be written more conveniently by computing the coefficients after the second term in accordance with property 6. We illustrate this procedure in the following examples.

EXAMPLE 2 Expand the binomial $(b^2 + 3y)^5$ and simplify each term.

Solution: To obtain the expansion, we use formula (1) with $a = b^2$, $x = 3y$, and $n = 5$. In this example, however, we compute the coefficients after the second term by the use of property 6. Thus,

$$(b^2+3y)^5 = (b^2)^5 + 5(b^2)^4(3y) + 10(b^2)^3(3y)^2 + 10(b^2)^2(3y)^3$$
$$+ 5(b^2)(3y)^4 + (3y)^5$$

Taking the indicated power of b^2 and $3y$ in each term, we have

$$(b^2 + 3y)^5 = b^{10} + 15b^8y + 90b^6y^2 + 270b^4y^3 + 405b^2y^4 + 243y^5$$

EXAMPLE 3 Expand $(a - x)^7$ and simplify each term.

Solution: Noting that $a - x = a + (-x)$, we have

$$[a + (-x)]^7 = a^7 + 7a^6(-x) + 21a^5(-x)^2 + 35a^4(-x)^3 + 35a^3(-x)^4 + 21a^2(-x)^5 + 7a(-x)^6 + (-x)^7$$

Hence,

$$(a - x)^7 = a^7 - 7a^6x + 21a^5x^2 - 35a^4x^3 + 35a^3x^4 - 21a^2x^5 + 7ax^6 - x^7$$

The signs alternate here, as will always be the case when the binomial is the difference of two quantities.

EXAMPLE 4 Write and simplify the seventh term of the expansion of $(2x - y)^{11}$.

Solution: Using formula (4) with $n = 11$ and $r = 7$, we obtain

$$\text{seventh term} = \frac{11 \cdot 10 \cdot 9 \cdot 8 \cdot 7 \cdot 6}{6!}(2x)^5(-y)^6$$

$$= 462(2x)^5(-y)^6 = 14{,}784x^5y^6$$

EXAMPLE 5 Find the term involving x^9 in the expansion of $[x^2 + (1/x)]^{12}$.

Solution: Using $n = 12$, we find that the rth term of the expansion, except for the numerical coefficient, is given by

$$(x^2)^{12-r+1}\left(\frac{1}{x}\right)^{r-1} = x^{26-2r} \cdot x^{1-r} = x^{27-3r}$$

Then $27 - 3r$ must be equal to 9, and therefore $r = 6$. Using this value of r and formula (4), we have

$$\frac{12 \cdot 11 \cdot 10 \cdot 9 \cdot 8}{5 \cdot 4 \cdot 3 \cdot 2 \cdot 1}(x^2)^7\left(\frac{1}{x}\right)^5 = 792x^9$$

EXERCISE 16-1

Write the expansion of each binomial and simplify the terms.

1. $(2a - y)^3$
2. $(3a + 2y)^4$
3. $(a + 2x)^5$
4. $(2m + 3n)^5$
5. $(3m - n^2)^5$
6. $(1 - 2x)^6$
7. $(a - x)^8$
8. $(n^2 - 3m)^6$
9. $(x + y)^8$
10. $\left(\frac{x}{2} + \frac{2}{x}\right)^6$
11. $\left(\frac{a}{b} - \frac{b}{a}\right)^6$
12. $\left(x + \frac{1}{x}\right)^7$
13. $\left(\frac{x}{3} + \frac{y}{2}\right)^5$
14. $\left(\frac{y}{x} + \frac{1}{y}\right)^6$
15. $\left(\frac{a^2}{4} + \frac{2}{a}\right)^6$

Find the specified term in each of the following powers. Do not find the preceding terms.

16. $(a - x)^8$; fifth term
17. $(a + x)^7$; fourth term
18. $(m + 3)^7$; fourth term
19. $(x + 1)^{11}$; eighth term
20. $(2x + y)^{10}$; seventh term
21. $(2a^2 - 1)^{12}$; ninth term
22. $(2x^2 + y^2)^{13}$; tenth term
23. $(\frac{1}{2}a + \frac{1}{3}b)^{10}$; middle term
24. $(a - \frac{1}{2}b)^{10}$; middle term
25. $\left(a - \frac{b}{2}\right)^{12}$; middle term
26. $(x^2 + y^2)^{14}$; middle term
27. $(3x^2 - 2y)^8$; middle term
28. $\left(x + \frac{1}{x}\right)^8$; term free of x
29. $\left(x^2 - \frac{1}{x}\right)^9$; term free of x
30. $\left(3x + \frac{1}{x^2}\right)^9$; term free of x
31. $\left(x^3 - \frac{1}{x}\right)^{13}$; term involving x^3
32. $\left(x^2 - \frac{1}{x^3}\right)^{10}$; term involving x^5
33. $\left(x^3 - \frac{1}{x^2}\right)^{10}$; term involving x^5

Use enough terms of the binomial expansion to evaluate each of the following expressions to three decimal places.

34. $(1.01)^7 = (1 + .01)^7$
35. $(1.02)^8$
36. $(0.99)^5$

37. $(0.98)^9$ **38.** $(1.01)^{10}$ **39.** $(0.98)^8$

40. Prove that the total number of combinations of n things taken successively 1 at a time, 2 at a time, 3 at a time, and so on, to n at a time, is equal to $2^n - 1$.
Hint: Expand $(1 + 1)^n$, expressing the binomial coefficients as in formula (3).

The binomial coefficients of the expansions of $(a + x)^n$ from $n = 0$ to $n = 6$ are arranged below in a triangular form known as Pascal's triangle. The first and last numbers of each row is 1; and each of the other numbers, starting with the third row, is equal to the sum of the two nearest numbers in the row just above.

$(a + x)^0$	1
$(a + x)^1$	1 1
$(a + x)^2$	1 2 1
$(a + x)^3$	1 3 3 1
$(a + x)^4$	1 4 6 4 1
$(a + x)^5$	1 5 10 10 5 1
$(a + x)^6$	1 6 15 20 15 6 1
	. . . .

41. Form a Pascal triangle of 10 rows. Use the numbers of the tenth row to write the expansion of $(a + x)^9$.

16-2 THE BINOMIAL SERIES

Thus far, we have considered binomials whose exponents are positive integers. We now wish to let n be a number other than a positive integer. Expanding in the usual way, we have

$$(1 + x)^n = 1 + nx + \frac{n(n-1)}{2!}x^2 + \cdots + \frac{n(n-1)(n-2)\cdots(n-r+1)}{r!}x^r + \cdots$$

Observing the coefficient of the $(r + 1)$st term here, we see that the factor $(n - r + 1)$ is equal to 1 when $r = n$, provided n is a positive integer, and is equal to 0 when $r = n + 1$. Hence, the expansion ends with the $(n + 1)$st term. But no binomial coefficient could have 0 as a factor if n is different from a positive integer. In this case the expansion does not terminate, and it is called a *binomial series.* We can, however, write as many terms as we like of such a series. There are mathematical situations in which the first few terms of a binomial series yield useful results, provided that the values of x are chosen between -1 and 1. For any of these values an approximation to $(1 + x)^n$ can be obtained to any desired degree of accuracy by taking a sufficient number of terms of the expansion. And, more generally, a satisfactory approximation of $(a + b)^n$ can be obtained if the value of b/a is between -1 and 1.

EXAMPLE 1 Find the first four terms of the expansion of $(2 - x)^{1/2}$.

Solution: Applying the binomial formula and property 6 in the process, we get

$$\begin{aligned}
&(2 - x)^{1/2} \\
&= 2^{1/2} + \tfrac{1}{2}(2)^{-1/2}(-x) - \tfrac{1}{8}(2)^{-3/2}(-x)^2 + \tfrac{1}{16}(2)^{-5/2}(-x)^3 - \cdots \\
&= \sqrt{2} - \frac{x}{2\sqrt{2}} - \frac{x^2}{8 \cdot 2\sqrt{2}} - \frac{x^3}{16 \cdot 4\sqrt{2}} - \cdots \\
&= \sqrt{2}\left(1 - \frac{x}{4} - \frac{x^2}{32} - \frac{x^3}{128} - \cdots\right)
\end{aligned}$$

EXAMPLE 2 Find an approximation to $\sqrt[4]{17}$.

Solution: We first write $\sqrt[4]{17} = (16 + 1)^{1/4}$. Then

$$\begin{aligned}
(16 + 1)^{1/4} &= 16^{1/4} + \tfrac{1}{4}(16)^{-3/4}(1) - \tfrac{3}{32}(16)^{-7/4}(1)^2 + \cdots \\
&= 2 + \tfrac{1}{32} - \tfrac{3}{4096} + \cdots \\
&= 2 + 0.031 - 0.001 + \cdots \\
&= 2.03 \qquad \text{to two decimal places}
\end{aligned}$$

Notice that we obtained this approximation from two terms of the expansion. The terms farther on in the expansion decrease rapidly toward zero. The fourth term rounded off to five decimal places is 0.00003 and the fifth term rounded off to six decimal places is -0.000002. So it seems that we are safe in taking 2.03 as the best approximation, to two decimal places, in this example.

EXERCISE 16-2

Expand each binomial to four terms and simplify the coefficients.

1. $(1 + x)^{-1}$ **2.** $(1 - y)^{1/2}$ **3.** $(1 - y)^{-1}$

4. $(1 - x)^{-2}$ **5.** $(1 + x)^{-3}$ **6.** $(1 + y)^{-1/2}$

7. $(1 + 3x)^{-3/2}$ **8.** $(1 - 2y)^{2/3}$ **9.** $(1 - 4x)^{-4/3}$

10. $(x + y)^{-2}$ **11.** $(a - 2x)^{-3}$ **12.** $(3 - y)^{-4}$

13. $(2 + x)^{1/4}$ **14.** $(4 + x)^{-5}$ **15.** $(2 - y)^{-2/3}$

Approximate to two decimal places, using the necessary number of terms of each expansion.

16. $\sqrt{26} = (25 + 1)^{1/2}$ **17.** $\sqrt[3]{26}$ **18.** $\sqrt[4]{82}$

19. $\sqrt{1.02}$ **20.** $24^{-1/2}$ **21.** $31^{-1/5}$

22. $(0.99)^{-4}$ **23.** $(1.01)^{-3}$ **24.** $(0.97)^{-5}$

16-3 MATHEMATICAL INDUCTION

Certain mathematical theorems and formulas can be proved by a process of reasoning called *mathematical induction.* This method of proof is based on the following axiom:

Axiom 11 *The mathematical induction axiom:* Let A stand for a set of positive integers such that

1. The integer 1 belongs to A.
2. If k belongs to A, it follows that $k + 1$ belongs to A.

Then the set A consists of all positive integers.

As this axiom indicates, there are two essential parts in the proof of a theorem by mathematical induction. The two parts are fittingly illustrated by the ladder analogy. Suppose we seek conditions which will ensure that a climber can ascend to any height of an infinitely long ladder. The following two conditions are sufficient to convince us that the climber can ascend to any chosen height:

1. Suppose we know he can get on the first rung.
2. Having reached the first rung or any other rung of the ladder, suppose he can get on the next higher rung.

By condition 1, he can get on the first rung; then by repeated applications of condition 2, we see that he can get on successive rungs, the second, third, fourth, and so on.

The following examples are designed to aid the student in understanding the rather difficult concept of mathematical induction. Each example should be studied carefully.

EXAMPLE 1 Prove that the sum of the first n positive odd integers is n^2; that is, show that

$$1 + 3 + 5 + \cdots + (2n - 1) = n^2$$

Proof: The proof has two separate parts.

Part 1. For the formula to be true, it must hold in particular for the smallest integer which may be assigned to n. The smallest integer for which the formula would have meaning is $n = 1$. Then checking the formula for $n = 1$ is the first part of the proof. To illustrate the meaning of the formula, we substitute $n = 2$ and $n = 3$.

For $n = 1$:	$1 = 1^2$	or	$1 = 1$
For $n = 2$:	$1 + 3 = 2^2$	or	$4 = 4$
For $n = 3$:	$1 + 3 + 5 = 3^2$	or	$9 = 9$

Thus, we see that the formula is true for $n = 1$, $n = 2$, and $n = 3$. But this does not prove the formula for other values of n, and further checking would make the proof only for the particular values used. To complete the proof, we proceed to part 2.

Part 2. This part of the proof consists of showing that if the formula holds for any value of n, say $n = k$, then consequently it also holds for $n = k + 1$. That is, if k^2 is the sum of the first k positive odd integers, then it must be shown that $(k + 1)^2$ is the sum of the first $k + 1$ positive odd integers. Hence, substituting k for n, we *assume* that

$$1 + 3 + 5 + \cdots + (2k - 1) = k^2$$

Now if we add $2k + 1$ (the next consecutive odd integer) to both sides of the equation, we have

$$1 + 3 + 5 + \cdots + (2k - 1) + (2k + 1) = k^2 + (2k + 1)$$

The left member of this equation is the sum of the first $k + 1$ positive odd integers. The right member, when factored, is $(k + 1)^2$. Hence, the given formula is true for $n = k + 1$, provided it is true for $n = k$.

Having established the two parts of the proof, let us notice what conclusions can be drawn. By part 1 the formula holds for $n = 1$. Then by applying part 2 repeatedly, we see that the formula holds for $n = 2, 3, 4$, and so on. Hence, we conclude (Axiom 11) that the formula holds for n any positive integer.

We should notice that both parts of a proof by mathematical induction are essential. The failure of either part in a supposed formula means that the formula holds for no value of n or, at most, for only some values of n. For example, the relation

$$1 + 2 + 3 + \cdots + n = 2n - 1$$

holds for $n = 1$ and for $n = 2$. It does not hold for any other value of n; and hence part 2 could not be established.

Let us next test the formula

$$5 + 10 + 15 + \cdots + 5n = \tfrac{5}{2}n(n + 1) - 1$$

Assuming that the formula holds for $n = k$, we have

$$5 + 10 + 15 + \cdots + 5k = \tfrac{5}{2}k(k + 1) - 1$$

Add $5k + 5$ to both sides of the equation:

$$\begin{aligned} 5 + 10 + 15 + \cdots + 5k + (5k + 5) \\ &= \tfrac{5}{2}k(k + 1) - 1 + (5k + 5) \\ &= \tfrac{5}{2}(k + 1)(k + 2) - 1 \end{aligned}$$

The right member of this equation is just what we would get by substituting $n = k + 1$ in the given formula. Thus, part 2 of the proof is established. But if we try part 1 by substituting $n = 1$ in the supposed formula, we have

$$5 \neq \tfrac{5}{2}(1)(2) - 1 = 5 - 1 = 4$$

Hence, part 1 fails; the relation as given holds for no value of n.

EXAMPLE 2 Prove by mathematical induction that

$$1^2 + 2^2 + 3^2 + \cdots + n^2 = \tfrac{1}{6}n(n + 1)(2n + 1)$$

Proof: We must establish the two parts of the proof.

Part 1. If we try the formula for $n = 1$, we have

$$1^2 = \tfrac{1}{6}(1)(1 + 1)(2 + 1) = \tfrac{1}{6}(2)(3) = 1$$

Thus, we see that the formula holds for $n = 1$. Although this constitutes the demonstration of part 1, a few other small values of n may be tried if one so desires.

Part 2. We next assume that the formula is true for $n = k$; that is,

$$1^2 + 2^2 + 3^2 + \cdots + k^2 = \tfrac{1}{6}k(k + 1)(2k + 1)$$

Now adding the square of $k + 1$ to both sides of this equation, we obtain

$$\begin{aligned} 1^2 + 2^2 + 3^2 + \cdots + k^2 + (k + 1)^2 & \\ &= \tfrac{1}{6}k(k + 1)(2k + 1) + (k + 1)^2 \\ &= \tfrac{1}{6}(k + 1)(k + 2)(2k + 3) \end{aligned}$$

The left member of this equation is the sum of the squares of the first $k + 1$ positive integers. And the expression in the right member is exactly that which the given formula yields for $n = k + 1$; that is,

$$\begin{aligned} \tfrac{1}{6}n(n + 1)(2n + 1) &= \tfrac{1}{6}(k + 1)(k + 1 + 1)[2(k + 1) + 1] \\ &= \tfrac{1}{6}(k + 1)(k + 2)(2k + 3) \end{aligned}$$

Hence, by Axiom 11, we conclude that the given formula is true.

EXERCISE 16-3

Prove that the following formulas are true for all positive integral values of n. Use mathematical induction.

1. $1 + 2 + 3 + \cdots + n = \dfrac{n(n + 1)}{2}$

2. $2 + 4 + 6 + \cdots + 2n = n(n + 1)$

3. $1^2 + 3^2 + 5^2 + \cdots + (2n - 1)^2 = \dfrac{n(2n - 1)(2n + 1)}{3}$

4. $1^3 + 2^3 + 3^3 + \cdots + n^3 = \dfrac{n^2(n+1)^2}{4}$

5. $2^2 + 4^2 + 6^2 + \cdots + (2n)^2 = \dfrac{2n(n+1)(2n+1)}{3}$

6. $1^3 + 3^3 + 5^3 + \cdots + (2n-1)^3 = n^2(2n^2 - 1)$

7. $\dfrac{1}{1\cdot 2} + \dfrac{1}{2\cdot 3} + \dfrac{1}{3\cdot 4} + \cdots + \dfrac{1}{n(n+1)} = \dfrac{n}{n+1}$

8. $\dfrac{1}{1\cdot 3} + \dfrac{1}{3\cdot 5} + \dfrac{1}{5\cdot 7} + \cdots + \dfrac{1}{(2n-1)(2n+1)} = \dfrac{n}{2n+1}$

9. $a + (a+d) + (a+2d) + \cdots$

$$+ [a + (n-1)d] = \frac{n[2a + (n-1)d]}{2}$$

10. $a + ar + ar^2 + \cdots + ar^{n-1} = \dfrac{a - ar^n}{1 - r}$

11. Prove that $2^{3n} - 1$ is divisible by 7 if n is any positive integer.

Proof:

(*a*) When $n = 1$, the expression $2^{3n} - 1 = 7$.

(*b*) Assume that $2^{3k} - 1 = 7x$, when x is some positive integer. Then, multiplying both members of this equation by 8, we obtain

$$\begin{aligned} 2^3 \cdot 2^{3k} - 8 &= 56x \\ 2^{3(k+1)} - 1 &= 56x + 7 \\ &= 7(8x + 1) \end{aligned}$$

In the following problems, assume that n stands for any positive integer.

12. Prove that $2^{2n} - 1$ is divisible by 3.

13. Prove that $3^n - 1$ is divisible by 2.

14. Prove that $7^n - 1$ is divisible by 6.

15. Prove that $3^{2n} - 1$ is divisible by 8.

16. Prove that $3^{3n} - 1$ is divisible by 26.

17. Prove that $x^n - y^n$ is divisible by $x - y$, where $x - y \neq 0$.

Hint: $x^{k+1} - y^{k+1} = x^{k+1} - xy^k + xy^k - y^{k+1}$

$$= x(x^k - y^k) + y^k(x - y)$$

18. Prove that $x^{2n+1} + y^{2n+1}$ is divisible by $x + y$, where $x + y \neq 0$.

16-4 PROOF OF THE BINOMIAL THEOREM BY MATHEMATICAL INDUCTION

We now apply mathematical induction to obtain a second proof of the binomial theorem. Since the theorem has been tested for $n = 1$, we proceed to the second part of the proof.

Part 2. We assume that the formula is true for $n = k$. Then we have

$$\begin{aligned}(a + x)^k = a^k &+ ka^{k-1}x \\ &+ \cdots + \frac{k(k-1)\cdots(k-r+2)}{(r-1)!}a^{k-r+1}x^{r-1} \\ &+ \frac{k(k-1)\cdots(k-r+2)(k-r+1)}{r!}a^{k-r}x^r \\ &+ \cdots + kax^{k-1} + x^k \qquad (1)\end{aligned}$$

To obtain an expression for $(a + x)^{k+1}$, we multiply both members of Eq. (1) by $a + x$. This gives

$$\begin{aligned}(a + x)^{k+1} = a^{k+1} &+ (k+1)a^kx + \cdots \\ &+ \frac{(k+1)(k)(k-1)\cdots(k-r+2)}{r!}a^{k-r+1}x^r \\ &+ \cdots + x^{k+1}\end{aligned}$$

The student should verify this multiplication. It is easy to check the first, second, and last terms. The term containing $a^{k-r+1}x^r$ is obtained by multiplying the rth term of the expansion in (1) by x and the $(r + 1)$st term by a and then adding these two products. Thus, the coefficient of $a^{k-r+1}x^r$ is the sum of the coefficients of the rth term and the $(r + 1)$st term of the right member of (1); that is,

$$\begin{aligned}&\frac{k(k-1)\cdots(k-r+2)}{(r-1)!} + \frac{k(k-1)\cdots(k-r+2)(k-r+1)}{r!} \\ &\quad = \frac{rk(k-1)\cdots(k-r+2) + k(k-1)\cdots(k-r+2)(k-r+1)}{r!} \\ &\quad = \frac{k(k-1)\cdots(k-r+2)[r + (k-r+1)]}{r!} \\ &\quad = \frac{(k+1)(k)(k-1)\cdots(k-r+2)}{r!}\end{aligned}$$

The expression which we have obtained for $(a + x)^{k+1}$ is exactly what the binomial formula yields by substituting $n = k + 1$. Hence for n any positive integer, we have

$$(a + x)^n = a^n + na^{n-1}x + \frac{n(n-1)}{2!}a^{n-2}x^2 + \cdots + \frac{n(n-1)\cdots(n-r+2)}{(r-1)!}a^{n-r+1}x^{r-1} + \cdots + x^n$$

17

THEORY OF EQUATIONS

17-1 POLYNOMIALS AND POLYNOMIAL EQUATIONS

An expression of the form

$$a_0x^n + a_1x^{n-1} + a_2x^{n-2} + \cdots + a_{n-1}x + a_n$$

where n is a positive integer and $a_0, a_1, a_2, \ldots, a_n$ are any constants with $a_0 \neq 0$, is called a *polynomial,* or *rational integral expression,* of the nth degree in x. Denoting the polynomial by $f(x)$ and writing $f(x) = 0$, we form the corresponding polynomial equation. A value of x which satisfies the equation is called a *root* or *solution* of the equation and also a *zero* of the polynomial.

In this chapter, we shall study certain properties of polynomials with particular reference to their zeros or the roots of the corresponding polynomial equations. We have developed methods for solving polynomial equations of the first and second degrees. Finding roots of equations of higher degrees is, in general, less simple. In fact, the problem is much too involved for a complete investigation at this stage. We shall, however, establish certain theorems relating to polynomials and the roots of the corresponding equations.

17-2 THE REMAINDER THEOREM AND THE FACTOR THEOREM

If we divide $f(x) = 2x^3 - 3x^2 + x - 1$ by $x - 2$, we get the quotient $2x^2 + x + 3$ with 5 as the remainder. Now substituting 2 for x in the dividend, we have

$$f(2) = 2(2)^3 - 3(2)^2 + 2 - 1 = 5$$

Thus, we obtain the remainder 5 in two ways. This is an illustration of a theorem which, as we shall see, has important consequences.

Theorem 17-1 *The remainder theorem:* If a polynomial $f(x)$ is divided by $x - r$ until a remainder free of x is obtained, the resulting remainder is equal to $f(r)$.

Proof: Let $Q(x)$ stand for the quotient when $f(x)$ is divided by $x - r$, and let R stand for the remainder. Then

$$f(x) = (x - r)Q(x) + R$$

$$\text{(dividend} = \text{divisor} \times \text{quotient} + \text{remainder)}$$

Since this is an identity, true for all values of x, we may substitute r for x, and obtain

$$f(r) = (r - r)Q(r) + R$$
$$f(r) = 0 \cdot Q(r) + R$$
$$f(r) = R$$

This establishes the remainder theorem.

Theorem 17-2 *The factor theorem and its converse:* If r is a root of the polynomial equation $f(x) = 0$, then $x - r$ is a factor of $f(x)$; and conversely, if $x - r$ is a factor of $f(x)$, then r is a root of the equation $f(x) = 0$.

Proof: From the remainder theorem, we have

$$f(x) = (x - r)Q(x) + R = (x - r)Q(x) + f(r)$$

If r is a root of the equation $f(x) = 0$, then $f(r) = 0$. Hence, we may present $f(x)$ in the factored form

$$f(x) = (x - r)Q(x)$$

To prove the converse of the theorem we need but to notice that with $x - r$ given as a factor, we may write the identity

$$f(x) = (x - r)Q(x)$$

which is true for all the values of x. Hence, replacing x by r, we have

$$f(r) = (r - r)Q(r) = 0 \cdot Q(r) = 0$$

EXAMPLE 1 Show that $x + 3$ is a factor of $x^3 + 2x^2 - 7x - 12$.

Solution: To determine if $x + 3$ is a factor, we need only to see if -3 is a root of the equation

$$x^3 + 2x^2 - 7x - 12 = 0$$

Substituting -3 for x, we obtain

$$(-3)^3 + 2(-3)^2 - 7(-3) - 12 = -27 + 18 + 21 - 12 = 0$$

The equation has -3 as a root and, according to the factor theorem, $x + 3$ is a factor of the given polynomial.

EXAMPLE 2 Form a cubic equation $f(x) = 0$ whose roots are 3, -2, and $\frac{1}{5}$.

Solution: From the factor theorem, we know that $x - 3$, $x + 2$, and $x - \frac{1}{5}$ are factors of $f(x)$. Hence, we write the equation

$$(x - 3)(x + 2)(x - \tfrac{1}{5}) = 0$$

This equation has the required roots. To avoid fractional coefficients, however, we replace $x - \frac{1}{5}$ with $5x - 1$ and obtain an equivalent equation; that is, either

$$(x - 3)(x + 2)(5x - 1) = 0 \qquad \text{or} \qquad 5x^3 - 6x^2 - 29x + 6 = 0$$

EXAMPLE 3 Determine if $x - a$ and $x + a$ are factors of $x^{2n} - a^{2n}$, where n is a positive integer.

Solution: If a and $-a$ are roots of the equation $f(x) = x^{2n} - a^{2n} = 0$, then, by the factor theorem, $x - a$ and $x + a$ are factors of the given expression. We see at once that

$$f(a) = a^{2n} - a^{2n} = 0 \qquad \text{and} \qquad f(-a) = (-a)^{2n} - a^{2n} = 0$$

and consequently, $x - a$ and $x + a$ are factors of $x^{2n} - a^{2n}$.

EXERCISE 17-1

Find the remainder when the indicated division is performed. Use both long division and the remainder theorem.

1. $(2x^2 - 4x + 8) \div (x - 2)$
2. $(3x^2 + x - 7) \div (x - 1)$
3. $(3x^3 + 4x^2 - 7x + 5) \div (x + 1)$
4. $(x^3 - 5x^2 - 3x + 15) \div (x + 2)$
5. $(2x^4 + 6x^3 + 3x - 4) \div (x + 3)$
6. $(x^4 - 10x^2 - 9x - 20) \div (x - 4)$

Using the remainder theorem, find what the remainder would be if the first expression were divided by the second.

7. $2x^3 + x^2 + 2x - 1, \quad x + 2$
8. $x^3 + 5x^2 - 7x + 6, \quad x - 2$
9. $x^3 - 3x^2 - 2x - 2, \quad x + 3$
10. $4x^3 + 4x^2 + x + 3, \quad x + 3$
11. $2x^3 - 5x^2 - 8x + 3, \quad x - \frac{1}{2}$
12. $6x^3 + 10x^2 + 5x - 3, \quad x - \frac{1}{3}$
13. $x^4 - x^3 - x^2 + x - 1, \quad x - 3$
14. $x^4 + 2x^3 - 3x - 5, \quad x + 2$
15. $3x^4 + 10x^3 - 9x^2 + 8, \quad x + 4$
16. $2x^4 + 7x^2 - 6x + 6, \quad x + 3$
17. $4x^4 + 2x^3 + 8x - 1, \quad x + \frac{1}{2}$
18. $x^5 - 2x^4 + x^2 - 9, \quad x - 2$

By means of the factor theorem, show that the second expression is a factor of the first.

19. $2x^3 - 3x^2 + 2x - 8, \quad x - 2$

20. $x^3 - 4x^2 + 3x + 8, \quad x + 1$

21. $3x^3 + 5x^2 - 6x + 18, \quad x + 3$

22. $2x^3 - 3x^2 + x - 6, \quad x - 2$

23. $x^3 + 8x^2 + 8x - 32, \quad x + 4$

24. $x^3 + 2x^2 - 6x - 72, \quad x - 4$

25. $2x^3 - 7x^2 - 3x - 4, \quad x - 4$

26. $3x^3 + 11x^2 - 2x + 8, \quad x + 4$

27. $3x^4 - x^3 - 3x^2 + \frac{1}{3}, \quad x - \frac{1}{3}$

28. $2x^4 + x^3 + x - \frac{3}{4}, \quad x - \frac{1}{2}$

29. $x^4 - 81, \quad x + 3$

30. $x^4 - 256, \quad x + 4$

31. $x^5 - 1, \quad x - 1$

32. $x^5 + 1, \quad x + 1$

33. $x^7 + a^7, \quad x + a$

34. $x^6 - a^6, \quad x - a$

35. $x^9 + a^9, \quad x + a$

36. $x^n - a^n, \quad x - a,$ n a positive integer

37. $x^n - a^n, \quad x + a,$ n an even positive integer

38. $x^n + a^n, \quad x + a,$ n an odd positive integer

Write equations with integral coefficients which have the following numbers, and no others, as roots.

39. $1, 2, -3$

40. $1, -2, 3$

41. $-1, -2, -3$

42. $0, 1, -4$

43. $-4, 3, 0$

44. $\frac{1}{2}, 1, -2$

45. $\frac{1}{3}, 1, -1$

46. $\frac{1}{3}, \frac{2}{3}, -1$

47. $\sqrt{3}, -\sqrt{3}, 2$

17-3 SYNTHETIC DIVISION

A much-used operation in finding the roots of polynomial equations is that of dividing a polynomial $f(x)$ by a linear expression of the form $x - r$. This kind of division can be done rapidly by a process called *synthetic division*. The process is simply an abbreviated form of the usual long division. An explanation can best be made by an example.

EXAMPLE 1 Divide $2x^3 + x^2 - 18x + 7$ by $x - 2$.

Solution: We first use the long division method.

$$
\begin{array}{rrrrl}
2x^3 + & x^2 - & 18x + & 7 & \underline{|x - 2} \quad \text{Divisor} \\
\underline{2x^3 -} & \underline{4x^2} & & & |\underline{2x^2 + 5x - 8} \quad \text{Quotient} \\
 & 5x^2 - & 18x & & \\
 & \underline{5x^2 -} & \underline{10x} & & \\
 & - & 8x + & 7 & \\
 & \underline{-} & \underline{8x +} & \underline{16} & \\
 & & - & 9 & \text{Remainder}
\end{array}
$$

In the steps of the division the terms $-18x$ and 7 were rewritten. Now let us leave these terms in the first line and write beneath them the terms to be subtracted. The division then appears as

$$
\begin{array}{rrrrl}
2x^3 + & x^2 - & 18x + & 7 & \underline{|x - 2} \\
\underline{2x^3 -} & \underline{4x^2 -} & \underline{10x +} & \underline{16} & |\underline{2x^2 + 5x - 8} \\
 & 5x^2 - & 8x - & 9 &
\end{array}
$$

The third line has in order the coefficients 5, -8, and -9. Placing the first coefficient of the first line before these, we have

$$2, 5, -8, -9$$

These values, in proper order, are the coefficients of the quotient $2x^2 + 5x - 8$ and the remainder -9.

The powers of x are not essential in forming the coefficients of the second and third lines. So we next delete all the x's and bring the coefficient of x^3 in the second line down to the third line. The division then takes the form

$$
\begin{array}{rrrrl}
2 + & 1 - & 18 + & 7 & \underline{|-2} \\
 & \underline{- 4 -} & \underline{10 +} & \underline{16} & \\
2 + & 5 - & 8 - & 9 &
\end{array}
$$

The numbers in the third line, after the first, are obtained by subtraction. If we change -2 of the divisor to 2, the signs of the numbers in the second line will be reversed. Then addition, rather than subtraction, completes the numbers of the third line. Making these changes, we finally present the division in the abbreviated form

$$\begin{array}{rrrr|l} 2 + 1 & - 18 & + 7 & & \underline{2} \\ + 4 & + 10 & - 16 & & \\ \hline 2 + 5 & - 8 & - 9 & & \end{array}$$

The quotient may be written, by supplying the proper powers of x, as

$$2x^2 + 5x - 8$$

with -9 as the remainder.

Following the process above, we may formulate directions for dividing a polynomial $f(x)$ by $x - r$.

1. Arrange the coefficients of $f(x)$ in a line in order of descending powers of x, supplying zero as the coefficient of each missing power of x. At the end of this line write r.
2. Rewrite the first coefficient of this line underneath as the first number of the third line. Multiply this number by r and place the product immediately beneath the second number of the first line, and write the sum of these two second position numbers as the second number of the third line. Continue this process until the second and third lines are completed to the position of the last number of the first line, which is the constant term of $f(x)$.
3. The numbers of the third line give the coefficients of the quotient in order of descending powers of x, the last number being the remainder. The degree of the quotient is, of course, one less than the degree of the dividend.

EXAMPLE 2 Divide $3x^4 + 8x^3 + 7x^2 - 80$ by $x + 3$.

Solution: In this example $x - r = x + 3$, so that $r = -3$. Using this value for r and placing zero in the first line as the coefficient of the missing power of x, we have

$$\begin{array}{l|l} 3 + 8 + 7 + 0 - 80 & \underline{-3} \\ \quad - 9 + 3 - 30 + 90 & \\ \hline 3 - 1 + 10 - 30 + 10 & \end{array}$$

Thus the quotient is $3x^3 - x^2 + 10x - 30$, and the remainder is 10.

EXAMPLE 3 Find the value of $f(x) = 4x^3 - 7x^2 + 13x$ when x is replaced by 4.

Solution: Recalling that when $f(x)$ is divided by $x - r$ the remainder is $f(r)$, we may use synthetic division to find the value of a polynomial corresponding to a given value of x. Hence,

$$\begin{array}{rrrr|l} 4 - & 7 + & 13 + & 0 & \underline{4} \\ & + 16 + & 36 + & 196 & \\ \hline 4 + & 9 + & 49 + & 196 & \end{array}$$

The remainder is 196, and therefore $f(4) = 196$.

EXERCISE 17-2

Find the quotient and remainder in each problem by the use of synthetic division.

1. $(x^3 - x^2 + 9x + 13) \div (x + 1)$
2. $(x^3 - 5x^2 + 5x - 6) \div (x - 1)$
3. $(3x^3 + 3x^2 - 4x - 13) \div (x - 2)$
4. $(3x^3 - 3x^2 + 7x + 9) \div (x + 2)$
5. $(2x^3 + 2x^2 + x + 46) \div (x + 3)$
6. $(2x^3 - 9x^2 + 6x + 1) \div (x - 3)$
7. $(5x^3 + 8x^2 - 9x + 11) \div (x + 4)$
8. $(4x^3 + 5x^2 - 8x + 30) \div (x + 3)$
9. $(x^4 + x^3 - 3x + 10) \div (x + 2)$
10. $(x^4 - 9x^2 - 8x - 40) \div (x - 4)$
11. $(2x^4 + 7x^3 + x + 12) \div (x + 3)$
12. $(x^4 + 4x^3 - 7x^2 + 3) \div (x + 4)$
13. $(x^5 - x^4 - x^2 + 4) \div (x - 2)$
14. $(x^5 + x^3 + x - 1) \div (x + 2)$
15. $(x^5 + 2x^4 - 5x^3 + 2) \div (x + 4)$
16. $(x^5 - 5x^3 + 4x^2 + 1) \div (x - 2)$
17. $(x^5 + 1) \div (x + 1)$

18. $(x^5 - 1) \div (x - 1)$

19. $(x^5 + 32) \div (x + 2)$

20. $(x^5 - 32) \div (x - 2)$

21. $(2x^4 - 5x^3 - x - 4) \div (x - \frac{1}{2})$

22. $(4x^4 + 3x^2 - 4x + 3) \div (x - \frac{1}{2})$

23. $(4x^4 + 3x^2 - 2x + 3) \div (x + \frac{1}{2})$

24. $(6x^4 - 8x^3 + 5x^2) \div (x - \frac{1}{3})$

25. $(8x^3 + 2x^2 + 5x - 3) \div (x + \frac{3}{4})$

26. $(9x^3 + 6x^2 - 2x - 3) \div (x - \frac{2}{3})$

Evaluate by synthetic division.

27. If $f(x) = x^3 - 7x^2 + 12x - 3$, find $f(-2)$, $f(3)$, $f(4)$

28. If $f(x) = x^4 - 7x^3 + 7x - 17$, find $f(2)$, $f(-2)$, $f(-3)$

29. If $f(x) = 2x^4 + 6x^3 + 7x^2 - x + 4$, find $f(-1)$, $f(-2)$, $f(3)$

30. If $f(x) = x^3 - 4x^2 - 20x + 55$, find $f(5)$, $f(6)$, $f(7)$

17-4 THEOREMS CONCERNING ROOTS

Having found roots of certain polynomial equations, particularly equations of the first and second degrees, we might expect that every equation of this kind has a root. In this connection, we state the following theorem:

Theorem 17-3 *The fundamental theorem of algebra:* Every polynomial equation of degree $n \geq 1$ has at least one root, real or imaginary.

This important theorem was first proved in 1797 by a German mathematician named Gauss. The theorem cannot be established by elementary algebraic methods, and consequently we do not include a proof. We shall, however, assume the theorem to be true and from it establish other theorems.

Theorem 17-4 Every polynomial $f(x)$ of degree $n \geq 1$ can be expressed as the product of n linear factors.

Proof: Let $f(x) = a_0x^n + a_1x^{n-1} + \cdots + a_{n-1}x + a_n$, with $a_0 \neq 0$. From the fundamental theorem, $f(x) = 0$ has at least one root.

Denoting this root by r_1 and recalling the factor theorem, we conclude that $x - r_1$ is a factor of $f(x)$. Hence,

$$f(x) = (x - r_1)Q_1(x)$$

where the factor $Q_1(x)$ is a polynomial of degree $n - 1$, its highest degree term being a_0x^{n-1}. Using the fundamental theorem and the factor theorem again, we have

$$Q_1(x) = (x - r_2)Q_2(x)$$

Hence,

$$f(x) = (x - r_1)(x - r_2)Q_2(x)$$

where r_2 is a root of $Q_1(x) = 0$, and $Q_2(x)$ is a polynomial with a_0x^{n-2} its highest degree term. This process may be continued until n linear factors are obtained, the last factor having the term a_0x. Then factoring the constant a_0 from the last factor, we may write it as $a_0(x - r_n)$ and have

$$f(x) = a_0(x - r_1)(x - r_2) \cdots (x - r_n)$$

The numbers $r_1, r_2, \ldots, r_n$ need not all be distinct, or even real. That is, certain factors may be repeated and some, or all, may contain imaginary constants.

Theorem 17-5 Every polynomial equation $f(x) = 0$ of degree n has exactly n roots.

Proof: By the preceding theorem, we may factor $f(x)$ and write the equation as

$$a_0(x - r_1)(x - r_2) \cdots (x - r_n) = 0$$

The roots of the equation are $r_1, r_2, \ldots, r_n$ because any one of these values when substituted for x makes a factor equal to zero, and hence the product of the factors is equal to zero. A value of x different from these quantities will obviously make no factor equal to zero. Hence, the equation has only the n roots $r_1, r_2, \ldots, r_n$.

If no two factors of $f(x)$ are alike, the n roots are all different. Each root is then called a *single,* or *simple, root.* If the same factor occurs twice, a root occurs twice. A root which occurs twice is called a *double root.* If a factor occurs three times, there is a *triple root.* In general, if a factor occurs m times, the corresponding root is said to be of *multiplicity m.* It is customary to count a double root as two roots, a triple root as three roots, and a root of multiplicity m as m roots. Counting the roots in this way, we say that a polynomial equation of degree n has exactly n roots.

EXAMPLE 1 The equation $(x + 1)(x - 2)(x + 5) = 0$ is of degree 3, and it has three distinct roots $-1, 2, -5$.

EXAMPLE 2 The equation $(x + 4)^2(x - 3)(x + 2)^4 = 0$ is of degree 7. It has -4 as a double root, 3 as a simple root, and -2 as a root of multiplicity four. The total number of roots, giving each a count according to its multiplicity, is seven.

As we have seen, a quadratic equation with real coefficients which has the complex number $a + bi$, $b \neq 0$, as a root, also has $a - bi$ as a root. That is, conjugate complex roots occur in pairs. We now prove a general theorem concerning roots of this kind.

Theorem 17-6 If the complex number $a + bi$, $b \neq 0$, is a root of a polynomial equation with real coefficients, then the complex number $a - bi$ is also a root.

Proof: Designating the equation by $f(x) = 0$, we have $f(a + bi) = 0$ since $(a + bi)$ is a root of the equation. Then $(x - a - bi)$ is a factor of $f(x)$. We shall show that $(x - a + bi)$ is also a factor by showing that the product

$$(x - a - bi)(x - a + bi) = x^2 - 2ax + (a^2 + b^2)$$

is a factor of $f(x)$. Let $D(x)$ stand for this product. Now $f(x)$ can be divided by the quadratic expression $D(x)$ until a remainder of degree lower than two is obtained. Then indicating the quotient by $Q(x)$ and the remainder by $Rx + S$, we may write

$$f(x) = D(x)Q(x) + Rx + S$$

If we substitute $a + bi$ for x in this identity, we get

$$f(a + bi) = D(a + bi)Q(a + bi) + R(a + bi) + S$$

But, by hypothesis, $f(a + bi) = 0$; and $D(a + bi) = 0$ since $(x - a - bi)$ is a factor of $D(x)$. Thus, we obtain

$$0 = 0 \cdot Q(a + bi) + R(a + bi) + S$$

or

$$(Ra + S) + Rbi = 0$$

A complex number is equal to zero if, and only if, the real and imaginary components are each equal to zero. Hence, we must have

$$Ra + S = 0 \qquad \text{and} \qquad Rb = 0$$

Since $a + bi$ is an imaginary number, $b \neq 0$; and therefore $R = 0$. Substituting $R = 0$ in $Ra + S = 0$ gives $S = 0$. Hence, the remainder $Rx + S$ is equal to zero, and

$$f(x) = D(x)Q(x) = (x - a - bi)(x - a + bi)Q(x)$$

Then $(x - a + bi)$ is a factor of $f(x)$, and consequently $a - bi$ is a root of $f(x) = 0$.

It is quite easy to see that the interchange of $a + bi$ and $a - bi$ in the steps of the preceding proof reveals that $a + bi$ is a root of $f(x) = 0$ if $a - bi$ is a root. We conclude, then, that a polynomial equation with real coefficients which has a complex root has also the conjugate complex number as a root.

We next state a theorem quite similar to the preceding one. The proof of the theorem is much like that for conjugate complex roots and is left for the student.

Theorem 17-7 Let a and b be rational numbers with b positive and $\sqrt{b}$ irrational and let $f(x)$ be a polynomial with rational coefficients. Then if $a + \sqrt{b}$ is a root of $f(x) = 0$, the number $a - \sqrt{b}$ is also a root.

EXAMPLE 1 The complex number $3 - 2i$ is a root of the equation $x^4 - 6x^3 + 10x^2 + 18x - 39 = 0$. Find the other roots.

Solution: Since $(3 - 2i)$ is a root, $(3 + 2i)$ is also a root. Then the product

$$(x - 3 + 2i)(x - 3 - 2i) = x^2 - 6x + 13$$

is a factor of the left member of the given equation. The other factor, obtainable by division, is $x^2 - 3$. The roots of $x^2 - 3 = 0$ are $\pm\sqrt{3}$. Hence, the solution set in this example is

$$\{-\sqrt{3}, \sqrt{3}, 3 - 2i, 3 + 2i\}$$

EXAMPLE 2 Form an equation with integral coefficients and of lowest possible degree which has $2 - \sqrt{2}$ and $4 + i$ as roots.

Solution: In addition to the given roots, $2 + \sqrt{2}$ and $4 - i$ must also be roots. Hence, the required equation is

$$(x - 2 + \sqrt{2})(x - 2 - \sqrt{2})(x - 4 - i)(x - 4 + i) = 0$$

or, multiplying,

$$(x^2 - 4x + 2)(x^2 - 8x + 17) = 0$$
$$x^4 - 12x^3 + 51x^2 - 84x + 34 = 0$$

EXERCISE 17-3

Find the roots of each equation which are not given.

1. $x^3 - 4x^2 + x - 4 = 0,\quad i$ is a root
2. $x^3 + 3x^2 + 9x - 13 = 0,\quad -2 + 3i$ is a root
3. $x^3 - 8x^2 + 22x - 20 = 0,\quad 3 + i$ is a root
4. $3x^3 + 7x^2 - x - 1 = 0,\quad -1 + \sqrt{2}$ is a root
5. $x^4 + 4x^3 + x^2 - 28x - 56 = 0,\quad \sqrt{7}$ is a root
6. $x^4 - 8x^3 - 13x^2 + 40x - 40 = 0,\quad \sqrt{5}$ is a root
7. $x^4 + 2x^3 + 2x^2 - 4x - 8 = 0,\quad -1 + i\sqrt{3}$ is a root
8. $x^4 - 4x^3 + 2x^2 - 4x + 1 = 0,\quad 2 + \sqrt{3}$ is a root
9. $x^4 + 6x^3 + 17x^2 + 36x + 66 = 0,\quad i\sqrt{6}$ is a root
10. $x^5 - x^4 - x^3 + x^2 - 2x + 2 = 0,\quad i$ and $\sqrt{2}$ are roots
11. $x^5 + x^4 + 5x^3 + 5x^2 + 4x + 4 = 0,\quad i$ and -1 are roots

Find the equation of lowest possible degree and with integral coefficients which has the given numbers among its roots.

12. $1 - i, 2$ **13.** $4 + 3i, -3$ **14.** $i, 3 + i$

15. $5 + 2i, -\sqrt{2}$ **16.** $4i, 3 + \sqrt{2}$ **17.** $2 - 4i, 4 + 3i$

18. $\sqrt{5}, 3 - \sqrt{2}$ **19.** $2 - i, 3 + 2i$ **20.** $1 - \sqrt{2}, 1 - i\sqrt{2}$

21. Prove Theorem 17-7.

22. Prove that a cubic equation with real coefficients has either three real roots or one real root and a pair of conjugate imaginary roots.

23. Prove that a fourth-degree equation with real coefficients has four real roots, or four imaginary roots, or two real roots and two imaginary roots.

24. Prove that a polynomial equation of the fifth degree with real coefficients has at least one real root, and state the possibilities for the nature of the remaining roots.

25. State the possibilities for the distribution of the real and imaginary roots of a sixth-degree polynomial equation with real coefficients.

17-5 RATIONAL ROOTS

A rational number which is the root of an equation is called a *rational root.* We shall consider a quite simple method of finding the rational roots, if any, of a polynomial equation with rational coefficients.

Theorem 17-8 If a rational number b/c in its lowest terms is a root of the equation

$$a_0x^n + a_1x^{n-1} + \cdots + a_{n-1}x + a_n = 0$$

where the coefficients $a_0, a_1, a_2, \ldots, a_n$ are all integers with $a_0 \neq 0$, then b is a factor of a_n and c is a factor of a_0.

Proof: Assuming that b/c is a root, we have

$$a_0\left(\frac{b}{c}\right)^n + a_1\left(\frac{b}{c}\right)^{n-1} + \cdots + a_{n-1}\left(\frac{b}{c}\right) + a_n = 0 \qquad (1)$$

Multiplying this equation by c^n/b and transposing the last term gives

$$a_0b^{n-1} + a_1b^{n-2}c + a_2b^{n-3}c^2 + \cdots + a_{n-1}c^{n-1} = -\frac{a_nc^n}{b} \qquad (2)$$

The left member of Eq. (2), being made up of terms whose factors are integers, has an integral value. Hence, the right member

$$-\frac{a_nc^n}{b}$$

likewise yields an integer. By hypothesis b and c have no common integral factor other than ± 1; then it must follow that a_n is divisible by b.

Now returning to Eq. (1), we multiply by c^{n-1}, transpose the first term, and obtain

$$a_1 b^{n-1} + a_2 b^{n-2} c + a_3 b^{n-3} c^2 + \cdots + a_{n-1} b c^{n-2} + a_n c^{n-1} = -\frac{a_0 b^n}{c}$$

The left member of this equation, having an integral value, means that the fraction in the right member is expressible as an integer. Since b and c have no common integral factor other than ± 1, a_0 is divisible by c.

For the case in which $a_0 = 1$, the rational root b/c is an integer. This is true because c, being a factor of a_0, would be equal to 1 or -1. Hence, we have the following corollary.

Corollary Any rational root of the equation

$$x^n + a_1 x^{n-1} + a_2 x^{n-2} + \cdots + a_{n-1} x + a_n = 0$$

where the coefficients are integers, is an integer and a factor of a_n.

EXAMPLE 1 Find all the rational roots of

$$f(x) = x^4 + x^3 + 4x^2 + 6x - 12 = 0$$

Solution: The rational roots, if any, must be factors of -12; and the possibilities are

$$\pm 1 \qquad \pm 2 \qquad \pm 3 \qquad \pm 4 \qquad \pm 6 \qquad \pm 12$$

Testing 1 by synthetic division, we find $f(1) = 0$.

$$\begin{array}{rrrrr|l} 1 + & 1 + & 4 + & 6 - & 12 & \underline{1} \\ & 1 + & 2 + & 6 + & 12 & \\ \hline 1 + & 2 + & 6 + & 12 + & 0 & \end{array}$$

Then $f(x) = (x - 1)(x^3 + 2x^2 + 6x + 12)$. Additional roots of $f(x) = 0$ may more readily be found from the equation

$$Q_1(x) = x^3 + 2x^2 + 6x + 12 = 0$$

$Q_1(x) = 0$ is called the *first depressed equation.* This equation has no positive roots; for a positive value for x makes all the terms on the left positive and their sum, of course, not equal to zero. Trying negative values, and starting with -1, we find $Q_1(-1) = 7$. Passing to -2, we get

$$\begin{array}{r|l} 1 + 2 + 6 + 12 & \underline{-2} \\ -\ 2 + 0 - 12 & \\ \hline 1 + 0 + 6 + \ \ 0 & \end{array}$$

Hence $Q_1(-2) = 0$, and consequently

$$Q_1(x) = (x + 2)(x^2 + 6)$$

The equation $Q_2(x) = x^2 + 6 = 0$ is called the *second depressed equation.* Its roots are $\pm\sqrt{-6}$, which are not real. We conclude, then, that -2 and 1 are the rational roots of the given equation.

EXAMPLE 2 Find the rational roots, if any, of the equation

$$f(x) = 2x^4 - 3x^3 + 10x^2 - 13x + 9 = 0$$

Solution: The signs of the coefficients alternate here, and consequently all terms become positive for any negative value of x. Hence any real root of the equation must be positive. The possibilities for rational roots, then, are

$$\frac{1}{2} \qquad 1 \qquad \frac{3}{2} \qquad 3 \qquad \frac{9}{2} \qquad 9$$

We find $f(\frac{1}{2}) = -\frac{19}{4}$ and $f(1) = 5$. Testing $\frac{3}{2}$, we have

$$\begin{array}{r|l} 2 - 3 + 10 - 13 + \ \ 9 & \underline{\tfrac{3}{2}} \\ +\ 3 + \ \ 0 + 15 + \ \ 3 & \\ \hline 2 + 0 + 10 + \ \ 2 + 12 & \end{array}$$

We see that $\frac{3}{2}$ is not a root. Furthermore, we see that testing a number greater than $\frac{3}{2}$ will yield a number greater than 12 in the last number of the third line. Therefore, we need not test the values 3, $\frac{9}{2}$, and 9. The given equation has no rational root.

As illustrated in this example, the following theorem may in some instances be applied to advantage in testing for rational roots:

Theorem 17-9 If in the synthetic division of a polynomial $f(x)$ by $x - r$, where r is positive, each term of the third line is positive (some may be zero), then r is an upper limit for the real roots of the equation $f(x) = 0$. If r is negative and the terms of the third line are alternately positive and negative, then r is a lower limit for the roots.

The truth of the theorem may be observed immediately. If a positive r produces positive terms in the third line in the division process, a number larger than r would increase all terms of the line after the first. The last term of the line could not then be zero. Consequently, a value larger than r is not a root of the equation.

By quite similar reasoning, the second part of the theorem can be established.

EXAMPLE 3 Find upper and lower limits for the roots of the equation

$$2x^3 - 5x^2 - 7x + 4 = 0$$

Solution: We show the tests for $r = 3$ and $r = 4$

$$\begin{array}{rrrr|l} 2 & -5 & -7 & +4 & \underline{3} \\ & +6 & +3 & -12 & \\ \hline 2 & +1 & -4 & -8 & \end{array} \qquad \begin{array}{rrrr|l} 2 & -5 & -7 & +4 & \underline{4} \\ & +8 & +12 & +20 & \\ \hline 2 & +3 & +5 & +24 & \end{array}$$

These tests show that the smallest integral upper limit, as determined by the theorem, is 4.

We next try negative integers, starting with -1

$$\begin{array}{rrrr|l} 2 & -5 & -7 & +4 & \underline{-1} \\ & -2 & +7 & +0 & \\ \hline 2 & -7 & +0 & +4 & \end{array} \qquad \begin{array}{rrrr|l} 2 & -5 & -7 & +4 & \underline{-2} \\ & -4 & +18 & -22 & \\ \hline 2 & -9 & +11 & -18 & \end{array}$$

Since -2 makes the terms of the third line alternate in signs, this number is a lower limit. Hence, all the real roots of the given equation lie between -2 and 4.

EXERCISE 17-4

Find all the rational roots, if any, of each equation. Solve completely if the last depressed equation is quadratic.

1. $x^3 - x^2 - 8x + 12 = 0$
2. $x^3 - 2x^2 - 7x - 4 = 0$

3. $x^3 - 7x^2 + 17x - 15 = 0$

4. $3x^3 + x^2 + 27x + 9 = 0$

5. $4x^3 - x^2 + 16x - 4 = 0$

6. $8x^3 - 4x^2 - 2x + 1 = 0$

7. $8x^3 + 4x^2 - 2x - 1 = 0$

8. $2x^3 + 7x^2 + 16x + 15 = 0$

9. $3x^3 + 8x^2 + 19x + 10 = 0$

10. $x^4 + 2x^3 + x^2 + x - 2 = 0$

11. $x^4 - 2x^3 + x^2 - 3x + 2 = 0$

12. $x^4 - 8x^2 + 5x + 6 = 0$

13. $6x^4 - x^3 + 5x^2 - x - 1 = 0$

14. $6x^4 - 5x^3 - 5x - 6 = 0$

15. $2x^4 + 13x^3 + 30x^2 + 28x + 8 = 0$

16. $2x^4 - 13x^3 + 36x^2 - 70x + 75 = 0$

17. $3x^4 + 4x^3 - 8x^2 - 9x - 2 = 0$

18. $12x^4 + 7x^3 + 13x^2 + 7x + 1 = 0$

19. $x^5 + 9x^4 + 31x^3 + 51x^2 + 40x + 12 = 0$

20. $2x^5 + 9x^4 + 21x^3 + 28x^2 + 21x + 9 = 0$

21. $2x^5 - 3x^4 + 8x^3 - 12x^2 + 6x - 9 = 0$

22. $x^5 + 20x^4 + 70x^3 + 100x^2 + 65x + 16 = 0$

Factor the following polynomials.

23. $6x^3 + 5x^2 - 7x - 4$

24. $4x^3 + 4x^2 - 3x + 10$

25. $2x^3 + 7x^2 + 18x - 11$

26. $4x^4 - 4x^3 - 17x^2 + 9x + 18$

27. $x^4 - 4x^3 + 3x^2 - 4x - 4$

28. $x^4 + 4x^3 + x^2 - 16x - 20$

29. $x^5 - x^4 - 5x^3 + x^2 + 8x + 4$

30. $x^5 - x^4 - 5x^3 + 5x^2 + 4x - 4$

17-6 THE GRAPH OF A POLYNOMIAL FUNCTION

In Sec. 5-3 we introduced graphs of functions and relations and in Secs. 12-8 and 12-9 we constructed graphs of quadratic functions, or polynomial functions of the second degree. From the graphs of quadratic functions we estimated the real zeros of the functions, or the roots of the corresponding quadratic equations. The graphical method may be used to obtain approximately the real roots, both rational and irrational, of a polynomial equation of a degree higher than the second degree. Computing functional values in the case of higher degree equations may be somewhat tedious, but the work can be facilitated by the use of synthetic division. Here, as previously, we shall assume the following theorem to be true.

Theorem 17-10 If $f(x)$ denotes a polynomial with real coefficients and if a and b are real numbers such that $f(a)$ and $f(b)$ are of opposite signs, then the equation $f(x) = 0$ has at least one real root between a and b.

Interpreted geometrically, this theorem means that the graph of $y = f(x)$ crosses the x axis at least once between any two of its points which are on opposite sides of the x axis.

EXAMPLE Estimate, to one decimal place, the roots of the next equation

$$x^3 - 4x^2 + x + 3 = 0$$

Solution: To draw the graph of the function defined by $y = x^3 - 4x^2 + x + 3$, we prepare the following table. The functional values should be checked by synthetic division. When $x = 3.5$, for example, we have

$$\begin{array}{rrrr|l} 1 & -\ 4.0 & +\ 1.00 & +\ 3.000 & \underline{3.5} \\ & +\ 3.5 & -\ 1.75 & -\ 2.625 & \\ \hline 1 & -\ 0.5 & -\ 0.75 & +\ 0.375 & \end{array}$$

Hence, $y = 0.375$ when $x = 3.5$. We round off, however, and take $y = 0.4$.

x	-1	-0.5	0	1.5	2	3	3.5
y	-3	1.4	3	-1.1	-3	-3	0.4

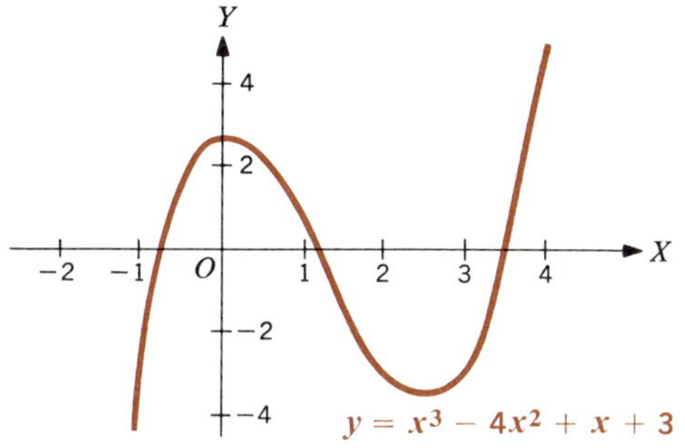

Figure 17-1

From the graph (Fig. 17-1), we estimate the zeros of the function, or the roots of the given equation to be -0.7, 1.2, and 3.5.

EXERCISE 17-5

Solve for the real roots by graphing. Estimate the roots to one decimal place.

1. $x^3 - 3x^2 - 1 = 0$

2. $x^3 + x^2 - 5x + 6 = 0$

3. $x^3 - 3x - 4 = 0$

4. $x^3 - 4x^2 + x + 4 = 0$

5. $x^3 - 7x + 5 = 0$

6. $x^3 + 3x^2 - 3 = 0$

7. $x^4 - 2x^2 - 3 = 0$

8. $x^4 - 4x^2 + 3 = 0$

9. $x^4 + 3x + 5 = 0$

10. $x^4 - 4x^3 + 16x + 5 = 0$

17-7 A METHOD OF APPROXIMATING IRRATIONAL ROOTS

There are several processes by which the irrational roots of a polynomial equation may be found to any desired degree of accuracy. We shall consider a method which is called the solution by *successive enlargements.* This method, as in the previous section, is based on the principle that if $f(a)$ and $f(b)$ have opposite signs, the equation $f(x) = 0$ has a root between a and b. The process consists in finding pairs of values of x closer and closer together for which $f(x)$ has opposite signs. We shall illustrate with some examples.

EXAMPLE 1 Find an approximation of the real root of the equation

$$f(x) = x^3 - 2x^2 + x - 1 = 0$$

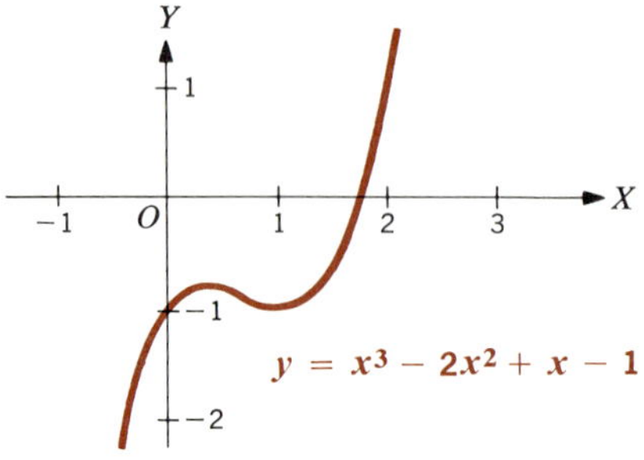

Figure 17-2

Solution: The graph (Fig. 17-2) of $y = f(x)$ shows that the given equation has a root between $x = 1$ and $x = 2$. The next step is to find consecutive tenths of units between which the curve crosses the x axis. Computing, we find $f(1.6) = -0.424$. Next, we try $x = 1.7$ and $x = 1.8$:

$$\begin{array}{rrrrl} 1 & -\ 2.0 & +\ 1.00 & -\ 1.000 & \underline{|1.7} \\ & 1.7 & -\ 0.51 & +\ 0.833 & \\ \hline 1 & -\ 0.3 & +\ 0.49 & -\ 0.167 & \end{array}$$

$$\begin{array}{rrrrl} 1 & -\ 2.0 & +\ 1.00 & -\ 1.000 & \underline{|1.8} \\ & 1.8 & -\ 0.36 & +\ 1.152 & \\ \hline 1 & -\ 0.2 & +\ 0.64 & +\ 0.152 & \end{array}$$

We see that $f(1.7) = -0.167$ and $f(1.8) = 0.152$. Hence, the points $(1.7, -0.167)$ and $(1.8, 0.152)$ are on the graph of $y = f(x)$, one below the x axis and the other above the x axis. Consequently, the desired root is between 1.7 and 1.8.

We next locate the root between consecutive hundredths of units. For this purpose, we plot the points of the graph for which $x = 1.7$ and $x = 1.8$ on an enlarged scale (Fig. 17-3). The line

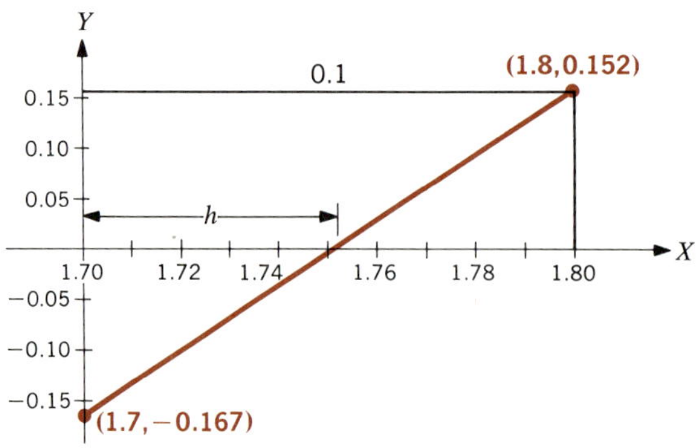

Figure 17-3

segment joining these points crosses the x axis at about $x = 1.75$, and the graph likely crosses the x axis at a point quite near. By synthetic division, as here shown, we find $f(1.75) = -0.01526$ and $f(1.76) = 0.016576$.

$$\begin{array}{rrrr|l} 1 & -\,2.00 & +\,1.0000 & -\,1.000000 & \underline{1.75} \\ & 1.75 & -\,0.4375 & +\,0.984375 & \\ \hline 1 & -\,0.25 & +\,0.5625 & -\,0.015625 & \end{array}$$

$$\begin{array}{rrrr|l} 1 & -\,2.000 & +\,1.0000 & -\,1.000000 & \underline{1.76} \\ & 1.76 & -\,0.4224 & +\,1.016576 & \\ \hline 1 & -\,0.24 & +\,0.5776 & +\,0.016576 & \end{array}$$

Thus, we see that the root is between 1.75 and 1.76, but we do not know at this stage which of these values is the better approximation. We could continue the process by next locating the root between consecutive thousandths of units. The computations, of course, would become increasingly tedious. It is easy, however, to estimate the root to one more digit. Figure 17-4 shows the points on the graph for which $x = 1.75$ and $x = 1.76$. The line connecting these points crosses the x axis at about $x = 1.755$. Hence, this value is approximately equal to a root of the given equation.

In this example, we used Fig. 17-3 to estimate the third digit of the desired root, and Fig. 17-4 to estimate the fourth digit. We could, however, obtain these estimates algebraically. Thus, from the similar triangles in Fig. 17-3, we have the proportion

$$\frac{h}{0.1} = \frac{0.167}{0.167 + 0.152}$$

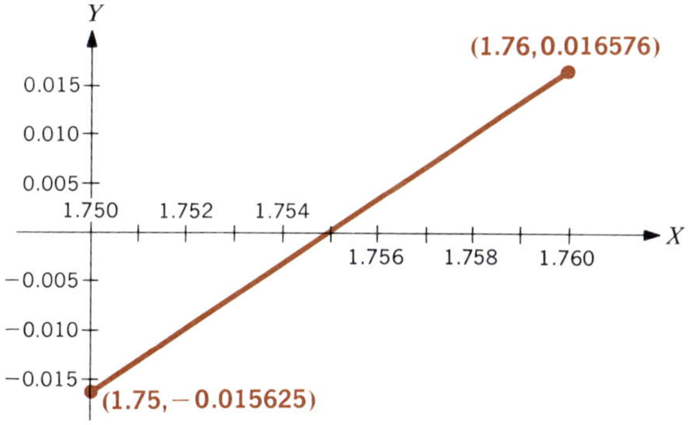

Figure 17-4

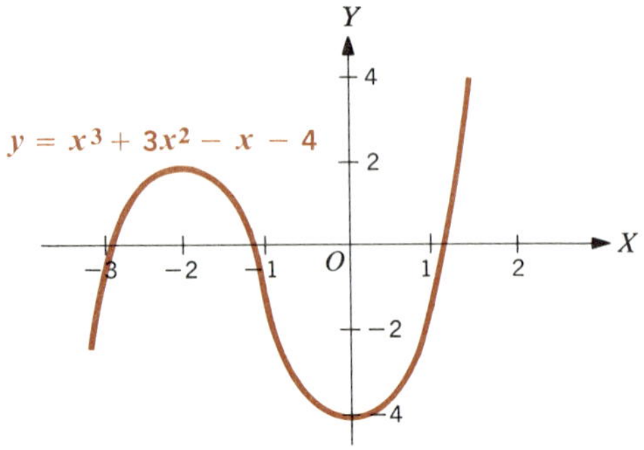

Figure 17-5

which yields $h = 0.05$, approximately, and therefore $x = 1.75$ is an approximation of the root. We note that this proportion, obtainable without the use of a figure, may be employed in lieu of a carefully constructed figure.

EXAMPLE 2 Find to two decimal places each of the real roots of

$$f(x) = x^3 + 3x^2 - x - 4 = 0$$

Solution: A rough graph of $y = f(x)$ may be made by plotting a few points corresponding to integral values of x. The graph (Fig. 17-5) shows there is a root between -3 and -2, another between -2 and -1, and a third root between 1 and 2. The desired accuracy in the approximations to these roots may be obtained by considering each separately. We show the necessary computations for the smallest of the three roots.

$$\begin{array}{rrrrl} 1 & +\,3.0 & -\,1.00 & -\,4.000 & \underline{|-2.9} \\ & -\,2.9 & -\,0.29 & +\,3.741 & \\ \hline 1 & +\,0.1 & -\,1.29 & -\,0.259 & \end{array}$$

$$\begin{array}{rrrrl} 1 & +\,3.0 & -\,1.00 & -\,4.000 & \underline{|-2.8} \\ & -\,2.8 & -\,0.56 & +\,4.368 & \\ \hline 1 & +\,0.2 & -\,1.56 & +\,0.368 & \end{array}$$

The root lies between $x = -2.9$ and $x = -2.8$. Figure 17-6 shows the points of the graph corresponding to these values of x. The line segment connecting the points crosses the x axis at about $x = -2.86$. We are not certain, however, that 6 is the best digit for the second decimal place. But, computing, we find $f(-2.86) = 0.005144$ and $f(-2.85) = -0.068375$. These results reveal that the root is between $x = -2.85$ and $x = -2.86$, with the root surely closer to -2.86.

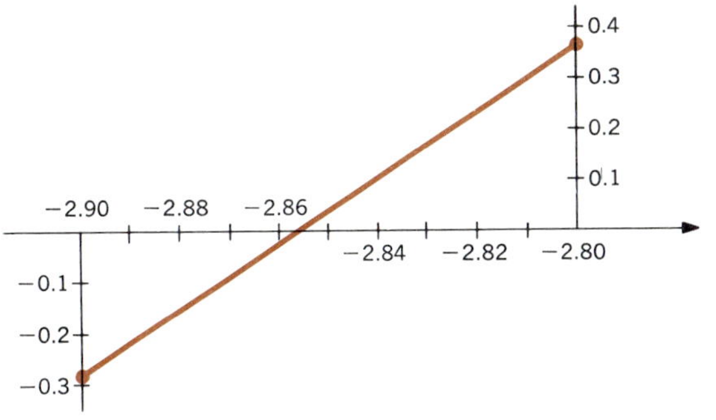

Figure 17-6

The other two roots to two decimal places, as the student should verify, are $x = -1.25$ and $x = 1.12$.

Illustrating with the equation of the preceding example, we describe an alternate procedure for approximating negative roots. If we replace x by $-x$ in the $x^3 + 3x^2 - x - 4 = 0$, we get

$$-x^3 + 3x^2 + x - 4 = 0 \qquad \text{or} \qquad x^3 - 3x^2 - x + 4 = 0$$

The roots of this new equation are the negatives of the roots of the original equation. Hence, the approximations to the negative roots of the original equation may be found by reversing the signs of the corresponding positive roots of the new equation.

In the two examples we have found approximations to the real roots of cubic equations. The process, of course, is applicable to equations of higher degrees. Furthermore, a root can be obtained to as many digits as desired. It is evident, however, that the computations increase in tediousness with the degree of the equation and with the accuracy required in the result.

EXERCISE 17-6

Find to two decimal places the real root of each of the following equations.

1. $x^3 + 3x - 8 = 0$

2. $x^3 + x - 14 = 0$

3. $x^3 - x^2 - 2 = 0$

4. $x^3 - 3x^2 + 4x - 3 = 0$

5. $x^3 + 3x^2 + 3x - 27 = 0$

6. $x^3 + 2x - 5 = 0$

7. $x^3 + 3x^2 + 5x - 2 = 0$

8. $x^3 - 2x^2 - 8 = 0$

9. $x^3 + 3x^2 + 3x - 8 = 0$

10. $2x^3 + 6x^2 + 6x - 3 = 0$

11. $3x^3 + 4x^2 - x - 14 = 0$

12. $2x^3 - 3x^2 - 3x - 6 = 0$

Find three roots to two decimal places.

13. $x^3 - 3x^2 - x + 4 = 0$

14. $x^3 - 3x^2 - 2x + 1 = 0$

15. $x^3 - 6x^2 - 8x + 1 = 0$

16. $x^3 - 5x^2 + 3x + 8 = 0$

17. $x^3 - 4x^2 + 3x + 1 = 0$

18. $x^3 - 9x^2 + 24x - 17 = 0$

19. $4x^3 - 12x^2 + 4x + 5 = 0$

20. $3x^3 - 5x^2 - 4x + 3 = 0$

Find the indicated root to two decimal places.

21. $x^4 - x - 3 = 0$; the root between 1 and 2

22. $x^4 + 2x^3 - 2 = 0$; the root between 0 and 1

23. $x^4 + x^3 - 6x - 15 = 0$; the root between 2 and 3

24. $x^4 - x^3 + x^2 - 3x - 4 = 0$; the root between 1 and 2

18

GRAPHICAL REPRESENTATION OF THE TRIGONOMETRIC FUNCTIONS

18-1 VARIATION OF THE TRIGONOMETRIC FUNCTIONS

The values of the trigonometric functions of an angle can be quite simply represented by directed line segments. We use Figs. 18-1 to 18-4 to provide a scheme for the representation. Each figure has a unit circle and a tangent line at the point Q. The terminal side of angle θ cuts the circle at the point P and the tangent line at R. Then, we have

$$\sin\theta = MP \qquad \cos\theta = OM \qquad \tan\theta = QR$$

A mental picture of the values which $\sin\theta$, $\cos\theta$, and $\tan\theta$ take as θ increases from 0 to 2π may be formed by focusing the attention on the resulting line segments MP, OM, and QR.

As θ increases from 0 to $\pi/2$, the directed distance MP, or $\sin\theta$, increases from 0 to 1; as θ increases from $\pi/2$ to π, $\sin\theta$ decreases from 1 to 0. For quadrants III and IV, the directed distance MP is negative. Hence as θ increases from π to $3\pi/2$, $\sin\theta$ varies from

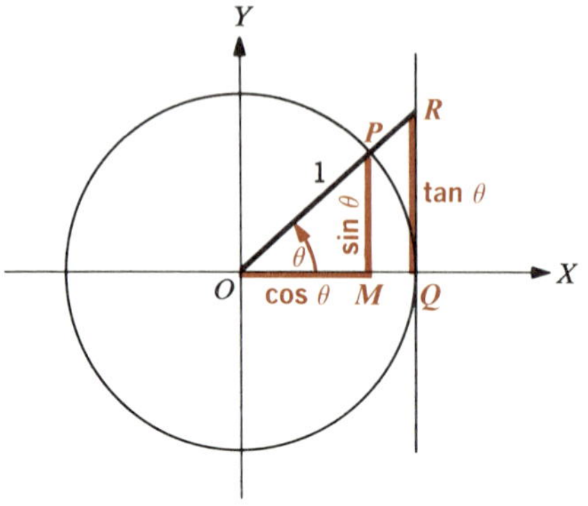

Figure 18-1

0 to -1; and finally, as θ increases from $3\pi/2$ to 2π, $\sin\theta$ varies from -1 to 0.

In a similar way the variation of $\cos\theta$ is revealed by the behavior of the line segment *OM*. For quadrants I and IV, the distance *OM* is positive and for quadrants II and III, the distance is negative. Consequently, the variation of $\cos\theta$ as θ increases from 0 to 2π, is, in order of quadrants, 1 to 0, 0 to -1, -1 to 0, 0 to 1.

Figures 18-1 to 18-4 also enable us to visualize the variation of $\tan\theta$ as θ increases. If θ starts at 0 and increases, the point *R* starts at *Q* and moves upward along the tangent line. The angle may be taken close enough to $\pi/2$ to make its tangent as large as any chosen positive number. For example, if the terminal side of θ passes through the point (1,1000), $\tan\theta$ is 1000. We conclude then that as θ increases from 0 to $\pi/2$, $\tan\theta$ starts at 0 and assumes all positive values. The terminal side of $\pi/2$ does not intersect the tangent line. Hence, there is no line representation of $\tan\pi/2$.

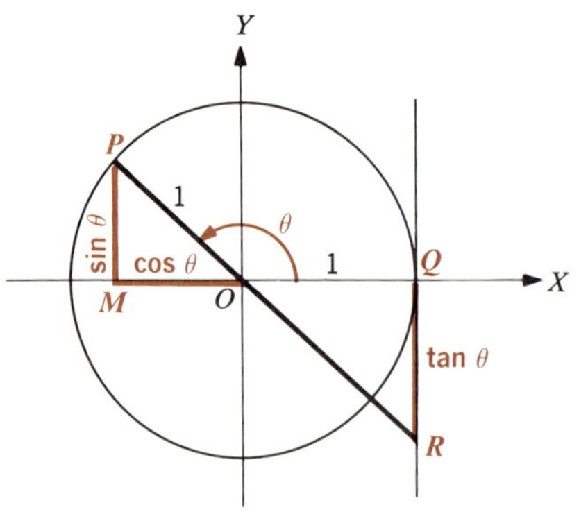

Figure 18-2

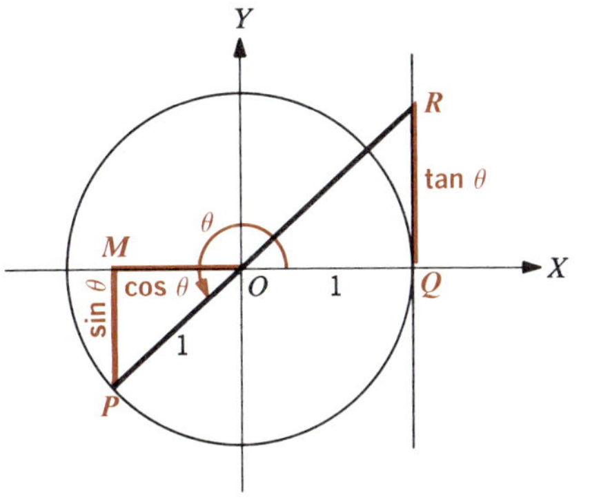

Figure 18-3

This agrees with an earlier observation that $\pi/2$ has no tangent (Sec. 8-4). We indicate the behavior of $\tan\theta$ in the first quadrant by employing the infinity symbol and writing $0 \to +\infty$.

When θ is in the second quadrant, $\tan\theta$ is negative and we extend the terminal side of θ through the origin to obtain the line segment QR. This line segment, directed downward, is negative and has the proper length to represent $\tan\theta$. As θ increases from $\pi/2$ to π, the point R moves upward and reaches Q at $\theta = \pi$. Hence, in this process, $\tan\theta$ assumes all negative values and zero. We indicate this variation by the symbolism $-\infty \to 0$.

As θ increases from π to 2π, $\tan\theta$ repeats its behavior in quadrants III and IV exactly as in quadrants I and II, respectively.

The variations of $\sin\theta$, $\cos\theta$, and $\tan\theta$ are indicated in the following table. We suggest that the student prepare a similar table for the reciprocals $\cot\theta$, $\sec\theta$, and $\csc\theta$.

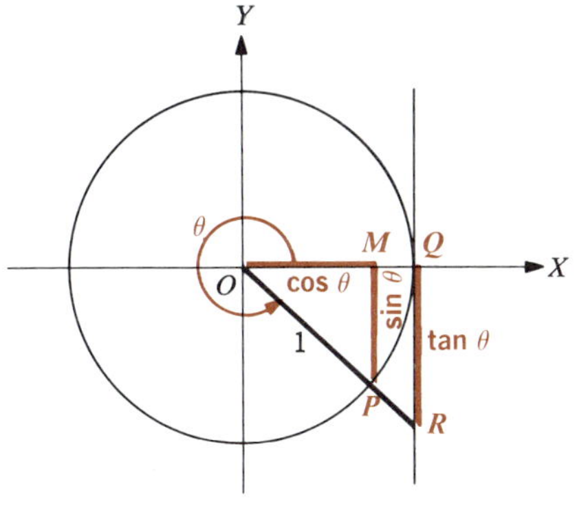

Figure 18-4

As θ varies from	$\sin\theta$ varies from	$\cos\theta$ varies from	$\tan\theta$ varies from
0 to $\frac{\pi}{2}$	0 to 1	1 to 0	$0 \to +\infty$
$\frac{\pi}{2}$ to π	1 to 0	0 to -1	$-\infty \to 0$
π to $\frac{3\pi}{2}$	0 to -1	-1 to 0	$0 \to +\infty$
$\frac{3\pi}{2}$ to 2π	-1 to 0	0 to 1	$-\infty \to 0$

18-2 PERIODS OF THE TRIGONOMETRIC FUNCTIONS

A trigonometric function of an angle is not changed in value when the angle is changed by any integral multiple of 2π radians. For this reason, the functions are said to be *periodic.* The sine function, for example, satisfies the equations

$$\sin\theta = \sin(\theta + 2\pi) = \sin(\theta + n \cdot 2\pi)$$

where θ is the radian measure of an angle and n is any integer. Thus, the values of the sine of angles recur in intervals of 2π radians. The variation of the sine function is such that the recurrence does not take place in smaller intervals (Sec. 18-1), and consequently 2π radians is called the *period* of the function. The cosecant, cosine, and the secant functions also have a period of 2π radians. To find the period of the tangent and cotangent functions, we set $A = \theta$ and $B = \pi$ in formula (13) of Sec. 10-5 and obtain

$$\tan(\theta + \pi) = \tan\theta$$

Hence, the values of the tangent and cotangent of angles recur in intervals of π radians. Since the recurrence does not take place in a smaller interval, the period of these functions is π radians.

The property of periodicity makes the trigonometric functions useful in the study of numerous problems where phenomena are of a recurrent nature. As examples, the functions are employed in the study of sound waves, light waves, vibrating strings, the pendulum, and alternating currents and voltages.

An understanding of the periods of the trigonometric functions is helpful in constructing their graphs. If the graph of the tangent or cotangent function is obtained over an interval of π radians, then this part of the graph can be reproduced in other intervals of

π radians. The same procedure is applicable to the other four functions except that the period is 2π radians.

Next, we proceed to discuss the graphs of the trigonometric functions. In this discussion, however, we shall let the domains of the functions be real numbers. In Sec. 8-6 we pointed out the close relation between the trigonometric functions of angles and the trigonometric functions of real numbers, and discovered that either kind of domain seemed equally reasonable and logical. In this connection, we recall that the usual tables of trigonometric functions serve to find the values of the functions of real numbers. Thus, illustrating with a real number x and the sine function, we have

$$\sin x = \sin (x \text{ radians}) = \sin \left(\frac{180x}{\pi}\right)^0$$

18-3 THE GRAPHS OF THE SINE AND COSINE FUNCTIONS

To plot the graph of the sine function we use the equation $y = \sin x$. We first assign several real values to x in the interval 0 to 2π, choosing integral multiples of $\pi/6$. The sines of these numbers, equal to the sines of multiples of 30°, may be found by referring to Table 2. The values of x and the corresponding values of $\sin x$, or y, are shown in the table.

x	0	$\frac{\pi}{6}$	$\frac{\pi}{3}$	$\frac{\pi}{2}$	$\frac{2\pi}{3}$	$\frac{5\pi}{6}$	π	$\frac{7\pi}{6}$	$\frac{4\pi}{3}$	$\frac{3\pi}{2}$	$\frac{5\pi}{3}$	$\frac{11\pi}{6}$	2π
$y = \sin x$	0	0.50	0.87	1	0.87	0.50	0	−0.50	−0.87	−1	−0.87	−0.50	0

We next lay off the same unit of length along the x axis and the y axis and take a point on the x axis about 3.14 units from the origin to correspond to the number π.* The corresponding values of x and y are plotted and a smooth curve is drawn through the points thus determined (Fig. 18-5). This part of the graph, which comprises one interval, may be duplicated in intervals of 2π to the right and left. The curve is called the *sine curve.*

By proceeding in just the same way as with the sine function,

* The following plan, however, is reasonably accurate and easier to use. Choose a convenient number of spaces of the graph paper to stand for the length $\pi/3$ (equals 1.05, approximately) on the x axis, and let the same number of spaces be the unit length on the y axis.

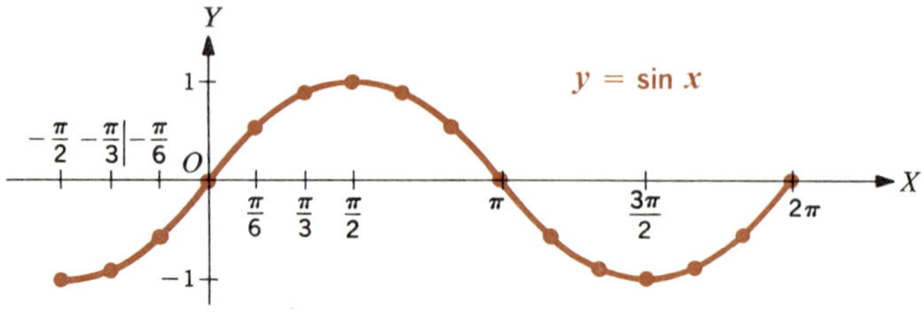

Figure 18-5

we can draw the graph of $y = \cos x$. The part of the cosine curve from $x = -\pi/2$ to $x = 5\pi/2$ is shown in Fig. 18-6.

An examination of the sine curve and the cosine curve reveals that the values of these two functions vary from -1 to 1. These are called the *extreme values.* Each graph crosses the x axis in every selected interval of length π. The extreme values and the zero values correspond to integral multiples of $\pi/2$. Moreover, the sine and cosine curves are shaped alike. Indeed, the identity $\sin(x + \frac{1}{2}\pi) = \cos x$ shows that the two curves drawn on the same coordinate axes would coincide if the sine curve were shifted $\pi/2$ units to the left.

18-4 THE GRAPHS OF THE OTHER FUNCTIONS

The period of the tangent function is π. Hence we may construct the graph of $y = \tan x$ for x between $-\frac{1}{2}\pi$ and $\frac{1}{2}\pi$ and then reproduce the curve in intervals of π to the right and left. The tabulated values for x and $\tan x$ may be verified by referring to Table 3 in the Appendix. To obtain the functional values for negative values of x, we recall the identity $\tan(-x) = -\tan x$ (Sec. 10-4).

x	0	0.4	0.8	1	1.2	1.3
$\tan x$	0	0.42	1.0	1.6	2.6	3.6

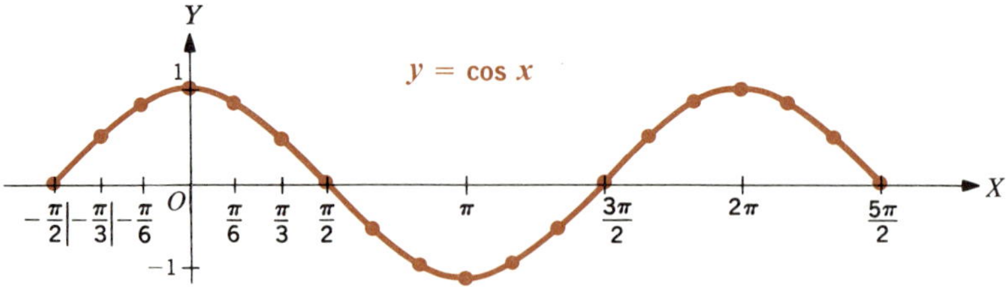

Figure 18-6

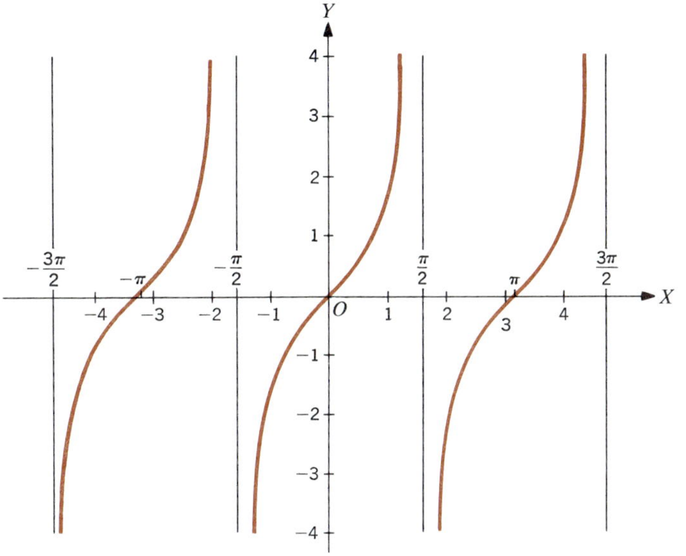

Figure 18-7 $y = \tan x$

The graph of the function (Fig. 18-7) is separated into *branches* at values of x which are odd integral multiples of $\pi/2$, where the tangent function does not exist. And the branches cross the x axis at integral multiples of π.

The graphs of the cotangent, secant, and cosecant functions are shown in Figs. 18-8 to 18-10. The graph of each function is separated into branches at values of x for which the function is not defined.

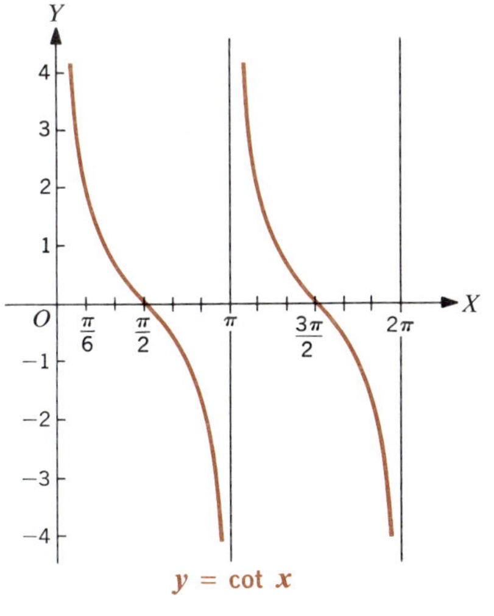

Figure 18-8 $y = \cot x$

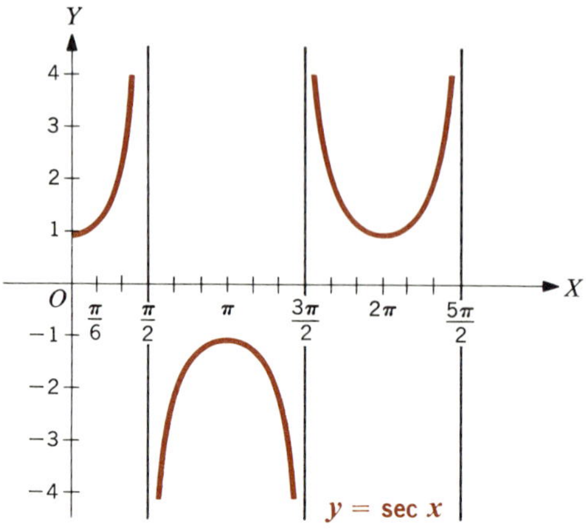

Figure 18-9

18-5 GRAPHS OF MORE GENERAL FUNCTIONS

Having constructed the graphs of the trigonometric functions of the variable x, we next consider certain generalizations of the sine and cosine functions. Thus for the sine function, we introduce the equations

$$y = a \sin bx \qquad \text{and} \qquad y = a \sin (bx + c)$$

where a, b, and c are constants. Since $\sin bx$ and $\sin (bx + c)$ vary from -1 to 1, the extreme values of the functions defined by the

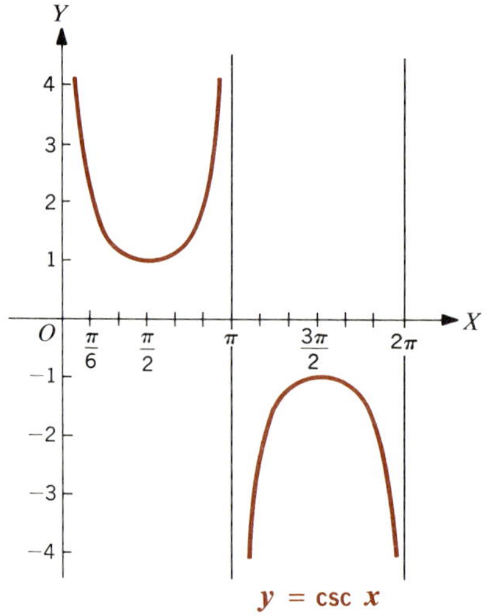

Figure 18-10

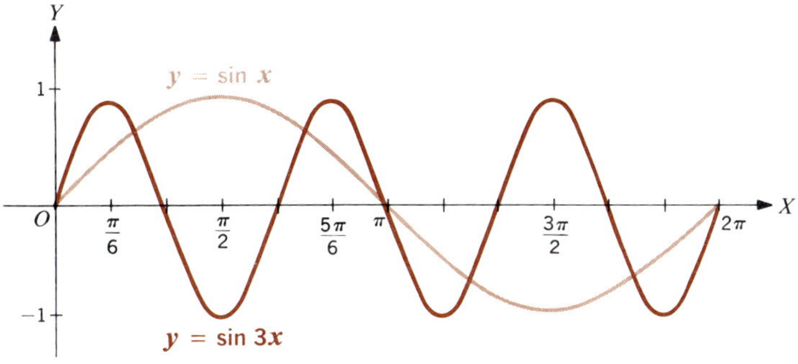

Figure 18-11

two equations are $-|a|$ and $|a|$. The period of $\sin bx$ may be obtained by determining how much x changes in producing a change of 2π in bx. As x varies from 0 to $2\pi/b$, the number bx varies from 0 to 2π if $b > 0$ and from 0 to -2π if $b < 0$. Accordingly, the period is $2\pi/b$. This is also the period of $\sin(bx + c)$.

The graphs of the equations $y = a \sin bx$ and $y = a \sin(bx + c)$, on the same coordinate axes, are alike except for position. This is true because the two values of y become equal if x is replaced by $x - c/b$ in the second equation. This tells us that a shift of c/b units along the x axis will bring the graph of the second equation into coincidence with the graph of the first equation. The shift is to the right or left depending on whether b and c have like or unlike signs.

Our discussion of the above equations applies in a similar way to the equations $y = a \cos bx$ and $y = a \cos(bx + c)$. We proceed now to construct graphs of special cases of the extended sine and cosine functions.

EXAMPLE 1 Draw the graph of the equation $y = \sin 3x$.

Solution: The function defined by this equation has extreme values of -1 and 1 and a period of $2\pi/3$. Hence, $\sin 3x$ and $\sin x$ have the same extreme values, but the period of $\sin 3x$ is one-third the period of $\sin x$. These facts are shown in the graphs of $y = \sin x$ and $y = \sin 3x$ (Fig. 18-11).

EXAMPLE 2 Construct the graph of $y = 2 \cos \frac{1}{2}x$.

Solution: The extreme values are -2 and 2 and the period, 2π divided by $\frac{1}{2}$, is 4π. The graph constructed from the accompanying table of values is shown in Fig. 18-12. Each y value is twice the corresponding value of the graph of $y = \cos \frac{1}{2}x$, which is included for comparison.

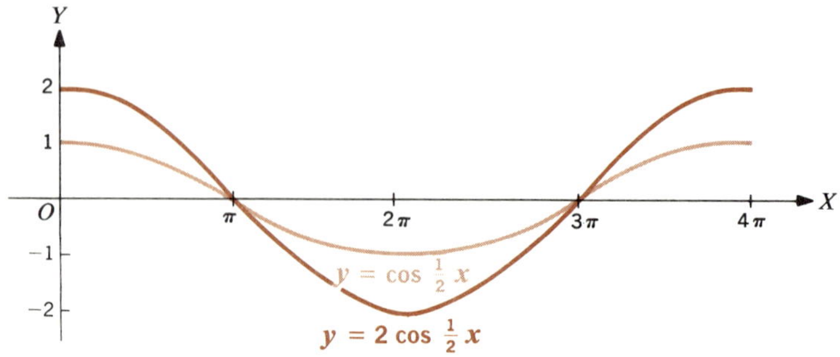

Figure 18-12

x	0	$\frac{\pi}{2}$	π	$\frac{3\pi}{2}$	2π	$\frac{5\pi}{2}$	3π	$\frac{7\pi}{2}$	4π
$\frac{1}{2}x$	0	$\frac{\pi}{4}$	$\frac{\pi}{2}$	$\frac{3\pi}{4}$	π	$\frac{5\pi}{4}$	$\frac{3\pi}{2}$	$\frac{7\pi}{4}$	2π
$2 \cos \frac{1}{2}x$	2	1.4	0	−1.4	−2	−1.4	0	1.4	2

EXAMPLE 3 Draw the graph of the equation $y = 1.5 \sin 2x$.

Solution: The extreme values are −1.5 and 1.5, and the period is $2\pi/2 = \pi$. The graph (Fig. 18-13) resembles the graph of $y = \sin x$; comparatively, it is compressed horizontally and stretched vertically.

EXAMPLE 4 Construct the graphs of the equations

$$y = 1.5 \sin 2x \qquad \text{and} \qquad y = 1.5 \sin (2x + \tfrac{1}{3}\pi)$$

Solution: According to our discussion above, the graphs of the

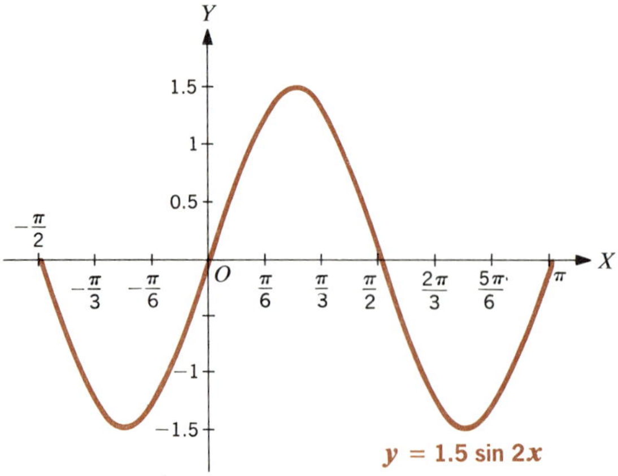

Figure 18-13

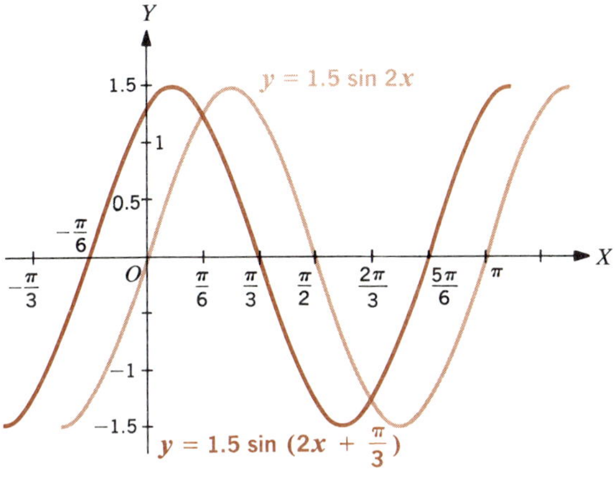

Figure 18-14

two equations are alike except for position (Fig. 18-14). And the curves would coincide if the graph of $y = 1.5 \sin (2x + \frac{1}{3}\pi)$ were moved $\frac{1}{6}\pi$ units to the right.

EXAMPLE 5 Draw the graphs of the equations

$$y = \cos 3x \qquad \text{and} \qquad y = \cos\left(3x + \frac{2\pi}{3}\right)$$

Solution: The graphs are drawn through two periods in Fig. 18-15. We note that the intersection points are at $x = 2\pi/9$, $5\pi/9$, $8\pi/9$, and $11\pi/9$. The graphs would coincide if the graph of the second equation were shifted $2\pi/9$ units to the right.

We remark that a rough drawing of each graph in Figs. 18-5 to 18-15 could be made without plotting points. The extreme values,

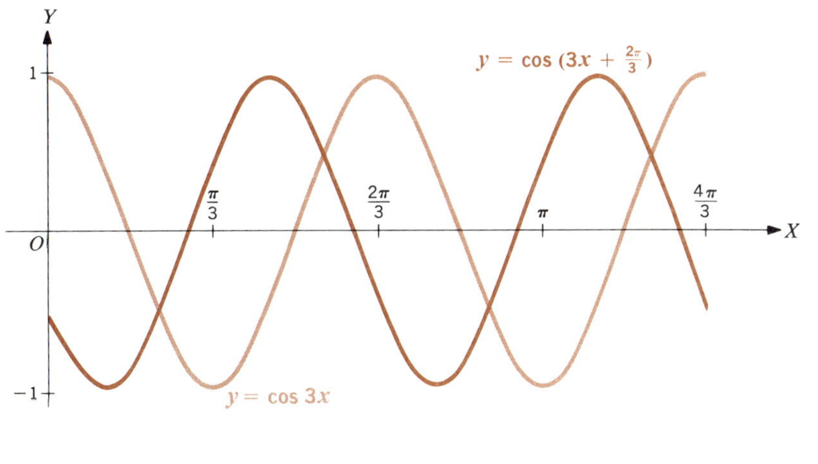

Figure 18-15

the period, and a knowledge of the way in which the function varies suggest about where the curve should be drawn. This method of graphing is called *sketching*. In many problems a sketch serves as well as an accurately drawn curve.

18-6 GRAPHING BY ADDITION OF ORDINATES

Frequently, trigonometric functions of sums can be conveniently graphed by a process called *addition of ordinates*. As an illustration of this method, we shall construct the graph of the equation

$$y = 2 \cos x + \sin 3x$$

This plan requires that the graphs of $y_1 = 2 \cos x$ and $y_2 = \sin 3x$ be drawn on the same coordinate axes (Fig. 18-16). Then for any x, the corresponding ordinate of $y = 2 \cos x + \sin 3x$ can be had from the two graphs. For example, the ordinate AD is the sum of the ordinates AB and AC. The addition of ordinates in this manner must be algebraic. Thus MN is negative, and the point Q is obtained by measuring downward from P, so that $PQ = MN$. By plotting a sufficient number of points, in the way indicated, the desired graph can be drawn. This process of graphing is especially helpful when the graphs of the separate terms can be drawn easily.

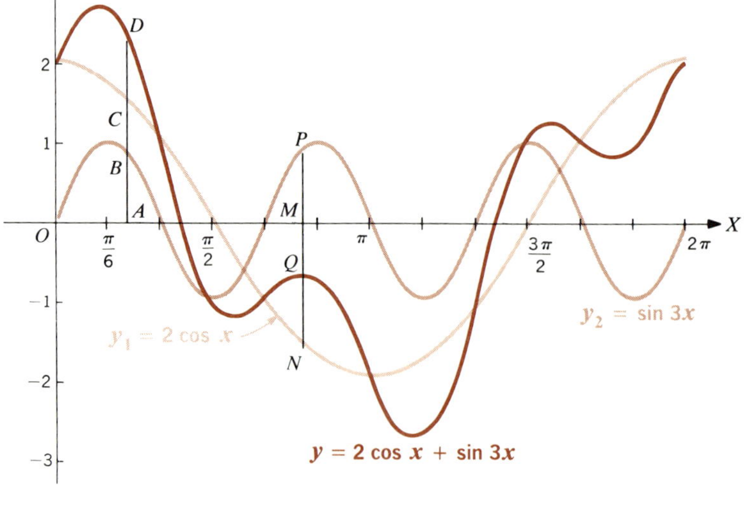

Figure 18-16

EXERCISE 18-1

Find the period of the graph of each equation. Give the extreme values of those involving the sine and cosine.

1. $y = \cos 4x$
2. $y = 3 \cos \frac{1}{2}x$
3. $y = -\cos \frac{1}{3}x$
4. $y = 4 \tan 6x$
5. $y = \frac{1}{2} \text{tin}\ 5x$
6. $y = \sin \frac{2}{3}x$
7. $y = \tan 2x$
8. $y = \tan \frac{1}{2}x$
9. $y = \cot \frac{3}{4}x$
10. $y = \sec \frac{3}{2}x$
11. $y = \sec 7x$
12. $y = \csc 6x$

Construct one period of the graph of each equation.

13. $y = \sin 2x$
14. $y = \cos 3x$
15. $y = \frac{1}{2} \tan 2x$
16. $y = 2 \cos 2x$
17. $y = 2 \sin \frac{1}{2}x$
18. $y = \tan \frac{1}{2}x$

Sketch one period of the graph of each pair of equations on the same coordinate axes.

19. $y = \sin x, \quad y = 2 \sin x$
20. $y = \sin x, \quad y = \sin 2x$
21. $y = \cos x, \quad y = 3 \cos x$
22. $y = \cos x, \quad y = \cos \frac{1}{2}x$
23. $y = \sin x, \quad y = \sin \left(x + \dfrac{\pi}{4}\right)$
24. $y = \cos x, \quad y = \cos \left(x - \dfrac{\pi}{6}\right)$
25. $y = \sin x, \quad y = \sin \left(x - \dfrac{\pi}{6}\right)$
26. $y = 2 \cos x, \quad y = 2 \cos \left(x + \dfrac{\pi}{3}\right)$
27. $y = 1.5 \cos 2x, \quad y = 1.5 \cos \left(2x + \dfrac{\pi}{3}\right)$
28. $y = \sin 3x, \quad y = \sin \left(3x + \dfrac{2\pi}{3}\right)$
29. $y = \frac{1}{2} \tan x, \quad y = \frac{1}{2} \tan \left(x + \dfrac{\pi}{4}\right)$
30. $y = 2 \cos \dfrac{x}{2}, \quad y = 2 \cos \left(\dfrac{x}{2} + \dfrac{\pi}{4}\right)$

Sketch one period of the graph of each of the following equations by the method of addition of ordinates.

32. $y = \sin x + \cos x$

33. $y = \sin x + \cos 2x$

34. $y = \sin x + \sin 2x$

35. $y = \sin x + \cos 3x$

36. $y = 2 \sin x + \sin 2x$

37. $y = 2 \cos x - \cos 2x$

19

TRIGONOMETRIC EQUATIONS AND INVERSE FUNCTIONS

19-1 SOLUTIONS OF TRIGONOMETRIC EQUATIONS

An equation involving one or more trigonometric functions of a variable is called a *trigonometric equation.* In Chap. 10 trigonometric equations of a particular kind, namely, trigonometric identities, were introduced. We shall now consider conditional trigonometric equations—equations which hold for only some of the angles (or numbers) for which the members are defined. Our problem will be to find solutions of such equations.

The simplest type of trigonometric equation is that in which a function of an angle is equal to a constant. Equations of this kind were used in finding unknown angles of a triangle (Chap. 9). The equation $\sin x = \frac{1}{2}$ is an example of a simple trigonometric equation. The equation is satisfied by $x = 30°$ and $x = 150°$, and all angles which differ from these by any integral multiple of 360°; that is, for all integers n, the solutions are

$$x = 30° + n \cdot 360° \qquad \text{and} \qquad x = 150° + n \cdot 360°$$

When a trigonometric equation is not of the simplest type, the procedure is to derive two or more simple equations which yield all the solutions of the given equation. The tools for this procedure are algebraic operations and trigonometric identities.

At the outset, however, we note that the solutions of a trigonometric equation may be expressed by

1. angles in degree measure
2. angles in radian measure
3. real numbers (Sec. 8-6)

We expressed the solutions of $\sin x = \frac{1}{2}$ in degree measure. Henceforth, however, we shall express solutions in terms of real numbers unless it is necessary to use a table. When a table is needed, we shall refer to Table 2 and use degree measure. Also, for brevity, we shall give only the solutions in the interval $0 \leq x < 2\pi$ or the interval $0 \leq x < 360°$.

EXAMPLE 1 Find all values of x in the interval $0 \leq x < 2\pi$ which satisfy the equation

$$2 \cos^2 x - 5 \cos x + 2 = 0$$

Solution: Factoring the left member gives

$$(\cos x - 2)(2 \cos x - 1) = 0$$

By setting each factor equal to zero, we obtain the simple equations

$$\cos x - 2 = 0 \qquad \text{and} \qquad 2 \cos x - 1 = 0$$

The first of these equations has no solution because the value of a cosine never exceeds 1. From the second equation, we have $\cos x = \frac{1}{2}$, and the solutions are

$$x = \frac{\pi}{3} \qquad \text{and} \qquad x = \frac{5\pi}{3}$$

Each of these values may be changed by any integral multiple of 2π, but we are giving only the solutions in the interval $0 \leq x < 2\pi$. We check our results by substituting in the given equation. Thus,

$$2 \cos^2 \frac{\pi}{3} - 5 \cos \frac{\pi}{3} + 2 = 2\left(\frac{1}{2}\right)^2 - 5\left(\frac{1}{2}\right) + 2 = 0$$

$$2 \cos^2 \frac{5\pi}{3} - 5 \cos \frac{5\pi}{3} + 2 = 2\left(\frac{1}{2}\right)^2 - 5\left(\frac{1}{2}\right) + 2 = 0$$

The two values satisfy the given equation, and therefore the solution set is

$$\left\{\frac{\pi}{3}, \frac{5\pi}{3}\right\}$$

EXAMPLE 2 Find the solution set in the interval $0 \leq x < 360°$ of the equation

$$\tan^2 x - 3 \tan x - 2 = 0$$

Solution: This is a quadratic equation in $\tan x$, and applying the quadratic formula, we have

$$\tan x = \frac{3 \pm \sqrt{9 + 8}}{2} = \frac{3 \pm \sqrt{17}}{2} = 3.562, \ -0.5616$$

We use Table 2 and find a solution of $\tan x = 3.562$. Then the angle in the third quadrant having the solution thus found as a reference angle is also a solution. Similarly, we find an angle whose tangent is $+0.5616$. The angles in the second and fourth quadrants having this angle as a reference angle are solutions of $\tan x = -0.5616$ (Sec. 9-4). Hence,

$$\begin{aligned} \tan x &= 3.562 & \tan x &= -0.5616 \\ x &= 74°\ 19' & x &= 180° - 29°\ 19' \\ & & &= 150°\ 41' \\ x &= 180° + 74°\ 19' & x &= 360° - 29°\ 19' \\ &= 254°\ 19' & &= 330°\ 41' \end{aligned}$$

These four results are solutions to the nearest minute. They check approximately in the given equation. Hence, we express the solution set by

$$\{74°\ 19', 150°\ 41', 254°\ 19', 330°\ 41'\}$$

EXAMPLE 3 Find all values of x in the interval $0 \leq x < 2\pi$ which satisfy the equation

$$\sin 2x \sec^2 x = 2 \sin 2x$$

Solution: We express the equation with one member equal to zero and then factor the other member. Thus,

$$\sin 2x \sec^2 x - 2 \sin 2x = 0$$

$$\sin 2x(\sec^2 x - 2) = 0$$

Equating the first factor to zero, we have

$$\sin 2x = 0$$

$$2x = 0, \pi, 2\pi, 3\pi$$

$$x = 0, \frac{\pi}{2}, \pi, \frac{3\pi}{2}$$

From the second factor, $\sec^2 x - 2$, we obtain

$$\sec x = \sqrt{2} \qquad \sec x = -\sqrt{2}$$

$$x = \frac{\pi}{4}, \frac{7\pi}{4} \qquad x = \frac{3\pi}{4}, \frac{5\pi}{4}$$

All the values which we have found for x will satisfy the given equation except $\pi/2$ and $3\pi/2$. These values must be rejected because they have no secant. Hence, the solution set is

$$\left\{0, \frac{\pi}{4}, \frac{3\pi}{4}, \pi, \frac{5\pi}{4}, \frac{7\pi}{4}\right\}$$

We make two important observations concerning this problem. First, we did not divide the members of the given equation by $\sin 2x$. Such a division would have resulted in the loss of 0 and π as solutions. Second, in order to find the solutions of $\sin 2x = 0$ for $0 \leq x < 2\pi$, it is necessary to find the solutions for $0 \leq 2x < 4\pi$.

EXERCISE 19-1

Find the solution set of each equation in the interval $0 \leq x < 2\pi$ *or, when a table is used, the interval* $0 \leq x < 360°$.

1. $2 \cos x + 1 = 0$

2. $\sin x + 1 = 0$

3. $\cot x = 1$

4. $\tan x + \sqrt{3} = 0$

5. $2 \cos x + \sqrt{3} = 0$

6. $\csc x = 2$

7. $\sec x - \sqrt{2} = 0$

8. $2 \sin x - \sqrt{3} = 0$

9. $\sin 2x = 0$

10. $\cot \frac{1}{2}x - 1 = 0$

11. $\sqrt{2} \cos 2x = 1$

12. $\sqrt{3} \csc 2x = 2$

13. $3 \cos \frac{1}{2}x - 1 = 0$

14. $\tan \frac{1}{2}x - 1 = 0$

15. $4 \sin \frac{1}{2}x = 1$

16. $\cos 3x = 0$

17. $2 \cot x - 3 = 0$

18. $4 \cos x = 3$

19. $4 \sin^2 x - 1 = 0$

20. $3 \tan^2 x - 1 = 0$

21. $4 \sin^2 x = 3$

22. $\sin^2 x + \sin x = 0$

23. $\sin x \cos x - \cos x = 0$

24. $\sqrt{3} \tan x \cot x = \tan x$

25. $\sqrt{3} \sec x = 2 \sin x \sec x$

26. $\cot x \sec x - \sqrt{3} \sec x = 0$

27. $2 \cos^2 x + \cos x - 1 = 0$

28. $2 \sin^2 x + \sin x = 1$

29. $2 \sin^2 x + 5 \sin x + 2 = 0$

30. $3 \cos^2 x + 2 \cos x = 1$

31. $4 \cos^2 x + 3 \cos x = 1$

32. $\sin^2 x + \sin x - 6 = 0$

33. $2 \csc^2 x - \csc x = 1$

34. $3 \cot^2 x + 2 \cot x = 3$

35. $2 \tan^2 x - \tan x = 3$

36. $2 \sec^2 x + 3 \sec x = 2$

37. $\cot^2 x - 2 \cot x = 3$

38. $6 \cos^2 x + 5 \cos x = 6$

39. $3 \sin^2 x - 7 \sin x = 6$

40. $\cot^2 x + \cot x + 1 = 0$

41. $\cos^2 x + 5 \cos x + 2 = 0$

42. $\sin^2 x - 3 \cos x + 1 = 0$

19-2 IDENTITIES USED IN SOLVING EQUATIONS

We now consider trigonometric equations whose solutions can be found through use of certain of the identities (1) to (30) of Chap. 10. Here, as in the preceding section, the plan consists in finding simple equations (each containing one function of an angle or number) which have all the roots of the equation to be solved. When an operation is performed which might introduce extraneous roots, such as squaring both members of an equation or multiplying both members by a trigonometric expression, it is essential that all results be tested in the given equation.

EXAMPLE 1 Solve the equation $\sin x = \cos 2x$.

Solution: This equation has a function of x and a function of $2x$. But we obtain an equation with a function of x only by applying identity (16), $\cos 2x = 1 - 2\sin^2 x$. Thus,

$$\sin x = 1 - 2\sin^2 x$$

$$2\sin^2 x + \sin x - 1 = 0$$

$$(2\sin x - 1)(\sin x + 1) = 0$$

The solution set, obtained by setting each factor equal to zero, is

$$\left\{\frac{\pi}{6}, \frac{5\pi}{6}, \frac{3\pi}{2}\right\}$$

EXAMPLE 2 Solve the equation $2\sin x + \cot x - \csc x = 0$.

Solution: First multiplying both members of the equation by $\sin x$, we get

$$2\sin^2 x + \cos x - 1 = 0$$

$$2(1 - \cos^2 x) + \cos x - 1 = 0$$

$$2\cos^2 x - \cos x - 1 = 0$$

$$(2\cos x + 1)(\cos x - 1) = 0$$

The first factor in the left member of the last equation yields $x = 2\pi/3, 4\pi/3$, and the second factor yields $x = 0$. The values $2\pi/3$ and $4\pi/3$ will check in the given equation. But we reject $x = 0$ because neither the cotangent nor the cosecant is defined for this number. The extraneous root was introduced by multiplying the members of the given equation by $\sin x$.

EXAMPLE 3 Solve the equation $1 - \sin x = \cos x$.

Solution: To obtain an equation with only one function of x, we square the members of the given equation and substitute for

$\cos^2 x$. Thus, we get

$$\begin{aligned} 1 - 2\sin x + \sin^2 x &= \cos^2 x \\ &= 1 - \sin^2 x \\ 2\sin^2 x - 2\sin x &= 0 \\ 2\sin x(\sin x - 1) &= 0 \end{aligned}$$

The factors of the left side give $x = 0, \pi/2, \pi$. These results are solutions of the equation obtained by squaring the members of the original equation. Because this operation may have introduced extraneous roots, it is essential that each result be tested. Substituting in each side of the given equation, we find that

$$\begin{array}{lll} x = 0 & \text{gives} & 1 - 0 = 1 \\ x = \dfrac{\pi}{2} & \text{gives} & 1 - 1 = 0 \\ x = \pi & \text{gives} & 1 - 0 \neq -1 \end{array}$$

We see that 0 and $\pi/2$ are roots but π is not a root. Hence, the solution set is

$$\left\{0, \frac{\pi}{2}\right\}$$

EXAMPLE 4 Solve the equation $2\sin 3x - 2\sin x + 3\cos 2x = 0$.

Solution: By identity (28), $\sin 3x - \sin x = 2\cos 2x \sin x$. Then from the given equation, we get

$$\begin{aligned} 4\cos 2x \sin x + 3\cos 2x &= 0 \\ \cos 2x(4\sin x + 3) &= 0 \end{aligned}$$

Setting the first factor equal to zero gives

$$\begin{aligned} \cos 2x &= 0 \\ 2x &= 90°, 270°, 450°, 630° \\ x &= 45°, 135°, 225°, 315° \end{aligned}$$

Equating the second factor to zero, we have $\sin x = -0.75$. Solutions of this equation have terminal sides in the third and fourth quadrants with the angle whose sine is $+0.75$ as a reference angle. The reference angle is $48° \, 35'$, and consequently, we obtain

$$\sin x = -0.75$$
$$x = 180° + 48° \, 35' = 228° \, 35'$$
$$x = 360° - 48° \, 35' = 311° \, 25'$$

Combining the foregoing results, we have the solution set

$$\{45°, 135°, 225°, 315°, 228° \, 35', 311° \, 25'\}$$

Notice that we found all solutions of $\cos 2x = 0$ for $0° \leq 2x < 720°$. This enabled us to obtain all solutions of $\cos 2x = 0$ for which $0° \leq x < 360°$.

The operations which we have used in the preceding examples are often helpful in finding solutions of trigonometric equations. For emphasis, we list the operations in the form of directions:

1. If two or more angles appear in an equation, use the fundamental identities (1) to (30), Sec. 10-9, to express the equation in terms of functions of a single angle.
2. If different functions appear in the equation, use the fundamental identities to express the equation in terms of a single function.
3. Write the equation with one member equal to zero and the other member in factored form with each factor containing a single function, using the fundamental identities if necessary.
4. Simplify the equation by multiplying each member by a suitable trigonometric expression. Check to discover extraneous solutions.
5. Square both members of the equation. Check for extraneous solutions.

EXERCISE 19-2

Solve each of the following equations. Check the results when there is a possibility of an extraneous solution.

1. $\cos 2x - \sin x = 0$
2. $\sin \frac{1}{2}x - \cos x = 0$
3. $\cot^2 x - \csc x - 1 = 0$
4. $2 \sin^2 x + \cos x + 4 = 0$
5. $\cos^2 x - \sin x + 5 = 0$
6. $\sec^2 x - \tan x - 3 = 0$
7. $4 \cos x + \sec x - 4 = 0$
8. $\csc x + 4 \sin x + 4 = 0$

9. $\csc x + \sin x + 2 = 0$

10. $\tan x + \sec x = -\cos x$

11. $\cot x + \csc x = -\sin x$

12. $\cot x + 2 \sin x = \csc x$

13. $\tan x + 2 \cos x = \sec x$

14. $1 - \sin x - \cos x = 0$

15. $2 \cos x + 1 = \sin x$

16. $\cos x + \sqrt{3} \sin x = 1$

17. $3 \sin x - 2 = \cos x$

18. $2 \cos x - \sin x = 1$

19. $\cos x - \sqrt{2} \sin x = 1$

20. $2 \cot x - \csc x = 0$

21. $\sin 3x + \sin x = 0$

22. $\cos 3x - \cos x = 0$

23. $\cos 4x + \cos x = \cos 2x$

24. $\sin 3x + \sin x = 0$

25. $\sin 3x + \sin x = \sin 2x$

26. $\cot^3 x - 8 = 0$

27. $\cos^3 x \csc^3 x = 4 \cot^2 x$

28. $\sin^2 x + \sec^2 x = \cos^2 x$

29. $\sin 2x + 2 \sin^2 \frac{1}{2}x = 1$

30. $2 \cos^2 \frac{1}{2}x + \sin 2x = 1$

31. $\cot x + \csc x = 1$

32. $\tan 2x - \cos x = 0$

33. $\tan \frac{1}{2}x + 2 \sin 2x = \csc x$

34. $\tan 2x + \cot x = 0$

35. $\sec x \tan x - \sqrt{2} = 0$

36. $\tan 2x + \tan x = 0$

37. $2 \cos^4 x - \sin^2 x = 0$

38. $3 \sin^4 x - 4 \cos^4 x = 0$

39. $4 \sin^4 x - \cos^2 2x = 0$

40. $\sin^2 x \csc x + \csc x = 1$

41. $\cos x + \sin x = \frac{1}{2}\sqrt{6}$

42. $\cos x - \sin x = \frac{1}{2}\sqrt{2}$

43. $2 \cot x + \csc x = 1$

44. $2 \tan x + \sec x = 1$

45. $2 \cos^2 x + \sin x \cos x - \sin^2 x = 0$

46. $\sin 2x - 4 \sin x - \cos x + 2 = 0$

19-3 THE INVERSE TRIGONOMETRIC FUNCTIONS

Consider the equations $y = \sin x$ and $x = \sin y$. The graph of $y = \sin x$ winds along the x axis (Fig. 18-5). Clearly, then, the graph of $x = \sin y$ is of the same form but winds along the y axis (Fig. 19-1). If we assign x a value in the domain of $y = \sin x$, we obtain one definite corresponding value of y. If, on the other hand, we assign x any value from -1 to 1 in $x = \sin y$, we see there are infinitely many corresponding values of y. These facts are exhibited in the graphs of the equations. Setting $x = 0$ in $x = \sin y$, for example, we have the resulting equation $\sin y = 0$. The solutions of this equation are $y = n\pi$ for all integral values of n.

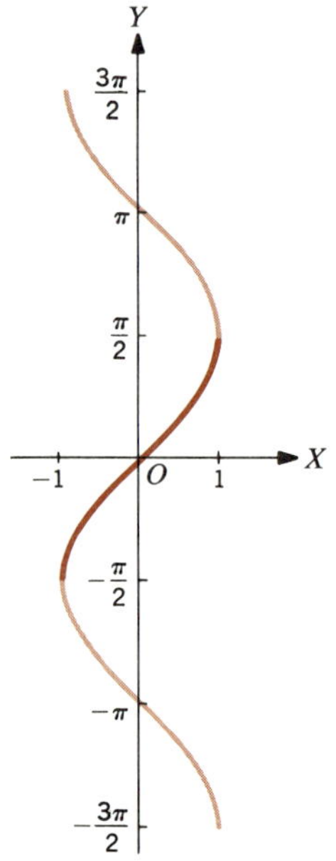

Figure 19-1 $x = \sin y$

If we restrict the y values so that $-\pi/2 \leq y \leq \pi/2$, then the equation $x = \sin y$ has only one y value for each x value. To indicate this restriction on y, we use a capital S and write $x = \text{Sin}\, y$. The graph of this equation is the heavy part of the curve in Fig. 19-1. A y value in the equation $x = \text{Sin}\, y$ is the number whose sine is x. To indicate this relation of x and y, we use the notation $y = \text{Sin}^{-1} x$. Hence,

$$y = \text{Sin}^{-1} x \qquad \text{is equivalent to} \qquad x = \text{Sin}\, y$$

Then either of these equations may be read "y is the number whose sine is x." We caution that $\text{Sin}^{-1} x$ *does not mean* $1/\text{Sin}\, x$.

The function defined by $y = \text{Sin}^{-1} x$ is called an *inverse trigonometric function*, or, specifically, the *inverse sine function*. This designation comes from the way in which $y = \text{Sin}^{-1} x$ is related to $y = \sin x$ with x restricted to the values $-\pi/2$ to $\pi/2$. We indicate this restriction on the domain of the sine function by using a capital S and writing $y = \text{Sin}\, x$. Now any ordered pair (x, y) belonging to $y = \text{Sin}\, x$ has the ordered pair (y, x) belonging to $y = \text{Sin}^{-1} x$. This is true because the inverse sine function is obtained by first interchanging the variables in the equation $y = \sin x$. Consequently, the sets $\{-\pi/2 \leq x \leq \pi/2\}$ and $\{-1 \leq y \leq 1\}$ which constitute the domain and range of $y = \text{Sin}\, x$ are interchanged to become the range and domain of $y = \text{Sin}^{-1} x$.

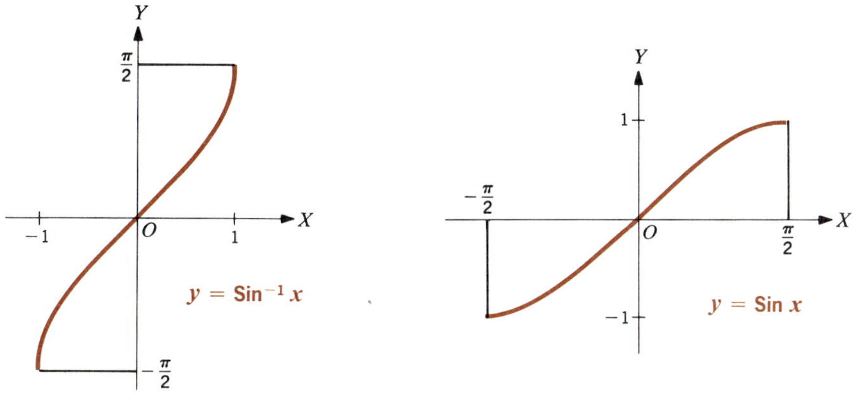

Figure 19-2

Each of the other trigonometric functions has its inverse with notation similar to that for the inverse sine. Thus the inverse cosine, the inverse tangent, and the inverse cotangent functions are defined as follows:

$y = \text{Cos}^{-1}\, x$ is equivalent to $x = \cos y$ $\quad 0 \leq y \leq \pi$

$y = \text{Tan}^{-1}\, x$ is equivalent to $x = \tan y$ $\quad -\frac{\pi}{2} < y < \frac{\pi}{2}$

$y = \text{Cot}^{-1}\, x$ is equivalent to $x = \cot y$ $\quad 0 < y < \pi$

The graphs of the four inverse functions which we have defined and the restricted trigonometric functions of which they are inverses are shown in Figs. 19-2 to 19-5. These inverse functions are used more frequently than are the inverse secant and the inverse cosecant functions. Consequently, we shall not discuss the latter two functions.

Figure 19-3

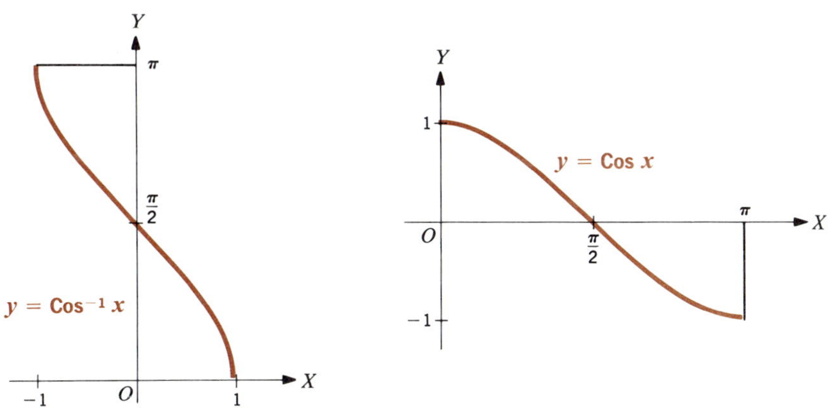

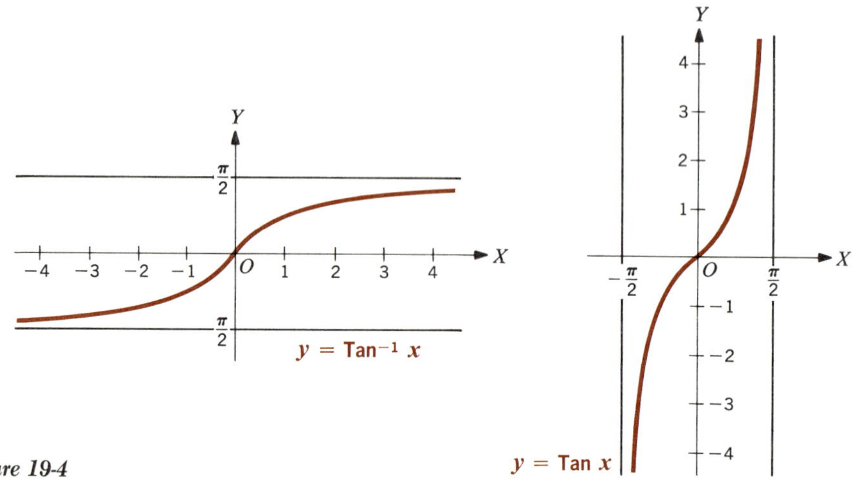

Figure 19-4

For emphasis and convenience, we list here the chosen domains and the resulting ranges of the four defined inverse functions:

Equation	Domain	Range
$y = \text{Sin}^{-1} x$	$-1 \leq x \leq 1$	$-\frac{\pi}{2} \leq \text{Sin}^{-1} x \leq \frac{\pi}{2}$
$y = \text{Cos}^{-1} x$	$-1 \leq x \leq 1$	$0 \leq \text{Cos}^{-1} x \leq \pi$
$y = \text{Tan}^{-1} x$	$-\infty < x < \infty$	$-\frac{\pi}{2} < \text{Tan}^{-1} x < \frac{\pi}{2}$
$y = \text{Cot}^{-1} x$	$-\infty < x < \infty$	$0 < \text{Cot}^{-1} x < \pi$

The notation for the inverse trigonometric functions, as written above, is widely used. There is, however, an equally acceptable notation in which Arcsin x is used instead of $\text{Sin}^{-1} x$, and similarly for the other functions.

EXAMPLES

$$\text{Sin}^{-1}\left(-\frac{1}{2}\right) = -\frac{\pi}{6} \qquad \text{Cos}^{-1}\left(\frac{-1}{\sqrt{2}}\right) = \frac{3\pi}{4}$$

$$\text{Arctan } \sqrt{3} = \frac{\pi}{3} \qquad \text{Arccot } (-\sqrt{3}) = \frac{5\pi}{6}$$

These results, as we have indicated in our discussion of the inverse trigonometric functions, are real numbers. We could, however, interpret the results as angles expressed in radian measure.

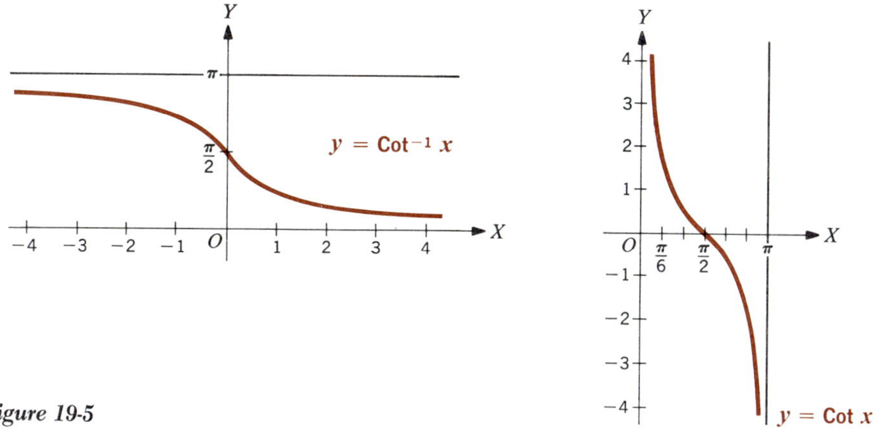

Figure 19-5

For example, $\text{Sin}^{-1}(-\frac{1}{2})$ means the *number* whose sine is $-\frac{1}{2}$ or the *angle* whose sine is $-\frac{1}{2}$. Hence, the result is the number $-\pi/6$ or the angle of measure $-\pi/6$ radian.

In some situations a trigonometric function of an inverse trigonometric function value is required. The following examples illustrate the procedure for handling a problem of this kind:

EXAMPLE 1 Find the value of $\tan \text{Cos}^{-1}(-2/\sqrt{13})$.

Solution: If we let $x = \text{Cos}^{-1}(-2/\sqrt{13})$, then by definition $\cos x = -2/\sqrt{13}$. Furthermore, the range of the inverse cosine function requires that x have a value between 0 and π, and therefore

$$\sin x = +\sqrt{1 - \cos^2 x} = \sqrt{1 - \frac{4}{13}} = \frac{3}{\sqrt{13}}$$

Finally, using $\tan x = (\sin x)/(\cos x)$, we have

$$\tan\left[\text{Cos}^{-1}\left(\frac{-2}{\sqrt{13}}\right)\right] = -\frac{3}{2}$$

EXAMPLE 2 Find the value of $\sin[\text{Arctan}(-0.7291)]$.

Solution: Referring to Table 3, we find $\text{Arctan}(+0.7291) = 0.63$, and therefore $\text{Arctan}(-0.7291) = -0.63$. Then, from the table, we find that

$$\sin[\text{Arctan}(-0.7291)] = \sin(-0.63) = -0.5891$$

EXAMPLE 3 Find the value of $\cos(2\,\text{Sin}^{-1}\frac{3}{5})$.

Solution: Let $x = \text{Sin}^{-1}\frac{3}{5}$ or $\sin x = \frac{3}{5}$. We then need the value of $\cos 2x$. So, using the identity of $\cos 2x = 1 - 2\sin^2 x$, we get

$$\cos(2\,\text{Sin}^{-1}\tfrac{3}{5}) = 1 - 2(\tfrac{3}{5})^2 = 1 - \tfrac{18}{25} = \tfrac{7}{25}$$

EXERCISE 19-3

Evaluate the following expressions. Use Table 3 when necessary.

1. $\text{Cos}^{-1} 0$ **2.** $\text{Sin}^{-1} 1$ **3.** $\text{Tan}^{-1} 1$

4. $\text{Sin}^{-1}(-1)$ **5.** $\text{Cos}^{-1}(-1)$ **6.** $\text{Cot}^{-1}(-1)$

7. $\text{Cos}^{-1}\dfrac{\sqrt{3}}{2}$ **8.** $\text{Sin}^{-1}\frac{1}{2}$ **9.** $\text{Cos}^{-1}\frac{1}{2}$

10. $\text{Sin}^{-1}\left(-\dfrac{\sqrt{3}}{2}\right)$ **11.** $\text{Cos}^{-1}(-\frac{1}{2})$ **12.** $\text{Sin}^{-1}\left(-\dfrac{\sqrt{2}}{2}\right)$

13. $\text{Tan}^{-1}\dfrac{1}{\sqrt{3}}$ **14.** $\text{Cot}^{-1}(-\sqrt{3})$

15. $\text{Tan}^{-1}\left(-\dfrac{\sqrt{3}}{3}\right)$ **16.** $\text{Sin}^{-1} 0.3986$

17. $\text{Cos}^{-1} 0.3153$ **18.** $\text{Tan}^{-1} 2.733$

19. $\text{Cos}^{-1}(-0.8776)$ **20.** $\text{Tan}^{-1}(-0.6131)$

21. $\text{Sin}^{-1}(-0.9915)$ **22.** $\sin(\text{Arccos}\,\frac{1}{2})$

23. $\tan[\text{Arcsin}\,(-\frac{1}{2})]$ **24.** $\cot\left[\text{Arccos}\left(-\dfrac{\sqrt{2}}{2}\right)\right]$

25. $\sin(\text{Arccos}\,\frac{2}{3})$ **26.** $\tan(\text{Arctan}\,\sqrt{3})$

27. $\cos[\text{Arccos}\,(-\frac{1}{3})]$ **28.** $\sin[\text{Arccot}\,(-4)]$

29. $\cos[\text{Arcsin}\,(-\frac{3}{4})]$ **30.** $\text{Arctan}\left(\tan\dfrac{\pi}{3}\right)$

31. $\text{Arccos}\left(\cos\dfrac{\pi}{4}\right)$ **32.** $\tan(2\,\text{Arcsin}\,\frac{1}{2})$

33. $\cos(2\,\text{Arccot}\,\frac{4}{3})$ **34.** $\tan(\frac{1}{2}\,\text{Arcsin}\,\frac{2}{3})$

35. $\cos(2\,\text{Arccos}\,\frac{1}{3})$ **36.** $\sin(2\,\text{Tan}^{-1}\frac{1}{2})$

37. $\sin(\frac{1}{2}\,\text{Cos}^{-1}\frac{2}{3})$ **38.** $\sin[\text{Cos}^{-1}(-0.9523)]$

39. $\cos[\text{Sin}^{-1}(-0.6816)]$ **40.** $\tan[\text{Sin}^{-1}(-0.7104)]$

41. $\tan[\text{Cos}^{-1}(-0.9780)]$

19-4 INVERSE TRIGONOMETRIC EQUATIONS

In this section, we shall consider certain expressions and equations involving inverse function values.

EXAMPLE 1 Verify the equality of the members of the equation

$$\text{Cos}^{-1}\left(-\tfrac{12}{13}\right) - \text{Sin}^{-1}\tfrac{4}{5} = \text{Cos}^{-1}\left(-\tfrac{16}{65}\right)$$

Solution: We observe that the right member of the equation is a number between $\pi/2$ and π, and that its cosine is $-\frac{16}{65}$. So we test the validity of the equation by finding the cosine of the left member. Setting

$$A = \text{Cos}^{-1}\left(-\tfrac{12}{13}\right) \qquad \text{and} \qquad B = \text{Sin}^{-1}\tfrac{4}{5}$$

we have

$$\cos A = -\tfrac{12}{13} \qquad \sin A = \tfrac{5}{13} \qquad \sin B = \tfrac{4}{5} \qquad \cos B = \tfrac{3}{5}$$

$$\begin{aligned}\cos(A - B) &= \cos A \cos B + \sin A \sin B \\ &= -\tfrac{12}{13}\cdot\tfrac{3}{5} + \tfrac{5}{13}\cdot\tfrac{4}{5} = -\tfrac{16}{65}\end{aligned}$$

We now see that the cosines of the members of the equation are equal, and consequently the equation is true. Showing that the sines of the members of the equation are equal would not definitely establish its validity. Tell why.

EXAMPLE 2 Solve the equation $\text{Sin}^{-1} 3x + \text{Sin}^{-1} x = \pi/2$.

Solution: We add $-\text{Sin}^{-1} x$ to both members of the given equation and then take the cosine of the members of the resulting equation. Thus, we get

$$\begin{aligned}\text{Sin}^{-1} 3x &= \frac{\pi}{2} - \text{Sin}^{-1} x \\ \cos(\text{Sin}^{-1} 3x) &= \cos\left(\frac{\pi}{2} - \text{Sin}^{-1} x\right) \\ \sqrt{1 - 9x^2} &= \sin(\text{Sin}^{-1} x) \\ 1 - 9x^2 &= x^2 \\ x &= \pm\frac{1}{\sqrt{10}}\end{aligned}$$

It may be verified that $x = 1/\sqrt{10}$ satisfies the given equation. But the negative value for x must be rejected; it makes the left side of the given equation negative, and therefore not equal to $\pi/2$.

EXAMPLE 3 Find the solution set of the equation

$$2 \text{ Cot}^{-1} x = \text{Sin}^{-1} \frac{2x}{x^2 + 1}$$

Solution: Recalling the ranges of the inverse cotangent function and the inverse sine function, we see that:

1. If x is a negative number or equal to zero, the left member of the equation is positive but the right member is not positive.
2. If x is positive and less than 1, the left member exceeds $\pi/2$ but the right member cannot exceed $\pi/2$.
3. If x is equal to 1, each member of the equation is equal to $\pi/2$.
4. If x is a number greater than 1, the value of each member is between 0 and $\pi/2$.

And so we test the validity of the equation for values of x greater than 1 by taking the sines of the members of the equation. The sine of the right member is $2x/(x^2 + 1)$ and, for the left member, we find

$$\begin{aligned} \sin(2 \text{ Cot}^{-1} x) &= 2 \sin(\text{Cot}^{-1} x) \cos(\text{Cot}^{-1} x) \\ &= 2 \cdot \frac{1}{\sqrt{x^2 + 1}} \cdot \frac{x}{\sqrt{x^2 + 1}} \end{aligned}$$

The sines of the members are equal, and therefore the solution set of the given equation is $\{x \geq 1\}$.

EXERCISE 19-4

Draw the graphs of the following equations.

1. $y = \text{Arcsin } \frac{1}{2}x$ **2.** $y = \frac{1}{2} \text{ Arccos } x$

3. $y = \text{Arctan } \frac{1}{2}x$ **4.** $y = 2 \text{ Arccot } x$

Verify the following equations without using a table.

5. $\text{Arctan } 2 + \text{Arctan } 3 = \dfrac{3\pi}{4}$

6. $\text{Arctan}\,\dfrac{3}{4} + \text{Arctan}\,\dfrac{1}{7} = \dfrac{\pi}{4}$

7. $\text{Arctan}\,(-3) + \text{Arctan}\,2 = \text{Arctan}\,(-\frac{1}{7})$

8. $\text{Arcsin}\,\frac{3}{5} - \text{Arccos}\,(-\frac{5}{13}) = \text{Arctan}\,(-\frac{63}{16})$

9. $\text{Arccos}\,\dfrac{3}{\sqrt{10}} - \text{Arcsin}\left(-\dfrac{1}{\sqrt{5}}\right) = \dfrac{\pi}{4}$

10. $\text{Arctan}\,\dfrac{3}{2} + \text{Arcsin}\,\dfrac{3}{\sqrt{13}} = \text{Arccos}\left(-\dfrac{5}{13}\right)$

Evaluate each of the following expressions.

11. $\cos\,(\text{Sin}^{-1}\,\frac{4}{5} + \text{Tan}^{-1}\,\frac{5}{12})$

12. $\sin\,(\text{Cos}^{-1}\,\frac{5}{13} + \text{Tan}^{-1}\,\frac{12}{5})$

13. $\tan\,(\text{Cos}^{-1}\,\frac{4}{5} - \text{Sin}^{-1}\,\frac{12}{13})$

14. $\sin\,(\text{Cos}^{-1}\,\frac{1}{2} + \text{Sin}^{-1}\,\frac{3}{5})$

15. $\cos\,[\text{Sin}^{-1}\,(-\frac{3}{5}) + \text{Cos}^{-1}\,(-\frac{3}{5})]$

16. $\cos\,(\text{Sin}^{-1}\,x + \text{Cos}^{-1}\,x)$

17. $\sin\,(\text{Cos}^{-1}\,x - \text{Cos}^{-1}\,y)$

18. $\tan\,(\text{Tan}^{-1}\,x - \text{Tan}^{-1}\,2y)$

19. $\cos\,(2\,\text{Cos}^{-1}\,x + \text{Sin}^{-1}\,x)$

20. $\sin\,(2\,\text{Tan}^{-1}\,x + \text{Cot}^{-1}\,x)$

Solve each of the following equations for x. Check the result.

21. $\text{Cos}^{-1}\,x - \text{Sin}^{-1}\,x = 0$

22. $\text{Tan}^{-1}\,x - \text{Cot}^{-1}\,x = 0$

23. $\text{Sin}^{-1}\,2x - \text{Cos}^{-1}\,x = 0$

24. $\text{Sin}^{-1}\,2\sqrt{x} = \text{Cos}^{-1}\,x$

25. $\text{Tan}^{-1}\,2x + \text{Tan}^{-1}\,3x = \dfrac{\pi}{4}$

26. $\text{Cos}^{-1}\,x + \text{Cos}^{-1}\,2x = \dfrac{\pi}{2}$

27. $\text{Cos}^{-1}\,x + \text{Sin}^{-1}\,(1 - x) = \dfrac{\pi}{2}$

28. $\text{Sin}^{-1}\,\frac{1}{2}x = \text{Cos}^{-1}\,x$

29. $\text{Cos}^{-1}\,x = \text{Cos}^{-1}\,\dfrac{1}{2x}$

30. $\text{Sin}^{-1}\,x + \text{Cos}^{-1}\,(1 - x) = \dfrac{\pi}{2}$

Find the solution set of each equation.

31. $\cos\,(\text{Sin}^{-1}\,x) = \sqrt{1 - x^2}$

32. $\sin\,(\text{Cos}^{-1}\,x) = \sqrt{1 - x^2}$

33. $\tan(\text{Sin}^{-1} x) = \dfrac{x}{\sqrt{1 - x^2}}$

34. $\cos(\text{Tan}^{-1} x) = \dfrac{1}{\sqrt{1 + x^2}}$

35. $\cot(\text{Cos}^{-1} x) = \dfrac{x}{\sqrt{1 - x^2}}$

36. $\sin(\text{Tan}^{-1} x) = \dfrac{x}{\sqrt{1 + x^2}}$

37. $\cos(2\ \text{Cot}^{-1} x) = \dfrac{x^2 - 1}{x^2 + 1}$

38. $\cos\left(\dfrac{1}{2}\ \text{Cos}^{-1} x\right) = \sqrt{\dfrac{1 + x}{2}}$

39. $\sin(2\ \text{Tan}^{-1} x) = \dfrac{2x}{1 + x^2}$

40. $\text{Sin}^{-1} x + \text{Cos}^{-1} x = \text{Sin}^{-1} 1$

41. $\text{Tan}^{-1} x + \text{Cot}^{-1} x = \text{Cos}^{-1} 0$

42. $2\ \text{Sin}^{-1} x = \text{Cos}^{-1}(1 - 2x^2)$

43. $\text{Tan}^{-1} x + \text{Tan}^{-1}(-x) = 0$

44. $\dfrac{1}{2}\ \text{Cos}^{-1} x = \text{Cot}^{-1} \dfrac{\sqrt{1 - x^2}}{1 - x}$

45. $2\ \text{Tan}^{-1} x = \text{Cos}^{-1} \dfrac{1 - x^2}{1 + x^2}$

46. $2\ \text{Tan}^{-1} \dfrac{1}{x} = \text{Sin}^{-1} \dfrac{2x}{x^2 + 1}$

47. $2\ \text{Tan}^{-1} x = \text{Sin}^{-1} \dfrac{2x}{x^2 + 1}$

48. $2\ \text{Cot}^{-1} x = \text{Cos}^{-1} \dfrac{x^2 - 1}{x^2 + 1}$

20

LOGARITHMS

20-1 THE LOGARITHMIC FUNCTION

Logarithms are of much value in mathematics. With their use many numerical computations can be made with surprising simplicity. Of greater import, however, are the theoretical applications of logarithms in calculus and several other areas of more advanced mathematics. As we shall see, exponents lie at the basis of the study of logarithms. For this reason, the student might well review the laws of exponents, which are listed in Chap. 7.

Definition 20-1 If a is a positive number other than 1 and y is a real number in the equation $a^y = x$, then y is called the *logarithm* of x to the base a; that is, the logarithm of x is the exponent which the base a must have to yield a value equal to x.

We write $y = \log_a x$ to mean that y is the logarithm of x to the base a. Hence, the equations

$$a^y = x \qquad \text{and} \qquad y = \log_a x$$

express the same relation among the numbers a, y, and x. The first equation is in *exponential form*, and the second in *logarithmic form*.

In the definition of a logarithm we specify that the base $a > 0$ and that the exponent y is a real number (positive, negative, or zero). As a consequence, a^y is a positive number; hence we shall consider logarithms of positive numbers only. We assume (but do not prove) that if a and x are any positive numbers except $a = 1$, then there is one and only one real value of y which will satisfy the equation $a^y = x$. That is, every positive number has a unique logarithm with respect to a given base. We reject unity as a base because 1 with any exponent is still 1.

Since a positive number has a unique logarithm, the equation $y = \log_a x$ defines a function with the set of positive numbers as the domain and the set of real numbers as the range. Hence, the function consists of the set of number pairs

$$\{(x, \log_a x) | x > 0\}$$

Suppose we choose $a = 3$ and construct the graph of the equation $y = \log_3 x$. We use the equivalent equation $x = 3^y$, assign integral values to y, and obtain the corresponding values of x as exhibited in the following table.

x	$\frac{1}{9}$	$\frac{1}{3}$	1	3	9
y	-2	-1	0	1	2

The graph (Fig. 20-1) is shown along with the graph of $y = \log_2 x$. Comparatively, we observe that $|\log_3 x| < |\log_2 x|$ except at $x = 1$. Both graphs cross the x axis at $x = 1$, as would be true for any base a because $a^0 = 1$.

As additional illustrations of the meaning of logarithms, we list some logarithmic equations and the corresponding equivalent exponential equations:

Logarithmic Form	Exponential Form
$\log_2 8 = 3$	$2^3 = 8$
$\log_9 27 = \frac{3}{2}$	$9^{3/2} = 27$
$\log_5 \frac{1}{25} = -2$	$5^{-2} = \frac{1}{25}$
$\log_a a = 1$	$a^1 = a$

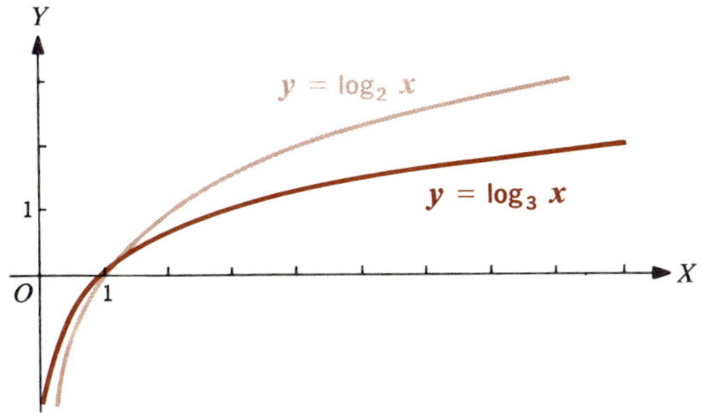

Figure 20-1

EXAMPLE 1 Find x if $\log_5 x = -4$.

Solution: Using the equivalent exponential form, we have

$$x = 5^{-4} = \frac{1}{5^4} = \frac{1}{625}$$

EXAMPLE 2 Find a if $\log_a 36 = \frac{2}{3}$.

Solution: The exponential form is $a^{2/3} = 36$. We take the $\frac{3}{2}$ power of both members of this equation and obtain

$$(a^{2/3})^{3/2} = 36^{3/2} \qquad \text{and} \qquad a = 216$$

EXAMPLE 3 Find the value of $\log_7 \frac{1}{49}$.

Solution: We let $y = \log_7 \frac{1}{49}$. Then changing to the exponential form, we get

$$7^y = \tfrac{1}{49} = 7^{-2} \qquad \text{and} \qquad y = -2$$

EXERCISE 20-1

Give the logarithmic form of each of the following equations.

1. $2^5 = 32$	**2.** $4^3 = 64$	**3.** $6^{-1} = \frac{1}{6}$
4. $9^1 = 9$	**5.** $8^0 = 1$	**6.** $4^{3/2} = 8$
7. $16^{1/4} = 2$	**8.** $4^{-5/2} = \frac{1}{32}$	**9.** $10^3 = 1000$
10. $(\frac{2}{3})^{-2} = \frac{9}{4}$	**11.** $(\frac{3}{4})^{-3} = \frac{64}{27}$	**12.** $(\frac{1}{9})^{-3/2} = 27$

Give the exponential form of the following equations.

13. $\log_2 1 = 0$ **14.** $\log_{10} 10 = 1$

15. $\log_5 25 = 2$ **16.** $\log_{10} 1 = 0$

17. $\log_2 16 = 4$ **18.** $\log_5 125 = 3$

19. $\log_{1/2} 8 = -3$ **20.** $\log_3 \frac{1}{9} = -2$

21. $\log_{1/5} 125 = -3$ **22.** $\log_{1/7} 49 = -2$

23. $\log_7 \frac{1}{49} = -2$ **24.** $\log_{2/3} \frac{16}{81} = 4$

Find x, a, or y in each of the following equations.

25. $\log_{10} x = 3$ **26.** $\log_3 x = 1$

27. $\log_{11} x = -1$ **28.** $\log_{16} x = -\frac{3}{4}$

29. $\log_{10} x = 0$ **30.** $\log_8 x = \frac{5}{3}$

31. $\log_{2/3} x = -3$ **32.** $\log_{4/9} x = \frac{3}{2}$

33. $\log_{1/3} x = -3$ **34.** $\log_a 25 = 2$

35. $\log_a 1 = 0$ **36.** $\log_a 8 = -3$

37. $\log_a 125 = -\frac{3}{2}$ **38.** $\log_a \frac{1}{5} = -1$

39. $\log_a \frac{9}{4} = -\frac{2}{5}$ **40.** $\log_a \frac{9}{16} = \frac{2}{3}$

41. $y = \log_4 256$ **42.** $y = \log_{16} 4$

43. $y = \log_5 \frac{1}{25}$ **44.** $y = \log_{243} \frac{1}{7}$

45. $y = \log_4 \frac{1}{256}$ **46.** $y = \log_{1/3} 9$

47. $y = \log_{5/2} \frac{8}{125}$ **48.** $y = \log_{2/3} \frac{16}{81}$

20-2 PROPERTIES OF LOGARITHMS

We shall state and prove three properties of logarithms. These properties follow readily from the laws of exponents.

Theorem 20-1 The logarithm of a product is equal to the sum of the logarithms of the factors; that is,

$$\log_a MN = \log_a M + \log_a N$$

Proof: Let $\log_a M = x$ and $\log_a N = y$. Then $M = a^x$, $N = a^y$,

$$MN = a^x \cdot a^y = a^{x+y} \qquad \text{and} \qquad \log_a MN = x + y$$

Substituting for x and y, we obtain

$$\log_a MN = \log_a M + \log_a N$$

For a product of three factors, as MNP, we have

$$\begin{aligned}\log_a MNP &= \log_a (MN)(P) \\ &= \log_a MN + \log_a P \\ &= \log_a M + \log_a N + \log_a P\end{aligned}$$

Similarly, the theorem may be extended to any number of factors.

EXAMPLE 1

$$\log_{10} (784)(92.7) = \log_{10} 784 + \log_{10} 92.7$$

Theorem 20-2 The logarithm of a quotient is equal to the logarithm of the dividend minus the logarithm of the divisor. That is,

$$\log_a \frac{M}{N} = \log_a M - \log_a N$$

Proof: If $\log_a M = x$ and $\log_a N = y$, then $M = a^x$ and $N = a^y$. Hence,

$$\frac{M}{N} = \frac{a^x}{a^y} = a^{x-y} \qquad \text{and} \qquad \log_a \frac{M}{N} = x - y$$

By substituting for x and y, we obtain

$$\log_a \frac{M}{N} = \log_a M - \log_a N$$

EXAMPLE 2

$$\log_5 \tfrac{1078}{5234} = \log_5 1078 - \log_5 5234$$

Theorem 20-3 The logarithm of a power of a number is equal to the exponent times the logarithm of the number; that is,

$$\log_a M^p = p \log_a M$$

Proof: We let $\log_a M = x$, or $M = a^x$. Then

$$M^p = (a^x)^p = a^{px} \qquad \text{and} \qquad \log_a M^p = px$$

Substituting for x gives

$$\log_a M^p = p \log_a M$$

EXAMPLE 3

$$\log_4 7842^3 = 3 \log_4 7842$$

EXAMPLE 4

$$\log_{10} 597^{-5} = -5 \log_{10} 597$$

EXAMPLE 5

$$\log_3 \sqrt{95} = \log_3 95^{1/2} = \tfrac{1}{2} \log_3 95$$

EXAMPLE 6 In this example we use Theorems 20-1, 20-2, and 20-3 to express a logarithm as the algebraic sum of logarithms. Thus,

$$\begin{aligned} \log_2 \frac{28^2(6.7)^{1/2}}{(8.4)(17)} &= \log_2 28^2(6.7)^{1/2} - \log_2 (8.4)(17) \\ &= 2 \log_2 28 + \tfrac{1}{2} \log_2 6.7 - \log_2 8.4 - \log_2 17 \end{aligned}$$

EXAMPLE 7 Express $\log_6 5 - 3 \log_6 2 + \frac{2}{3} \log_6 28$ as a single logarithm.

Solution:

$$\begin{aligned} &\log_6 5 - 3 \log_6 2 + \tfrac{2}{3} \log_6 28 \\ &\qquad = \log_6 5 - \log_6 2^3 + \log_6 28^{2/3} \\ &\qquad = \log_6 \frac{5 \cdot 28^{2/3}}{2^3} \end{aligned}$$

EXERCISE 20-2

Write each of the following expressions as the algebraic sum of logarithms. A base is not indicated. Any positive number except 1 may be used.

1. $\log (251)(46)(18)$ **2.** $\log (34)^2(2.7)$

3. $\log (24)^{1/2}(35)^3$ **4.** $\log \dfrac{(83)(41)}{29}$

5. $\log \dfrac{(49)(65)}{(71)(86)}$

6. $\log \dfrac{(2.7)^2(58)^{1/3}}{(75)(89)^2}$

7. $\log \sqrt{\dfrac{(87)(28)}{15}}$

8. $\log \sqrt{\dfrac{125}{(37)(44)}}$

9. $\log \dfrac{58\sqrt[3]{28}}{34\sqrt{56}}$

Write each expression as a single logarithm.

10. $\log a + \log b + \log c$

11. $\log x + \log y - \log z$

12. $2 \log x - 3 \log y + \log z$

13. $\log \pi + 2 \log r + \log h$

14. $3 \log x - \log (x - 2)$

15. $\frac{1}{2} \log 52 - \frac{2}{3} \log 5 - 4 \log 3$

16. $\log 48 + \frac{1}{2} \log 913 - \log 82 - \frac{1}{2} \log 397$

17. $\frac{1}{2}(\log 82 + \log 44 + \log 21 - \log 798)$

Given $\log_{10} 2 = 0.3010$ *and* $\log_{10} 3 = 0.4771$, *find the value of each of the following quantities.*

18. $\log_{10} 8$

19. $\log_{10} 9$

20. $\log_{10} 6$

21. $\log_{10} 54$

22. $\log_{10} \sqrt{12}$

23. $\log_{10} \sqrt[3]{18}$

24. $\log_{10} \frac{64}{9}$

25. $\log_{10} \sqrt[4]{72}$

26. $\log_{10} \frac{27}{16}$

20-3 COMMON LOGARITHMS

There are two systems of logarithms which are used extensively in mathematics. In one of these the base is an irrational number, designated by e, whose value is approximately 2.718. Logarithms to this base are called *natural*, or *Naperian, logarithms*. This system is convenient in many theoretical considerations and is used in calculus and higher mathematics. The other system employs 10 as a base. Logarithms to the base 10 are called *common*, or *Briggs, logarithms*. Common logarithms are especially suited for making computations.

We shall be interested in logarithms to the base 10; and henceforth, for brevity, this base will not be indicated in logarithmic expressions. Thus, we shall write $\log_{10} N$ simply as $\log N$.

The common logarithms of the integral powers of 10 are integers.

This is illustrated in the following list of powers of 10 and their logarithms:

$$\begin{aligned} 10^3 &= 1000 & \log 1000 &= 3 \\ 10^2 &= 100 & \log 100 &= 2 \\ 10^1 &= 10 & \log 10 &= 1 \\ 10^0 &= 1 & \log 1 &= 0 \\ 10^{-1} &= 0.1 & \log 0.1 &= -1 \\ 10^{-2} &= 0.01 & \log 0.01 &= -2 \\ 10^{-3} &= 0.001 & \log 0.001 &= -3 \end{aligned}$$

Noticing that $\log 1 = 0$, we may conclude that numbers greater than 1 have positive logarithms and that numbers between 0 and 1 have negative logarithms. Furthermore, since the logarithm of an integral power of 10 is an integer, we conclude that the logarithms of other positive numbers are not integers. For example, 63 is between 10 and 10^2, and therefore its logarithm is between 1 and 2, or 1 plus a fraction (between 0 and 1). In general, then, a logarithm consists of two parts, an integer and a fraction.

20-4 THE CHARACTERISTIC AND MANTISSA

When a logarithm is expressed as an integer plus a decimal (between 0 and 1), the integer is called the *characteristic* and the decimal is called the *mantissa*. Finding the logarithm of a number, then, requires the determination of both the characteristic and the mantissa. We shall consider first the problem of finding the characteristic. Since $\log 1 = 0$ and $\log 10 = 1$, we assume (and it seems reasonable) that the logarithm of a number between 1 and 10 is zero plus a fraction between 0 and 1. In order to determine the characteristics corresponding to numbers between 0 and 1 and numbers greater than 10, we introduce the idea of the *reference position* of the decimal point.

Definition 20-2 The reference position of the decimal point comes just after the first nonzero digit of the number.

The arrow indicates the reference position of the decimal point of each of the following numbers.

$$\underset{\uparrow}{3}.741 \qquad 5\underset{\uparrow}{}78.3 \qquad 0.4\underset{\uparrow}{}56 \qquad 0.0001\underset{\uparrow}{}25$$

The decimal points of these numbers are in the reference position, 2 places to the right of the reference position, 1 place to the left, and 4 places to the left, respectively.

Let us now find out how the characteristic of the logarithm of a number is related to the reference position of the decimal point of the number. To illustrate the procedure, we first determine the characteristic of the logarithm of each of the numbers 7832 and 0.07832. Expressing these numbers in scientific notation (Sec. 9-1), we have

$$7832 = 7.832 \times 10^3$$

$$0.07832 = 7.832 \times 10^{-2}$$

and, by Theorems 20-1 and 20-3, we have

$$\log 7832 = \log (7.832 \times 10^3) = 3 + \log 7.832$$

$$\log 0.07832 = \log (7.832 \times 10^{-2}) = -2 + \log 7.832$$

Thus, we see that the characteristic of the logarithm of each of the numbers is the exponent of 10 when the number is expressed in scientific notation.

Consider now the general case in which a number A is expressed in scientific notation as $A = B \times 10^n$, where $1 \leq B < 10$ and n is an integer. Taking the logarithm of each member, we write

$$\log A = \log (B \times 10^n) = \log B + n$$

The mantissa of $\log A$ is equal to $\log B$ because $0 \leq \log B < 1$, and the characteristic is the integer n. If A is greater than 10, the characteristic is a positive integer; if A is between 0 and 1, the characteristic is a negative integer.

We note also that the characteristic of the logarithm of a number depends solely on the location of the decimal point. Furthermore, to determine the characteristic, it is sufficient to notice what the exponent of 10 would be if the number were expressed in scientific notation. The mantissa, however, is independent of the position of the decimal point; it depends only on the sequence of digits in the number. Accordingly, we state the following:

Theorem 20-4 If the decimal point of a number is in the reference position, the characteristic of its logarithm is zero. If the decimal point is n places to the right of the reference position, the characteristic is n. If the decimal point is n places to the left of the reference position, the characteristic is $-n$.

The mantissa is independent of the position of the decimal point.

EXAMPLE 1 Find the characteristic of the logarithm of each of the numbers 0.0003405, 3.405, and 3405.

Solution: We count the number of places which the decimal point is removed from the reference position in each case. Thus, we find that the characteristics are respectively -4, 0, and 3.

0.0003405	3.405	3405
↑____↑	↑	↑_↑
-4	0	3

EXAMPLE 2 Given $\log 5.826 = 0.7654$, write the logarithms of 582.6, 58,260, and 0.05826.

Solution: The mantissas of the logarithms of these numbers are all the same as the given mantissa. Relative to the reference position, the decimal points of 582.6, 58,260, and 0.05826 are, respectively, 2 places to the right, 4 places to the right, and 2 places to the left. Hence, the required characteristics are 2, 4, and -2, and we have

$$\log 582.6 = 2.7654$$
$$\log 58{,}260 = 4.7654$$
$$\log 0.05826 = -2 + 0.7654$$

We have noticed that the logarithm of a number between 0 and 1 is negative. That is, the characteristic is a negative integer, and the mantissa is either zero or a positive fraction. The characteristic may be combined with the mantissa to yield a single negative quantity. For example,

$$\log 0.05826 = -2 + 0.7654 = -1.2346$$

For most computations, however, it is preferable to express a logarithm with the fractional part positive.

The logarithm $-2 + 0.7654$ could not be written as -2.7654 since this would indicate that the fractional part is also negative. A negative characteristic is customarily expressed as the difference of two integers. Thus, since $-2 = 2 - 4$ and also equal to $8 - 10$, we have

$$\begin{aligned}\log 0.05826 &= -2 + 0.7654\\ &= 2.7654 - 4\\ &= 8.7654 - 10\end{aligned}$$

A negative characteristic may obviously be written as the difference of two integers in an unlimited number of ways. The usual practice, however, is to use a positive integer minus 10 or an integral multiple of 10.

20-5 TABLES OF LOGARITHMS

We have shown how to determine immediately the characteristic of the logarithm of a number. Finding mantissas, however, is not simple. But methods are developed in more advanced mathematics which permit the computation of a mantissa to any desired number of decimal places. The mantissas corresponding to many numbers have been computed and arranged in tabular form. Such an arrangement is called a *table of logarithms*, or a *table of mantissas*. Since most mantissas are unending decimal fractions, their values are approximated to two, three, or more decimal places. Accordingly, a table is called a *two-place* table, a *three-place* table, and so on, depending on the number of decimal places used.

A four-place table (Table 4) is given in the Appendix. This table gives the mantissas of the numbers from 1.00 to 9.99 in steps of 0.01 of a unit. The first column, headed by N, contains the first two digits of a number and the third digit is to the right in the first line of the page. The mantissa corresponding to a number is in the horizontal line of the first two digits of the number and in the column headed by the third digit. It is customary to omit decimal points in the table. A decimal point belongs in front of each mantissa. Since the characteristic of the logarithms of 1 and numbers between 1 and 10 is zero, the table gives the logarithms of all three-digit numbers from 1.00 to 9.99. In order to use the table for a positive number outside this range, it is necessary to supply the proper nonzero characteristic in accordance with the discussion in Sec. 20-4. We shall explain with examples how to use the table in finding logarithms of numbers and in finding numbers corresponding to given logarithms.

EXAMPLE 1 Find (a) log 4.58; (b) log 0.00458.

Solution: (a) Since the decimal point is in the reference position, the characteristic of the logarithm is 0. To find the mantissa, look at Table 4 in the Appendix and go down the first column to 45, the first two digits in 4.58. Then look along the horizontal line from

45, stopping at the column which has 8 at the top, the third digit in 4.58. The mantissa .6609 is thus located. Hence,

$$\log 4.58 = 0.6609$$

(*b*) The decimal point of 0.00458 is three places to the left of the reference position, and this makes the characteristic -3. Although the characteristic is determined from the position of the decimal point, the mantissa is independent of the decimal point. Hence, the numbers 4.58 and 0.00458 have the same mantissa, and therefore

$$\log 0.00458 = 7.6609 - 10$$

EXAMPLE 2 Find log 2047.

Solution: This number has four significant digits, and its mantissa cannot be read directly from the table. Hence, we resort to interpolation, a process explained in Sec. 9-3. The number 2047 is 0.7 of the way from 2040 and 2050. The mantissas corresponding to 2040 and 2050 are the same as those for 204 and 205 and can be read directly from the table. These mantissas are 3096 and 3118, except for the decimal points. To obtain the number 0.7 of the way from the smaller to the larger mantissa, we compute 0.7(3118 − 3096). Our answer is 15, to the nearest integer, and hence 3096 + 15 = 3111 is the required mantissa when the decimal point is supplied. The quantities involved in this interpolation process are tabulated as a further aid in understanding the steps:

		Number	*Mantissa*			
10	7	2040	3096	22	$0.7 \times 22 = 15.4$	
		2047	?		$= 15$	Rounded off
		2050	3118		$3096 + 15 = 3111$	

Since the required characteristic is 3, we have

$$\log 2047 = 3.3111$$

In beginning to find logarithms of numbers requiring interpolation, the student may wish to tabulate the quantities involved similar to that used in this example. Having done a few problems, however, he should notice that most, if not all, of the operations can be done mentally. Emphasizing the process, we list the three essential steps:

1. A mantissa is to be subtracted from the one immediately following in the table.

2. This difference is to be multiplied by an integral multiple of tenths (0.1 or more up to 0.9).

3. The product, rounded off, is to be added to the smaller mantissa.

EXERCISE 20-3

Give the characteristic of the logarithm of each of the following numbers.

1. 83	**2.** 50	**3.** 428.3
4. 9.75	**5.** 400.3	**6.** 8.00
7. 30.50	**8.** 9753	**9.** 43,870
10. 0.392	**11.** 0.072	**12.** 0.0047
13. 0.0007	**14.** 0.00001	**15.** 1.800
16. 6.34×10^3	**17.** 2.78×10^{-2}	**18.** 7.58×10^{-1}
19. 1.70×10	**20.** 3.65×10^{-4}	**21.** 1.03×10^8

Find the common logarithm of each of the following numbers.

22. 31.4	**23.** 5.72	**24.** 4320
25. 57,500	**26.** 90.9	**27.** 0.812
28. 0.0095	**29.** 0.085	**30.** 0.0432
31. 670	**32.** 700	**33.** 0.00078
34. 0.49	**35.** 756,000	**36.** 0.0208
37. 3.142	**38.** 64.55	**39.** 7582
40. 457.1	**41.** 0.7944	**42.** 0.009127
43. 84,230	**44.** 612,300	**45.** 0.004711
46. 82.45	**47.** 585.3	**48.** 908.3
49. 10.01	**50.** 0.1118	**51.** 102.4
52. 994.9	**53.** 0.0001118	**54.** 0.01001
55. 0.0000745	**56.** 745,500	**57.** 507.9
58. 6.449×10^{-5}	**59.** 3.446×10^5	**60.** 1.008×10^6
61. 9.105×10^{-7}	**62.** 4.403×10^{-9}	**63.** 7.766×10^8

20-6 FINDING ANTILOGARITHMS FROM TABLES

The number which corresponds to a logarithm is called the *antilogarithm* (abbreviated antilog); that is, if $\log N = x$, then $N =$ antilog x. The following examples illustrate the use of Table 4 in finding antilogarithms.

EXAMPLE 1 Find antilog $(8.8407 - 10)$.

Solution: We have the logarithm of a number and wish to find the number, or antilogarithm. Since the table has only the mantissas of logarithms, we disregard the characteristic for the moment and look in the columns of mantissas for 8407. The mantissas increase to the right along each row and from row to row. It is easy, therefore, to locate any mantissa which appears in the table. Having found the entry 8407, we look along its row to the left and find 69 in the N column. These are the first two digits of the antilogarithm. The third, found at the top of the column containing the given mantissa, is 3. Hence, 693 is the sequence of digits corresponding to the mantissa .8407. To place the decimal point, we notice that the characteristic $8 - 10$ is equal to -2, and consequently the decimal point belongs two places to the left of the reference position. Thus,

$$\text{antilog } (8.8407 - 10) = 0.0693$$

EXAMPLE 2 If $\log N = 4.7556$, find N.

Solution: The mantissa is not in the table, and we shall interpolate to find the antilogarithm. The mantissa 7556 is between the consecutive mantissas 7551 and 7559, and five-eighths of the way across. The numbers 569 and 570 or, equivalently, 5690 and 5700 correspond to the consecutive mantissas. Hence, we find the number five-eighths of the way from 5690 to 5700. This gives 5696, which, except for the decimal point, is the required antilogarithm. Placing the decimal point in accordance with the given characteristic 4, we have

$$N = 56{,}960$$

The zero in this result serves to locate the decimal point; it is not a significant digit.

The numbers involved in this example are presented in the following tabular form.

	Number	*Mantissa*		
	5690	7551 ⎤ 5 ⎤	$\frac{5}{8} \times 10 = 6$	Nearest integer
10	?	7556 ⎦ ⎥ 8	$5690 + 6 = 5696$	
	5700	7559 ⎦	$N = 56{,}960$	

Again, the student should observe that in most cases the computations required in finding antilogarithms can be done mentally.

EXERCISE 20-4

Use Table 4 and find the antilogarithm of each of the following logarithms.

1. 2.1004	**2.** 0.2742	**3.** 1.5011
4. 3.5224	**5.** 9.7860 − 10	**6.** 8.8710 − 10
7. 5.4314	**8.** 6.0212	**9.** 7.0170 − 10
10. 0.6236	**11.** 1.3471	**12.** 2.7155
13. 3.6109	**14.** 0.8009	**15.** 9.9706 − 10
16. 6.9932 − 10	**17.** 7.8612 − 10	**18.** 8.0846 − 10
19. 9.1120 − 10	**20.** 8.3410 − 10	**21.** 6.2491 − 10
22. 0.3221	**23.** 2.7111	**24.** 3.9709
25. 1.6004	**26.** 3.3330	**27.** 5.7130
28. 4.1445	**29.** 6.5435	**30.** 5.2316

20-7 LOGARITHMS USED IN COMPUTATIONS

We are now ready to compute products, quotients, powers, and roots of numbers, or any combination of these operations, by means of logarithms. The logarithmic theorems of Sec. 20-2 serve for this purpose. We shall illustrate their use in some examples:

EXAMPLE 1 Compute N by means of logarithms, given

$$N = \frac{(32.41)(81.93)}{(1.854)(0.7949)}$$

Solution: Applying Theorem 20-2, we write

$$\log N = \log (32.41)(81.93) - \log (1.854)(0.7949)$$

And by Theorem 20-1,

$$\log N = \log 32.41 + \log 81.93 - (\log 1.854 + \log 0.7949)$$

To help in avoiding errors and also to speed the work, the student should first make an outline form in which the characteristics of the given numbers are written and spaces are left for the quantities to be found. We suggest a form for this problem:

Numerator	Denominator
$\log 32.41 = 1$	$\log 1.854 = 0$
$(+)\log 81.93 = 1$	$(+)\log 0.7949 = 9 \qquad -10$
$\log \text{num} =$	$\log \text{denom} =$
$(-)\log \text{denom} =$	
$\log N =$	

Having made the form, the tables should then be used and the spaces filled in. We show the completed form:

$$\begin{aligned}
\log 32.41 &= 1.5106 \\
(+)\log 81.93 &= 1.9135 \\
\log \text{num} &= 3.4241 \\
(-)\log \text{denom} &= 0.1684 \\
\log N &= 3.2557 \\
N &= 1802
\end{aligned}
\qquad
\begin{aligned}
\log 1.854 &= 0.2681 \\
(+)\log 0.7949 &= 9.9003 - 10 \\
\log \text{denom} &= 10.1684 - 10 \\
&= 0.1684
\end{aligned}$$

EXAMPLE 2 Evaluate $\sqrt[3]{321.4}/\sqrt{208.7}$.

Solution: Letting N stand for the given fraction, we have

$$N = \frac{\sqrt[3]{321.4}}{\sqrt{208.7}} = \frac{(321.4)^{1/3}}{(208.7)^{1/2}}$$

and

$$\log N = \tfrac{1}{3}\log 321.4 - \tfrac{1}{2}\log 208.7$$

$$\log 321.4 = 2.5071 \qquad \log 208.7 = 2.3195$$

$$\tfrac{1}{3}\log 321.4 = 0.8357 \qquad \tfrac{1}{2}\log 208.7 = 1.1598$$

The logarithm of the denominator is greater than the logarithm of the numerator. We wish to obtain the difference, or log N, in the form of a negative integer plus a positive fraction (mantissa). The table may then be used to find N. To obtain log N in this usable form, we first express the characteristic of the logarithm of the numerator as 10 − 10. Thus, we write

$$\begin{aligned} \tfrac{1}{3}\log 321.4 &= 10.8357 - 10 \\ (-)\ \tfrac{1}{2}\log 208.7 &= \underline{\ \ 1.1598\qquad\quad} \\ \log N &= \ \ 9.6759 - 10 \\ N &= 0.4741 \end{aligned}$$

EXAMPLE 3 Compute $\sqrt[3]{0.8316}$.

Solution: We first write $N = \sqrt[3]{0.8316} = (0.8316)^{1/3}$. Then

$$\log N = \tfrac{1}{3}\log 0.8316$$

$$\log 0.8316 = 9.9199 - 10$$

Dividing 9.9199 − 10 by 3, we get the result, 3.3066 − 3.3333, in an undesirable form. A usable form is obtained at once if the characteristic 9 − 10 is replaced by the equal quantity 29 − 30. Thus, we get

$$\begin{aligned} \log 0.8316 &= 29.9199 - 30 \\ \tfrac{1}{3}\log 0.8316 &= \ \ 9.9733 - 10 \\ N &= \ \ 0.9404 \end{aligned}$$

EXERCISE 20-5

Use logarithms and perform the indicated operations.

1. (5.13)(0.837)

2. (3.82)(61.2)

3. 44.8 ÷ 81.3

4. (0.0081)(723)

5. 9.07 ÷ 75.6

6. 224 ÷ 68.7

7. (10.7)(8.16)(70.7)

8. (5.01)(3.77)(9.12)

9. (90.92)(0.141)(6.683)

10. (771.3)(31.25)

11. (12.36)(357.9)(0.167)

12. (7.103)(23.09)(1.783)

13. $\dfrac{(7531)(3456)}{(1813)(1339)}$

14. $\dfrac{(314)(671)}{(544)(837)}$

15. $(2.718)^2$

16. $\dfrac{(404)(505)}{(707)(114)}$

17. $(3.318)^3$

18. $(0.07184)^4$

19. $\sqrt[3]{8.082}$

20. $\sqrt[3]{75.66}$

21. $\sqrt[4]{0.04623}$

22. $\sqrt{0.6118}$

23. $\sqrt[3]{66.78}$

24. $\sqrt[4]{17.81}$

25. $(0.1776)^{3/4}$

26. $(21.08)^{2/3}$

27. $(11.08)^{-4/3}$

28. $(1.414)^{3/5}$

29. $(1004)^{-5/2}$

30. $(1881)^{-4/5}$

31. $\dfrac{\sqrt{1357}}{(5.635)^3}$

32. $\dfrac{\sqrt[3]{8143}}{\sqrt{5625}}$

33. $\sqrt[3]{\dfrac{(4881)(2321)}{87.91}}$

34. $\sqrt{\dfrac{(57.81)(6.183)}{19.78}}$

35. $\dfrac{25.73\sqrt{5.183}}{\sqrt[3]{893.2}}$

36. $\sqrt[4]{\dfrac{(19.03)(30.91)}{(1.456)(2.531)}}$

20-8 EXPONENTIAL EQUATIONS

An equation in which the unknown appears in an exponent is called an *exponential equation.* Some quite simple equations of this kind may be solved by inspection. Certain others are readily solved by the use of logarithms. The following examples illustrate methods of solving exponential equations.

EXAMPLE 1 Solve $5^x = 625^2$.

Solution: First express 625 as 5^4, then $5^x = (5^4)^2 = 5^8$. With the bases equal, the exponents must be the same; hence, $x = 8$.

EXAMPLE 2 Solve $2^x = 29$.

Solution: Since equal quantities have equal logarithms, we may write

$$\log 2^x = \log 29$$

Hence,

$$x \log 2 = \log 29$$

and

$$x = \frac{\log 29}{\log 2} = \frac{1.4624}{0.3010} = 4.858$$

The value of x is obtained by dividing log 29 by log 2. This is the quotient of two logarithms; it should not be confused with the logarithm of a quotient.

EXAMPLE 3 Solve the equation $5(3)^{2x+1} = 7^{1-x}$.

Solution: Taking the logarithm of each member, we get

$$\log 5(3)^{2x+1} = \log 7^{1-x}$$

and

$$\log 5 + (2x + 1) \log 3 = (1 - x) \log 7$$

This is a linear equation in x and may be solved in the usual way. Thus,

$$(2 \log 3 + \log 7)x = \log 7 - \log 5 - \log 3$$

and

$$x = \frac{\log 7 - \log 5 - \log 3}{2 \log 3 + \log 7} = \frac{-0.3310}{1.7993} = -0.1840$$

EXERCISE 20-6

Solve the following exponential equations without the use of tables.

1. $5^x = 125$	**2.** $3^{x-1} = 9$	**3.** $e^{2x} = e^{10}$
4. $4^{x+2} = 16^x$	**5.** $10^{3x-3} = 1000$	**6.** $25^x = 5$
7. $4^{2x+3} = 4$	**8.** $6^x = \frac{1}{36}$	**9.** $3^{2x} = 3^0$

Solve the following exponential equations.

10. $4^x = 7$	**11.** $9^x = 15$	**12.** $(1.27)^x = 2.73$
13. $9^x = 7$	**14.** $6^{2x-3} = 4$	**15.** $(8.13)^x = 68$
16. $4^x = 5^x$	**17.** $11^{2x-1} = 7^x$	**18.** $20^x = 4^x$
19. $100^x = 3^x$	**20.** $(0.3)^x = 0.5$	**21.** $(0.182)^x = 0.93$
22. $(0.718)^x = 1.07$	**23.** $3^x = 0.13$	**24.** $17^x = 0.19$
25. $6 \cdot 10^x = 11^x$	**26.** $9^x = 4 \cdot 19^{x+1}$	**27.** $5 \cdot 7^x = 6 \cdot 8^x$

20-9 LOGARITHMS TO DIFFERENT BASES

It is sometimes desirable to express the logarithm of a number to one base in terms of its logarithm to another base. In particular, a change from a common logarithm to a natural logarithm, or vice versa, is advantageous in certain calculus problems. We shall derive the relation between the logarithms of a number to any two bases.

Let a and b stand for two positive numbers different from 1, and let

$$\log_a N = x \qquad \text{or} \qquad N = a^x$$

Taking the logarithm to the base b of both members of the exponential equation, we obtain

$$\log_b N = x \log_b a$$

and

$$x = \frac{\log_b N}{\log_b a}$$

Since $x = \log_a N$, we obtain

$$\log_a N = \frac{\log_b N}{\log_b a}$$

This formula can be used to change from one base to another. By substituting the natural base e (2.718 approximately) for a and 10 for b, we have

$$\log_e N = \frac{\log_{10} N}{\log_{10} e}$$

The values of $\log_{10} e$ and $1/\log_{10} e$, to four decimal places, are 0.4343 and 2.3026, respectively. Making these substitutions, we get the relations

$$\log_{10} N = 0.4343 \log_e N$$

and

$$\log_e N = 2.3026 \log_{10} N$$

EXAMPLE 1

$$\log_e 46 = 2.3026 \log_{10} 46 = (2.3026)(1.6628) = 3.829$$

EXAMPLE 2

$$\log_4 15 = \frac{\log_{10} 15}{\log_{10} 4} = \frac{1.1761}{0.6021} = 1.953$$

EXERCISE 20-7

Find the following logarithms.

1. $\log_e 100$	**2.** $\log_e 158$	**3.** $\log_e 0.478$
4. $\log_2 54$	**5.** $\log_3 175$	**6.** $\log_4 204$
7. $\log_2 0.416$	**8.** $\log_5 0.067$	**9.** $\log_7 62.4$
10. $\log_{11} 150$	**11.** $\log_e 0.041$	**12.** $\log_{12} 247$
13. $\log_8 9.85$	**14.** $\log_9 0.009$	**15.** $\log_e \sqrt{2}$

20-10 LOGARITHMS OF TRIGONOMETRIC FUNCTIONS

Table 5 in the Appendix gives to four decimal places the logarithms of the values of the sine, cosine, tangent, and cotangent of angles, in 10′ steps from 0° to 90°. The arrangement of the table is like that of Table 2. Because the values of the sine and cosine of angles between 0° and 90° are between 0 and 1, their logarithms are negative. The tangent of angles between 0° and 45° and the cotangent of angles between 45° and 90° also have negative logarithms. The characteristic of a negative logarithm is usually written as a positive integer minus 10. In Table 5, however, the −10 is omitted from every logarithmic entry and must be supplied by the user.

EXAMPLE 1 Find the log tan 67° 27′.

Solution: The angle 67° 27′ is 0.7 of the way from 67° 20′ to 67° 30′. We assume that this relation holds for their tangents. From the table we obtain the following figures:

$$\left.\begin{aligned} \log\tan 67^\circ\,20' &= 10.3792 - 10 \\ \log\tan 67^\circ\,30' &= 10.3828 - 10 \end{aligned}\right] 36$$

We add 0.7 of the tabular difference, 36, to the smaller logarithm. Thus 3792 + 0.7(36) = 3817, and hence

$$\log\tan 67^\circ\,27' = 0.3817$$

EXAMPLE 2 Find log cos 15° 8′.

Solution: Since the cosine decreases as the angle increases from 0° to 90°, the correction must be subtracted in the interpolation process. We find

$$\log \cos 15° = 9.9849 - 10$$

with 3 the tabular difference. Then $0.8 \times 3 = 2.4$ (2 rounded off); therefore, we get

$$\log \cos 15°\ 8' = 9.9847 - 10$$

EXAMPLE 3 Find the angle A between 0° and 90° if $\log \sin A = 9.7556 - 10$.

Solution: The necessary computations follow:

$$10' \left[\begin{array}{ll} \log \sin 34°\ 40' & = 9.7550 - 10 \\ \log \sin A & = 9.7556 - 10 \\ \log \sin 34°\ 50' & = 9.7568 - 10 \end{array} \right] \begin{array}{l} 6 \\ 18 \end{array}$$

Since $\frac{6}{18} \times 10' = 3'$, rounded off, we have $A = 34°\ 43'$.

EXERCISE 20-8

Find the following logarithms.

1. log sin 26° 20′
2. log tan 73° 40′
3. log cos 39° 10′
4. log cot 16° 30′
5. log sin 31° 12′
6. log cos 41° 32′
7. log cos 3° 7′
8. log sin 80° 4′
9. log tan 45° 2′
10. log cos 74° 25′

Find angle A between 0° and 90° in each problem.

11. $\log \sin A = 9.9340 - 10$
12. $\log \cos A = 9.7559 - 10$
13. $\log \cos A = 9.8640 - 10$
14. $\log \sin A = 9.9381 - 10$
15. $\log \tan A = 9.8396 - 10$
16. $\log \cot A = 0.5109$
17. $\log \sin A = 8.9538 - 10$
18. $\log \cos A = 8.9249 - 10$
19. $\log \tan A = 0.0058$
20. $\log \cot A = 9.9998 - 10$

21

THE LAW OF SINES AND THE LAW OF COSINES

21-1 OBLIQUE TRIANGLES

In Chap. 9, we discussed the solution of right triangles. We advance now to a consideration of oblique triangles—triangles with no right angle. Unlike the right triangle, the ratio of two sides of an oblique triangle does not in general yield a function of any of the angles. We shall, however, derive relations involving the sides and angles. These new relations, called *trigonometric laws,* will enable us to solve for the unknown parts of a triangle when there are sufficient known parts to determine the triangle. We shall take up the following cases of given parts of a triangle:

Case I	A side and two angles
Case II	Two sides and the angle opposite one of them
Case III	Two sides and the included angle
Case IV	Three sides

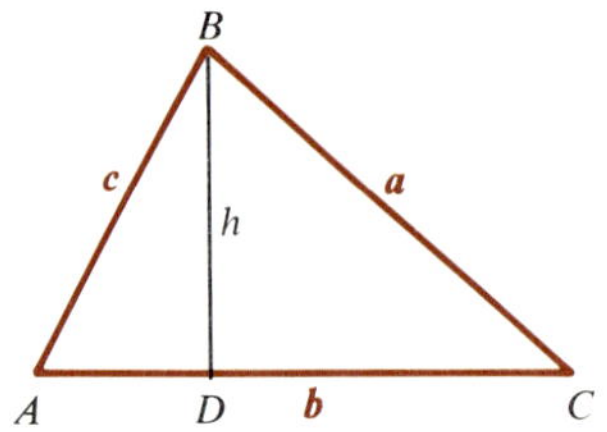

Figure 21-1

21-2 THE LAW OF SINES

Theorem 21-1 In any triangle the sides a, b, c are proportional to the sines of the opposite angles A, B, C. That is,

$$\frac{a}{\sin A} = \frac{b}{\sin B} = \frac{c}{\sin C}$$

Proof: In each of the triangles constructed in Figs. 21-1 and 21-2, h is an altitude drawn from the vertex of angle B. From either figure, we have at once

$$\frac{h}{c} = \sin A \qquad \text{and} \qquad \frac{h}{a} = \sin C$$

Solving each of these equations for h, we obtain

$$h = c \sin A \qquad \text{and} \qquad h = a \sin C$$

and, consequently,

$$a \sin C = c \sin A$$

Dividing by $\sin A \sin C$, we reduce the last equation to

$$\frac{a}{\sin A} = \frac{c}{\sin C}$$

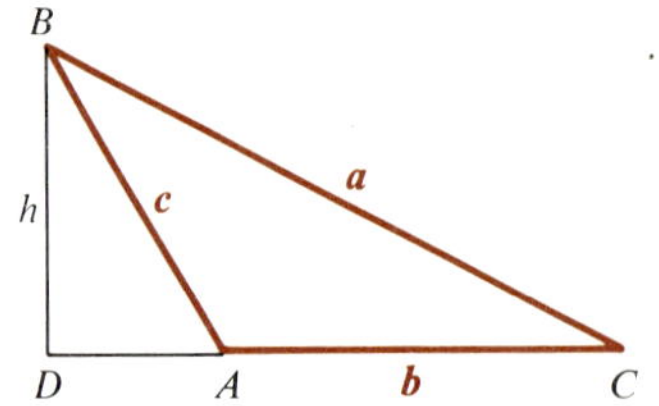

Figure 21-2

By drawing the altitude from the vertex of C, we can show in the same way that

$$\frac{a}{\sin A} = \frac{b}{\sin B}$$

Combining the two results, we present the relation of the sides and angles in the form

$$\frac{a}{\sin A} = \frac{b}{\sin B} = \frac{c}{\sin C}$$

We may see at once that the law of sines furnishes a method for solving Case I. If two angles are known, the third angle can be found from the fact that the sum of the three angles is 180°. Then if b stands for the known side, the equations

$$\frac{a}{\sin A} = \frac{b}{\sin B} \qquad \text{and} \qquad \frac{c}{\sin C} = \frac{b}{\sin B}$$

may be solved separately to yield a and c. If two sides and the angle opposite one of them are known, as a, b, and B, the equation

$$\frac{a}{\sin A} = \frac{b}{\sin B}$$

may be solved for its one unknown A.

We shall use the law of sines first in solving triangles in which a side and two angles are known.

EXAMPLE 1 Solve the triangle which has $c = 231.7$, $A = 52° 17'$, and $B = 69° 53'$.

Solution: The triangle is drawn to scale in Fig. 21-3. Since the sum of the three angles is 180°, we have

$$C = 180° - (52° 17' + 69° 53') = 57° 50'$$

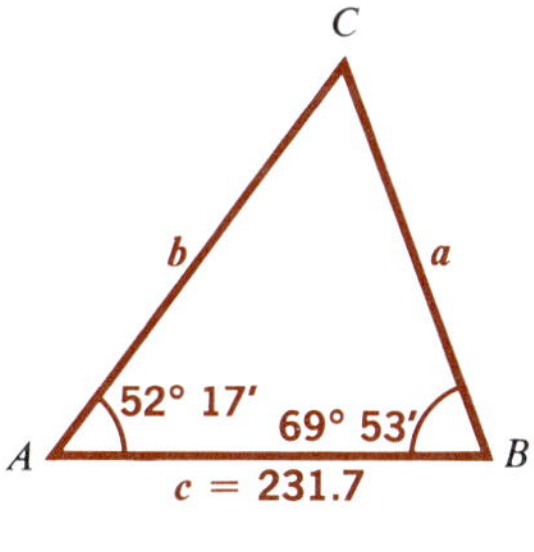

Figure 21-3

By the law of sines,

$$\frac{a}{\sin 52^\circ\ 17'} = \frac{231.7}{\sin 57^\circ\ 50'}$$

and

$$a = \frac{231.7 \sin 52^\circ\ 17'}{\sin 57^\circ\ 50'}$$

Hence,

$$\log a = \log 231.7 + \log \sin 52^\circ\ 17' - \log 57^\circ\ 50'$$

Using the tables, we obtain

$$\begin{aligned} \log 231.7 &= 2.3649 \\ \log \sin 52^\circ\ 17' &= \underline{9.8982 - 10} \\ &\ 12.2631 - 10 \\ \log \sin 57^\circ\ 50' &= \underline{9.9276 - 10} \\ \log a &= 2.3355 \\ a &= 216.5 \end{aligned}$$

Using the law of sines again, we get

$$\frac{b}{\sin 69^\circ\ 53'} = \frac{231.7}{\sin 57^\circ\ 50'}$$

and

$$\log b = \log 231.7 + \log \sin 69^\circ\ 53' - \log \sin 57^\circ\ 50'$$

$$\begin{aligned} \log 231.7 &= 2.3649 \\ \log \sin 69^\circ\ 53' &= \underline{9.9727 - 10} \\ &\ 12.3376 - 10 \\ \log \sin 57^\circ\ 50' &= \underline{9.9276 - 10} \\ \log b &= 2.4100 \\ b &= 257.1 \end{aligned}$$

The computed parts are $C = 57^\circ\ 50'$, $a = 216.5$, and $b = 257.1$.

EXAMPLE 2 A tree stands vertically on a hillside which makes an angle of 21° with the horizontal. From a point 50 feet directly down the hill from the base of the tree the angle of elevation of the top of the tree is 44°. Find the height of the tree.

Solution: We let a in Fig. 21-4 represent the height of the tree. Then from the triangle ABC, we have

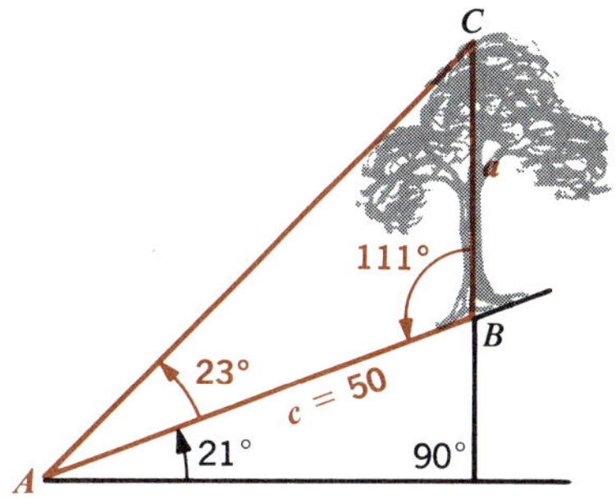

Figure 21-4

$$A = 44^\circ - 21^\circ = 23^\circ$$

$$B = 90^\circ + 21^\circ = 111^\circ$$

$$C = 180^\circ - (A + B) = 46^\circ$$

Applying the law of sines, we obtain

$$\frac{a}{\sin 23^\circ} = \frac{50}{\sin 46^\circ}$$

$$a = \frac{50 \sin 23^\circ}{\sin 46^\circ}$$

$$= \frac{50(0.391)}{0.719} = 27 \text{ feet}$$

EXAMPLE 3 A pilot heads his plane in the direction 60°, but a wind blowing in the direction 210° causes his course to be in the direction 70°. If he travels on this course 160 miles in 1 hour, find the airspeed of the plane and the speed of the wind.

Solution: Recalling the discussion of vectors in Sec. 9-10, we construct triangle ABC (Fig. 21-5). The vector **AB** represents the

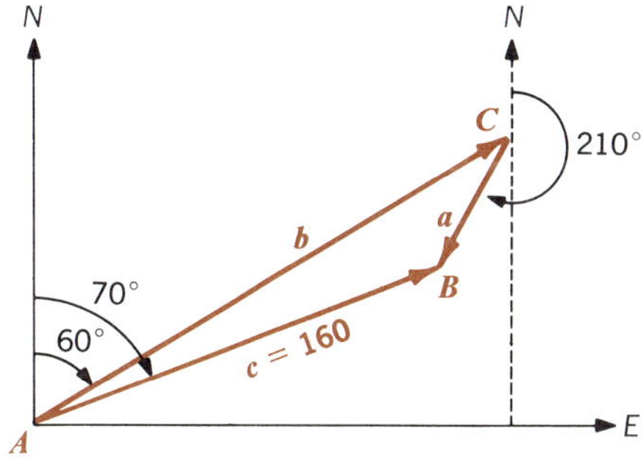

Figure 21-5

groundspeed and direction of travel, **CB** the speed and direction of the wind, and **AC** the airspeed and direction of heading. From the given directions, we find the angles of triangle ABC to be $A = 10°$, $C = 30°$, and $B = 140°$. Hence, by the law of sines,

$$\frac{a}{\sin 10°} = \frac{160}{\sin 30°} \quad \text{and} \quad \frac{b}{\sin 140°} = \frac{160}{\sin 30°}$$

These equations yield $a = 56$ miles per hour for the windspeed and $b = 210$ miles per hour for the airspeed.

EXERCISE 21-1

Draw each triangle approximately to scale. Then solve for the unknown parts without the use of logarithms. Give final results rounded off in accordance with the discussions in Secs. 9-1 and 9-8.

1. $c = 80, \quad A = 46°, \quad B = 64°$

2. $b = 40, \quad A = 42°, \quad C = 67°$

3. $b = 3.7, \quad A = 31°, \quad C = 68°$

4. $a = 9.1, \quad B = 51°, \quad C = 29°$

5. $a = 19, \quad B = 21°, \quad C = 106°$

6. $c = 0.60, \quad B = 27°, \quad C = 38°$

7. $b = 300, \quad B = 50°\ 10', \quad C = 71°\ 00'$

8. $c = 400, \quad A = 53°\ 40', \quad C = 81°\ 10'$

9. $a = 0.422, \quad A = 98°\ 00', \quad B = 24°\ 30'$

Use logarithms in solving problems 10 through 15.

10. $b = 4.85, \quad B = 69°\ 10', \quad C = 40°\ 50'$

11. $a = 734.6, \quad B = 61°\ 18', \quad C = 83°\ 39'$

12. $b = 6761, \quad A = 11°\ 10', \quad B = 38°\ 47'$

13. $c = 16.73, \quad B = 84°\ 11', \quad C = 48°\ 7'$

14. $a = 800.4, \quad B = 55°\ 1', \quad C = 66°\ 19'$

15. $b = 1784, \quad A = 40°\ 13', \quad B = 70°\ 9'$

16. To determine the distance between two points A and B on opposite sides of a river, we measure the distance from A to C

to be 215 feet, the angle CAB to be 100° 30′, and the angle ACB to be 42° 20′. Find the distance AB.

17. A tree stands vertically on a hillside which makes an angle of 18° with the horizontal. From a point 48 feet directly down the hill from the base of the tree the angle of elevation of the top of the tree is 45°. Find the height of the tree.

18. Two men, 2200 feet apart, sight a balloon which is between them and in their vertical plane. The angle of elevation of the balloon as measured by one of the men is 38° and by the other is 54°. Find the distance of the balloon from each observer and the height of the balloon.

19. Solve the preceding problem if the men are on the same side of the balloon, the other data being unchanged.

20. A tree stands vertically on a hillside which makes an angle of 18° with the horizontal. From a point 180 feet up the hill from the tree the angle of depression of the top of the tree is 9°. Find the height of the tree.

21. In crossing a river, a boat, headed in the direction N 36° E, reached a point on the opposite bank N 55° E of the starting point. If the rate of the boat in still water is 4.0 miles per hour and if the river flows due south, find the speed of the current and the actual speed of the boat in crossing.

22. A pilot, flying with an airspeed of 280 miles per hour, heads his plane in the direction 150°. If the wind is blowing due east and the flight course is in the direction 140°, find the wind velocity and the groundspeed of the plane.

23. A pilot heads his plane in the direction 80°. A 45-mile-per-hour wind blowing in the direction 45° causes the plane to move in the direction 75°. Find the airspeed and the groundspeed of the plane.

24. Two men are carrying a 320-pound weight. One lifts in a direction 25° from the vertical, and the other lifts in a direction 20° from the vertical. Find the force which each man exerts.

25. A force of 176 pounds makes an angle of 78° 00′ with a second force. The resultant makes an angle of 32° 00′ with the second force. Find the magnitude of the second force and of the resultant.

26. Derive the Mollweide equation

$$\frac{a-b}{c} = \frac{\sin \frac{1}{2}(A-B)}{\cos \frac{1}{2}C}$$

21-3 THE AMBIGUOUS CASE

As we have seen, a unique triangle is determined by a side and two angles. Case II, however, presents a new situation; there may be no triangle, one triangle, or two triangles, depending on the given parts. Because two triangles are sometimes constructible from two sides and an angle opposite one of them, Case II is called the *ambiguous case.* The various possibilities for the ambiguous case are pictured in Figs. 21-6 to 21-11, in which the given parts are designated by a, b, and A.

If $A \geq 90°$ and $a \leq b$, no triangle exists; a is too short to complete a triangle (Fig. 21-6). But if $a > b$, one triangle can be constructed (Fig. 21-7).

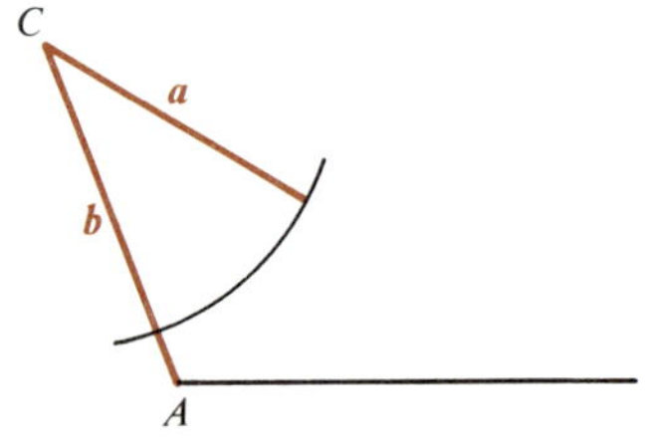

Figure 21-6

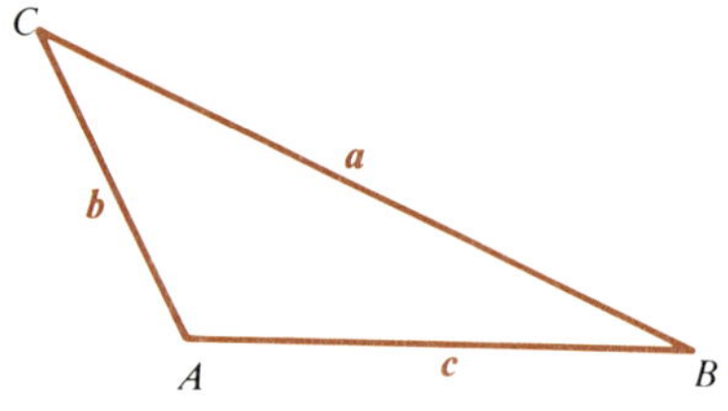

Figure 21-7

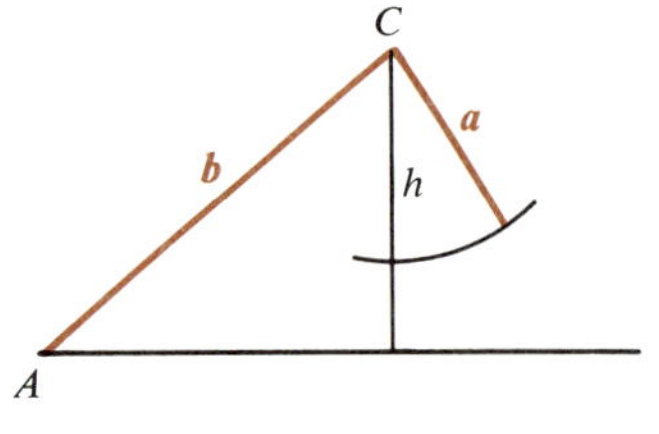

Figure 21-8

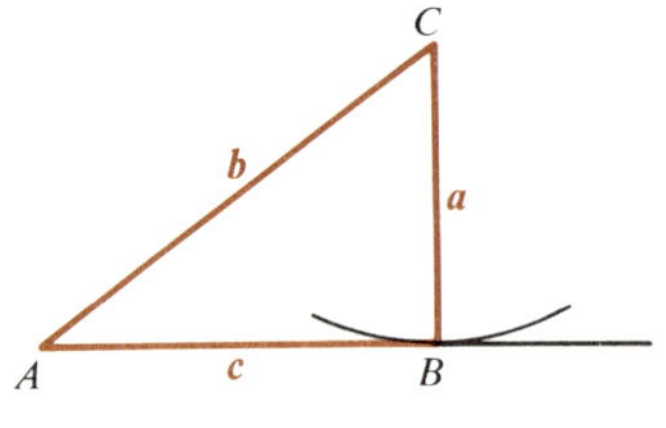

Figure 21-9

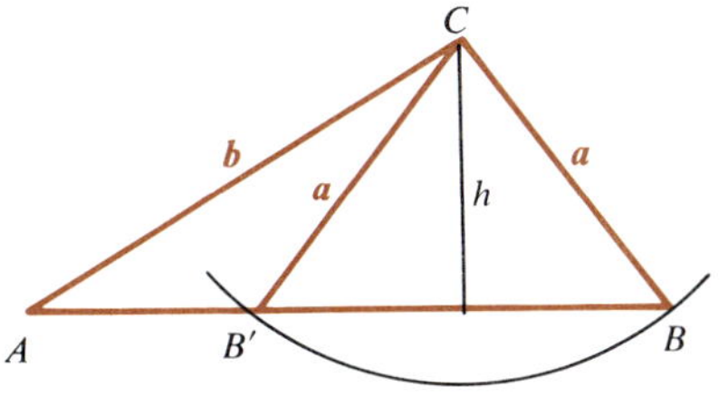

Figure 21-10

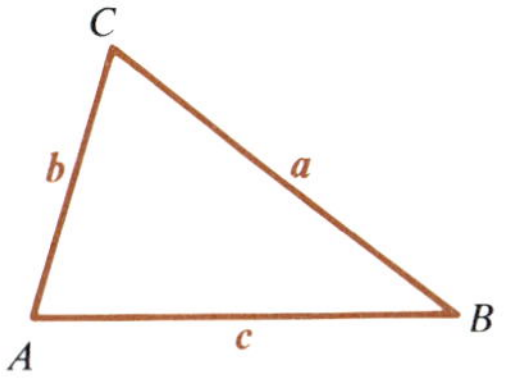

Figure 21-11

If A is an acute angle, we have the following possibilities:

1. If a is less than the perpendicular h, a triangle cannot be formed (Fig. 21-8).
2. If $a = h$, there is one triangle, a right triangle (Fig. 21-9).
3. If a is greater than h and less than b ($h < a < b$), two triangles are possible (Fig. 21-10).
4. If $a \geq b$, one triangle can be formed (Fig. 21-11).

The figures reveal that there is uncertainty only when A is acute and $a < b$. The decision in this instance can be made by applying the law of sines and computing sin B. Thus,

$$\sin B = \frac{b \sin A}{a} = \frac{h}{a}$$

If the computed value for $\sin B > 1$, then $a < h$ and no triangle is possible (Fig. 21-8). If $\sin B = 1$, then $a = h$ and a right triangle can be formed (Fig. 21-9). If $\sin B < 1$, then $a > h$ and two triangles are constructible (Fig. 21-10).

If logarithms are used in computing, $\log \sin B > 0$ means no triangle, $\log \sin B = 0$ means one triangle, and $\log \sin B < 0$ means two triangles.

EXAMPLE 1 Given $a = 17$, $b = 28$, and $B = 108°$, find angle A.

Solution: It is easy to see that there is only one solution. Applying the inverted form of the law of sines, which places the unknown in the numerator, we have

$$\frac{\sin A}{17} = \frac{\sin 108^\circ}{28}$$

$$\sin A = \frac{17 \sin 108^\circ}{28}$$

$$= 0.577$$

$$A = 35^\circ \qquad \text{Nearest degree}$$

EXAMPLE 2 Determine if a triangle can be formed with the parts $a = 41$, $b = 62$, and $A = 43^\circ$.

Solution: Substituting in the law of sines gives

$$\frac{\sin B}{62} = \frac{\sin 43^\circ}{41}$$

$$\sin B = 1.031$$

Since the sine of an angle can never exceed 1, we conclude that a triangle cannot be formed with the given parts. This situation is shown in Fig. 21-8.

EXAMPLE 3 Solve the triangle whose given parts are $a = 182$, $b = 243$, and $A = 43^\circ\ 20'$.

Solution: Since A is acute and $a < b$, there may be no, one, or two solutions. By constructing angle A and taking a distance along one side to represent b, we surmise that a may be drawn in either of two positions to complete a triangle. And there are two solutions because, as we shall see, $\sin B < 1$. We solve first the triangle made by placing a so that angle B is acute (Fig. 21-12). We have

$$\frac{\sin B}{243} = \frac{\sin 43^\circ\ 20'}{182}$$

$$\sin B = \frac{243 \sin 43^\circ\ 20'}{182} = \frac{243(0.6862)}{182} = 0.9162$$

$$B = 66^\circ\ 22'$$

$$C = 180^\circ - (43^\circ\ 20' + 66^\circ\ 22') = 70^\circ\ 18'$$

To find c, we write

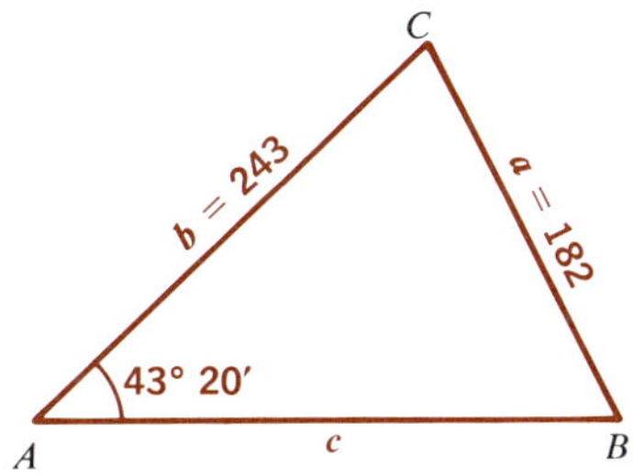

Figure 21-12

$$\frac{c}{\sin 70^\circ\ 18'} = \frac{182}{\sin 43^\circ\ 20'}$$

$$c = \frac{182 \sin 70^\circ\ 18'}{\sin 43^\circ\ 20'} = \frac{182(0.9415)}{0.6862} = 249.7$$

Having found B, C, and c, we next solve the triangle formed by placing side a in the position shown in Fig. 21-13. We designate the unknown parts of this triangle by B', C', and c'. Here angle B' is obtuse, but $\sin B'$, expressed in terms of the given parts, is exactly the same as $\sin B$ for the first triangle. We conclude therefore that B' and B are supplementary. Hence, B' and C' are readily obtainable. Thus,

$$B' = 180^\circ - B = 180^\circ - 66^\circ\ 22' = 113^\circ\ 38'$$

$$C' = 180^\circ - (43^\circ\ 20' + 113^\circ\ 38') = 23^\circ\ 2'$$

The law of sines now yields c'.

$$\frac{c'}{\sin 23^\circ\ 2'} = \frac{182}{\sin 43^\circ\ 20'}$$

$$c' = \frac{182 \sin 23^\circ\ 2'}{\sin 43^\circ\ 20'} = \frac{182(0.3912)}{0.6862} = 103.8$$

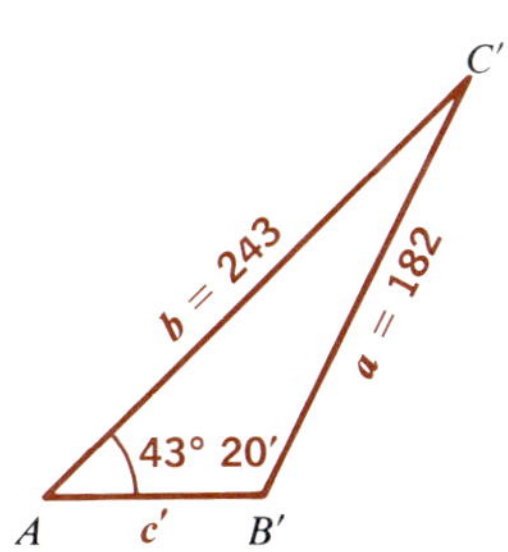

Figure 21-13

Summarizing and rounding off the computed parts, we have

$$B = 66^\circ\ 20', \qquad C = 70^\circ\ 20', \qquad c = 250$$
$$B' = 113^\circ\ 40', \qquad C' = 23^\circ\ 00', \qquad c' = 104$$

EXERCISE 21-2

Draw each triangle approximately to scale. Then solve for the unknown parts without the use of logarithms. Give the results properly rounded off.

1. $a = 40, \quad b = 24, \quad A = 100^\circ$
2. $a = 30, \quad b = 18, \quad A = 96^\circ$
3. $b = 70, \quad c = 80, \quad B = 105^\circ$
4. $b = 53, \quad c = 53, \quad C = 90^\circ$
5. $a = 29, \quad b = 25, \quad A = 52^\circ$
6. $a = 38, \quad b = 48, \quad A = 37^\circ$
7. $b = 3.6, \quad c = 2.4, \quad C = 28^\circ$
8. $b = 9.1, \quad c = 8.5, \quad C = 73^\circ$
9. $a = 51, \quad b = 64, \quad A = 54^\circ$
10. $b = 33, \quad c = 46, \quad B = 35^\circ$

Use logarithms in solving problems 11 through 16.

11. $b = 186, \quad c = 233, \quad C = 101^\circ\ 10'$
12. $a = 64.0, \quad c = 50.9, \quad C = 53^\circ\ 40'$
13. $a = 2518, \quad b = 3116, \quad B = 51^\circ\ 23'$
14. $b = 0.7010, \quad c = 0.9119, \quad C = 23^\circ\ 37'$
15. $b = 128.3, \quad c = 160.3, \quad B = 40^\circ\ 3'$
16. $a = 68.51, \quad b = 80.32, \quad A = 46^\circ\ 17'$
17. To find the distance between two points A and B, separated by a lake, a surveyor measures a distance of 247 feet from A to point C such that angle BAC is $47^\circ\ 30'$. He then measures CB as 274 feet. Find the distance AB.
18. A pole leans southward and makes an angle of 60° with the horizontal. A prop 11 feet long supports the pole. If the

base of the prop is 12 feet south of the pole, find the greater acute angle which the prop may make with the horizontal.

19. In crossing a river, a boat, whose rate in still water is 8.0 miles per hour, has a groundspeed of 5.6 miles per hour. The boat is headed in the direction N 30° E, and the river flows due south. Find the bearing of the path of the boat and the smaller of the two possible rates of flow.

20. A pilot wishes to travel in the direction 90°. His airspeed is 150 miles per hour, and a 35-mile-per-hour wind is blowing in the direction 38°. Find the direction he should head his plane, and find his speed relative to the ground.

21. A pilot flying with an airspeed of 560 miles per hour travels in the direction 140°. If the wind is blowing due east at 56 miles per hour, find the heading and groundspeed of the plane.

21-4 THE LAW OF COSINES

The law of sines is not directly applicable to Cases III and IV. But the law of cosines, which we shall now derive, is applicable to these cases.

Theorem 21-2 In any triangle the square of any side is equal to the sum of the squares of the other two sides minus twice the product of those sides and the cosine of the included angle. That is,

$$a^2 = b^2 + c^2 - 2bc \cos A$$

$$b^2 = a^2 + c^2 - 2ac \cos B$$

$$c^2 = a^2 + b^2 - 2ab \cos C$$

Proof: Consider the triangle ABC (Fig. 21-14) where we have introduced a coordinate system with the origin at the vertex of

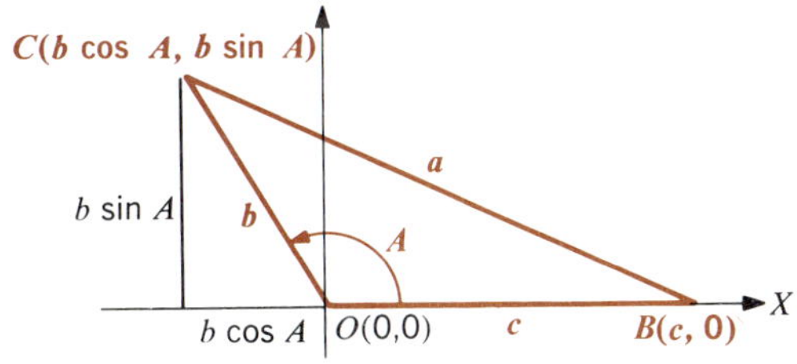

Figure 21-14

angle A and the x axis along side c. Note that the coordinates of the vertices of the triangle are

$$O(0, 0), \qquad B(c, 0), \qquad C(b \cos A, b \sin A)$$

Although angle A is obtuse in the figure, these coordinates would be unchanged if A were an acute angle or a right angle. We now use the distance formula (Sec. 5-4) to find the square of length a. Thus,

$$\begin{aligned} a^2 &= (b \cos A - c)^2 + (b \sin A - 0)^2 \\ &= b^2 \cos^2 A - 2bc \cos A + c^2 + b^2 \sin^2 A \\ &= b^2(\cos^2 A + \sin^2 A) - 2bc \cos A + c^2 \\ &= b^2 + c^2 - 2bc \cos A \end{aligned}$$

Clearly, we could place the origin of coordinates at each of the other vertices and obtain relations involving $\cos B$ and $\cos C$.

It is immediately evident that the third side of a triangle may be computed by the law of cosines if two sides and their included angle are known. The formulas show also that any angle may be found if three sides of the triangle are given. Hence, the law of cosines is applicable to Case III and to Case IV. The cosine formulas, however, involving addition and subtraction, are not well suited to logarithmic computation. Because of this disadvantage, other laws, later to be derived, are usually preferable for handling Cases III and IV. Still the law of cosines is of much importance, for it has numerous applications in situations where computations are not involved.

EXAMPLE 1 Two sides and the included angle of a triangle are measured to be 11 feet, 20 feet, and 112°. Find the length of the third side.

Solution: We let b, c, and A stand for the given parts as indicated in Fig. 21-15. Then using the law of cosines, we obtain

$$\begin{aligned} a^2 &= b^2 + c^2 - 2bc \cos A \\ &= (11)^2 + (20)^2 - 2(11)(20) \cos 112^\circ \\ &= 121 + 400 - 440(-0.3746) \\ &= 521 + 164.8 = 685.8 \end{aligned}$$

$$a = 26 \qquad \text{Rounded off to two digits}$$

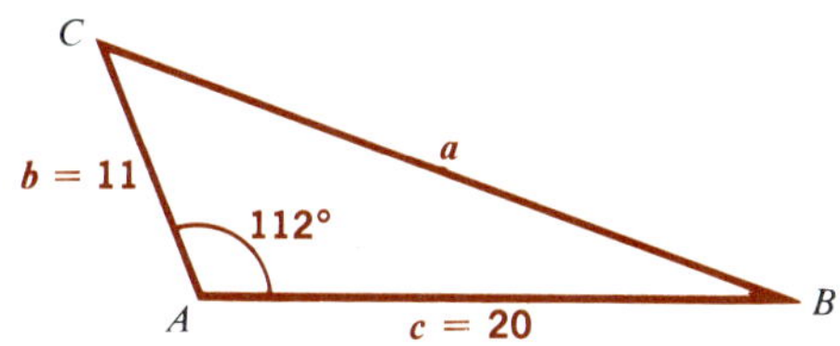

Figure 21-15

Notice that the reference angle of 112° is 68°, and therefore cos 112° = −cos 68° (Sec. 9-4). We remark also that the square root of a number may be found by the use of logarithms.

We can express the law of cosines in other useful forms by solving for the cosines of the angles. Thus,

$$\cos A = \frac{b^2 + c^2 - a^2}{2bc}$$

$$\cos B = \frac{a^2 + c^2 - b^2}{2ac}$$

$$\cos C = \frac{a^2 + b^2 - c^2}{2ab}$$

These formulas are convenient when the angles are to be found from the three known sides.

EXAMPLE 2 Given $a = 15.2$, $b = 20.3$, and $c = 30.0$, find A, B, and C.

Solution: We shall find each of the angles by use of the law of cosines. We could, of course, employ the law of sines after one angle is found.

$$\cos A = \frac{(20.3)^2 + (30.0)^2 - (15.2)^2}{2(20.3)(30.0)} = \frac{1081}{1218} = 0.8875$$

and $A = 27° \ 30'$.

$$\cos B = \frac{(15.2)^2 + (30.0)^2 - (20.3)^2}{2(15.2)(30.0)} = \frac{719.0}{912.0} = 0.7884$$

and $B = 38° \ 00'$.

$$\cos C = \frac{(15.2)^2 + (20.3)^2 - (30.0)^2}{2(15.2)(20.3)} = \frac{-256.9}{617.1} = -0.4163$$

The cosine is negative; hence C is obtuse. We find the angle whose cosine is $+.4163$ and take its supplement to get C. Thus,

$$C = 180° - 65° 20' = 114° 40'$$

Check:

$$A + B + C = 27° 30' + 38° 00' + 114° 40' = 180° 10'$$

The check is satisfactory. The extra $10'$ in the sum of the angles is not surprising since the value of each angle is computed to the nearest integral multiple of $10'$.

EXERCISE 21-3

Find the unknown parts of the triangles in problems 1 to 15.

1. $a = 3.0, \quad b = 4.0, \quad C = 52°$
2. $b = 5.0, \quad c = 8.0, \quad A = 60°$
3. $a = 11, \quad c = 16, \quad B = 96°$
4. $a = 1.4, \quad b = 2.3, \quad C = 113°$
5. $b = 12, \quad c = 23, \quad A = 35°$
6. $a = 20, \quad c = 31, \quad B = 45°$
7. $b = 0.26, \quad c = 0.41, \quad A = 131°$
8. $b = 0.60, \quad c = 0.30, \quad A = 70°$
9. $a = 6.0, \quad b = 9.0, \quad c = 8.0$
10. $a = 32, \quad b = 30, \quad c = 40$
11. $a = 21, \quad b = 25, \quad c = 22$
12. $a = 20, \quad b = 22, \quad c = 25$
13. $a = 18.0, \quad b = 23.0, \quad c = 30.0$
14. $a = 2.00, \quad b = 1.20, \quad c = 1.50$
15. $a = 240, \quad b = 304, \quad c = 400$
16. Two sides of a parallelogram make an angle of 53°. The lengths of the two sides are 20 and 32 feet. Find the length of each diagonal.

17. Two points A and B are on opposite sides of a pond. The distance from A to a third point C is measured to be 20.3 feet and the distance from C to B, 27.6 feet. Find the distance from A to B if angle ACB is $53° 20'$.

18. Two boats leave port at the same time. One travels 28.0 miles in the direction N $50° 30'$ E while the other travels 24.0 miles in the direction S $80° 10'$ E. What is the distance between the new positions of the boats?

19. A force of 50 pounds makes an angle of $67°$ with a force of 70 pounds. Find the magnitude of the resultant of the forces.

20. A plane with an airspeed of 170 miles per hour heads in the direction $67°$. It is carried from this course by a 42-mile-per-hour wind blowing in the direction $0°$. Find the direction in which the plane travels and its speed relative to the ground.

21. Referring to Fig. 21-1 (Sec. 21-2), find two expressions for h^2 from the right triangles ABD and BCD. Equate the expressions and proceed to derive the law of cosines. Also make the derivation from Fig. 21-2 where A is an obtuse angle.

21-5 THE AREA OF A TRIANGLE

We shall now derive formulas for the area S of a triangle in terms of

1. Two sides and the included angle
2. One side and three angles
3. Three sides

Theorem 21-3 The area of a triangle is equal to half the product of two sides times the sine of the included angle; that is,

$$S = \tfrac{1}{2}bc \sin A, \qquad S = \tfrac{1}{2}ac \sin B, \qquad S = \tfrac{1}{2}ab \sin C$$

Proof: The area of a triangle is equal to half the product of the base and the altitude. Hence for either triangle of Fig. 21-16 with h the altitude, we have

$$S = \tfrac{1}{2}hc = \tfrac{1}{2}bc \sin A$$

since $h = b \sin A$. The other two forms for the area evidently hold because any of the three angles could be labeled A.

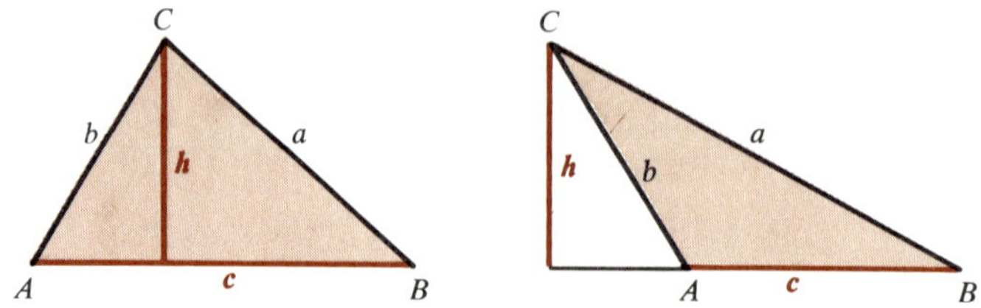

Figure 21-16

Theorem 21-4 The area of a triangle is equal to half the square of a side times the product of the sines of the adjacent angles divided by the sine of the opposite angle; that is,

$$S = \frac{a^2 \sin B \sin C}{2 \sin A} = \frac{b^2 \sin C \sin A}{2 \sin B} = \frac{c^2 \sin A \sin B}{2 \sin C}$$

Proof: If only one side and two (therefore three) angles are known, a second side is obtainable from the law of sines; that is,

$$\frac{b}{\sin B} = \frac{a}{\sin A} \qquad \text{and} \qquad b = \frac{a \sin B}{\sin A}$$

Then substituting for b in $S = \frac{1}{2}ab \sin C$ gives

$$S = \frac{a^2 \sin B \sin C}{2 \sin A}$$

The other two forms above for the area may be obtained by replacing a in this formula, in turn, by

$$\frac{b \sin A}{\sin B} \qquad \text{and} \qquad \frac{c \sin A}{\sin C}$$

Theorem 21-5 The area of a triangle in terms of its sides is given by

$$S = \sqrt{s(s - a)(s - b)(s - c)}$$

where $s = \frac{1}{2}(a + b + c)$.

Proof: For this case, we start with $S = \frac{1}{2}ab \sin C$. Then

$$\begin{aligned}
S^2 &= \tfrac{1}{4}a^2b^2 \sin^2 C \\
&= \tfrac{1}{4}a^2b^2(1 - \cos^2 C) \\
&= \tfrac{1}{4}a^2b^2\left[1 - \left(\frac{a^2 + b^2 - c^2}{2ab}\right)^2\right] \qquad \text{By law of cosines} \\
&= \tfrac{1}{4}a^2b^2\left[\frac{4a^2b^2 - (a^2 + b^2 - c^2)^2}{4a^2b^2}\right] \\
&= \tfrac{1}{16}[2ab + (a^2 + b^2 - c^2)][2ab - (a^2 + b^2 - c^2)] \\
&= \tfrac{1}{16}[(a + b)^2 - c^2][c^2 - (a - b)^2] \\
&= \tfrac{1}{16}(a + b + c)(a + b - c)(c + a - b)(c - a + b)
\end{aligned}$$

To simplify the right member, we set

$$a + b + c = 2s$$

Then subtracting $2a$, $2b$, and $2c$, in turn, from both sides gives

$$\begin{aligned}
b + c - a &= 2(s - a) \\
a + c - b &= 2(s - b) \\
a + b - c &= 2(s - c)
\end{aligned}$$

We may now write

$$S^2 = \frac{2s \cdot 2(s - c) \cdot 2(s - b) \cdot 2(s - a)}{16}$$

and hence,

$$S = \sqrt{s(s - a)(s - b)(s - c)}$$

None of the formulas which we have derived gives the area of a triangle in terms of two sides and the angle opposite one of them. For these known parts the angles can first be found and then the area is obtainable through two sides and the included angle. This is the ambiguous case, however, and there is a possibility of two solutions.

EXAMPLE 1 If $a = 200$, $b = 300$, and $C = 40° 10'$, find the area of the triangle to three significant figures.

Solution: Using $s = \frac{1}{2}ab \sin C$, we have

$$\begin{aligned} S &= \tfrac{1}{2}(200)(300) \sin 40^\circ\ 10' \\ &= \tfrac{1}{2}(200)(300)(0.6450) \\ &= 19{,}400 \qquad \text{Rounded off} \end{aligned}$$

EXAMPLE 2 Find the area of the triangle with the sides $a = 20.55$, $b = 32.06$, and $c = 41.27$.

Solution: We use $S = \sqrt{s(s-a)(s-b)(s-c)}$.

$$\begin{aligned} a &= 20.55 & s &= 46.94 \\ b &= 32.06 & s - a &= 26.39 \\ c &= \underline{41.27} & s - b &= 14.88 \\ 2s &= 93.88 & s - c &= 5.67 \end{aligned}$$

$$\begin{aligned} \log s &= 1.6716 & \log S &= 2.5096 \\ \log (s - a) &= 1.4214 & & \\ \log (s - b) &= 1.1726 & S &= 323.3 \\ \log (s - c) &= \underline{0.7536} & & \\ \log S^2 &= 5.0192 & & \end{aligned}$$

EXERCISE 21-4

Find the area of each triangle. Use logarithms only where the computation will be facilitated.

1. $a = 10, \quad b = 15, \quad C = 150^\circ$
2. $a = 10, \quad b = 15, \quad C = 30^\circ$
3. $b = 40, \quad c = 30, \quad A = 46^\circ$
4. $b = 2.2, \quad c = 3.0, \quad A = 113^\circ$
5. $a = 70, \quad B = 64^\circ, \quad C = 79^\circ$
6. $b = 60, \quad A = 75^\circ, \quad C = 95^\circ$
7. $a = 51.3, \quad B = 64^\circ\ 10', \quad C = 79^\circ\ 40'$
8. $b = 7.14, \quad A = 31^\circ\ 20', \quad C = 110^\circ\ 30'$
9. $a = 6.0, \quad b = 5.0, \quad A = 62^\circ$
10. $b = 13.2, \quad c = 20.5, \quad C = 100^\circ\ 10'$
11. $a = 58.4, \quad b = 70.1, \quad A = 28^\circ\ 50'$

12. $b = 20.6, \quad c = 12.2, \quad B = 99° \, 20'$

13. $a = 17, \quad b = 26, \quad c = 39$

14. $a = 8.00, \quad b = 9.18, \quad c = 6.34$

15. $a = 0.7557, \quad b = 0.8813, \quad c = 0.5813$

16. Denoting the length of each of the equal sides of an isosceles triangle by a, and each of the equal angles by A, show that the area is given by the formula

$$S = \tfrac{1}{2}a^2 \sin 2A$$

17. Verify each of the formulas for areas of triangles for equilateral triangles, letting a denote the length of a side.

22

THE LAW OF TANGENTS AND THE HALF-ANGLE FORMULAS

22-1 THE LAW OF TANGENTS

In Chap. 21 the law of sines and the law of cosines were used in solving the oblique triangle for the various cases of known parts. The law of sines, as we saw, is in a suitable form for the application of logarithms. But the law of cosines is not well suited for logarithmic computations. In this chapter, we shall return to Cases III and IV (Sec. 21-1) and derive other laws which apply to them. These new laws are readily adaptable to the use of logarithms.

Theorem 22-1 In any triangle the difference of two sides divided by their sum is equal to the tangent of half the difference of the opposite angles divided by the tangent of half their sum; that is,

$$\frac{a-b}{a+b}=\frac{\tan\frac{1}{2}(A-B)}{\tan\frac{1}{2}(A+B)}$$

$$\frac{b - c}{b + c} = \frac{\tan \frac{1}{2}(B - C)}{\tan \frac{1}{2}(B + C)}$$

$$\frac{c - a}{c + a} = \frac{\tan \frac{1}{2}(C - A)}{\tan \frac{1}{2}(C + A)}$$

Proof: We begin by writing the laws of sines in the form

$$\frac{a}{b} = \frac{\sin A}{\sin B}$$

This equation may be expressed in two different forms by subtracting 1 from both members and adding 1 to both members. Thus, we have

$$\frac{a}{b} - 1 = \frac{\sin A}{\sin B} - 1 \qquad \text{and} \qquad \frac{a}{b} + 1 = \frac{\sin A}{\sin B} + 1$$

Hence, combining, we obtain

$$\frac{a - b}{b} = \frac{\sin A - \sin B}{\sin B} \qquad \text{and} \qquad \frac{a + b}{b} = \frac{\sin A + \sin B}{\sin B}$$

These two equations, by division, yield

$$\frac{a - b}{a + b} = \frac{\sin A - \sin B}{\sin A + \sin B}$$

Then applying formulas (27) and (28) of Sec. 10-8, we have

$$\frac{a - b}{a + b} = \frac{2 \cos \frac{1}{2}(A + B) \sin \frac{1}{2}(A - B)}{2 \sin \frac{1}{2}(A + B) \cos \frac{1}{2}(A - B)}$$

$$= \cot \tfrac{1}{2}(A + B) \tan \tfrac{1}{2}(A - B)$$

Hence,

$$\frac{a - b}{a + b} = \frac{\tan \frac{1}{2}(A - B)}{\tan \frac{1}{2}(A + B)}$$

We note that the second of the above forms for the law of tangents may be obtained by the same steps by starting with b and c in the law of sines and the third form, by starting with sides c and a.

The law of tangents is applicable to Case III in which two sides and the included angle are known. Suppose we indicate the two known sides by b and c and the included angle by A. With A given, half the sum of B and C can be readily obtained. Then the equation

$$\frac{b-c}{b+c} = \frac{\tan \frac{1}{2}(B-C)}{\tan \frac{1}{2}(B+C)}$$

contains $\tan \frac{1}{2}(B-C)$ as an unknown and consequently may be solved for $\frac{1}{2}(B-C)$. With $\frac{1}{2}(B+C)$ and $\frac{1}{2}(B-C)$ known, B and C can be found individually. Finally, side a is obtainable by the law of sines.

EXAMPLE Solve the triangle ABC if $A = 46° 20'$, $b = 173$, and $c = 283$.

Solution: Since $c > b$, it is convenient to use the law of tangents in the form

$$\frac{c-b}{c+b} = \frac{\tan \frac{1}{2}(C-B)}{\tan \frac{1}{2}(C+B)}$$

The given values yield $\frac{1}{2}(C+B) = \frac{1}{2}(180° - A) = 66° 50'$, $c - b = 110$, and $c + b = 456$. Then substituting in the chosen form of the law of tangents gives

$$\tan \tfrac{1}{2}(C-B) = \frac{110 \tan 66° 50'}{456}$$

and

$$\log \tan \tfrac{1}{2}(C-B) = \log 110 + \log \tan 66° 50' - \log 456$$

Computing, we find

$$\begin{aligned} \log 110 &= 2.0414 \\ \log \tan 66° 50' &= \underline{10.3686 - 10} \\ & 12.4100 - 10 \\ \log 456 &= \underline{2.6590} \\ \log \tan \tfrac{1}{2}(C-B) &= 9.7510 - 10 \\ \tfrac{1}{2}(C-B) &= 29° 24' \end{aligned}$$

The sum and the difference of the corresponding members of the equations $\frac{1}{2}(C+B) = 66° 50'$ and $\frac{1}{2}(C-B) = 29° 24'$ yields $C = 96° 14'$ and $B = 37° 26'$. Then side a, obtainable by the law of sines, is 205.6. Rounding off, we give the final results as $B = 37° 30'$, $C = 96° 10'$, and $a = 206$.

EXERCISE 22-1

Solve by the law of tangents and the law of sines. Logarithms may be used to advantage in some of the problems.

1. $a = 18, \quad b = 32, \quad C = 59°$
2. $a = 74, \quad b = 48, \quad C = 27°$
3. $b = 81, \quad c = 67, \quad A = 133°$
4. $b = 44, \quad c = 36, \quad A = 161°$
5. $a = 63, \quad c = 42, \quad B = 46°$
6. $a = 90, \quad c = 71, \quad B = 75°$
7. $b = 90.4, \quad c = 60.4, \quad A = 91° \, 40'$
8. $a = 415, \quad b = 715, \quad C = 89° \, 10'$
9. $a = 124, \quad c = 150, \quad B = 38° \, 30'$
10. $b = 57.0, \quad c = 65.1, \quad A = 143° \, 20'$
11. $b = 7.071, \quad c = 4.032, \quad A = 84° \, 53'$
12. $a = 671.2, \quad b = 981.7, \quad C = 107° \, 30'$
13. $a = 0.3408, \quad b = 0.2003, \quad C = 118° \, 34'$
14. $b = 1842, \quad c = 2450, \quad A = 32° \, 35'$
15. The angle made by the directions of two forces is 56°. If the magnitudes of the forces are 430 and 660 pounds, find the magnitude of their resultant.
16. Two forces, acting at a point, have magnitudes of 220 and 140 pounds. If the directions of the two forces make an angle of 28°, find the magnitude of the resultant.
17. The diagonals of a parallelogram are of lengths 27 and 48 inches. Find the lengths of the sides if the diagonals make an angle of 34°.
18. Two adjacent sides of a parallelogram are of lengths 45 and 80 and make an angle of 38°. Find the length of each diagonal.
19. An airplane heads in the direction 46° and has an airspeed of 240 miles per hour. If the wind is blowing from the east at 35 miles per hour, find the direction of the course and the groundspeed of the plane.

20. An airplane heads in the direction 142° with an airspeed of 210 miles per hour. If the wind is blowing in the direction 38° at 30 miles per hour, find the direction of the course and the groundspeed of the plane.

22-2 THE HALF-ANGLE FORMULAS

We shall now derive formulas by which the angles of a triangle may be computed from the lengths of the sides.

Theorem 22-2 If r is the radius of the inscribed circle of a triangle and s is half the sum of the lengths of the sides, then

$$\tan \tfrac{1}{2}A = \frac{r}{s-a}, \qquad \tan \tfrac{1}{2}B = \frac{r}{s-b}, \qquad \tan \tfrac{1}{2}C = \frac{r}{s-c}$$

Proof: The bisectors of angle A and angle B (Fig. 22-1) meet at O. The point O, being on the bisector of A, is equally distant from the sides b and c. Likewise, O is equally distant from the sides a and c. Hence, O is the same distance from the three sides, and therefore a circle can be inscribed in the triangle. The center, of course, is at O and the radius r is the perpendicular distance to the sides.

We see readily from the figure that

$$\tan \tfrac{1}{2}A = \frac{r}{AF}, \qquad \tan \tfrac{1}{2}B = \frac{r}{BD}, \qquad \tan \tfrac{1}{2}C = \frac{r}{CE}$$

To complete the derivation, we need to express r, AF, BD, and CE in terms of the sides a, b, and c. The area of triangle ABC is equal to the sum of the areas of triangles BOC, COA, and AOB. Thus,

$$S = \tfrac{1}{2}ar + \tfrac{1}{2}br + \tfrac{1}{2}cr = \tfrac{1}{2}r(a+b+c) = rs$$

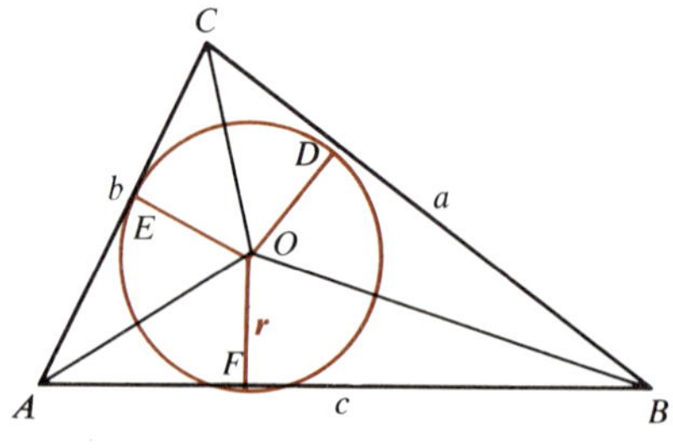

Figure 22-1

where $a + b + c = 2s$. Hence, the area of a triangle is equal to the radius of the inscribed circle multiplied by one-half the sum of the sides. If we equate rs to the formula for the area of a triangle, as found in Sec. 21-5, we obtain

$$rs = \sqrt{s(s-a)(s-b)(s-c)}$$

and

$$r = \sqrt{\frac{(s-a)(s-b)(s-c)}{s}} \tag{1}$$

The triangles AFO and AEO are congruent, and hence $AF = AE$. Similarly, $BD = BF$ and $CD = CE$. Since the sum of these six segments is equal to $a + b + c$, or $2s$, we have

$$2AF + 2BD + 2CD = 2s$$

$$AF = s - (BD + CD) = s - a$$

In a similar way, $BD = s - b$ and $CE = s - c$. Now, we may write

$$\tan \tfrac{1}{2}A = \frac{r}{s-a} \qquad \tan \tfrac{1}{2}B = \frac{r}{s-b} \qquad \tan \tfrac{1}{2}C = \frac{r}{s-c}$$

These are the *half-angle formulas.* If the three angles of a triangle are computed by these formulas, then the sum of the angles furnishes a quick check.

EXAMPLE The sides of a triangle are $a = 14$, $b = 15$, and $c = 20$. Find A, B, and C.

Solution: The quantities s, $s - a$, $s - b$, and $s - c$ need to be computed first. From these values we can find $\log r$ and the logarithms of the denominators in the half-angle formulas. Thus, taking the logarithms of the members of formula (1) above, we have

$$\log r = \tfrac{1}{2}[\log (s-a) + \log (s-b) + \log (s-c) - \log s]$$

we find that $s = \frac{1}{2}(a + b + c) = 24.5$. Then we proceed with the following computations.

$$\begin{array}{rr} s = 24.5 & \log (s-a) = 1.0212 \\ s - a = 10.5 & (+) \log (s-b) = 0.9777 \\ s - b = 9.5 & (+) \log (s-c) = \underline{0.6532} \\ s - c = 4.5 & 2.6521 \\ & (-) \log s = \underline{1.3892} \\ & \log r^2 = 1.2629 \\ & \log r = 0.6314 \end{array}$$

We now have the necessary logarithms for applying the half-angle formulas. Thus, we get

$$\log \tan \tfrac{1}{2}A = \log r - \log (s - a) = 9.6102 - 10$$

$$\tfrac{1}{2}A = 22° \, 10' \qquad A = 44° \, 20'$$

$$\log \tan \tfrac{1}{2}B = \log r - \log (s - b) = 9.6537 - 10$$

$$\tfrac{1}{2}B = 24° \, 20' \qquad B = 48° \, 40'$$

$$\log \tan \tfrac{1}{2}C = \log r - \log (s - c) = 9.9782 - 10$$

$$\tfrac{1}{2}C = 43° \, 30' \qquad C = 87° \, 00'$$

Rounding off to the nearest degrees, we find $A = 44°$, $B = 49°$, and $C = 87°$.

EXERCISE 22-2

Solve the following triangles.

1. $a = 16, \quad b = 25, \quad c = 37$
2. $a = 62, \quad b = 78, \quad c = 86$
3. $a = 13, \quad b = 15, \quad c = 21$
4. $a = 11, \quad b = 13, \quad c = 20$
5. $a = 7.6, \quad b = 6.2, \quad c = 9.8$
6. $a = 0.23, \quad b = 0.31, \quad c = 0.34$
7. $a = 19.0, \quad b = 21.4, \quad c = 27.6$
8. $a = 409, \quad b = 437, \quad c = 532$
9. $a = 520, \quad b = 700, \quad c = 902$
10. $a = 360, \quad b = 210, \quad c = 217$
11. $a = 785.3, \quad b = 564.2, \quad c = 343.1$
12. $a = 40.09, \quad b = 50.17, \quad c = 70.01$
13. A force of 520 pounds just balances two forces of 313 and 412 pounds. Find the angles between the directions of the forces.
14. Each of three circles of radii 42, 60, and 72 is tangent externally to the other two. Find the angles of the triangle whose vertices are the centers of the circles.

15. An airplane travels due east at 200 miles per hour. If the airspeed of the plane is 170 miles per hour and the wind is blowing at 35 miles per hour, find the heading of the plane.

16. Notice in Fig. 22-1 that

$$\sin \tfrac{1}{2}A = \frac{r}{\sqrt{r^2 + (s - a)^2}} \quad \text{and} \quad \cos \tfrac{1}{2}A = \frac{s - a}{\sqrt{r^2 + (s - a)^2}}$$

Simplify the right members of these equations and obtain

$$\sin \tfrac{1}{2}A = \sqrt{\frac{(s - b)(s - c)}{bc}} \quad \text{and} \quad \cos \tfrac{1}{2}A = \sqrt{\frac{s(s - a)}{bc}}$$

Then write, without deriving, the corresponding formulas for these functions of $\frac{1}{2}B$ and $\frac{1}{2}C$.

23

MATRICES AND DETERMINANTS

23-1 MATRICES

In this chapter, we shall give a brief introduction to matrices and determinants, and apply these concepts in treating systems of linear equations.

Definition 23-1 A rectangular array of numbers forming m rows and n columns is called a *matrix*. Each number of a matrix is called an *element*.

Here, we list three examples of matrices, denoting each with a capital letter:

$$A = \begin{bmatrix} 2 & 4 \\ -3 & 7 \end{bmatrix}, \qquad B = \begin{bmatrix} 5 & 2 & 1 \\ 3 & 0 & 6 \end{bmatrix}, \qquad C = \begin{bmatrix} 1 & 2 & 4 \\ 0 & -5 & 6 \\ 7 & 8 & 0 \end{bmatrix}$$

It is customary to place brackets, parentheses, or double vertical bars around the elements of a matrix. Matrix A has 2 rows and 2 columns, B has 2 rows and 3 columns, and C has 3 rows and 3 columns. Accordingly they are said to be 2 by 2, 2 by 3, and 3 by 3 matrices, respectively. In general, a matrix of m rows and n columns is an m by n matrix, the number of rows given first.

We may represent the elements of a matrix by letters with one subscript or by a single letter with two subscripts. These methods of representation are illustrated by the following matrices.

$$A = \begin{bmatrix} a_1 & b_1 & c_1 \\ a_2 & b_2 & c_2 \\ a_3 & b_3 & c_3 \end{bmatrix} \qquad B = \begin{bmatrix} b_{11} & b_{12} & b_{13} \\ b_{21} & b_{22} & b_{23} \\ b_{31} & b_{32} & b_{33} \end{bmatrix}$$

The row of each element of A is given by the subscript and the column by a letter. The row of each element of B is given by the first subscript and the column by the second subscript.

Definition 23-2 Two matrices of the same dimensions (same number of rows and same number of columns) are said to be *equal* if and only if the corresponding elements are equal.

For example,

$$\begin{bmatrix} a_1 & b_1 & c_1 \\ a_2 & b_2 & c_2 \end{bmatrix} = \begin{bmatrix} 3 & 0 & -2 \\ 5 & 7 & -4 \end{bmatrix}$$

if and only if

$$a_1 = 3, \quad b_1 = 0, \quad c_1 = -2, \quad a_2 = 5, \quad b_2 = 7, \quad c_2 = -4$$

Definition 23-3 The *sum* of two matrices of the same dimensions is the matrix obtained by adding each element of one of the matrices to the corresponding element of the other matrix.

Thus,

$$\begin{bmatrix} 2 & -3 \\ 4 & 0 \end{bmatrix} + \begin{bmatrix} 9 & 3 \\ 6 & 2 \end{bmatrix} = \begin{bmatrix} 2+9 & -3+3 \\ 4+6 & 0+2 \end{bmatrix} = \begin{bmatrix} 11 & 0 \\ 10 & 2 \end{bmatrix}$$

Definition 23-4 The *product* of a number k and a matrix is the matrix obtained by multiplying each element of the matrix by k.

For example,

$$(-3)\begin{bmatrix} 8 & 7 & -2 & 1 \\ 2 & 5 & -8 & 9 \end{bmatrix} = \begin{bmatrix} -24 & -21 & 6 & -3 \\ -6 & -15 & 24 & -27 \end{bmatrix}$$

Definition 23-5 A matrix whose elements are all zero is called a *zero matrix.*

Thus, the matrices

$$[0], \qquad \begin{bmatrix} 0 \\ 0 \end{bmatrix}, \qquad \begin{bmatrix} 0 & 0 & 0 \\ 0 & 0 & 0 \end{bmatrix}$$

are zero matrices of dimensions 1 by 1, 2 by 1, and 2 by 3, respectively.

23-2 PRODUCT OF TWO MATRICES

We denote a matrix of m rows and n columns by the array

$$\begin{bmatrix} a_{11} & a_{12} & a_{13} & \cdots & a_{1n} \\ a_{21} & a_{22} & a_{23} & \cdots & a_{2n} \\ \cdot & \cdot & \cdot & \cdot & \cdot \\ a_{m1} & a_{m2} & a_{m3} & \cdots & a_{mn} \end{bmatrix}$$

The subscripts of a indicate the position of an element. Thus, a_{23} represents the element in the second row and third column and, more generally, a_{ij} represents the element in the ith row and the jth column. If $m = n$, the matrix is called a *square matrix of order n.* The elements $a_{11}, a_{22}, a_{33}, \ldots, a_{nn}$ extending from the upper left to the lower right corner of a square matrix form the *principal diagonal.*

Definition 23-6 If A is an m by n matrix and B is an n by p matrix, then the product AB is the m by p matrix C ($C = AB$) whose element c_{ij} in the ith row and jth column is given by

$$c_{ij} = a_{i1}b_{1j} + a_{i2}b_{2j} + a_{i3}b_{3j} + \cdots + a_{in}b_{nj}$$

According to this definition, we "multiply" the ith row of A by the jth column of B to find the element in the ith row and jth column of the product AB. Hence the number of columns in A must be equal to the number of rows in B. Matrix multiplication is not defined when this condition fails.

EXAMPLE 1 Find the product of the matrices

$$A = \begin{bmatrix} a_{11} & a_{12} \\ a_{21} & a_{22} \end{bmatrix} \quad \text{and} \quad B = \begin{bmatrix} b_{11} & b_{12} & b_{13} \\ b_{21} & b_{22} & b_{23} \end{bmatrix}$$

Solution: Applying the definition of a product, we have

$$C = \begin{bmatrix} a_{11} & a_{12} \\ a_{21} & a_{22} \end{bmatrix} \begin{bmatrix} b_{11} & b_{12} & b_{13} \\ b_{21} & b_{22} & b_{23} \end{bmatrix}$$

$$\begin{bmatrix} a_{11}b_{11} + a_{12}b_{21} & a_{11}b_{12} + a_{12}b_{22} & a_{11}b_{13} + a_{12}b_{23} \\ a_{21}b_{11} + a_{22}b_{21} & a_{21}b_{12} + a_{22}b_{22} & a_{21}b_{13} + a_{22}b_{23} \end{bmatrix}$$

EXAMPLE 2

$$\begin{bmatrix} 1 & -2 \\ 0 & 6 \\ -4 & 5 \end{bmatrix} \begin{bmatrix} 3 & 6 & -1 & 4 \\ 5 & 2 & 0 & 7 \end{bmatrix} = \begin{bmatrix} -7 & 2 & -1 & -10 \\ 30 & 12 & 0 & 42 \\ 13 & -14 & 4 & 19 \end{bmatrix}$$

The elements $-7, 2, -1, -10$ of the first row of the product of the two matrices come from the expressions

$$1 \cdot 3 + (-2)5, \quad 1 \cdot 6 + (-2)2, \quad 1(-1) + (-2)0, \quad 1 \cdot 4 + (-2)7$$

That is, we go across the first row of the first matrix and successively down the first, second, and third columns of the second matrix to obtain the elements of the first row of the product of the matrices. Then in a similar way we use the second and third rows of the first given matrix and columns of the second matrix to find the second and third rows of the product of the matrices.

Notice that the first matrix in this example has dimensions 3 by 2 and the second 2 by 3, making the number of columns in the first matrix equal to the number of rows in the second matrix. If we interchange the positions of the matrices, however, the first matrix would then have 4 columns and the second 3 rows, which means the product is not defined. Even though the products AB and BA

of two matrices exists, the results may or may not be the same. The next example illustrates the fact that matrix multiplication is, in general, not commutative.

EXAMPLE 3

$$\begin{bmatrix} 5 & -1 \\ 6 & 7 \end{bmatrix}\begin{bmatrix} 2 & 1 \\ 3 & 4 \end{bmatrix} = \begin{bmatrix} 5\cdot 2 + (-1)3 & 5\cdot 1 + (-1)4 \\ 6\cdot 2 + 7\cdot 3 & 6\cdot 1 + 7\cdot 4 \end{bmatrix} = \begin{bmatrix} 7 & 1 \\ 33 & 34 \end{bmatrix}$$

$$\begin{bmatrix} 2 & 1 \\ 3 & 4 \end{bmatrix}\begin{bmatrix} 5 & -1 \\ 6 & 7 \end{bmatrix} = \begin{bmatrix} 2\cdot 5 + 1\cdot 6 & 2(-1) + 1\cdot 7 \\ 3\cdot 5 + 4\cdot 6 & 3(-1) + 4\cdot 7 \end{bmatrix} = \begin{bmatrix} 16 & 5 \\ 39 & 25 \end{bmatrix}$$

EXERCISE 23-1

Given the matrices

$$A = \begin{bmatrix} 1 & 3 \\ 4 & 2 \end{bmatrix} \quad \text{and} \quad B = \begin{bmatrix} -2 & 5 \\ 4 & 3 \end{bmatrix}$$

find the matrices resulting from the indicated operations.

1. $A + B$ **2.** $B + A$ **3.** $A - B$

4. $2A + B$ **5.** AB **6.** BA

Compute the indicated products

7. $\begin{bmatrix} 1 & -2 \\ 3 & 2 \end{bmatrix}\begin{bmatrix} 1 & 2 & 3 \\ 3 & 2 & 1 \end{bmatrix}$

8. $\begin{bmatrix} 2 & 3 & 4 \\ 4 & 1 & 5 \end{bmatrix}\begin{bmatrix} 1 & -1 \\ 0 & -2 \\ 2 & 4 \end{bmatrix}$

9. $\begin{bmatrix} 1 & -1 \\ 0 & -2 \\ 2 & 4 \end{bmatrix}\begin{bmatrix} 2 & 3 & 4 \\ 4 & 1 & 5 \end{bmatrix}$

10. $\begin{bmatrix} 1 & -2 \\ 4 & -3 \\ 0 & 5 \end{bmatrix}\begin{bmatrix} 1 & 2 & 3 & 4 \\ 4 & 3 & 2 & 1 \end{bmatrix}$

11. $\begin{bmatrix} 3 \\ 2 \\ 1 \end{bmatrix}\begin{bmatrix} 1 & 3 & 5 & 6 \end{bmatrix}$

12. $\begin{bmatrix} -5 & -2 & 1 & 4 \end{bmatrix}\begin{bmatrix} 2 \\ 4 \\ 5 \\ 7 \end{bmatrix}$

13. $\begin{bmatrix} 1 & 2 & 4 \\ 2 & 3 & 5 \\ 1 & 0 & 1 \end{bmatrix} \begin{bmatrix} 2 & -1 & 5 \\ 4 & 0 & 1 \\ 3 & -3 & 0 \end{bmatrix}$ **14.** $\begin{bmatrix} 2 & -1 & 5 \\ 4 & 0 & 1 \\ 3 & -3 & 0 \end{bmatrix} \begin{bmatrix} 1 & 2 & 4 \\ 2 & 3 & 5 \\ 1 & 0 & 1 \end{bmatrix}$

Given the matrices

$$A = \begin{bmatrix} a & b \\ c & d \end{bmatrix} \qquad B = \begin{bmatrix} e & f \\ g & h \end{bmatrix} \qquad C = \begin{bmatrix} i & j \\ k & l \end{bmatrix}$$

prove that the following equations are true.

15. $A + B = B + A$

16. $(A + B) + C = A + (B + C)$

17. $A(B + C) = AB + AC$

18. $A(B - C) = AB - AC$

19. $k(A + B) = kA + kB$

20. $(m + n)A = mA + nA$

23-3 DETERMINANTS OF ORDERS TWO AND THREE

A matrix of n rows and n columns is called a square matrix of order n. With a square matrix we shall associate a number called the *determinant* of the matrix. The determinant of a matrix is said to be of the same order as the matrix. We define the determinant of the second-order matrix

$$\begin{bmatrix} a_1 & b_1 \\ a_2 & b_2 \end{bmatrix}$$

to be the number $a_1b_2 - a_2b_1$. To symbolize this definition, we write

$$\det \begin{bmatrix} a_1 & b_1 \\ a_2 & b_2 \end{bmatrix} = \begin{vmatrix} a_1 & b_1 \\ a_2 & b_2 \end{vmatrix} = a_1b_2 - a_2b_1$$

It is customary to indicate the determinant of a matrix by enclosing the elements of the matrix by vertical bars instead of brackets.

EXAMPLE 1

$$\begin{vmatrix} 3 & 4 \\ -2 & 5 \end{vmatrix} = 3(5) - (-2)(4) = 15 + 8 = 23$$

We next show how to express the solution of a system of linear equations by the use of determinants. The system

$$\begin{aligned} a_1x + b_1y &= c_1 \\ a_2x + b_2y &= c_2 \end{aligned} \tag{1}$$

may be solved by either method of Sec. 6-6 to yield

$$x = \frac{b_2c_1 - b_1c_2}{a_1b_2 - a_2b_1} \qquad y = \frac{a_1c_2 - a_2c_1}{a_1b_2 - a_2b_1} \tag{2}$$

The numerator and denominator of each of these fractions are the numbers associated with matrices of order two. Hence, we may express the solution in the symbolic form

$$x = \frac{\begin{vmatrix} c_1 & b_1 \\ c_2 & b_2 \end{vmatrix}}{\begin{vmatrix} a_1 & b_1 \\ a_2 & b_2 \end{vmatrix}} \qquad y = \frac{\begin{vmatrix} a_1 & c_1 \\ a_2 & c_2 \end{vmatrix}}{\begin{vmatrix} a_1 & b_1 \\ a_2 & b_2 \end{vmatrix}} \tag{3}$$

These formulas may be applied to find the solution of any consistent system of the form (1), that is, systems for which $a_1b_2 - a_2b_1 \neq 0$.

EXAMPLE 2 Solve the system of equations

$$\begin{aligned} 3x + 2y &= 3 \\ 2x - 5y &= -17 \end{aligned}$$

Solution: Taking $a_1 = 3$, $b_1 = 2$, $c_1 = 3$ and $a_2 = 2$, $b_2 = -5$, $c_2 = -17$ and substituting in formulas (3), we get

$$x = \frac{\begin{vmatrix} 3 & 2 \\ -17 & -5 \end{vmatrix}}{\begin{vmatrix} 3 & 2 \\ 2 & -5 \end{vmatrix}} = \frac{-15 + 34}{-15 - 4} = \frac{19}{-19} = -1$$

$$y = \frac{\begin{vmatrix} 3 & 3 \\ 2 & -17 \end{vmatrix}}{\begin{vmatrix} 3 & 2 \\ 2 & -5 \end{vmatrix}} = \frac{-51 - 6}{-15 - 4} = \frac{-57}{-19} = 3$$

We define the determinant of a third-order matrix by the equation

$$\begin{vmatrix} a_1 & b_1 & c_1 \\ a_2 & b_2 & c_2 \\ a_3 & b_3 & c_3 \end{vmatrix} = a_1 \begin{vmatrix} b_2 & c_2 \\ b_3 & c_3 \end{vmatrix} - b_1 \begin{vmatrix} a_2 & c_2 \\ a_3 & c_3 \end{vmatrix} + c_1 \begin{vmatrix} a_2 & b_2 \\ a_3 & b_3 \end{vmatrix}$$

In this equation, we have a symbol for the determinant of a third-order matrix and symbols for the determinants of three second-order matrices. Symbols of this kind, though denoting values of determinants of matrices, are often referred to simply as determinants.

EXAMPLE 3 Find the value of the determinant

$$\begin{vmatrix} 2 & 5 & 2 \\ 1 & -3 & 1 \\ -4 & 1 & -5 \end{vmatrix}$$

Solution: Employing the definition of a third-order determinant and evaluating the resulting second-order determinants, we find

$$\begin{vmatrix} 2 & 5 & 2 \\ 1 & -3 & 1 \\ -4 & 1 & -5 \end{vmatrix} = 2 \begin{vmatrix} -3 & 1 \\ 1 & -5 \end{vmatrix} - 5 \begin{vmatrix} 1 & 1 \\ -4 & -5 \end{vmatrix} + 2 \begin{vmatrix} 1 & -3 \\ -4 & 1 \end{vmatrix}$$

$$= 2(15 - 1) - 5(-5 + 4) + 2(1 - 12)$$

$$= 28 + 5 - 22$$

$$= 11$$

EXERCISE 23-2

Evaluate each of the following determinants.

1. $\begin{vmatrix} 3 & 4 \\ 9 & -7 \end{vmatrix}$ **2.** $\begin{vmatrix} 5 & -6 \\ 4 & 3 \end{vmatrix}$ **3.** $\begin{vmatrix} 4 & -8 \\ 7 & 11 \end{vmatrix}$

4. $\begin{vmatrix} a & 5 \\ 3a & 2 \end{vmatrix}$ 5. $\begin{vmatrix} 5 & 20 \\ 0 & -1 \end{vmatrix}$ 6. $\begin{vmatrix} x & y \\ 5 & 6 \end{vmatrix}$

Solve the following systems of equations for x and y, using determinants.

7. $4x + 3y = 1$
   $2x + 5y = 11$

8. $2x - 4y = -3$
   $4x + 2y = 9$

9. $3x - 4y = 7$
   $5x - 5y = 8$

10. $2x = 7y + 17$
    $4x = 5y + 25$

11. $7x = 4 - 3y$
    $4x = 3 - 2y$

12. $5x + 7y = -2$
    $4x + 6y = -3$

13. $ax - 3by = c$
    $2ax + 2by = 9c$

14. $ax - by = 2b$
    $3ax + 4by = 7a - b$

15. $3x + ay = 4$
    $2x + ay = 2$

Evaluate each of the third-order determinants.

16. $\begin{vmatrix} 3 & 4 & 5 \\ 2 & 1 & 4 \\ 3 & 2 & 1 \end{vmatrix}$ 17. $\begin{vmatrix} 1 & -3 & 2 \\ 4 & -1 & 2 \\ 3 & 5 & 2 \end{vmatrix}$ 18. $\begin{vmatrix} 0 & 2 & 6 \\ 1 & 5 & 0 \\ 3 & 7 & 1 \end{vmatrix}$

19. $\begin{vmatrix} -3 & 5 & 1 \\ -4 & 1 & 3 \\ 3 & 1 & 0 \end{vmatrix}$ 20. $\begin{vmatrix} 6 & 7 & 0 \\ 0 & -9 & 2 \\ 4 & 0 & -5 \end{vmatrix}$ 21. $\begin{vmatrix} 0 & 3 & 4 \\ 2 & 0 & 1 \\ 5 & 6 & 0 \end{vmatrix}$

Solve the following equations for x.

22. $\begin{vmatrix} x & 14 \\ 3 & 11 \end{vmatrix} = 2$ 23. $\begin{vmatrix} 4 & -10 \\ x & -3 \end{vmatrix} = 8$

24. $\begin{vmatrix} 2 - x & 3 \\ 2 + x & 4 \end{vmatrix} = 0$ 25. $\begin{vmatrix} 6 & -x \\ x - 4 & -2 \end{vmatrix} = 0$

26. $\begin{vmatrix} 2 & x & 1 \\ 1 & 0 & 1 \\ 3 & 4 & 2 \end{vmatrix} = 0$ 27. $\begin{vmatrix} 1 & 4 & -3 \\ 4 & x & -5 \\ 2 & 0 & -3 \end{vmatrix} = 0$

23-4 DETERMINANTS OF ORDER n

We have defined determinants of orders two and three. We wish now to give a general definition of determinants which applies to those of orders one, two, three, and higher orders. First, however, let us explain the notation of inversions which is helpful in defining and studying the properties of determinants. In a sequence of positive integers, an *inversion* occurs whenever a larger number precedes a smaller number. Thus, the sequence 2143 has two inversions because 2 precedes 1, and 4 precedes 3. But 5143 has four inversions; 5 preceding 1, 4, and 3 furnishes three inversions, and 4 preceding 3 gives another inversion.

Definition 23-7 If A is an n by n matrix, the determinant of the matrix is denoted symbolically by

$$\begin{vmatrix} a_{11} & a_{12} & a_{13} & \cdots & a_{1n} \\ a_{21} & a_{22} & a_{23} & \cdots & a_{2n} \\ a_{31} & a_{32} & a_{33} & \cdots & a_{3n} \\ \vdots & \vdots & \vdots & \vdots & \vdots \\ a_{n1} & a_{n2} & a_{n3} & \cdots & a_{nn} \end{vmatrix}$$

The subscripts here indicate the row and column of each number, the first subscript giving the row and the second the column. Thus, a_{34} stands for the number in the third row and fourth column, and a_{ij} stands for the number in the ith row and jth column.

The determinant stands for the number which is the algebraic sum of all possible products obtained:

1. By taking as factors one and only one number from each row and each column, and
2. By giving to each product a plus or minus sign according to whether there is an even or odd number of inversions of the second (the column) subscripts when the factors of the product are written so that the first (the row) subscripts have no inversions.

Each of the numbers in the square array is called an *element* of the determinant. The elements along the line from the upper left to the lower right corner form the *principal diagonal.* Each of the signed products described in the definition is called a *term* of the determinant and the sum of the terms is the *expansion,* or *value,* of the determinant. There are, of course, n factors in each term since one and only one factor is taken from each row and each column.

According to the preceding definition, the expansion of a third-order determinant is given by the equation

$$\begin{vmatrix} a_{11} & a_{12} & a_{13} \\ a_{21} & a_{22} & a_{23} \\ a_{31} & a_{32} & a_{33} \end{vmatrix} = a_{11}a_{22}a_{33} - a_{11}a_{23}a_{32} - a_{12}a_{21}a_{33} + a_{12}a_{23}a_{31} + a_{13}a_{21}a_{32} - a_{13}a_{22}a_{31}$$

In this expansion each product has exactly one factor from each row and each column, and the first subscripts are in the order 1, 2, 3. The inversions of the second subscripts of the factors of the terms in the order written are respectively zero, one, one, two, two, three. The signs of the terms with an even number of inversions are positive and those with an odd number of inversions are negative.

23-5 PROPERTIES OF DETERMINANTS

From the definition of determinants a number of interesting and useful properties can be established. In this section, we state some of the elementary properties as theorems, including those which lead to a practical method for evaluating determinants. Because the proofs would be somewhat lengthy we simply illustrate, rather than establish, Theorems 23-2 and 23-4.

Theorem 23-1 The expansion of a determinant of order n has $n!$ terms.

Proof: If the n factors of a term are written with the first subscripts in order of magnitude, then any arrangement of the integers from 1 to n, inclusive, as the second subscripts yield the factors of one term. But the n subscripts can be assigned in $n!$ different ways (Sec. 15-2). Since there is a term for each arrangement of the second subscripts, the expansion has $n!$ terms.

Theorem 23-2 The value of a determinant is not altered if corresponding rows and columns are interchanged.

We illustrate this theorem by the equation

$$\begin{vmatrix} 1 & 4 & 8 \\ 2 & 0 & 9 \\ 3 & 5 & 6 \end{vmatrix} = \begin{vmatrix} 1 & 2 & 3 \\ 4 & 0 & 5 \\ 8 & 9 & 6 \end{vmatrix}$$

Since the interchange of corresponding rows and columns of a determinant does not change its value, a property of a determinant involving its rows is also a property involving its columns, and vice versa. Hence, the proof of a property pertaining to rows or columns need be made for only one of the two.

Theorem 23-3 If each of the elements of a row (or column) of a determinant is zero, then the value of the determinant is zero.

Proof: This property follows at once from the fact that each term of the expansion has a factor from every row and every column of the determinant. A product in which zero is a factor is, of course, zero.

Theorem 23-4 If two columns (or rows) of a determinant are interchanged, the sign of the determinant is changed.

Thus, interchanging the first and third columns of the determinant on the left, we have the equation

$$\begin{vmatrix} 3 & 8 & 2 \\ 1 & 0 & 4 \\ 5 & 9 & 7 \end{vmatrix} = -\begin{vmatrix} 2 & 8 & 3 \\ 4 & 0 & 1 \\ 7 & 9 & 5 \end{vmatrix}$$

Theorem 23-5 If two columns (or rows) of a determinant are identical, the value of the determinant is zero.

Proof: If the two identical columns are interchanged, the value of the determinant is unchanged. But, by Theorem 23-4, the sign of the determinant is reversed. Hence, if D is the value of the determinant, the value of the determinant after the interchange of the identical columns is $-D$, then we have $D = -D$, $2D = 0$, and $D = 0$.

Theorem 23-6 If each of the elements of a column (or row) of a determinant is multiplied by a number m, the value of the determinant is multiplied by m.

Proof: Since each term of the expansion has one and only one factor from each row and each column, m is a factor of each term of the expansion of the new determinant. Then the expansion can be written as m times the expansion of the first determinant.

EXAMPLE

$$\begin{vmatrix} 2 & 5m & 1 \\ 4 & m & 2 \\ 3 & 8m & 7 \end{vmatrix} = m\begin{vmatrix} 2 & 5 & 1 \\ 4 & 1 & 2 \\ 3 & 8 & 7 \end{vmatrix}$$

Corollary If each element of a column (or row) of a determinant is a constant times the corresponding elements of another column (or row), the value of the determinant is zero.

We leave the proof of this corollary to the student.

Theorem 23-7 If each element of a column, the kth say, of a determinant is expressed as the sum of two terms, the determinant is equal to the sum of two determinants, where

1. The elements of each of the two determinants are identical to the corresponding elements of the given determinant except for the elements of the kth column.
2. The first terms of the elements of the kth column of the given determinant form the kth column of one of the two determinants and the second terms form the kth column of the other determinant.

Proof: Consider the determinants in the equation

$$\begin{vmatrix} a_{11}+b_{11} & a_{12} & a_{13} \\ a_{21}+b_{21} & a_{22} & a_{23} \\ a_{31}+b_{31} & a_{32} & a_{33} \end{vmatrix} = \begin{vmatrix} a_{11} & a_{12} & a_{13} \\ a_{21} & a_{22} & a_{23} \\ a_{31} & a_{32} & a_{33} \end{vmatrix} + \begin{vmatrix} b_{11} & a_{12} & a_{13} \\ b_{21} & a_{22} & a_{23} \\ b_{31} & a_{32} & a_{33} \end{vmatrix}$$

Each term of the expansion of the determinant on the left side contains the sum of two terms as a factor, which is equal to the sum of the corresponding terms of the determinants on the right side. For example, $(a_{11} + b_{11})a_{22}a_{33}$ is a term of the left side of the equation, and the corresponding terms, $a_{11}a_{22}a_{33}$ and $b_{11}a_{22}a_{33}$, appear on the right side. Hence, the expansion of the left side of the equation is equal to the sum of the expansions on the right side.

The reasoning which we have used here applies in the case of a determinant of any order. We conclude, therefore, that the theorem is true.

Theorem 23-8 The value of a determinant is not changed if each element of a column (or row) is multiplied by a number m and added to the corresponding elements of any column (or row).

Proof: Let us show that

$$D = \begin{vmatrix} a_{11} & a_{12} & a_{13} + ma_{11} \\ a_{21} & a_{22} & a_{23} + ma_{21} \\ a_{31} & a_{32} & a_{33} + ma_{31} \end{vmatrix} = \begin{vmatrix} a_{11} & a_{12} & a_{13} \\ a_{21} & a_{22} & a_{23} \\ a_{31} & a_{32} & a_{33} \end{vmatrix}$$

By Theorem 23-7, the determinant on the left may be written as the sum of two determinants; that is,

$$D = \begin{vmatrix} a_{11} & a_{12} & a_{13} \\ a_{21} & a_{22} & a_{23} \\ a_{31} & a_{32} & a_{33} \end{vmatrix} + \begin{vmatrix} a_{11} & a_{12} & ma_{11} \\ a_{21} & a_{22} & ma_{21} \\ a_{31} & a_{32} & ma_{31} \end{vmatrix}$$

But the second of these determinants is equal to zero since it may be expressed as m times a determinant with the first and third columns identical.

Inasmuch as the method of proof for a third-order determinant is applicable for the case of a determinant of any order, we conclude that the theorem is true.

EXAMPLE Without expanding the determinant find the root of the equation

$$\begin{vmatrix} 3 & 2 & 4 \\ 4 & -1 & -2 \\ 5 & x & 4 \end{vmatrix} = 0$$

Solution: If we replace x by 2, the elements of the third column become twice the corresponding elements of the second column, and the value of the determinant zero. Why zero? Hence, $x = 2$ is the root of the equation.

EXERCISE 23-3

Without expanding show that the value of each of the following determinants is zero.

1. $\begin{vmatrix} 3 & -1 & 3 \\ 2 & 4 & 2 \\ 5 & -3 & 5 \end{vmatrix}$ **2.** $\begin{vmatrix} 1 & 6 & -6 \\ 5 & 2 & -2 \\ 4 & -3 & 3 \end{vmatrix}$ **3.** $\begin{vmatrix} 1 & 2 & 3 \\ 2 & 4 & 6 \\ 0 & 2 & 1 \end{vmatrix}$

4. $\begin{vmatrix} 1 & 2 & -5 \\ 2 & 4 & -4 \\ 0 & 0 & -3 \end{vmatrix}$ **5.** $\begin{vmatrix} 0 & 1 & 0 \\ 5 & -6 & -7 \\ 0 & -4 & 0 \end{vmatrix}$ **6.** $\begin{vmatrix} 7 & 6 & 5 \\ 6 & 5 & 4 \\ 5 & 4 & 3 \end{vmatrix}$

Without expanding the determinants, show that the following equations are true.

7. $$\begin{vmatrix} 4 & 1 & 6 \\ 2 & 3 & 15 \\ 6 & 0 & 3 \end{vmatrix} = 6\begin{vmatrix} 2 & 1 & 2 \\ 1 & 3 & 5 \\ 3 & 0 & 1 \end{vmatrix}$$

8. $$\begin{vmatrix} 5 & -3 & 1 \\ 4 & 3 & 1 \\ 1 & -2 & 6 \end{vmatrix} = \begin{vmatrix} 9 & 0 & 2 \\ 4 & 3 & 1 \\ 1 & -2 & 6 \end{vmatrix}$$

9. $$\begin{vmatrix} 6 & 4 & 0 \\ 3 & 2 & 1 \\ 5 & 3 & 1 \end{vmatrix} = \begin{vmatrix} 2 & 4 & 4 \\ 1 & 2 & 3 \\ 2 & 3 & 4 \end{vmatrix}$$

10. $$\begin{vmatrix} 2 & 4 & 5 \\ -3 & 7 & 1 \\ 5 & 6 & 0 \end{vmatrix} + \begin{vmatrix} 3 & 4 & 5 \\ 5 & 7 & 1 \\ 4 & 6 & 0 \end{vmatrix} = \begin{vmatrix} 5 & 4 & 5 \\ 2 & 7 & 1 \\ 9 & 6 & 0 \end{vmatrix}$$

Without expanding the determinants, solve the following equations for x.

11. $\begin{vmatrix} 2 & 1 & 4 \\ 6 & x & 12 \\ 5 & 0 & 3 \end{vmatrix} = 0$ **12.** $\begin{vmatrix} 2 & x & -1 \\ 4 & x^2 & 1 \\ 5 & 5 & 5 \end{vmatrix} = 0$

13. $\begin{vmatrix} 2 & 4 & 8 \\ x & x^2 & x^3 \\ -2 & 4 & -8 \end{vmatrix} = 0$ **14.** $\begin{vmatrix} x & a & -3 \\ x^2 & a^2 & 9 \\ 2 & 2 & 2 \end{vmatrix} = 0$

23-6 EXPANSION BY COFACTORS

The definition of the determinant of a matrix (Definition 23-7) describes its expansion and hence provides a way for evaluating the determinant. But evaluating a determinant of order four or higher directly from the definition would be quite tedious. In this section we shall develop a practical method for finding the values of determinants. As a preliminary step, however, we need to introduce the idea of a minor and cofactor corresponding to an element of a determinant.

Definition 23-8 The *minor* of an element of a determinant of order n is the determinant of order $n - 1$ which remains after the row and column of the element are removed.

EXAMPLE 1 Denoting the minors if a_{23} and a_{41} of the determinant

$$\begin{vmatrix} a_{11} & a_{12} & a_{13} & a_{14} \\ a_{21} & a_{22} & a_{23} & a_{24} \\ a_{31} & a_{32} & a_{33} & a_{34} \\ a_{41} & a_{42} & a_{43} & a_{44} \end{vmatrix} \tag{1}$$

by A_{23} and A_{41}, we have

$$A_{23} = \begin{vmatrix} a_{11} & a_{12} & a_{14} \\ a_{31} & a_{32} & a_{34} \\ a_{41} & a_{42} & a_{44} \end{vmatrix} \quad \text{and} \quad A_{41} = \begin{vmatrix} a_{12} & a_{13} & a_{14} \\ a_{22} & a_{23} & a_{24} \\ a_{32} & a_{33} & a_{34} \end{vmatrix}$$

Definition 23-9 The *cofactor* of an element a_{ij} of a determinant is the product of the minor of a_{ij} and $(-1)^{i+j}$; that is, if the minor and cofactor of a_{ij} are denoted respectively by A_{ij} and C_{ij}, then $C_{ij} = (-1)^{i+j}A_{ij}$.

According to this definition, the cofactors of a_{23}, a_{41}, and a_{42} of the determinant above are respectively $C_{23} = -A_{23}$, $C_{41} = -A_{41}$, and $C_{42} = A_{42}$.

Theorem 23-9 The value of a determinant may be obtained by the following steps:

1. Multiply each element of a selected row (or column) by its cofactor.
2. Then take the algebraic sum of the resulting products.

Before proving this theorem, we illustrate its meaning by referring to the fourth-order determinant of Example 1. The value of the determinant may be expressed by

$$a_{11}C_{11} + a_{12}C_{12} + a_{13}C_{13} + a_{14}C_{14}$$

which is called the *expansion according to the elements of the first row.* Similarly the expansion according to the elements of the third column is

$$a_{13}C_{13} + a_{23}C_{23} + a_{33}C_{33} + a_{43}C_{43}$$

Proof: We first point out that the terms of a determinant D having a_{11} as a factor are given by $a_{11}A_{11}$, where A_{11} is the minor of a_{11}. This is true because each term of the expansion of D has one and only one factor from each row and each column, and consequently the terms of D involving a_{11} can have no other factor from the first row and first column. Now if each term of A_{11} is multiplied by a_{11}, no change is made in the number of inversions of the subscripts of any product. Hence, the product of a_{11} and the expansion of A_{11} are terms of D. Thus, we see that the product of the element in the first row and column and its minor are terms of D involving that element.

Next, we consider the product of an element and its minor if the element is in the ith row and jth column. This element a_{ij} can be transferred to the first row and first column in the following way:

1. First bring the ith row to row one by $i - 1$ interchanges of adjacent rows.
2. Then bring the jth column to column one by $j - 1$ interchanges of adjacent columns.

We should notice that these interchanges involve adjacent rows or adjacent columns, and therefore do not disturb the minor of a_{ij}. Furthermore, each interchange reverses the sign of the determinant (Theorem 23-4). Since the total number of interchanges is $i + j - 2$, the final determinant D' and the original determinant are related by the equation

$$D' = (-1)^{i+j-2}D = (-1)^{i+j}D$$

We conclude, then, that the terms of the expansion of D with a_{ij} as a factor are given by

$$a_{ij}(-1)^{i+j}A_{ij} = a_{ij}C_{ij}$$

where A_{ij} is the minor of a_{ij} and C_{ij} is the cofactor.

We recall that each term of the expansion of a determinant has as a factor one and only one element from each row and each column. Hence, the sum of the products obtained by multiplying each element of any selected row (or column) by its cofactor yields all the terms of the expansion. This completes the proof of the theorem.

Expanding by cofactors, we can express a determinant of order n in terms of determinants of order $n-1$. Each of the resulting determinants can then be expressed in terms of determinants of order $n-2$. A continuation of this process leads to determinants of order 2, whose expansions can be easily written. The evaluation of determinants, however, can be greatly facilitated by employing certain theorems which were established in Sec. 23-5. A particularly helpful device is to obtain a determinant whose elements, except one, of some row or column, are zero. We shall illustrate with examples.

EXAMPLE 2 Evaluate the determinant

$$D = \begin{vmatrix} 2 & -3 & 4 \\ 3 & 5 & -1 \\ 7 & 6 & -2 \end{vmatrix}$$

Solution: Let us obtain an equivalent determinant which has all the elements, except one, of some row or column equal to zero. The element -1 in the second row, third column, makes this end easily achieved. We multiply the elements of the second row by 4 and add the products to the corresponding elements of the first row. Next, we multiply the elements of the second row by -2 and add the products to the corresponding elements of the third row. These operations do not alter the value of the determinant (Theorem 23-8), and therefore

$$D = \begin{vmatrix} 2 & -3 & 4 \\ 3 & 5 & -1 \\ 7 & 6 & -2 \end{vmatrix} = \begin{vmatrix} 14 & 17 & 0 \\ 3 & 5 & -1 \\ 1 & -4 & 0 \end{vmatrix}$$

Taking advantage of the zero elements, we expand the new determinant according to the elements of the third column and have

$$D = 0 - (-1)\begin{vmatrix} 14 & 17 \\ 1 & -4 \end{vmatrix} + 0 = 14(-4) - 1(17) = -73$$

Notice that the sum of the row number and the column number of the element -1, given by $2 + 3 = 5$, is odd. Consequently, the cofactor of this element is the negative of the minor.

EXAMPLE 3 Evaluate the determinant

$$D = \begin{vmatrix} 2 & -12 & 7 & 5 \\ -4 & 6 & 4 & -3 \\ 10 & 0 & 1 & 2 \\ 6 & 9 & 5 & 0 \end{vmatrix}$$

Solution: We first factor 2 from the elements of column one and 3 from the elements of column two, thus getting

$$D = 6\begin{vmatrix} 1 & -4 & 7 & 5 \\ -2 & 2 & 4 & -3 \\ 5 & 0 & 1 & 2 \\ 3 & 3 & 5 & 0 \end{vmatrix}$$

To obtain additional zeros in the third row, we multiply the elements of the third column by 5 and subtract the products from the corresponding elements of the first column, and next multiply the elements of the third column by 2 and subtract the products from the corresponding elements of the fourth column. This gives the determinant

$$D = 6\begin{vmatrix} -34 & -4 & 7 & -9 \\ -22 & 2 & 4 & -11 \\ 0 & 0 & 1 & 0 \\ -22 & 3 & 5 & -10 \end{vmatrix}$$

We now expand according to the elements of the third row and have

$$D = 6\begin{vmatrix} -34 & -4 & -9 \\ -22 & 2 & -11 \\ -22 & 3 & -10 \end{vmatrix}$$

We next subtract the elements of the second row from the corresponding elements of the third row; and, in the resulting determinant, subtract the elements of the third column from the corresponding elements of the second column. These operations give

$$D = 6\begin{vmatrix} -34 & -4 & -9 \\ -22 & 2 & -11 \\ 0 & 1 & 1 \end{vmatrix} = 6\begin{vmatrix} -34 & 5 & -9 \\ -22 & 13 & -11 \\ 0 & 0 & 1 \end{vmatrix}$$

and

$$D = 6\begin{vmatrix} -34 & 5 \\ -22 & 13 \end{vmatrix} = 6(-442 + 110) = -1992$$

EXERCISE 23-4

Evaluate the following determinants.

1. $\begin{vmatrix} 1 & 2 & 3 \\ 6 & 4 & -5 \\ 4 & 0 & 1 \end{vmatrix}$

2. $\begin{vmatrix} 6 & -3 & 7 \\ 7 & 5 & -9 \\ 6 & 3 & 7 \end{vmatrix}$

3. $\begin{vmatrix} 5 & 0 & 1 \\ 3 & 4 & 2 \\ 6 & 1 & 3 \end{vmatrix}$

4. $\begin{vmatrix} 0 & 5 & 4 & 1 \\ 3 & 2 & 3 & -2 \\ 0 & 1 & 2 & 3 \\ 3 & 0 & 1 & 4 \end{vmatrix}$

5. $\begin{vmatrix} 6 & 8 & 7 & 2 \\ 3 & 2 & 4 & 1 \\ 2 & 3 & 8 & 2 \\ 5 & 4 & 2 & 1 \end{vmatrix}$

6. $\begin{vmatrix} 3 & 2 & 2 & 3 \\ 3 & 1 & 1 & 1 \\ 6 & 2 & 2 & 1 \\ 9 & 0 & 1 & 1 \end{vmatrix}$

7. $\begin{vmatrix} 2 & -1 & 2 & 0 \\ 0 & -2 & 1 & 6 \\ 6 & 0 & 0 & 1 \\ 4 & 3 & 4 & 2 \end{vmatrix}$

8. $\begin{vmatrix} 4 & 8 & 8 & 4 \\ 1 & 2 & 0 & 2 \\ 0 & 1 & 2 & 3 \\ 2 & 4 & 1 & 4 \end{vmatrix}$

9. $\begin{vmatrix} 3 & 2 & 3 & 0 \\ 2 & 3 & 0 & 2 \\ 1 & 0 & 1 & 1 \\ 0 & 1 & 2 & 3 \end{vmatrix}$

10. $\begin{vmatrix} 3 & -9 & 0 & 12 \\ -2 & 1 & 3 & 0 \\ 3 & -2 & 1 & -3 \\ 0 & 2 & -3 & 1 \end{vmatrix}$

11. $\begin{vmatrix} 6 & 6 & -1 & 1 \\ 3 & -1 & 1 & 1 \\ 3 & 1 & 6 & 1 \\ 6 & -1 & 1 & -6 \end{vmatrix}$

12. $\begin{vmatrix} 1 & -2 & 1 & -1 & 3 \\ 1 & 2 & 2 & 1 & 1 \\ 2 & 1 & -1 & -3 & 2 \\ 1 & 1 & 3 & 2 & 1 \\ 2 & -1 & 1 & 1 & 1 \end{vmatrix}$

13. $\begin{vmatrix} 1 & 2 & 2 & 1 & 3 \\ 2 & 1 & 1 & 1 & 1 \\ 3 & 2 & 1 & 2 & 1 \\ 1 & 3 & 2 & 1 & 1 \\ 1 & 1 & 3 & 3 & 1 \end{vmatrix}$

14. Let the vertices of a triangle, reading counterclockwise, be $A(x_1, y_1)$, $B(x_2, y_2)$, and $C(x_3, y_3)$, as in Fig. 23-1. Observe that *DECA*, *EFBC*, and *DFBA* are trapezoids. The sum of the areas of the first two trapezoids minus the area of the third trapezoid is equal to the area of triangle *ABC*. Recalling that the area of a trapezoid is equal to half the sum of

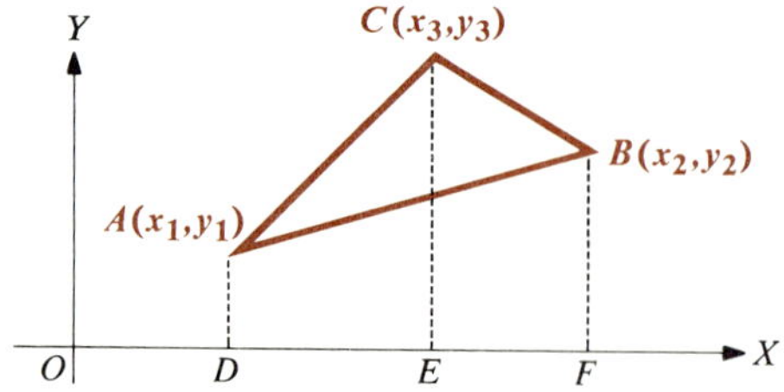

Figure 23-1

the parallel sides times the altitude, we have area $DECA = \frac{1}{2}(y_1 + y_3)(x_3 - x_1)$. With this start, show that the area S of the triangle is

$$S = \tfrac{1}{2}[x_1(y_2 - y_3) - y_1(x_2 - x_3) + (x_2y_3 - x_3y_2)]$$

By expanding the determinant, show that the area of the triangle can be expressed as

$$S = \frac{1}{2}\begin{vmatrix} x_1 & y_1 & 1 \\ x_2 & y_2 & 1 \\ x_3 & y_3 & 1 \end{vmatrix}$$

What change is made in this result if the vertices are numbered in the clockwise direction?

Using the determinant of problem 14, find the area of each triangle having the given vertices.

15. $A(0, 0)$, $B(6, 0)$, $C(4, 3)$

16. $A(-2, 4)$, $B(2, -6)$, $C(5, 4)$

17. $A(2, 7)$, $B(1, 1)$, $C(10, 8)$

18. $A(5, -1)$, $B(-1, 4)$, $C(3, 6)$

19. $A(1, 5)$, $B(6, 1)$, $C(8, 7)$

20. $A(-2, -3)$, $B(3, 2)$, $C(-1, -8)$

23-7 SOLUTION OF SYSTEMS OF LINEAR EQUATIONS BY DETERMINANTS

In Sec. 23-3 we solved systems of linear equations in two unknowns by the use of determinants. Determinants may also be used in solving systems of n linear equations in n unknowns. We shall establish the method for the case when $n = 4$. Consider the system

$$\begin{aligned} a_1x + b_1y + c_1z + d_1w &= e_1 \\ a_2x + b_2y + c_2z + d_2w &= e_2 \\ a_3x + b_3y + c_3z + d_3w &= e_3 \\ a_4x + b_4y + c_4z + d_4w &= e_4 \end{aligned}$$

We shall use D to stand for the determinant of the matrix of the coefficients of the unknowns x, y, z, and w. And N_x, N_y, N_z, and N_w will stand for the four determinants obtained from D by replacing the coefficients of x, y, z, and w, in turn, by the corresponding constant terms. Then the solution of the given system of equations, as we shall show, may be written as

$$x = \frac{N_x}{D}, \qquad y = \frac{N_y}{D}, \qquad z = \frac{N_z}{D}, \qquad w = \frac{N_w}{D}$$

if $D \neq 0$. If we multiply both members of

$$D = \begin{vmatrix} a_1 & b_1 & c_1 & d_1 \\ a_2 & b_2 & c_2 & d_2 \\ a_3 & b_3 & c_3 & d_3 \\ a_4 & b_4 & c_4 & d_4 \end{vmatrix}$$

by x, we get

$$Dx = \begin{vmatrix} a_1x & b_1 & c_1 & d_1 \\ a_2x & b_2 & c_2 & d_2 \\ a_3x & b_3 & c_3 & d_3 \\ a_4x & b_4 & c_4 & d_4 \end{vmatrix}$$

Now we multiply the second column of the determinant by y, the third column by z, and the fourth column by w and add each of the products to the first column to obtain

$$Dx = \begin{vmatrix} a_1x + b_1y + c_1z + d_1w & b_1 & c_1 & d_1 \\ a_2x + b_2y + c_2z + d_2w & b_2 & c_2 & d_2 \\ a_3x + b_3y + c_3z + d_3w & b_3 & c_3 & d_3 \\ a_4x + b_4y + c_4z + d_4w & b_4 & c_4 & d_4 \end{vmatrix}$$

The elements of the first column of this determinant are the left members of the given system of equations, and hence may be replaced by their equals as found in the right members. We have then

$$Dx = \begin{vmatrix} e_1 & b_1 & c_1 & d_1 \\ e_2 & b_2 & c_2 & d_2 \\ e_3 & b_3 & c_3 & d_3 \\ e_4 & b_4 & c_4 & d_4 \end{vmatrix} = N_x$$

and $x = N_x/D$.

Similarly, we may find

$$y = \frac{N_y}{D}, \qquad z = \frac{N_z}{D}, \qquad w = \frac{N_w}{D}$$

Then we have a solution of the given system of equations if these values for the unknowns x, y, z, and w satisfy each of the equations. Suppose that we check the first equation. The remaining equations could be checked in the same way. Substituting, we have

$$\frac{a_1N_x}{D} + \frac{b_1N_y}{D} + \frac{c_1N_z}{D} + \frac{d_1N_w}{D} = e_1$$

Now transposing the constant term to the left side and multiplying by D gives

$$a_1N_x + b_1N_y + c_1N_z + d_1N_w - e_1D = 0$$

If this expression on the left is equal to zero, the first equation of the given system is satisfied. The expression is the expansion of a determinant; that is,

$$a_1N_x + b_1N_y + c_1N_z + d_1N_w - e_1D = \begin{vmatrix} a_1 & e_1 & b_1 & c_1 & d_1 \\ a_1 & e_1 & b_1 & c_1 & d_1 \\ a_2 & e_2 & b_2 & c_2 & d_2 \\ a_3 & e_3 & b_3 & c_3 & d_3 \\ a_4 & e_4 & b_4 & c_4 & d_4 \end{vmatrix}$$

as may be verified by expanding the determinant according to the elements of the first row. But the determinant is equal to zero since the first and second rows are identical.

The steps which we used in solving the system of four equations may be applied in just the same way to solve a system of n linear equations in n unknowns. Accordingly, we shall state a theorem covering the general case. The theorem, known as Cramer's rule, is named after the Swiss mathematician Gabriel Cramer (1704–1752).

Theorem 23-10 Let a system of n linear equations in n unknowns be denoted by

$$\begin{aligned} a_{11}x_1 + a_{12}x_2 + a_{13}x_3 + \cdots + a_{1n}x_n &= k_1 \\ a_{21}x_1 + a_{22}x_2 + a_{23}x_3 + \cdots + a_{2n}x_n &= k_2 \\ &\vdots \\ a_{n1}x_1 + a_{n2}x_2 + a_{n3}x_3 + \cdots + a_{nn}x_n &= k_n \end{aligned}$$

Let D be the determinant of the coefficients of the unknowns $x_1, x_2, x_3, \ldots, x_n$, and N_i the determinant obtained from D by replacing the coefficients of x_i by the corresponding constant terms and leaving all other elements unchanged. Then, if $D \neq 0$, the solution of the system is given by

$$x_1 = \frac{N_1}{D}, \quad x_2 = \frac{N_2}{D}, \quad x_3 = \frac{N_3}{D}, \quad \ldots, \quad x_n = \frac{N_n}{D}$$

EXAMPLE Solve the system

$$\begin{aligned} 2x \qquad\;\; + z - \;\, w &= 1 \\ x + 2y - 2z + 3w &= 9 \\ 3x - 2y + 3z - 2w &= 2 \\ 4x - 4y - 2z - 3w &= 3 \end{aligned}$$

Solution: The solution may be had from the five determinants D, N_x, N_y, N_z, and N_w. We show how D may be evaluated.

$$D = \begin{vmatrix} 2 & 0 & 1 & -1 \\ 1 & 2 & -2 & 3 \\ 3 & -2 & 3 & -2 \\ 4 & -4 & -2 & -3 \end{vmatrix}$$

We add twice the fourth column to the first column, and add the fourth column to the third column, and then expand by the elements of the first row; whence,

$$D = \begin{vmatrix} 0 & 0 & 0 & -1 \\ 7 & 2 & 1 & 3 \\ -1 & -2 & 1 & -2 \\ -2 & -4 & -5 & -3 \end{vmatrix} = \begin{vmatrix} 7 & 2 & 1 \\ -1 & -2 & 1 \\ -2 & -4 & -5 \end{vmatrix}$$

$$= 2\begin{vmatrix} 7 & 1 & 1 \\ -1 & -1 & 1 \\ -2 & -2 & -5 \end{vmatrix}$$

Referring to the determinant on the right, we add the third column to the first and also to the second column and obtain

$$D = 2\begin{vmatrix} 8 & 2 & 1 \\ 0 & 0 & 1 \\ -7 & -7 & -5 \end{vmatrix} = -2\begin{vmatrix} 8 & 2 \\ -7 & -7 \end{vmatrix} = 84$$

By evaluating, we may verify that

$$N_x = \begin{vmatrix} 1 & 0 & 1 & -1 \\ 9 & 2 & -2 & 3 \\ 2 & -2 & 3 & -2 \\ 3 & -4 & -2 & -3 \end{vmatrix} = 168$$

$$N_y = \begin{vmatrix} 2 & 1 & 1 & -1 \\ 1 & 9 & -2 & 3 \\ 3 & 2 & 3 & -2 \\ 4 & 3 & -2 & -3 \end{vmatrix} = -84$$

$$N_z = \begin{vmatrix} 2 & 0 & 1 & -1 \\ 1 & 2 & 9 & 3 \\ 3 & -2 & 2 & -2 \\ 4 & -4 & 3 & -3 \end{vmatrix} = 0$$

$$N_w = \begin{vmatrix} 2 & 0 & 1 & 1 \\ 1 & 2 & -2 & 9 \\ 3 & -2 & 3 & 2 \\ 4 & -4 & -2 & 3 \end{vmatrix} = 252$$

Hence,

$$x = \frac{168}{84} = 2, \quad y = \frac{-84}{84} = -1, \quad z = \frac{0}{84} = 0, \quad w = \frac{252}{84} = 3$$

EXERCISE 23-5

Solve the following systems of equations by determinants and check your results.

1. $\begin{aligned} 2x + 3y + z &= 9 \\ x - 2y - 3z &= 1 \\ 5x + 4y + 6z &= 5 \end{aligned}$

2. $\begin{aligned} 5x + 4y + 3z &= 1 \\ 2x - y - z &= 7 \\ 3x + y &= 6 \end{aligned}$

3. $\begin{aligned} 2x + 3y + z &= 9 \\ 4x + y &= 7 \\ x - 3y - 7z &= 6 \end{aligned}$

4. $\begin{aligned} 3x + y + z &= 2 \\ x + y + z &= 0 \\ 5x - 2y + 3z &= -8 \end{aligned}$

5. $\begin{aligned} 3x - 5z &= -1 \\ 2x + 7y &= 6 \\ x + y + z &= 5 \end{aligned}$

6. $\begin{aligned} x + 5y - z &= 9 \\ 3x - 3y + 2z &= 7 \\ 2x - 4y + 3z &= 1 \end{aligned}$

7. $\begin{aligned} -x + 2y + 3z - 2w &= 2 \\ 3x + y + z - w &= 4 \\ 2x - y - z + 2w &= 0 \\ x + y + z + w &= 0 \end{aligned}$

8. $\begin{aligned} 2x + 2y - z - w &= 1 \\ 2x - 4y + 3z + 2w &= 8 \\ 3x + 2y - 2z - 3w &= 7 \\ x + y - z + w &= 0 \end{aligned}$

9. $\begin{aligned} 2x - 3z + w &= 1 \\ x - y + 2w &= 1 \\ -3y + z + w &= 1 \\ x + y + z &= 1 \end{aligned}$

10. $\begin{aligned} x - y + z - w &= 3 \\ x + y + z + w &= -5 \\ x - 3y - z - w &= 9 \\ x + y - z + w &= 1 \end{aligned}$

11. $\begin{aligned} x + y + 2z + 2w &= 3 \\ 2x - y - z + 2w &= 1 \\ 2x + 2y - 3z + w &= 2 \\ 4x - 2y + z + 3w &= 6 \end{aligned}$

12. $$\begin{aligned} x + y + 2z + 3w &= -2 \\ 2x + y - 3z - 2w &= -1 \\ 3x + 2y - z - w &= -5 \\ 2x + 3y - z - 2w &= 1 \end{aligned}$$

13. $$\begin{aligned} x + y + z + w &= 2 \\ x - 2y + 2z + 2w &= -6 \\ 2x + y - 2z + 2w &= -5 \\ 3x - y + 3z - 3w &= -3 \end{aligned}$$

14. $$\begin{aligned} 2x + 3y + z + w &= 2 \\ 2x - 2y + z - w &= 7 \\ x - y - 3z - w &= 2 \\ 3x + 2y + 2z + 2w &= 5 \end{aligned}$$

APPENDIX

TABLE 1 POWERS AND ROOTS

No.	Sq.	Sq. root	Cube	Cube root	No.	Sq.	Sq. root	Cube	Cube root
1	1	1.000	1	1.000	**51**	2,601	7.141	132,651	3.708
2	4	1.414	8	1.260	**52**	2,704	7.211	140,608	3.733
3	9	1.732	27	1.442	**53**	2,809	7.280	148,877	3.756
4	16	2.000	64	1.587	**54**	2,916	7.348	157,464	3.780
5	25	2.236	125	1.710	**55**	3,025	7.416	166,375	3.803
6	36	2.449	216	1.817	**56**	3,136	7.483	175,616	3.826
7	49	2.646	343	1.913	**57**	3,249	7.550	185,193	3.849
8	64	2.828	512	2.000	**58**	3,364	7.616	195,112	3.871
9	81	3.000	729	2.080	**59**	3,481	7.681	205,379	3.893
10	100	3.162	1,000	2.154	**60**	3,600	7.746	216,000	3.915
11	121	3.317	1,331	2.224	**61**	3,721	7.810	226,981	3.936
12	144	3.464	1,728	2.289	**62**	3,844	7.874	238,328	3.958
13	169	3.606	2,197	2.351	**63**	3,969	7.937	250,047	3.979
14	196	3.742	2,744	2.410	**64**	4,096	8.000	262,144	4.000
15	225	3.873	3,375	2.466	**65**	4,225	8.062	274,625	4.021
16	256	4.000	4,096	2.520	**66**	4,356	8.124	287,496	4.041
17	289	4.123	4,913	2.571	**67**	4,489	8.185	300,763	4.062
18	324	4.243	5,832	2.621	**68**	4,624	8.246	314,432	4.082
19	361	4.359	6,859	2.668	**69**	4,761	8.307	328,509	4.102
20	400	4.472	8,000	2.714	**70**	4,900	8.367	343,000	4.121
21	441	4.583	9,261	2.759	**71**	5,041	8.426	357,911	4.141
22	484	4.690	10,648	2.802	**72**	5,184	8.485	373,248	4.160
23	529	4.796	12,167	2.844	**73**	5,329	8.544	389,017	4.179
24	576	4.899	13,824	2.884	**74**	5,476	8.602	405,224	4.198
25	625	5.000	15,625	2.924	**75**	5,625	8.660	421,875	4.217
26	676	5.099	17,576	2.962	**76**	5,776	8.718	438,976	4.236
27	729	5.196	19,683	3.000	**77**	5,929	8.775	456,533	4.254
28	784	5.292	21,952	3.037	**78**	6,084	8.832	474,552	4.273
29	841	5.385	24,389	3.072	**79**	6,241	8.888	493,039	4.291
30	900	5.477	27,000	3.107	**80**	6,400	8.944	512,000	4.309
31	961	5.568	29,791	3.141	**81**	6,561	9.000	531,441	4.327
32	1,024	5.657	32,768	3.175	**82**	6,724	9.055	551,368	4.344
33	1,089	5.745	35,937	3.208	**83**	6,889	9.110	571,787	4.362
34	1,156	5.831	39,304	3.240	**84**	7,056	9.165	592,704	4.380
35	1,225	5.916	42,875	3.271	**85**	7,225	9.220	614,125	4.397
36	1,296	6.000	46,656	3.302	**86**	7,396	9.274	636,056	4.414
37	1,369	6.083	50,653	3.332	**87**	7,569	9.327	658,503	4.431
38	1,444	6.164	54,872	3.362	**88**	7,744	9.381	681,472	4.448
39	1,521	6.245	59,319	3.391	**89**	7,921	9.434	704,969	4.465
40	1,600	6.325	64,000	3.420	**90**	8,100	9.487	729,000	4.481
41	1,681	6.403	68,921	3.448	**91**	8,281	9.539	753,571	4.498
42	1,764	6.481	74,088	3.476	**92**	8,464	9.592	778,688	4.514
43	1,849	6.557	79,507	3.503	**93**	8,649	9.644	804,357	4.531
44	1,936	6.633	85,184	3.530	**94**	8,836	9.695	830,584	4.547
45	2,025	6.708	91,125	3.557	**95**	9,025	9.747	857,375	4.563
46	2,116	6.782	97,336	3.583	**96**	9,216	9.798	884,736	4.579
47	2,209	6.856	103,823	3.609	**97**	9,409	9.849	912,673	4.595
48	2,304	6.928	110,592	3.634	**98**	9,604	9.899	941,192	4.610
49	2,401	7.000	117,649	3.659	**99**	9,801	9.950	970,299	4.626
50	2,500	7.071	125,000	3.684	**100**	10,000	10.000	1,000,000	4.642

TABLE 2 NATURAL TRIGONOMETRIC FUNCTIONS

Degrees	Sin	Cos	Tan	Cot	Sec	Csc	
0° 00′	.0000	1.0000	.0000		1.000		**90° 00′**
10	.0029	1.0000	.0029	343.8	1.000	343.8	50
20	.0058	1.0000	.0058	171.9	1.000	171.9	40
30	.0087	1.0000	.0087	114.6	1.000	114.6	30
40	.0116	.9999	.0116	85.94	1.000	85.95	20
50	.0145	.9999	.0145	68.75	1.000	68.76	10
1° 00′	.0175	.9998	.0175	57.29	1.000	57.30	**89° 00′**
10	.0204	.9998	.0204	49.10	1.000	49.11	50
20	.0233	.9997	.0233	42.96	1.000	42.98	40
30	.0262	.9997	.0262	38.19	1.000	38.20	30
40	.0291	.9996	.0291	34.37	1.000	34.38	20
50	.0320	.9995	.0320	31.24	1.001	31.26	10
2° 00′	.0349	.9994	.0349	28.64	1.001	28.65	**88° 00′**
10	.0378	.9993	.0378	26.43	1.001	26.45	50
20	.0407	.9992	.0407	24.54	1.001	24.56	40
30	.0436	.9990	.0437	22.90	1.001	22.93	30
40	.0465	.9989	.0466	21.47	1.001	21.49	20
50	.0494	.9988	.0495	20.21	1.001	20.23	10
3° 00′	.0523	.9986	.0524	19.08	1.001	19.11	**87° 00′**
10	.0552	.9985	.0553	18.07	1.002	18.10	50
20	.0581	.9983	.0582	17.17	1.002	17.20	40
30	.0610	.9981	.0612	16.35	1.002	16.38	30
40	.0640	.9980	.0641	15.60	1.002	15.64	20
50	.0669	.9978	.0670	14.92	1.002	14.96	10
4° 00′	.0698	.9976	.0699	14.30	1.002	14.34	**86° 00′**
10	.0727	.9974	.0729	13.73	1.003	13.76	50
20	.0756	.9971	.0758	13.20	1.003	13.23	40
30	.0785	.9969	.0787	12.71	1.003	12.75	30
40	.0814	.9967	.0816	12.25	1.003	12.29	20
50	.0843	.9964	.0846	11.83	1.004	11.87	10
5° 00′	.0872	.9962	.0875	11.43	1.004	11.47	**85° 00′**
10	.0901	.9959	.0904	11.06	1.004	11.10	50
20	.0929	.9957	.0934	10.71	1.004	10.76	40
30	.0958	.9954	.0963	10.39	1.005	10.43	30
40	.0987	.9951	.0992	10.08	1.005	10.13	20
50	.1016	.9948	.1022	9.788	1.005	9.839	10
6° 00′	.1045	.9945	.1051	9.514	1.006	9.567	**84° 00′**
10	.1074	.9942	.1080	9.255	1.006	9.309	50
20	.1103	.9939	.1110	9.010	1.006	9.065	40
30	.1132	.9936	.1139	8.777	1.006	8.834	30
40	.1161	.9932	.1169	8.556	1.007	8.614	20
50	.1190	.9929	.1198	8.345	1.007	8.405	10
7° 00′	.1219	.9925	.1228	8.144	1.008	8.206	**83° 00′**
10	.1248	.9922	.1257	7.953	1.008	8.016	50
20	.1276	.9918	.1287	7.770	1.008	7.834	40
30	.1305	.9914	.1317	7.596	1.009	7.661	30
40	.1334	.9911	.1346	7.429	1.009	7.496	20
50	.1363	.9907	.1376	7.269	1.009	7.337	10
8° 00′	.1392	.9903	.1405	7.115	1.010	7.185	**82° 00′**
10	.1421	.9899	.1435	6.968	1.010	7.040	50
20	.1449	.9894	.1465	6.827	1.011	6.900	40
30	.1478	.9890	.1495	6.691	1.011	6.765	30
40	.1507	.9886	.1524	6.561	1.012	6.636	20
50	.1536	.9881	.1554	6.435	1.012	6.512	10
9° 00′	.1564	.9877	.1584	6.314	1.012	6.392	**81° 00′**
	Cos	Sin	Cot	Tan	Csc	Sec	Degrees

TABLE 2 NATURAL TRIGONOMETRIC FUNCTIONS (continued)

Degrees	Sin	Cos	Tan	Cot	Sec	Csc	
9° 00′	.1564	.9877	.1584	6.314	1.012	6.392	**81° 00′**
10	.1593	.9872	.1614	6.197	1.013	6.277	50
20	.1622	.9868	.1644	6.084	1.013	6.166	40
30	.1650	.9863	.1673	5.976	1.014	6.059	30
40	.1679	.9858	.1703	5.871	1.014	5.955	20
50	.1708	.9853	.1733	5.769	1.015	5.855	10
10° 00′	.1736	.9848	.1763	5.671	1.015	5.759	**80° 00′**
10	.1765	.9843	.1793	5.576	1.016	5.665	50
20	.1794	.9838	.1823	5.485	1.016	5.575	40
30	.1822	.9833	.1853	5.396	1.017	5.487	30
40	.1851	.9827	.1883	5.309	1.018	5.403	20
50	.1880	.9822	.1914	5.226	1.018	5.320	10
11° 00′	.1908	.9816	.1944	5.145	1.019	5.241	**79° 00′**
10	.1937	.9811	.1974	5.066	1.019	5.164	50
20	.1965	.9805	.2004	4.989	1.020	5.089	40
30	.1994	.9799	.2035	4.915	1.020	5.016	30
40	.2022	.9793	.2065	4.843	1.021	4.945	20
50	.2051	.9787	.2095	4.773	1.022	4.876	10
12° 00′	.2079	.9781	.2126	4.705	1.022	4.810	**78° 00′**
10	.2108	.9775	.2156	4.638	1.023	4.745	50
20	.2136	.9769	.2186	4.574	1.024	4.682	40
30	.2164	.9763	.2217	4.511	1.024	4.620	30
40	.2193	.9757	.2247	4.449	1.025	4.560	20
50	.2221	.9750	.2278	4.390	1.026	4.502	10
13° 00′	.2250	.9744	.2309	4.331	1.026	4.445	**77° 00′**
10	.2278	.9737	.2339	4.275	1.027	4.390	50
20	.2306	.9730	.2370	4.219	1.028	4.336	40
30	.2334	.9724	.2401	4.165	1.028	4.284	30
40	.2363	.9717	.2432	4.113	1.029	4.232	20
50	.2391	.9710	.2462	4.061	1.030	4.182	10
14° 00′	.2419	.9703	.2493	4.011	1.031	4.134	**76° 00′**
10	.2447	.9696	.2524	3.962	1.031	4.086	50
20	.2476	.9689	.2555	3.914	1.032	4.039	40
30	.2504	.9681	.2586	3.867	1.033	3.994	30
40	.2532	.9674	.2617	3.821	1.034	3.950	20
50	.2560	.9667	.2648	3.776	1.034	3.906	10
15° 00′	.2588	.9659	.2679	3.732	1.035	3.864	**75° 00′**
10	.2616	.9652	.2711	3.689	1.036	3.822	50
20	.2644	.9644	.2742	3.647	1.037	3.782	40
30	.2672	.9636	.2773	3.606	1.038	3.742	30
40	.2700	.9628	.2805	3.566	1.039	3.703	20
50	.2728	.9621	.2836	3.526	1.039	3.665	10
16° 00′	.2756	.9613	.2867	3.487	1.040	3.628	**74° 00′**
10	.2784	.9605	.2899	3.450	1.041	3.592	50
20	.2812	.9596	.2931	3.412	1.042	3.556	40
30	.2840	.9588	.2962	3.376	1.043	3.521	30
40	.2868	.9580	.2994	3.340	1.044	3.487	20
50	.2896	.9572	.3026	3.305	1.045	3.453	10
17° 00′	.2924	.9563	.3057	3.271	1.046	3.420	**73° 00′**
10	.2952	.9555	.3089	3.237	1.047	3.388	50
20	.2979	.9546	.3121	3.204	1.048	3.356	40
30	.3007	.9537	.3153	3.172	1.049	3.326	30
40	.3035	.9528	.3185	3.140	1.049	3.295	20
50	.3062	.9520	.3217	3.108	1.050	3.265	10
18° 00′	.3090	.9511	.3249	3.078	1.051	3.236	**72° 00′**
	Cos	Sin	Cot	Tan	Csc	Sec	Degrees

TABLE 2 NATURAL TRIGONOMETRIC FUNCTIONS (continued)

Degrees	Sin	Cos	Tan	Cot	Sec	Csc	
18° 00′	.3090	.9511	.3249	3.078	1.051	3.236	**72° 00′**
10	.3118	.9502	.3281	3.047	1.052	3.207	50
20	.3145	.9492	.3314	3.018	1.053	3.179	40
30	.3173	.9483	.3346	2.989	1.054	3.152	30
40	.3201	.9474	.3378	2.960	1.056	3.124	20
50	.3228	.9465	.3411	2.932	1.057	3.098	10
19° 00′	.3256	.9455	.3443	2.904	1.058	3.072	**71° 00′**
10	.3283	.9446	.3476	2.877	1.059	3.046	50
20	.3311	.9436	.3508	2.850	1.060	3.021	40
30	.3338	.9426	.3541	2.824	1.061	2.996	30
40	.3365	.9417	.3574	2.798	1.062	2.971	20
50	.3393	.9407	.3607	2.773	1.063	2.947	10
20° 00′	.3420	.9397	.3640	2.747	1.064	2.924	**70° 00′**
10	.3448	.9387	.3673	2.723	1.065	2.901	50
20	.3475	.9377	.3706	2.699	1.066	2.878	40
30	.3502	.9367	.3739	2.675	1.068	2.855	30
40	.3529	.9356	.3772	2.651	1.069	2.833	20
50	.3557	.9346	.3805	2.628	1.070	2.812	10
21° 00′	.3584	.9336	.3839	2.605	1.071	2.790	**69° 00′**
10	.3611	.9325	.3872	2.583	1.072	2.769	50
20	.3638	.9315	.3906	2.560	1.074	2.749	40
30	.3665	.9304	.3939	2.539	1.075	2.729	30
40	.3692	.9293	.3973	2.517	1.076	2.709	20
50	.3719	.9283	.4006	2.496	1.077	2.689	10
22° 00′	.3746	.9272	.4040	2.475	1.079	2.669	**68° 00′**
10	.3773	.9261	.4074	2.455	1.080	2.650	50
20	.3800	.9250	.4108	2.434	1.081	2.632	40
30	.3827	.9239	.4142	2.414	1.082	2.613	30
40	.3854	.9228	.4176	2.394	1.084	2.595	20
50	.3881	.9216	.4210	2.375	1.085	2.577	10
23° 00′	.3907	.9205	.4245	2.356	1.086	2.559	**67° 00′**
10	.3934	.9194	.4279	2.337	1.088	2.542	50
20	.3961	.9182	.4314	2.318	1.089	2.525	40
30	.3987	.9171	.4348	2.300	1.090	2.508	30
40	.4014	.9159	.4383	2.282	1.092	2.491	20
50	.4041	.9147	.4417	2.264	1.093	2.475	10
24° 00′	.4067	.9135	.4452	2.246	1.095	2.459	**66° 00′**
10	.4094	.9124	.4487	2.229	1.096	2.443	50
20	.4120	.9112	.4522	2.211	1.097	2.427	40
30	.4147	.9100	.4557	2.194	1.099	2.411	30
40	.4173	.9088	.4592	2.177	1.100	2.396	20
50	.4200	.9075	.4628	2.161	1.102	2.381	10
25° 00′	.4226	.9063	.4663	2.145	1.103	2.366	**65° 00′**
10	.4253	.9051	.4699	2.128	1.105	2.352	50
20	.4279	.9038	.4734	2.112	1.106	2.337	40
30	.4305	.9026	.4770	2.097	1.108	2.323	30
40	.4331	.9013	.4806	2.081	1.109	2.309	20
50	.4358	.9001	.4841	2.066	1.111	2.295	10
26° 00′	.4384	.8988	.4877	2.050	1.113	2.281	**64° 00′**
10	.4410	.8975	.4913	2.035	1.114	2.268	50
20	.4436	.8962	.4950	2.020	1.116	2.254	40
30	.4462	.8949	.4986	2.006	1.117	2.241	30
40	.4488	.8936	.5022	1.991	1.119	2.228	20
50	.4514	.8923	.5059	1.977	1.121	2.215	10
27° 00′	.4540	.8910	.5095	1.963	1.122	2.203	**63° 00′**
	Cos	Sin	Cot	Tan	Csc	Sec	Degrees

TABLE 2 NATURAL TRIGONOMETRIC FUNCTIONS (continued)

Degrees	Sin	Cos	Tan	Cot	Sec	Csc	
27° 00′	.4540	.8910	.5095	1.963	1.122	2.203	**63° 00′**
10	.4566	.8897	.5132	1.949	1.124	2.190	50
20	.4592	.8884	.5169	1.935	1.126	2.178	40
30	.4617	.8870	.5206	1.921	1.127	2.166	30
40	.4643	.8857	.5243	1.907	1.129	2.154	20
50	.4669	.8843	5280	1.894	1.131	2.142	10
28° 00′	.4695	.8829	.5317	1.881	1.133	2.130	**62° 00′**
10	.4720	.8816	.5354	1.868	1.134	2.118	50
20	.4746	.8802	.5392	1.855	1.136	2.107	40
30	.4772	.8788	.5430	1.842	1.138	2.096	30
40	.4797	.8774	.5467	1.829	1.140	2.085	20
50	.4823	.8760	.5505	1 816	1.142	2.074	10
29° 00′	.4848	.8746	.5543	1.804	1.143	2.063	**61° 00′**
10	.4874	.8732	.5581	1.792	1.145	2.052	50
20	.4899	.8718	.5619	1.780	1.147	2.041	40
30	.4924	.8704	.5658	1.767	1.149	2.031	30
40	.4950	.8689	.5696	1.756	1.151	2.020	20
50	.4975	.8675	.5735	1.744	1.153	2.010	10
30° 00′	.5000	.8660	.5774	1.732	1.155	2.000	**60° 00′**
10	.5025	.8646	.5812	1.720	1.157	1.990	50
20	.5050	.8631	.5851	1.709	1.159	1.980	40
30	.5075	.8616	.5890	1.698	1.161	1.970	30
40	.5100	.8601	.5930	1.686	1.163	1.961	20
50	.5125	.8587	.5969	1.675	1.165	1.951	10
31° 00′	.5150	.8572	.6009	1.664	1.167	1.942	**59° 00′**
10	.5175	.8557	.6048	1.653	1.169	1.932	50
20	.5200	.8542	.6088	1.643	1.171	1.923	40
30	.5225	.8526	.6128	1.632	1.173	1.914	30
40	.5250	.8511	.6168	1.621	1.175	1.905	20
50	.5275	.8496	.6208	1.611	1.177	1.896	10
32° 00′	.5299	.8480	.6249	1.600	1.179	1.887	**58° 00′**
10	.5324	.8465	.6289	1.590	1.181	1.878	50
20	.5348	.8450	.6330	1.580	1.184	1.870	40
30	.5373	.8434	.6371	1.570	1.186	1.861	30
40	.5398	.8418	.6412	1.560	1.188	1.853	20
50	.5422	.8403	.6453	1.550	1.190	1.844	10
33° 00′	.5446	.8387	.6494	1.540	1.192	1.836	**57° 00′**
10	.5471	.8371	.6536	1.530	1.195	1.828	50
20	.5495	.8355	.6577	1.520	1.197	1.820	40
30	.5519	.8339	.6619	1.511	1.199	1.812	30
40	.5544	.8323	.6661	1.501	1.202	1.804	20
50	.5568	.8307	.6703	1.492	1.204	1.796	10
34° 00′	.5592	.8290	.6745	1.483	1.206	1.788	**56° 00′**
10	.5616	.8274	.6787	1.473	1.209	1.781	50
20	.5640	.8258	.6830	1.464	1.211	1.773	40
30	.5664	.8241	.6873	1.455	1.213	1.766	30
40	.5688	.8225	.6916	1.446	1.216	1.758	20
50	.5712	.8208	.6959	1.437	1.218	1.751	10
35° 00′	.5736	.8192	.7002	1.428	1.221	1.743	**55° 00′**
10	.5760	.8175	.7046	1.419	1.223	1.736	50
20	.5783	.8158	.7089	1.411	1.226	1.729	40
30	.5807	.8141	.7133	1.402	1.228	1.722	30
40	.5831	.8124	.7177	1.393	1.231	1.715	20
50	.5854	.8107	.7221	1.385	1.233	1.708	10
36° 00′	.5878	.8090	.7265	1.376	1.236	1.701	**54° 00′**
	Cos	Sin	Cot	Tan	Csc	Sec	Degrees

TABLE 2 NATURAL TRIGONOMETRIC FUNCTIONS (continued)

Degrees	Sin	Cos	Tan	Cot	Sec	Csc	
36° 00′	.5878	.8090	.7265	1.376	1.236	1.701	**54° 00′**
10	.5901	.8073	.7310	1.368	1.239	1.695	50
20	.5925	.8056	.7355	1.360	1.241	1.688	40
30	.5948	.8039	.7400	1.351	1.244	1.681	30
40	.5972	.8021	.7445	1.343	1.247	1.675	20
50	.5995	.8004	.7490	1.335	1.249	1.668	10
37° 00′	.6018	.7986	.7536	1.327	1.252	1.662	**53° 00′**
10	.6041	.7969	.7581	1.319	1.255	1.655	50
20	.6065	.7951	.7627	1.311	1.258	1.649	40
30	.6088	.7934	.7673	1.303	1.260	1.643	30
40	.6111	.7916	.7720	1.295	1.263	1.636	20
50	.6134	.7898	.7766	1.288	1.266	1.630	10
38° 00′	.6157	.7880	.7813	1.280	1.269	1.624	**52° 00′**
10	.6180	.7862	.7860	1.272	1.272	1.618	50
20	.6202	.7844	.7907	1.265	1.275	1.612	40
30	.6225	.7826	.7954	1.257	1.278	1.606	30
40	.6248	.7808	.8002	1.250	1.281	1.601	20
50	.6271	.7790	.8050	1.242	1.284	1.595	10
39° 00′	.6293	.7771	.8098	1.235	1.287	1.589	**51° 00′**
10	.6316	.7753	.8146	1.228	1.290	1.583	50
20	.6338	.7735	.8195	1.220	1.293	1.578	40
30	.6361	.7716	.8243	1.213	1.296	1.572	30
40	.6383	.7698	.8292	1.206	1.299	1.567	20
50	.6406	.7679	.8342	1.199	1.302	1.561	10
40° 00′	.6428	.7660	.8391	1.192	1.305	1.556	**50° 00′**
10	.6450	.7642	.8441	1.185	1.309	1.550	50
20	.6472	.7623	.8491	1.178	1.312	1.545	40
30	.6494	.7604	.8541	1.171	1.315	1.540	30
40	.6517	.7585	.8591	1.164	1.318	1.535	20
50	.6539	.7566	.8642	1.157	1.322	1.529	10
41° 00′	.6561	.7547	.8693	1.150	1.325	1.524	**49° 00′**
10	.6583	.7528	.8744	1.144	1.328	1.519	50
20	.6604	.7509	.8796	1.137	1.332	1.514	40
30	.6626	.7490	.8847	1.130	1.335	1.509	30
40	.6648	.7470	.8899	1.124	1.339	1.504	20
50	.6670	.7451	.8952	1.117	1.342	1.499	10
42° 00′	.6691	.7431	.9004	1.111	1.346	1.494	**48° 00′**
10	.6713	.7412	.9057	1.104	1.349	1.490	50
20	.6734	.7392	.9110	1.098	1.353	1.485	40
30	.6756	.7373	.9163	1.091	1.356	1.480	30
40	.6777	.7353	.9217	1.085	1.360	1.476	20
50	.6799	.7333	.9271	1.079	1.364	1.471	10
43° 00′	.6820	.7314	.9325	1.072	1.367	1.466	**47° 00′**
10	.6841	.7294	.9380	1.066	1.371	1.462	50
20	.6862	.7274	.9435	1.060	1.375	1.457	40
30	.6884	.7254	.9490	1.054	1.379	1.453	30
40	.6905	.7234	.9545	1.048	1.382	1.448	20
50	.6926	.7214	.9601	1.042	1.386	1.444	10
44° 00′	.6947	.7193	.9657	1.036	1.390	1.440	**46° 00′**
10	.6967	.7173	.9713	1.030	1.394	1.435	50
20	.6988	.7153	.9770	1.024	1.398	1.431	40
30	.7009	.7133	.9827	1.018	1.402	1.427	30
40	.7030	.7112	.9884	1.012	1.406	1.423	20
50	.7050	.7092	.9942	1.006	1.410	1.418	10
45° 00′	.7071	.7071	1.000	1.000	1.414	1.414	**45° 00′**
	Cos	Sin	Cot	Tan	Csc	Sec	Degrees

TABLE 3 VALUES OF TRIGONOMETRIC FUNCTIONS (RADIANS)

Radians	Sin	Cos	Tan	Radians	Sin	Cos	Tan
.00	.0000	1.0000	.0000	**.40**	.3894	.9211	.4228
.01	.0100	1.0000	.0100	.41	.3986	.9171	.4346
.02	.0200	.9998	.0200	.42	.4078	.9131	.4466
.03	.0300	.9996	.0300	.43	.4169	.9090	.4586
.04	.0400	.9992	.0400	.44	.4259	.9048	.4708
.05	.0500	.9988	.0500	**.45**	.4350	.9004	.4831
.06	.0600	.9982	.0601	.46	.4439	.8961	.4954
.07	.0699	.9976	.0701	.47	.4529	.8916	.5080
.08	.0799	.9968	.0802	.48	.4618	.8870	.5206
.09	.0899	.9960	.0902	.49	.4706	.8823	.5334
.10	.0998	.9950	.1003	**.50**	.4794	.8776	.5463
.11	.1098	.9940	.1104	.51	.4882	.8727	.5594
.12	.1197	.9928	.1206	.52	.4969	.8678	.5726
.13	.1296	.9916	.1307	.53	.5055	.8628	.5859
.14	.1395	.9902	.1409	.54	.5141	.8577	.5994
.15	.1494	.9888	.1511	**.55**	.5227	.8525	.6131
.16	.1593	.9872	.1614	.56	.5312	.8473	.6269
.17	.1692	.9856	.1717	.57	.5396	.8419	.6410
.18	.1790	.9838	.1820	.58	.5480	.8365	.6552
.19	.1889	.9820	.1923	.59	.5564	.8309	.6696
.20	.1987	.9801	.2027	**.60**	.5646	.8253	.6841
.21	.2085	.9780	.2131	.61	.5729	.8196	.6989
.22	.2182	.9759	.2236	.62	.5810	.8139	.7139
.23	.2280	.9737	.2341	.63	.5891	.8080	.7291
.24	.2377	.9713	.2447	.64	.5972	.8021	.7445
.25	.2474	.9689	.2553	**.65**	.6052	.7961	.7602
.26	.2571	.9664	.2660	.66	.6131	.7900	.7761
.27	.2667	.9638	.2768	.67	.6210	.7838	.7923
.28	.2764	.9611	.2876	.68	.6288	.7776	.8087
.29	.2860	.9582	.2984	.69	.6365	.7712	.8253
.30	.2955	.9553	.3093	**.70**	.6442	.7648	.8423
.31	.3051	.9523	.3203	.71	.6518	.7584	.8595
.32	.3146	.9492	.3314	.72	.6594	.7518	.8771
.33	.3240	.9460	.3425	.73	.6669	.7452	.8949
.34	.3335	.9428	.3537	.74	.6743	.7385	.9131
.35	.3429	.9394	.3650	**.75**	.6816	.7317	.9316
.36	.3523	.9359	.3764	.76	.6889	.7248	.9505
.37	.3616	.9323	.3879	.77	.6961	.7179	.9697
.38	.3709	.9287	.3994	.78	.7033	.7109	.9893
.39	.3802	.9249	.4111	.79	.7104	.7038	1.009

TABLE 3 VALUES OF TRIGONOMETRIC FUNCTIONS (RADIANS) (continued)

Radians	Sin	Cos	Tan	Radians	Sin	Cos	Tan
.80	.7174	.6967	1.030	**1.20**	.9320	.3624	2.572
.81	.7243	.6895	1.050	1.21	.9356	.3530	2.650
.82	.7311	.6822	1.072	1.22	.9391	.3436	2.733
.83	.7379	.6749	1.093	1.23	.9425	.3342	2.820
.84	.7446	.6675	1.116	1.24	.9458	.3248	2.912
.85	.7513	.6600	1.138	**1.25**	.9490	.3153	3.010
.86	.7578	.6524	1.162	1.26	.9521	.3058	3.113
.87	.7643	.6448	1.185	1.27	.9551	.2963	3.224
.88	.7707	.6372	1.210	1.28	.9580	.2867	3.341
.89	.7771	.6294	1.235	1.29	.9608	.2771	3.467
.90	.7833	.6216	1.260	**1.30**	.9636	.2675	3.602
.91	.7895	.6137	1.286	1.31	.9662	.2579	3.747
.92	.7956	.6058	1.313	1.32	.9687	.2482	3.903
.93	.8016	.5978	1.341	1.33	.9711	.2385	4.072
.94	.8076	.5898	1.369	1.34	.9735	.2288	4.256
.95	.8134	.5817	1.398	**1.35**	.9757	.2190	4.455
.96	.8192	.5735	1.428	1.36	.9779	.2092	4.673
.97	.8249	.5653	1.459	1.37	.9799	.1995	4.913
.98	.8305	.5570	1.491	1.38	.9819	.1896	5.177
.99	.8360	.5487	1.524	1.39	.9837	.1798	5.471
1.00	.8415	.5403	1.557	**1.40**	.9854	.1700	5.798
1.01	.8468	.5319	1.592	1.41	.9871	.1601	6.165
1.02	.8521	.5234	1.628	1.42	.9887	.1502	6.581
1.03	.8573	.5148	1.665	1.43	.9901	.1403	7.055
1.04	.8624	.5062	1.704	1.44	.9915	.1304	7.602
1.05	.8674	.4976	1.743	**1.45**	.9927	.1205	8.238
1.06	.8724	.4889	1.784	1.46	.9939	.1106	8.989
1.07	.8772	.4801	1.827	1.47	.9949	.1006	9.887
1.08	.8820	.4713	1.871	1.48	.9959	.0907	10.98
1.09	.8866	.4625	1.917	1.49	.9967	.0807	12.35
1.10	.8912	.4536	1.965	**1.50**	.9975	.0707	14.10
1.11	.8957	.4447	2.014	1.51	.9982	.0608	16.43
1.12	.9001	.4357	2.066	1.52	.9987	.0508	19.67
1.13	.9044	.4267	2.120	1.53	.9992	.0408	24.50
1.14	.9086	.4176	2.176	1.54	.9995	.0308	32.46
1.15	.9128	.4085	2.234	**1.55**	.9998	.0208	48.08
1.16	.9168	.3993	2.296	1.56	.9999	.0108	92.62
1.17	.9208	.3902	2.360	1.57	1.000	.0008	1256.
1.18	.9246	.3809	2.427				
1.19	.9284	.3717	2.498				

TABLE 4 LOGARITHMS OF NUMBERS

N	0	1	2	3	4	5	6	7	8	9
10	0000	0043	0086	0128	0170	0212	0253	0294	0334	0374
11	0414	0453	0492	0531	0569	0607	0645	0682	0719	0755
12	0792	0828	0864	0899	0934	0969	1004	1038	1072	1106
13	1139	1173	1206	1239	1271	1303	1335	1367	1399	1430
14	1461	1492	1523	1553	1584	1614	1644	1673	1703	1732
15	1761	1790	1818	1847	1875	1903	1931	1959	1987	2014
16	2041	2068	2095	2122	2148	2175	2201	2227	2253	2279
17	2304	2330	2355	2380	2405	2430	2455	2480	2504	2529
18	2553	2577	2601	2625	2648	2672	2695	2718	2742	2765
19	2788	2810	2833	2856	2878	2900	2923	2945	2967	2989
20	3010	3032	3054	3075	3096	3118	3139	3160	3181	3201
21	3222	3243	3263	3284	3304	3324	3345	3365	3385	3404
22	3424	3444	3464	3483	3502	3522	3541	3560	3579	3598
23	3617	3636	3655	3674	3692	3711	3729	3747	3766	3784
24	3802	3820	3838	3856	3874	3892	3909	3927	3945	3962
25	3979	3997	4014	4031	4048	4065	4082	4099	4116	4133
26	4150	4166	4183	4200	4216	4232	4249	4265	4281	4298
27	4314	4330	4346	4362	4378	4393	4409	4425	4440	4456
28	4472	4487	4502	4518	4533	4548	4564	4579	4594	4609
29	4624	4639	4654	4669	4683	4698	4713	4728	4742	4757
30	4771	4786	4800	4814	4829	4843	4857	4871	4886	4900
31	4914	4928	4942	4955	4969	4983	4997	5011	5024	5038
32	5051	5065	5079	5092	5105	5119	5132	5145	5159	5172
33	5185	5198	5211	5224	5237	5250	5263	5276	5289	5302
34	5315	5328	5340	5353	5366	5378	5391	5403	5416	5428
35	5441	5453	5465	5478	5490	5502	5514	5527	5539	5551
36	5563	5575	5587	5599	5611	5623	5635	5647	5658	5670
37	5682	5694	5705	5717	5729	5740	5752	5763	5775	5786
38	5798	5809	5821	5832	5843	5855	5866	5877	5888	5899
39	5911	5922	5933	5944	5955	5966	5977	5988	5999	6010
40	6021	6031	6042	6053	6064	6075	6085	6096	6107	6117
41	6128	6138	6149	6160	6170	6180	6191	6201	6212	6222
42	6232	6243	6253	6263	6274	6284	6294	6304	6314	6325
43	6335	6345	6355	6365	6375	6385	6395	6405	6415	6425
44	6435	6444	6454	6464	6474	6484	6493	6503	6513	6522
45	6532	6542	6551	6561	6571	6580	6590	6599	6609	6618
46	6628	6637	6646	6656	6665	6675	6684	6693	6702	6712
47	6721	6730	6739	6749	6758	6767	6776	6785	6794	6803
48	6812	6821	6830	6839	6848	6857	6866	6875	6884	6893
49	6902	6911	6920	6928	6937	6946	6955	6964	6972	6981
50	6990	6998	7007	7016	7024	7033	7042	7050	7059	7067
51	7076	7084	7093	7101	7110	7118	7126	7135	7143	7152
52	7160	7168	7177	7185	7193	7202	7210	7218	7226	7235
52	7243	7251	7259	7267	7275	7284	7292	7300	7308	7316
54	7324	7332	7340	7348	7356	7364	7372	7380	7388	7396
N	0	1	2	3	4	5	6	7	8	9

TABLE 4 LOGARITHMS OF NUMBERS (continued)

N	0	1	2	3	4	5	6	7	8	9
55	7404	7412	7419	7427	7435	7443	7451	7459	7466	7474
56	7482	7490	7497	7505	7513	7520	7528	7536	7543	7551
57	7559	7566	7574	7582	7589	7597	7604	7612	7619	7627
58	7634	7642	7649	7657	7664	7672	7679	7686	7694	7701
59	7709	7716	7723	7731	7738	7745	7752	7760	7767	7774
60	7782	7789	7796	7803	7810	7818	7825	7832	7839	7846
61	7853	7860	7868	7875	7882	7889	7896	7903	7910	7917
62	7924	7931	7938	7945	7952	7959	7966	7973	7980	7987
63	7993	8000	8007	8014	8021	8028	8035	8041	8048	8055
64	8062	8069	8075	8082	8089	8096	8102	8109	8116	8122
65	8129	8136	8142	8149	8156	8162	8169	8176	8182	8189
66	8195	8202	8209	8215	8222	8228	8235	8241	8248	8254
67	8261	8267	8274	8280	8287	8293	8299	8306	8312	8319
68	8325	8331	8338	8344	8351	8357	8363	8370	8376	8382
69	8388	8395	8401	8407	8414	8420	8426	8432	8439	8445
70	8451	8457	8463	8470	8476	8482	8488	8494	8500	8506
71	8513	8519	8525	8531	8537	8543	8549	8555	8561	8567
72	8573	8579	8585	8591	8597	8603	8609	8615	8621	8627
73	8633	8639	8645	8651	8657	8663	8669	8675	8681	8686
74	8692	8698	8704	8710	8716	8722	8727	8733	8739	8745
75	8751	8756	8762	8768	8774	8779	8785	8791	8797	8802
76	8808	8814	8820	8825	8831	8837	8842	8848	8854	8859
77	8865	8871	8876	8882	8887	8893	8899	8904	8910	8915
78	8921	8927	8932	8938	8943	8949	8954	8960	8965	8971
79	8976	8982	8987	8993	8998	9004	9009	9015	9020	9025
80	9031	9036	9042	9047	9053	9058	9063	9069	9074	9079
81	9085	9090	9096	9101	9106	9112	9117	9122	9128	9133
82	9138	9143	9149	9154	9159	9165	9170	9175	9180	9186
83	9191	9196	9201	9206	9212	9217	9222	9227	9232	9238
84	9243	9248	9253	9258	9263	9269	9274	9279	9284	9289
85	9294	9299	9304	9309	9315	9320	9325	9330	9335	9340
86	9345	9350	9355	9360	9365	9370	9375	9380	9385	9390
87	9395	9400	9405	9410	9415	9420	9425	9430	9435	9440
88	9445	9450	9455	9460	9465	9469	9474	9479	9484	9489
89	9494	9499	9504	9509	9513	9518	9523	9528	9533	9538
90	9542	9547	9552	9557	9562	9566	9571	9576	9581	9586
91	9590	9595	9600	9605	9609	9614	9619	9624	9628	9633
92	9638	9643	9647	9652	9657	9661	9666	9671	9675	9680
93	9685	9689	9694	9699	9703	9708	9713	9717	9722	9727
94	9731	9736	9741	9745	9750	9754	9759	9763	9768	9773
95	9777	9782	9786	9791	9795	9800	9805	9809	9814	9818
96	9823	9827	9832	9836	9841	9845	9850	9854	9859	9863
97	9868	9872	9877	9881	9886	9890	9894	9899	9903	9908
98	9912	9917	9921	9926	9930	9934	9939	9943	9948	9952
99	9956	9961	9965	9969	9974	9978	9983	9987	9991	9996
N	0	1	2	3	4	5	6	7	8	9

TABLE 5 LOGARITHMS OF TRIGONOMETRIC FUNCTIONS

→	L Sin	L Tan	L Cot	L Cos	
0°00′				10.0000	**90°00′**
10′	7.4637	7.4637	12.5363	.0000	89°50′
20′	.7648	.7648	.2352	.0000	40′
30′	7.9408	7.9409	12.0591	.0000	30′
40′	8.0658	8.0658	11.9342	.0000	20′
0°50′	.1627	.1627	.8373	10.0000	10′
1°00′	8.2419	8.2419	11.7581	9.9999	**89°00′**
10′	.3088	.3089	.6911	.9999	88°50′
20′	.3668	.3669	.6331	.9999	40′
30′	.4179	.4181	.5819	.9999	30′
40′	.4637	.4638	.5362	.9998	20′
1°50′	.5050	.5053	.4947	.9998	10′
2°00′	8.5428	8.5431	11.4569	9.9997	**88°00′**
10′	.5776	.5779	.4221	.9997	87°50′
20′	.6097	.6101	.3899	.9996	40′
30′	.6397	.6401	.3599	.9996	30′
40′	.6677	.6682	.3318	.9995	20′
2°50′	.6940	.6945	.3055	.9995	10′
3°00′	8.7188	8.7194	11.2806	9.9994	**87°00′**
10′	.7423	.7429	.2571	.9993	86°50′
20′	.7645	.7652	.2348	.9993	40′
30′	.7857	.7865	.2135	.9992	30′
40′	.8059	.8067	.1933	.9991	20′
3°50′	.8251	.8261	.1739	.9990	10′
4°00′	8.8436	8.8446	11.1554	9.9989	**86°00′**
10′	.8613	.8624	.1376	.9989	85°50′
20′	.8783	.8795	.1205	.9988	40′
30′	.8946	.8960	.1040	.9987	30′
40′	.9104	.9118	.0882	.9986	20′
4°50′	.9256	.9272	.0728	.9985	10′
5°00′	8.9403	8.9420	11.0580	9.9983	**85°00′**
10′	.9545	.9563	.0437	.9982	84°50′
20′	.9682	.9701	.0299	.9981	40′
30′	.9816	.9836	.0164	.9980	30′
40′	8.9945	8.9966	11.0034	.9979	20′
5°50′	9.0070	9.0093	10.9907	.9977	10′
6°00′	9.0192	9.0216	10.9784	9.9976	**84°00′**
	L Cos	L Cot	L Tan	L Sin	←

TABLE 5 LOGARITHMS OF TRIGONOMETRIC FUNCTIONS (continued)

⟶	L Sin	L Tan	L Cot	L Cos	
6°00′	9.0192	9.0216	10.9784	9.9976	**84°00′**
10′	.0311	.0336	.9664	.9975	83°50′
20′	.0426	.0453	.9547	.9973	40′
30′	.0539	.0567	.9433	.9972	30′
40′	.0648	.0678	.9322	.9971	20′
6°50′	.0755	.0786	.9214	.9969	10′
7°00′	9.0859	9.0891	10.9109	9.9968	**83°00′**
10′	.0961	.0995	.9005	.9966	82°50′
20′	.1060	.1096	.8904	.9964	40′
30′	.1157	.1194	.8806	.9963	30′
40′	.1252	.1291	.8709	.9961	20′
7°50′	.1345	.1385	.8615	.9959	10′
8°00′	9.1436	9.1478	10.8522	9.9958	**82°00′**
10′	.1525	.1569	.8431	.9956	81°50′
20′	.1612	.1658	.8342	.9954	40′
30′	.1697	.1745	.8255	.9952	30′
40′	.1781	.1831	.8169	.9950	20′
8°50′	.1863	.1915	.8085	.9948	10′
9°00′	9.1943	9.1997	10.8003	9.9946	**81°00′**
10′	.2022	.2078	.7922	.9944	80°50′
20′	.2100	.2158	.7842	.9942	40′
30′	.2176	.2236	.7764	.9940	30′
40′	.2251	.2313	.7687	.9938	20′
9°50′	.2324	.2389	.7611	.9936	10′
10°00′	9.2397	9.2463	10.7537	9.9934	**80°00′**
10′	.2468	.2536	.7464	.9931	79°50′
20′	.2538	.2609	.7391	.9929	40′
30′	.2606	.2680	.7320	.9927	30′
40′	.2674	.2750	.7250	.9924	20′
10°50′	.2740	.2819	.7181	.9922	10′
11°00′	9.2806	9.2887	10.7113	9.9919	**79°00′**
10′	.2870	.2953	.7047	.9917	78°50′
20′	.2934	.3020	.6980	.9914	40′
30′	.2997	.3085	.6915	.9912	30′
40′	.3058	.3149	.6851	.9909	20′
11°50′	.3119	.3212	.6788	.9907	10′
12°00′	9.3179	9.3275	10.6725	9.9904	**78°00′**
	L Cos	L Cot	L Tan	L Sin	⟵

TABLE 5 LOGARITHMS OF TRIGONOMETRIC FUNCTIONS (continued)

→	L Sin	L Tan	L Cot	L Cos	
12°00′	9.3179	9.3275	10.6725	9.9904	**78°00′**
10′	.3238	.3336	.6664	.9901	77°50′
20′	.3296	.3397	.6603	.9899	40′
30′	.3353	.3458	.6542	.9896	30′
40′	.3410	.3517	.6483	.9893	20′
12°50′	.3466	.3576	.6424	.9890	10′
13°00′	9.3521	9.3634	10.6366	9.9887	**77°00′**
10′	.3575	.3691	.6309	.9884	76°50′
20′	.3629	.3748	.6252	.9881	40′
30′	.3682	.3804	.6196	.9878	30′
40′	.3734	.3859	.6141	.9875	20′
13°50′	.3786	.3914	.6086	.9872	10′
14°00′	9.3837	9.3968	10.6032	9.9869	**76°00′**
10′	.3887	.4021	.5979	.9866	75°50′
20′	.3937	.4074	.5926	.9863	40′
30′	.3986	.4127	.5873	.9859	30′
40′	.4035	.4178	.5822	.9856	20′
14°50′	.4083	.4230	.5770	.9853	10′
15°00′	9.4130	9.4281	10.5719	9.9849	**75°00′**
10′	.4177	.4331	.5669	.9846	74°50′
20′	.4223	.4381	.5619	.9843	40′
30′	.4269	.4430	.5570	.9839	30′
40′	.4314	.4479	.5521	.9836	20′
15°50′	.4359	.4527	.5473	.9832	10′
16°00′	9.4403	9.4575	10.5425	9.9828	**74°00′**
10′	.4447	.4622	.5378	.9825	73°50′
20′	.4491	.4669	.5331	.9821	40′
30′	.4533	.4716	.5284	.9817	30′
40′	.4576	.4762	.5238	.9814	20′
16°50′	.4618	.4808	.5192	.9810	10′
17°00′	9.4659	9.4853	10.5147	9.9806	**73°00′**
10′	.4700	.4898	.5102	.9802	72°50′
20′	.4741	.4943	.5057	.9798	40′
30′	.4781	.4987	.5013	.9794	30′
40′	.4821	.5031	.4969	.9790	20′
17°50′	.4861	.5075	.4925	.9786	10′
18°00′	9.4900	9.5118	10.4882	9.9782	**72°00′**
	L Cos	L Cot	L Tan	L Sin	←

TABLE 5 LOGARITHMS OF TRIGONOMETRIC FUNCTIONS (continued)

⟶	L Sin	L Tan	L Cot	L Cos	
18°00′	9.4900	9.5118	10.4882	9.9782	**72°00′**
10′	.4939	.5161	.4839	.9778	71°50′
20′	.4977	.5203	.4797	.9774	40′
30′	.5015	.5245	.4755	.9770	30′
40′	.5052	.5287	.4713	.9765	20′
18°50′	.5090	.5329	.4671	.9761	10′
19°00′	9.5126	9.5370	10.4630	9.9757	**71°00′**
10′	.5163	.5411	.4589	.9752	70°50′
20′	.5199	.5451	.4549	.9748	40′
30′	.5235	.5491	.4509	.9743	30′
40′	.5270	.5531	.4469	.9739	20′
19°50′	.5306	.5571	.4429	.9734	10′
20°00′	9.5341	9.5611	10.4389	9.9730	**70°00′**
10′	.5375	.5650	.4350	.9725	69°50′
20′	.5409	.5689	.4311	.9721	40′
30′	.5443	.5727	.4273	.9716	30′
40′	.5477	.5766	.4234	.9711	20′
20°50′	.5510	.5804	.4196	.9706	10′
21°00′	9.5543	9.5842	10.4158	9.9702	**69°00′**
10′	.5576	.5879	.4121	.9697	68°50′
20′	.5609	.5917	.4083	.9692	40′
30′	.5641	.5954	.4046	.9687	30′
40′	.5673	.5991	.4009	.9682	20′
21°50′	.5704	.6028	.3972	.9677	10′
22°00′	9.5736	9.6064	10.3936	9.9672	**68°00′**
10′	.5767	.6100	.3900	.9667	67°50′
20′	.5798	.6136	.3864	.9661	40′
30′	.5828	.6172	.3828	.9656	30′
40′	.5859	.6208	.3792	.9651	20′
22°50′	.5889	.6243	.3757	.9646	10′
23°00′	9.5919	9.6279	10.3721	9.9640	**67°00′**
10′	.5948	.6314	.3686	.9635	66°50′
20′	.5978	.6348	.3652	.9629	40′
30′	.6007	.6383	.3617	.9624	30′
40′	.6036	.6417	.3583	.9618	20′
23°50′	.6065	.6452	.3548	.9613	10′
24°00′	9.6093	9.6486	10.3514	9.9607	**66°00′**
	L Cos	L Cot	L Tan	L Sin	⟵

TABLE 5 LOGARITHMS OF TRIGONOMETRIC FUNCTIONS (continued)

→	L Sin	L Tan	L Cot	L Cos	
24°00′	9.6093	9.6486	10.3514	9.9607	**66°00′**
10′	.6121	.6520	.3480	.9602	65°50′
20′	.6149	.6553	.3447	.9596	40′
30′	.6177	.6587	.3413	.9590	30′
40′	.6205	.6620	.3380	.9584	20′
24°50′	.6232	.6654	.3346	.9579	10′
25°00′	9.6259	9.6687	10.3313	9.9573	**65°00′**
10′	.6286	.6720	.3280	.9567	64°50′
20′	.6313	.6752	.3248	.9561	40′
30′	.6340	.6785	.3215	.9555	30′
40′	.6366	.6817	.3183	.9549	20′
25°50′	.6392	.6850	.3150	.9543	10′
26°00′	9.6418	9.6882	10.3118	9.9537	**64°00′**
10′	.6444	.6914	.3086	.9530	63°50′
20′	.6470	.6946	.3054	.9524	40′
30′	.6495	.6977	.3023	.9518	30′
40′	.6521	.7009	.2991	.9512	20′
26°50′	.6546	.7040	.2960	.9505	10′
27°00′	9.6570	9.7072	10.2928	9.9499	**63°00′**
10′	.6595	.7103	.2897	.9492	62°50′
20′	.6620	.7134	.2866	.9486	40′
30′	.6644	.7165	.2835	.9479	30′
40′	.6668	.7196	.2804	.9473	20′
27°50′	.6692	.7226	.2774	.9466	10′
28°00′	9.6716	9.7257	10.2743	9.9459	**62°00′**
10′	.6740	.7287	.2713	.9453	61°50′
20′	.6763	.7317	.2683	.9446	40′
30′	.6787	.7348	.2652	.9439	30′
40′	.6810	.7378	.2622	.9432	20′
28°50′	.6833	.7408	.2592	.9425	10′
29°00′	9.6856	9.7438	10.2562	9.9418	**61°00′**
10′	.6878	.7467	.2533	.9411	60°50′
20′	.6901	.7497	.2503	.9404	40′
30′	.6923	.7526	.2474	.9397	30′
40′	.6946	.7556	.2444	.9390	20′
29°50′	.6968	.7585	.2415	.9383	10′
30°00′	9.6990	9.7614	10.2386	9.9375	**60°00′**
	L Cos	L Cot	L Tan	L Sin	←

TABLE 5 LOGARITHMS OF TRIGONOMETRIC FUNCTIONS (continued)

⟶	L Sin	L Tan	L Cot	L Cos	
30°00′	9.6990	9.7614	10.2386	9.9375	**60°00′**
10′	.7012	.7644	.2356	.9368	59°50′
20′	.7033	.7673	.2327	9361	40′
30′	.7055	.7701	.2299	.9353	30′
40′	.7076	.7730	.2270	.9346	20′
30°50′	.7097	.7759	.2241	.9338	10′
31°00′	9.7118	9.7788	10.2212	9.9331	**59°00′**
10′	.7139	.7816	.2184	.9323	58°50′
20′	.7160	.7845	.2155	.9315	40′
30′	.7181	.7873	.2127	.9308	30′
40′	.7201	.7902	.2098	.9300	20′
31°50′	.7222	.7930	.2070	.9292	10′
32°00′	9.7242	9.7958	10.2042	9.9284	**58°00′**
10′	.7262	.7986	.2014	.9276	57°50′
20′	.7282	.8014	.1986	.9268	40′
30′	.7302	.8042	.1958	.9260	30′
40′	.7322	.8070	.1930	.9252	20′
32°50′	.7342	.8097	.1903	.9244	10′
33°00′	9.7361	9.8125	10.1875	9.9236	**57°00′**
10′	.7380	.8153	.1847	.9228	56°50′
20′	.7400	.8180	.1820	.9219	40′
30′	.7419	.8208	.1792	.9211	30′
40′	.7438	.8235	.1765	.9203	20′
33°50′	.7457	.8263	.1737	.9194	10′
34°00′	9.7476	9.8290	10.1710	9.9186	**56°00′**
10′	.7494	.8317	.1683	.9177	55°50′
20′	.7513	.8344	.1656	.9169	40′
30′	.7531	.8371	.1629	.9160	30′
40′	.7550	.8398	.1602	.9151	20′
34°50′	.7568	.8425	.1575	.9142	10′
35°00′	9.7586	9.8452	10.1548	9.9134	**55°00′**
10′	.7604	.8479	.1521	.9125	54°50′
20′	.7622	.8506	.1494	.9116	40′
30′	.7640	.8533	.1467	.9107	30′
40′	.7657	.8559	.1441	.9098	20′
35°50′	.7675	.8586	.1414	.9089	10′
36°00′	9.7692	9.8613	10.1387	9.9080	**54°00′**
	L Cos	L Cot	L Tan	L Sin	⟵

TABLE 5 LOGARITHMS OF TRIGONOMETRIC FUNCTIONS (continued)

⟶	L Sin	L Tan	L Cot	L Cos	
36°00′	9.7692	9.8613	10.1387	9.9080	**54°00′**
10′	.7710	.8639	.1361	.9070	53°50′
20′	.7727	.8666	.1334	.9061	40′
30′	.7744	.8692	.1308	.9052	30′
40′	.7761	.8718	.1282	.9042	20′
36°50′	.7778	.8745	.1255	.9033	10′
37°00′	9.7795	9.8771	10.1229	9.9023	**53°00′**
10′	.7811	.8797	.1203	.9014	52°50′
20′	.7828	.8824	.1176	.9004	40′
30′	.7844	.8850	.1150	.8995	30′
40′	.7861	.8876	.1124	.8985	20′
37°50′	.7877	.8902	.1098	.8975	10′
38°00′	9.7893	9.8928	10.1072	9.8965	**52°00′**
10′	.7910	.8954	.1046	.8955	51°50′
20′	.7926	.8980	.1020	.8945	40′
30′	.7941	.9006	.0994	.8935	30′
40′	.7957	.9032	.0968	.8925	20′
38°50′	.7973	.9058	.0942	.8915	10′
39°00′	9.7989	9.9084	10.0916	9.8905	**51°00′**
10′	.8004	.9110	.0890	.8895	50°50′
20′	.8020	.9135	.0865	.8884	40′
30′	.8035	.9161	.0839	.8874	30′
40′	.8050	.9187	.0813	.8864	20′
39°50′	.8066	.9212	.0788	.8853	10′
40°00′	9.8081	9.9238	10.0762	9.8843	**50°00′**
10′	.8096	.9264	.0736	.8832	49°50′
20′	.8111	.9289	.0711	.8821	40′
30′	.8125	.9315	.0685	.8810	30′
40′	.8140	.9341	.0659	.8800	20′
40°50′	.8155	.9366	.0634	.8789	10′
41°00′	9.8169	9.9392	10.0608	9.8778	**49°00′**
10′	.8184	.9417	.0583	.8767	48°50′
20′	.8198	.9443	.0557	.8756	40′
30′	.8213	.9468	.0532	.8745	30′
40′	.8227	.9494	.0506	.8733	20′
41°50′	.8241	.9519	.0481	.8722	10′
42°00′	9.8255	9.9544	10.0456	9.8711	**48°00′**
	L Cos	L Cot	L Tan	L Sin	⟵

TABLE 5 LOGARITHMS OF TRIGONOMETRIC FUNCTIONS (continued)

→	L Sin	L Tan	L Cot	L Cos	
42°00′	9.8255	9.9544	10.0456	9.8711	**48°00′**
10′	.8269	.9570	.0430	.8699	47°50′
20′	.8283	.9595	.0405	.8688	40′
30′	.8297	.9621	.0379	.8676	30′
40′	.8311	.9646	.0354	.8665	20′
42°50′	.8324	.9671	.0329	.8653	10′
43°00′	9.8338	9.9697	10.0303	9.8641	**47°00′**
10′	.8351	.9722	.0278	.8629	46°50′
20′	.8365	.9747	.0253	.8618	40′
30′	.8378	.9772	.0228	.8606	30′
40′	.8391	.9798	.0202	.8594	20′
43°50′	.8405	.9823	.0177	.8582	10′
44°00′	9.8418	9.9848	10.0152	9.8569	**46°00′**
10′	.8431	.9874	.0126	.8557	45°50′
20′	.8444	.9899	.0101	.8545	40′
30′	.8457	.9924	.0076	.8532	30′
40′	.8469	.9949	.0051	.8520	20′
44°50′	.8482	.9975	.0025	.8507	10′
45°00′	9.8495	10.0000	10.0000	9.8495	**45°00′**
	L Cos	L Cot	L Tan	L Sin	←

ANSWERS TO ODD-NUMBERED PROBLEMS

EXERCISE 1-1 PAGE 4

1. $\{1, 2, 3, 4\}$, $\{x|x$ is an integer between 0 and 5$\}$
3. $\{$M, Tu, W, Th, F, S$\}$, $\{x|x$ is a day of the week$\}$
5. $\{$Washington, Adams, Jefferson$\}$, $\{x|x$ is one of the first three presidents of the United States$\}$
7. $\{$Genesis, Exodus, Leviticus, Numbers, Deuteronomy$\}$, $\{x|x$ is a Book of the Pentateuch$\}$
9. None of these
11. $S = T = \varnothing$
13. $a = 0, b = 1, c = 4$
15. $\varnothing, \{1\}, \{2\}, \{3\}, \{1, 2\}, \{1, 3\}, \{2, 3\}, \{1, 2, 3\}$

EXERCISE 1-2 PAGE 8

1. $\{0, 1, 2, 3, 4, 5\}$
3. $\{0, 1, 3, 4, 5, 6, 7\}$
5. $\{4, 5, 6, 7\}$
7. $\varnothing$

9. $\{0, 1, 2, 3, 4, 5\}$

11. $\{0, 1, 3, 4, 5\}$

13. $\{r, s\} = \{r, s\}$

15. $\varnothing \neq \{p, q, r, s, t, u\}$

25. $P \times Q = \{(1, 2), (1, 3), (2, 2), (2, 3)\}$
$Q \times P = \{(2, 1), (2, 2), (3, 1), (3, 2)\}$

EXERCISE 2-1 PAGE 14

1. Commutative axiom
3. Commutative axiom, identity element
5. Commutative axiom, identity element
7. Inverse element
9. Inverse element
11. Distributive axiom
13. Commutative axiom, identity element
15. $-3, \frac{1}{3}$
17. $-1, \frac{1}{1}$
19. $a, \dfrac{1}{-a}$
21. $-\dfrac{5x}{2y}, \dfrac{1}{5x/2y}$
23. $-[-(2x + 3y)], \dfrac{1}{-(2x + 3y)}$

EXERCISE 2-3 PAGE 27

1. $\frac{38}{100}$
3. $-1162/100$
5. $0.8333 \cdots$
7. $3.090909 \cdots$
9. $0.142857142857 \cdots$
15. $4 < 9$
17. $8 > 2$
19. $3 > -3$
21. $-5 < -2$

EXERCISE 3-1 PAGE 32

1. 6
3. -8
5. 0
7. 8
9. 2
11. 9
13. $-9y$
15. $-12x$
17. $-6x$
19. $5x + y - 3$
$-x + 15y - 11$
21. $11x + 2y + 6z$
$7x + 16y - 6z$
23. $6x + xy - 3yz - a$
$2x - xy - 3yz + 21a$
25. $3xy + 13yz - 2x$
$-5xy + 9yz + 4x$
27. $7rs - 3rt + 12st - 3r - 5s$
$7rs - 13rt + 24st - 3r + 5s$
29. $-8y$
31. $-x + y - 1$
33. $-x + 2y$
35. $1 - 2a - 2b$
37. $4x$
39. -9
41. $7xy + 5x - 2y$
43. $1 + x + 10y$
45. $3a + 10ab + 2b - 7$
47. $a + (3b - 2c - 4)$
$a - (-3b + 2c + 4)$
49. $2 + (3 - p + q)$
$2 - (-3 + p - q)$
51. $r + (-2s + 3t - st)$
$r - (2s - 3t + st)$

EXERCISE 3-2 PAGE 36

1. 60
3. -48
5. x^8
7. $4y^{30}$
9. $-12a^3b^3$
11. $-6a^2y^3$
13. a^2b^5
15. $-2x^4y^5$
17. $-6a^2 + 9ab$
19. $-24a^3 + 18a^2$
21. $4x^3y^3 - 2x^2y^2$
23. $2a^2 - ab - b^2$
25. $6x^2 + x - 12$
27. $3m^2 - 5mn + 2n^2$
29. $2a^3b - 6a^2b^2 + 6a^2b^4$
31. $14xy^2 - 21x^2y^2 + 7xy^3 - 35xy$
33. $-4mn + 10m^2n - 12m^2n^2 + 6mn^2$
35. $x^3 + y^3$
37. $y^3 + xy^2 - 3x^2y - 2x^3$
39. $3a^3 - 15a^2b + 22ab^2 - 12b^3$
41. $3x^4 - 5x^3 - 15x^2 + 35x - 42$
43. $6x^2 - 15x + xy - y^2 + 9$
45. $6x^4 - x^3 + 12x^2 + 11x - 28$
47. $2x^3 + 13x^2 + 17x - 12$
49. $32 - 8x - 8x^2 + 2x^3$

EXERCISE 3-3 PAGE 41

1. $3x^2$
3. $-3ab$
5. $-2ab^2$
7. $3 - 2x$
9. $1 - 2x^2y^2$
11. $m - 2 - m^2$
13. $x + 6$
15. $x - 1$
17. $2x + 3 + \frac{-2}{4x - 1}$
19. $a^2 + ab + b^2$
21. $x^2 - 6x + 24$
23. $x^2 - 3x - 1 + \frac{1}{2x + 6}$
25. $2x^2 + x + 1$
27. $3x^2 + x + 2 + \frac{-8x}{x^2 - 2x + 4}$
29. $x^2 + 2xy - y^2$
31. $2x^3 - 3x^2 - 6x + 1 + \frac{-3}{5x - 4}$
33. $x^3 - x^2 + x + 4$

EXERCISE 3-4 PAGE 45

1. $x^2 - 9$
3. $4x^2 - 9y^2$
5. $25 - 9a^4$
7. $a^2 - 8a + 16$
9. $9x^2 - 12xy + 4y^2$
11. $x^6 - 4x^3y^3 + 4y^6$
13. $x^2 - 2x + 1 + 2xy - 2y + y^2$
15. $9x^2 + 16y^2 + 4 - 24xy + 12x - 16y$
17. $x^2 + 7x + 10$
19. $2x^2 - 9x - 5$
21. $8a^2 - 10a - 3$
23. $15a^2 - 14ab - 8b^2$
25. $6r^4 + 5r^2s^2 - 6s^4$
27. $x^3 - 9x^2 + 27x - 27$
29. $125 - 75x + 15x^2 - x^3$
31. $x^6 + 3x^4y^2 + 3x^2y^4 + y^6$
33. $27m^3 - 108m^2xy + 144mx^2y^2 - 64x^3y^3$

35. $x^2 + 4y^2 + z^2 + 4xy - 2xz - 4yz$
37. $a^4 + 2a^2b^2 + b^4 + 2a^2 + 2b^2 + 1$
39. $4x^2 + 9y^2 + 4z^2 + 12xy + 8xz + 12yz$
41. $8x^3 - 12x^2y + 6xy^2 - y^3 - 12x^2 + 12xy - 3y^2 + 6x - 3y - 1$
43. $a^2 - 2ab + b^2 - 1$
45. $1 - 4m + 4m^2 - n^2$
47. $x^2 - 25y^2 + 10y - 1$
49. $x^2 + 2xy + y^2 - z^2 - 2z - 1$

EXERCISE 4-1 PAGE 49

1. $x(x^2 + xy - 3y^3)$
3. $3x(x + y - 2y^2)$
5. $(2a + b)(m - n)$
7. $(x + 3y)(x - 3y)$
9. $(1 + 5y)(1 - 5y)$
11. $2a(x + 4y)(x - 4y)$
13. $(a + 1 + b)(a + 1 - b)$
15. $(3xy + z - 1)(3xy - z + 1)$
17. $(x - 2)^2$
19. $(3a + b)^2$
21. $3(x - 3y)^2$
23. $(x + 2y)(x^2 - 2xy + 4y^2)$
25. $(2b - 1)(4b^2 + 2b + 1)$
27. $5xy(y - x)(y^2 + xy + x^2)$
29. $(2x + y - 2)(4x^2 - 8x + 4 - 2xy + 2y + y^2)$
31. $2x^3y(3x - y)(9x^2 + 3xy + y^2)$
33. $(1 + 9y^2)(1 + 3y)(1 - 3y)$
35. $(x + y)(x - y)(x^2 - xy + y^2)(x^2 + xy + y^2)$
37. $(4x^2 + y^2)(16x^4 - 4x^2y^2 + y^4)$
39. $2ab^2(9a^2 + 1)(3a + 1)(3a - 1)$
41. $(x + 2)^2(x - 2)^2$
43. $(2x - 3y + 5)^2$
45. $(4x - 3y + 1)(4x - 3y - 1)$
47. $(x + 1)(x - 1)(x^2 + 1)(x^4 + 1)$
49. $(x^4 + 4)(x^2 + 2)(x^2 - 2)$

EXERCISE 4-2 PAGE 51

1. $(x + 3)(x + 6)$
3. $(x + 4)(x + 5)$
5. $(x + 3)(x - 5)$
7. $(a - 3b)(a + 7b)$
9. $(a + 2b)(a + 9b)$
11. $(a - 2)(3a + 2)$
13. $(2x + 3)(5x - 1)$
15. $(4 - 3x)(6 + 7x)$
17. $(3xy + 8)(xy + 1)$
19. $x(4x + 1)(3x + 2)$
21. $y^2(x + 2y)(4x - 3y)$
23. $(m^2 + 2)(m^2 + 6)$
25. $(3m^2 + 2)(2m^2 - 3)$
27. $(m^2 + 4)(4m^2 + 1)$
29. $(2m + 1)(2m - 1)(m + 3)(m - 3)$
31. $(x + 2)(x - 1)(x^2 - 2x + 4)(x^2 + x + 1)$
33. $(x + 1)(x + 3)(x^2 - x + 1)(x^2 - 3x + 9)$
35. $(x + y - 2)(x + y + 1)$
37. $2x^2y(x + 2y)(2x - 5y)$

EXERCISE 4-3 PAGE 53

1. $(x - 3)(4 + y)$
3. $(x - y)(a + 2c)$
5. $(3x - a)(y + 3a)$
7. $2(2 + m)(1 + m^2)$
9. $(x - 1)(x - 1 - a)$
11. $(x + y)(x - y - 5)$
13. $(x - y + 1)(x - y - 1)$
15. $(a + 2b - 3)(a - 2b + 3)$
17. $(x + 1 + y - z)(x + 1 - y + z)$
19. $(x + y + a + b)(x + y - a - b)$
21. $(x + y + 6)(x - y + 2)$
23. $(a - b)(a^2 + ab + b^2 - a + b)$
25. $(8a^2 + 1 + 4a)(8a^2 + 1 - 4a)$
27. $(a^2 + b^2 + ab)(a^2 + b^2 - ab)$
29. $(a^2 + 4b^2 + ab)(a^2 + 4b^2 - ab)$
31. $(3x^2 + 2 + 7x)(3x^2 + 2 - 7x)$

EXERCISE 4-4 PAGE 56

1. $\frac{x}{3y}$
3. $\frac{3b^5}{a^2}$
5. $\frac{x}{a}$
7. $\frac{x + 2}{x - 2}$
9. $-\frac{1}{x}$
11. $\frac{b - a}{b + a}$
13. $\frac{a}{2}$
15. $\frac{2a - b}{a - b}$
17. $-\frac{4x + 1}{x + 2}$
19. $\frac{3 - 2x}{3 - x}$
21. $\frac{a^2 + ab + b^2}{a - b}$
23. $\frac{x - y}{x - y + 1}$
25. $\frac{y + 1}{y - 1}$
27. $\frac{c - 1}{c + 3}$

EXERCISE 4-5 PAGE 58

1. 4
3. $\frac{3}{2b}$
5. $\frac{6}{bx}$
7. $\frac{ab}{18}$
9. $\frac{y^3}{24}$
11. -2
13. -2
15. $2(x - y)$
17. $\frac{4 + 2x + x^2}{2 - x}$
19. $\frac{6 + x}{3 + x}$
21. $\frac{2a(x - 2)}{b(x + 7)}$
23. $\frac{(x - 2)(x - 3)}{(x + 2)(x + 7)}$
25. $(a - 3)(a - 1)$
27. $\frac{2(1 - x)}{x}$
29. 1
31. $\frac{(a + 2)(a + 5)}{(a - 2)(3a + 5)}$
33. $\frac{2(x + 2y)(3x - 2y)}{(3x + y)(x + 5y)}$

EXERCISE 4-6 PAGE 62

1. $180x^2y^3$
3. $12x(x + 2)(x - 2)$
5. $(3a - b)(a + 2b)(2a + b)$
7. $\dfrac{5a + b}{3c}$
9. $\dfrac{4}{a - 3}$
11. $\dfrac{bc + ac + ab}{abc}$
13. $\dfrac{x^2 - 2x + 10}{x(x - 5)}$
15. $\dfrac{-x - 39}{(x - 9)(x + 3)}$
17. $\dfrac{2a + 1}{(a + 2)(a - 2)}$
19. $\dfrac{1 + 4b^2 - y^2}{2y + b}$
21. $\dfrac{2x^3 - 4x^2 + 3x - 1}{x^2 + 1}$
23. $\dfrac{19a - 39}{(a + 3)(a - 5)}$
25. $\dfrac{16 - 13a + 3a^2}{(3 - a)(1 - a)(2 - a)}$
27. $\dfrac{m^3 + 10m^2 - 20}{(m - 1)(m^2 - 4)}$
29. $\dfrac{2m}{2m + 1}$
31. $\dfrac{m}{3 - 2m}$
33. $\dfrac{3 - x}{x^2 - x + 1}$
35. $\dfrac{3}{a + 2}$
37. $\dfrac{6x + 11}{(x - 1)(x + 3)}$
39. $\dfrac{3a^2 - 6a - 18}{(2a - 1)(a + 1)(2a + 3)(a - 2)}$
41. $\dfrac{-x^2 + x - 9}{(2x + 7)(x - 1)(x - 2)}$

EXERCISE 4-7 PAGE 66

1. $\frac{2}{31}$
3. $\frac{2}{3}$
5. $\dfrac{bd + ac}{ad - bc}$
7. $\dfrac{x - 3}{x + 3}$
9. $\dfrac{2(x + 1)}{x + 2}$
11. $\dfrac{x + 2}{x + 4}$
13. $\dfrac{1 - 4x}{3 + x}$
15. $\dfrac{x}{x - 9}$
17. 2
19. $\dfrac{(3x - 10)(x + 2)}{(2x + 3)(x - 2)}$
21. $\dfrac{x(x + 4)}{(x + 1)(x + 3)}$
23. $\dfrac{c - 2d}{2c + 3d}$
25. $\dfrac{2 + x}{2 - x}$

EXERCISE 5-1 PAGE 72

1. $\{3, -3\}$ domain, $\{4, -4\}$ range, a relation
3. $\{0, 1, 2, 3\}$ domain, $\{4, -1, 2, 5\}$ range, a function
5. $\{-3, -2, 0\}$ domain, $\{0, 1, 3, 4\}$ range, a relation
7. $\{(-1, -5), (0, -3), (2, 1), (3, 3)\}$
9. $\{(-1, 0), (0, 1), (3, 4), (8, 9), (15, 16)\}$
11. $-1, -2, 1, 3, 5$
13. 10, 5, 10, 2

15. $\frac{1}{2}, 0, \frac{4}{5}, \frac{3}{2}$
17. $s^4 + 1, x^2 - 4x + 5, x^4 - 4x^3 + 8x^2 - 8x + 4$
19. $t^2 + 2ht + h^2 - 3t - 3h, 2ht + h^2 - 3h$
21. $\{x|x \geq 0\}$ domain, $\{y|y \geq 0\}$ range
23. $\{x|-4 \leq x \leq 4\}$ domain, $\{y|y \geq 0\}$ range
25. $\{x|x \in R\}$ domain, $\{y|y \geq 1\}$ range

EXERCISE 5-3 PAGE 80

1. 5
3. $3\sqrt{2}$
5. 5
7. 3, 4, 5
9. $\sqrt{29}, \sqrt{13}, 2\sqrt{17}$
11. Isosceles
13. Isosceles
15. Straight line
17. $x^2 + y^2 + 12x - 4y + 15 = 0$
19. $x^2 + y^2 = 36$
21. $x^2 + y^2 - 6x = 0$
23. $x^2 + y^2 - 4x - 8y = 14$
25. $x^2 + y^2 + 8x - 2y + 16 = 0$
27. $x^2 + y^2 - 6x - 6y + 10 = 0$

EXERCISE 6-1 PAGE 88

1. $x = -1$
3. $x = 0$
5. $x = -5$
7. $x = -\frac{11}{4}$
9. $x = \dfrac{3 + 5a}{9 - 2a}$
11. $x = \dfrac{3a + 13}{2a + 3}$
13. $x = \dfrac{a - 11}{2b - a}$
15. $x = \dfrac{7 - 3a}{c - b - 3}$
17. $x = \frac{8}{5}$
19. $x = \frac{1}{8}$
21. $x = \frac{2}{3}$
23. $x = \frac{3}{4}$
25. No solution
27. No solution
29. No solution

EXERCISE 6-2 PAGE 91

1. 40, 46, 52
3. 10 by 14
5. $6\frac{2}{3}$ days
7. $6\frac{3}{7}$ hr
9. $71\frac{1}{7}$ oz
11. $10\frac{5}{9}$ oz gold, $4\frac{4}{9}$ oz silver
13. $\dfrac{-3}{1}$
15. \$4,000 at 4.75%, \$6,000 at 5.5%
17. 4.75%, 5.25%
19. \$1050
21. $1\frac{2}{7}$ hr, $1\frac{6}{7}$ hr
23. 4 mph

EXERCISE 6-3 PAGE 98

5. Dependent equations
11. Inconsistent equations

EXERCISE 6-4 PAGE 101

1. $\{(\frac{3}{2}, -\frac{1}{2})\}$ 3. $\{(3, -2)\}$ 5. Inconsistent equations
7. $\{(\frac{9}{5}, \frac{9}{5})\}$ 9. $\{(-1, -3)\}$ 11. $\{(-\frac{3}{5}, -\frac{11}{5})\}$
13. $\{(1, -3)\}$ 15. $\{(-3, 4)\}$ 17. $\{(\frac{3}{2}, \frac{3}{2})\}$
19. $\left\{\left(2, -\frac{2}{a}\right)\right\}$ 21. $\left\{\left(\frac{a+b}{a}, \frac{a-b}{b}\right)\right\}$ 23. $\{(-2, 0)\}$
25. $\{(\frac{5}{13}, 5)\}$ 27. $\{(\frac{1}{5}, -\frac{1}{6})\}$ 29. $\{(\frac{1}{3}, -\frac{1}{2})\}$

EXERCISE 6-5 PAGE 104

1. Calves \$43, sheep \$22 3. \$2,500 at 5%, \$3,000 at 6%
5. *A*, 30 days; *B*, 40 days 7. $\frac{1}{6}, \frac{1}{7}$
9. 215 mph, 55 mph 11. 5 mph, 3 mph
13. 15°, 75°, 90°
15. \$2,000 at $2\frac{1}{2}$%, \$6,000 at 3%, \$4,000 at $3\frac{1}{2}$%

EXERCISE 6-6 PAGE 107

1. $\{(-2, 2, -1)\}$ 3. $\{(0, 0, -5)\}$ 5. $\{(\frac{2}{9}, -\frac{1}{18}, -\frac{2}{3})\}$
7. $\{(4, 3, -2)\}$ 9. $\{(\frac{1}{8}, -\frac{1}{8}, \frac{5}{8})\}$ 11. $\left\{\left(\frac{9}{5a}, \frac{1}{5b}, \frac{3}{5c}\right)\right\}$
13. $\{(b + c, -a + b, -a + c)\}$ 15. $\left\{\left(-\frac{3}{a}, -\frac{2}{b}, \frac{3}{c}\right)\right\}$
17. $\{(a, -b, -1)\}$ 19. $\{(-1, 1, -3)\}$ 21. $\{(1, 1, -3)\}$

EXERCISE 6-7 PAGE 111

1. $\frac{24}{5}$ 3. $\frac{1}{9}$ 5. $\frac{32}{15}$ 7. $2, \frac{3}{2}$
9. $\frac{11}{3}, -\frac{121}{9}$ 11. ± 6 13. ± 12 15. 10
17. ± 6 19. ± 10 21. $x = \frac{8}{5}, y = \frac{15}{2}$
23. $x = \frac{9}{5}, y = \frac{6}{5}$ 25. 64, 80 27. \$30.75, \$41, \$51.25

EXERCISE 6-8 PAGE 114

1. $S = kr^2$ 3. $V = khr^2$ 5. $s = kt^2$
7. 16 lb, 44 lb 9. 81 ft 11. $40\frac{24}{25}$ lb, $28\frac{4}{9}$ lb, 16 lb
13. 75 cu in. 15. 12,960 lb 17. Ratio 9 to 16
19. 675 watts 21. Ratio 5 to 3

EXERCISE 7-1 PAGE 120

1. 64	**3.** $\frac{1}{5}$	**5.** 1	**7.** $\frac{1}{4}$	**9.** $\frac{1}{6}$
11. $\frac{1}{25}$	**13.** 7	**15.** $\frac{1}{9}$	**17.** 32	**19.** $\dfrac{2y}{3x}$
21. $\dfrac{b^2}{a^2}$	**23.** $\dfrac{3}{2a^2b}$	**25.** $3a^2b$	**27.** $\dfrac{4y^2}{3x^3z}$	
29. $\dfrac{3a^3c}{2b^5}$	**31.** $\dfrac{9y^2}{x^2}$	**33.** $\dfrac{p^6q^4}{r^2}$	**35.** $\dfrac{1}{p^3q^{15}}$	
37. $\frac{1}{3}$	**39.** $\frac{8}{5}$	**41.** $\dfrac{1}{xy^2 - x^2y}$	**43.** $\dfrac{y^3 - x^3}{y^3 + x^3}$	

EXERCISE 7-2 PAGE 124

1. 3	**3.** 125	**5.** 32	**7.** $\frac{1}{9}$	**9.** 16
11. $\frac{1}{4}$	**13.** 3	**15.** $\frac{1}{4}$	**17.** $\dfrac{1}{a^{1/5}}$	**19.** $3^{5/6}$
21. a	**23.** $\dfrac{81}{x^2}$	**25.** $\dfrac{a^{3/2}}{81}$	**27.** [illegible]^{5}	
29. $a + 2a^{1/2}b^{1/2} + b$	**31.** $\dfrac{y^2}{x^2}$	**33.** $27xy^{7/4}$		
35. $\dfrac{27}{32x^{3/2}y^{1/6}}$	**37.** $\dfrac{y^2}{x^{20}}$	**39.** $\dfrac{4}{3xy^{1/6}}$		

EXERCISE 7-3 PAGE 127

1. $2\sqrt{5}$	**3.** $3a\sqrt{2}$	**5.** $2y\sqrt[3]{4x^2y}$
7. $3xy\sqrt[3]{x^2y}$	**9.** $2x\sqrt[5]{x^3}$	**11.** $\dfrac{1}{3a}\sqrt{15a}$
13. $\frac{4}{3}\sqrt[3]{3}$	**15.** $\dfrac{1}{2x}\sqrt{6}$	**17.** $\dfrac{1}{3x^3}\sqrt{6x}$
19. $\dfrac{1}{2y}\sqrt[3]{12xy^2}$	**21.** $\dfrac{1}{3x}\sqrt[4]{27x^3}$	**23.** $\dfrac{1}{y^2z^2}\sqrt[3]{x^2y^2z^2}$
25. $a\sqrt{3a}$	**27.** $xy\sqrt{3x}$	**29.** $\dfrac{1}{x}\sqrt[3]{5x^2}$
31. $\dfrac{1}{3x}\sqrt[3]{9x^2(x-1)}$	**33.** $\dfrac{1}{bc}\sqrt{bc(a+1)}$	**35.** $\sqrt{25x^2y}$
37. $\sqrt[3]{(4-x)x}$	**39.** $\sqrt{2ab}$	**41.** $\sqrt[3]{15ax^2}$
43. $\sqrt[6]{2}$	**45.** $\sqrt[6]{243}$	**47.** $\sqrt[8]{320}$

EXERCISE 7-4 PAGE 129

1. $2\sqrt{2}$ 3. $11\sqrt{7}$ 5. $5\sqrt{7} - 5\sqrt{2}$ 7. $\frac{8}{3}\sqrt{3}$

9. $3\sqrt[3]{5} - 3\sqrt[3]{2}$ 11. $(8 - 2x)\sqrt{2x}$ 13. $(15x^3y - 2x^2 - x)\sqrt{3y}$

15. $(3b^2 - 2ab - 1)\sqrt[3]{3a^2}$ 17. $(1 + a)\sqrt[3]{2ab^2}$ 19. $-a\sqrt{3}$

21. $\dfrac{3 - x}{2x}\sqrt{2x}$ 23. $\dfrac{(a - 2b)\sqrt{a^2 - b^2}}{a^2 - b^2}$ 25. $\dfrac{4(b\sqrt{2ab} + a^2)}{b^2}$

EXERCISE 7-5 PAGE 132

1. $\sqrt{21}$ 3. $18\sqrt{2}$ 5. $15\sqrt{3}$ 7. $3xy^2\sqrt{6xy}$

9. $9(x + y)$ 11. $a\sqrt[3]{12}$ 13. $\sqrt[6]{72}$

15. $2\sqrt[4]{8}$ 17. $2\sqrt{3}$ 19. -1

21. $-5\sqrt{6}$ 23. $\sqrt{25 - a}$ 25. $8 + 5\sqrt{2}$

27. $\frac{1}{3}$ 29. $\dfrac{2}{x}\sqrt{x}$ 31. x

33. $\dfrac{1}{a}\sqrt[3]{3a}$ 35. $\sqrt[6]{5}$ 37. $\dfrac{1}{b}\sqrt[6]{ab^2}$

39. $\sqrt[4]{x^3}$ 41. $\sqrt{3} - \sqrt{7}$ 43. $\sqrt{3} - \sqrt{2}$

45. $\dfrac{5\sqrt{2} + 2}{23}$ 47. $\dfrac{6 + 5\sqrt{2}}{7}$ 49. $\dfrac{54 - 13\sqrt{14}}{10}$

EXERCISE 8-1 PAGE 139

1. $\dfrac{\pi}{4}, \dfrac{\pi}{3}, \dfrac{3\pi}{4}, \dfrac{5\pi}{6}$ 3. $\dfrac{5\pi}{4}, \dfrac{7\pi}{4}, -\dfrac{73\pi}{180}, -\dfrac{2\pi}{9}$

5. 0.8725, 0.1134, 1.406 7. $30°, 45°, 270°, -18°$

9. $135°, 150°, 22.5°, -12°$

11. $114° \, 35.5'$, $-120° \, 19.3'$, $34° \, 22.7'$, $14° \, 53.8'$

13. 25.6 in., 25.1 in., 41.9 in. 15. $\frac{1}{2}$ rad, $\frac{6}{5}$ rad, 1 rad

17. 12,400 miles, approximately 19. $\frac{2}{15}$ rad, 0.3 rad, $\frac{13}{30}$ rad

21. 2160 miles

EXERCISE 8-2 PAGE 141

5. II, IV, IV, I 7. IV, II, I, IV

9. $-320°, 400°$ 11. $-490°, -130°$ 13. $-35°, -395°$

15. $\dfrac{\pi}{2}, -\dfrac{3\pi}{2}$ 17. $-\dfrac{\pi}{2}, \dfrac{3\pi}{2}$ 19. $-130°, 590°$

21. $-35°, 685°$ 23. $30°, -330°$ 25. $120°, -240°$

27. $\frac{5\pi}{6}, -\frac{7\pi}{6}$

EXERCISE 8-3 PAGE 149

The values of the functions in problems 1 through 11 are given in the order sine, cosine, tangent, cotangent, secant, and cosecant.

1. $\frac{3}{5}, \frac{4}{5}, \frac{3}{4}, \frac{4}{3}, \frac{5}{4}, \frac{5}{3}$ 3. $\frac{12}{13}, -\frac{5}{13}, -\frac{12}{5}, -\frac{5}{12}, -\frac{13}{5}, \frac{13}{12}$

5. $-\frac{1}{2}, -\frac{\sqrt{3}}{2}, \frac{1}{\sqrt{3}}, \sqrt{3}, -\frac{2}{\sqrt{3}}, -2$

7. $-\frac{\sqrt{5}}{3}, \frac{2}{3}, -\frac{\sqrt{5}}{2}, -\frac{2}{\sqrt{5}}, \frac{3}{2}, -\frac{3}{\sqrt{5}}$

9. $-\frac{\sqrt{2}}{3}, -\frac{\sqrt{7}}{3}, \frac{\sqrt{2}}{\sqrt{7}}, \frac{\sqrt{7}}{\sqrt{2}}, -\frac{3}{\sqrt{7}}, -\frac{3}{\sqrt{2}}$

11. $\frac{\sqrt{11}}{4}, -\frac{\sqrt{5}}{4}, -\frac{\sqrt{11}}{\sqrt{5}}, -\frac{\sqrt{5}}{\sqrt{11}}, -\frac{4}{\sqrt{5}}, \frac{4}{\sqrt{11}}$

13. III and IV 15. I and III 17. II 19. III

21. Sine and cosecant positive; others negative
23. Cosine and secant positive; others negative
25. Tangent and cotangent positive; others negative
27. All positive
29. Tangent and cotangent positive; others negative

Except for the quadrantal angles, the values of the functions in problems 31 through 55 are given in the order sine, cosine, tangent. The reciprocals of these values give the values of the other functions.

31. $\frac{12}{13}, \frac{5}{13}, \frac{12}{5}$ 33. $-\frac{8}{17}, -\frac{15}{17}, \frac{8}{15}$

35. $\frac{1}{\sqrt{17}}, -\frac{4}{\sqrt{17}}, -\frac{1}{4}$

37. $\cos\theta = 0, \quad \cot\theta = 0, \quad \csc\theta = 1$
39. $\sin\theta = 0, \quad \cos\theta = 1, \quad \tan\theta = 0$
41. $\sin 3\pi = 0, \quad \cos 3\pi = -1, \quad \tan 3\pi = 0, \quad \sec 3\pi = -1$
43. $\sin 10\pi = 0, \quad \cos 10\pi = 1, \quad \sec 10\pi = 1, \quad \tan 10\pi = 0$
45. $\frac{9}{41}, \frac{40}{41}, \frac{9}{40}; \frac{9}{41}, -\frac{40}{41}, -\frac{9}{40}$ 47. $\frac{24}{25}, -\frac{7}{25}, -\frac{24}{7}; -\frac{24}{25}, \frac{7}{25}, -\frac{24}{7}$

49. $\frac{4}{\sqrt{17}}, -\frac{1}{\sqrt{17}}, -4; -\frac{4}{\sqrt{17}}, \frac{1}{\sqrt{17}}, -4$

51. $\frac{\sqrt{5}}{3}, -\frac{2}{3}, -\frac{\sqrt{5}}{2}; -\frac{\sqrt{5}}{3}, -\frac{2}{3}, \frac{\sqrt{5}}{2}$

53. $\frac{3}{\sqrt{34}}, \frac{5}{\sqrt{34}}, \frac{3}{5}; -\frac{3}{\sqrt{34}}, -\frac{5}{\sqrt{34}}, \frac{3}{5}$

55. $\sin\theta = 1, \quad \cot\theta = 0, \quad \csc\theta = 1; \quad \sin\theta = -1, \quad \cot\theta = 0,$ $\csc\theta = -1$

EXERCISE 9-1 PAGE 155

1. 604.5 ft, 605.5 ft **3.** 65.95 ft, 66.05 ft
5. 8.1695 cm, 8.1705 cm **7.** 8560, 8600, 9000
9. 4070, 4100, 4000 **11.** 1.00, 1.0, 1
13. 279,000, 280,000, 300,000
15. 6.00×10^3, 6.0×10^3, 6×10^3 **17.** 9.6 **19.** 8.6
21. 352 **23.** 100 **25.** 12 **27.** 28

EXERCISE 9-2 PAGE 157

1. 0.5712 **3.** 1.054 **5.** 0.7314
7. 0.4592 **9.** 26.45 **11.** 0.0698
13. 7° 10′ **15.** 7° 20′ **17.** 27° 50′
19. 33° 40′ **21.** 54° 20′ **23.** 65° 20′
25. 49° 30′

EXERCISE 9-3 PAGE 159

1. 0.3489 **3.** 0.7467 **5.** 0.8971
7. 0.9713 **9.** 2.295 **11.** 0.6116
13. 1.039 **15.** 1.425 **17.** 0.9998
19. 0.4631 **21.** 3.560 **23.** 31° 44′
25. 64° 26′ **27.** 37° 16′ **29.** 29° 9′
31. 66° 47′ **33.** 38° 48′ **35.** 45° 32′
37. 39° 38′ **39.** 2° 25′ **41.** 84° 47′

EXERCISE 9-4 PAGE 163

1. 27° **3.** 35° **5.** $\frac{\pi}{4}$ **7.** $\frac{\pi}{3}$
9. 42° **11.** 40° **13.** 0.6293 **15.** -1.600
17. -1.367 **19.** -0.1822 **21.** -1.269 **23.** 0.8660

EXERCISE 9-5 PAGE 166

The values of the functions in problems 1 through 10 are in the order sine, cosine, tangent. The reciprocals of these values give the values of the other functions.

1. a. A: $\frac{4}{5}, \frac{3}{5}, \frac{4}{3}$; B: $\frac{3}{5}, \frac{4}{5}, \frac{3}{4}$
   b. A: $\frac{15}{17}, \frac{8}{17}, \frac{15}{8}$; B: $\frac{8}{17}, \frac{15}{17}, \frac{8}{15}$
   c. A: $\frac{5}{\sqrt{74}}, \frac{7}{\sqrt{74}}, \frac{5}{7}$; B: $\frac{7}{\sqrt{74}}, \frac{5}{\sqrt{74}}, \frac{7}{5}$

3. A: $\frac{24}{25}, \frac{7}{25}, \frac{24}{7}$; B: $\frac{7}{25}, \frac{24}{25}, \frac{7}{24}$

5. A: $\frac{35}{37}, \frac{12}{37}, \frac{35}{12}$; B: $\frac{12}{37}, \frac{35}{37}, \frac{12}{35}$

7. A: $\frac{6}{\sqrt{85}}, \frac{7}{\sqrt{85}}, \frac{6}{7}$; B: $\frac{7}{\sqrt{85}}, \frac{6}{\sqrt{85}}, \frac{7}{6}$

9. $\sin 55°$ **11.** $\tan 37°$ **13.** $\sec \frac{\pi}{5}$ **15.** $\cos \frac{\pi}{10}$

17. $45°$ **19.** $27°$ **21.** $0°$ **23.** $\frac{5\pi}{36}$

EXERCISE 9-6 PAGE 169

13. $2 - \sqrt{3}$ **15.** 1 **17.** $\frac{\sqrt{6} - \sqrt{2}}{4}$

19. $-\frac{\sqrt{3}}{2}, -\frac{1}{2}, \sqrt{3}$ **21.** $\frac{1}{2}, \frac{\sqrt{3}}{2}, \frac{1}{\sqrt{3}}$ **23.** $-\frac{1}{2}, -\frac{\sqrt{3}}{2}, \frac{1}{\sqrt{3}}$

25. $\frac{1}{\sqrt{2}}, \frac{-1}{\sqrt{2}}, -1$ **27.** $-\frac{1}{\sqrt{2}}, -\frac{1}{\sqrt{2}}, 1$

29. $\frac{\sqrt{3}}{2}, -\frac{1}{2}, -\sqrt{3}$

EXERCISE 9-7 PAGE 174

1. $B = 53°$, $b = 57$, $c = 71$ **3.** $A = 61°$, $a = 5.6$, $b = 3.1$
5. $B = 39°$, $a = 56$, $c = 72$ **7.** $A = 52°$, $B = 38°$, $c = 110$
9. $A = 41°$, $B = 49°$, $b = 7.0$ **11.** $A = 30°$, $B = 60°$, $c = 97$
13. $B = 57° 30'$, $b = 397$, $c = 471$
15. $B = 38° 20'$, $a = 0.787$, $c = 1.00$
17. $A = 38° 00'$, $B = 52° 00'$, $c = 5.71$
19. $A = 51° 00'$, $B = 39° 00'$, $b = 33.9$
21. $B = 27° 30'$, $a = 5322$, $b = 2770$
23. $A = 37° 53'$, $B = 52° 7'$, $c = 6838$
25. $A = 29° 20'$, $b = 1.344$, $c = 1.542$
27. $A = 50° 49'$, $B = 39° 11'$, $c = 108.7$
29. $51° 30'$, $51° 30'$, 71.5 in., 71.5 in.
31. 19.7 **33.** $39°$, 17 in.

EXERCISE 9-8 PAGE 178

1. 58°, 15 ft **3.** 130 ft **5.** 22.9 ft
7. 58° **9.** 540 ft **11.** S 54° W, 44 miles
13. 22.4 ft **15.** 48 ft **17.** 800 ft, 13,000 ft
19. 959 ft, 121 ft **21.** 25°

EXERCISE 9-9 PAGE 184

1. 260 lb, 39°, 51° **3.** 263 mph, N 81° 30′ E
5. 164 lb, 281 lb **7.** 180 lb, 312 lb
9. Heading makes 36° with perpendicular to stream, 3.3 mph
11. 124 mph, 118 mph **13.** 21° 00′
15. 86 lb **17.** 120 lb, 260 lb

EXERCISE 10-1 PAGE 191

1. $\sin\theta = \tan\theta\cos\theta,\ \cos\theta = \dfrac{\sin\theta}{\tan\theta}$

3. $\sin\theta = \pm\sqrt{1-\cos^2\theta},\ \cos\theta = \pm\sqrt{1-\sin^2\theta}$

5. $\csc\theta = \pm\sqrt{1+\cot^2\theta},\ \cot\theta = \pm\sqrt{\csc^2\theta - 1}$

7. I, II **9.** II, IV

19. $\sin\theta = \pm\sqrt{1-\cos^2\theta},\ \tan\theta = \pm\dfrac{\sqrt{1-\cos^2\theta}}{\cos\theta},\ \sec\theta = \dfrac{1}{\cos\theta}$

$\csc\theta = \pm\dfrac{1}{\sqrt{1-\cos^2\theta}},\ \cot\theta = \pm\dfrac{\cos\theta}{\sqrt{1-\cos^2\theta}}$

EXERCISE 10-4 PAGE 203

1. $\dfrac{\sqrt{6}+\sqrt{2}}{4}$ **3.** $\dfrac{\sqrt{2}-\sqrt{6}}{4}$ **5.** $-\dfrac{\sqrt{6}+\sqrt{2}}{4}$

7. $\dfrac{\sqrt{2}-\sqrt{6}}{4}$ **13.** $-\sin\theta$ **15.** $-\cos\theta$

17. $-\cos\theta$ **27.** $\frac{63}{65}, \frac{33}{65}$ **29.** $\frac{13}{85}, -\frac{77}{85}$

EXERCISE 10-5 PAGE 206

1. $\dfrac{\sqrt{2}-\sqrt{6}}{4}, \sqrt{3}-2$ **3.** $\dfrac{\sqrt{6}+\sqrt{2}}{4}, -2-\sqrt{3}$

5. $\dfrac{\sqrt{2}-\sqrt{6}}{4}, 2-\sqrt{3}$ 7. $-\dfrac{\sqrt{6}+\sqrt{2}}{4}, 2+\sqrt{3}$

33. $\frac{3}{5}, -\frac{3}{4}; \frac{117}{125}, \frac{-117}{44}$ 35. $\dfrac{\sqrt{2}}{10}, \dfrac{1}{7}; \dfrac{7\sqrt{2}}{10}, -7$

EXERCISE 10-6 PAGE 210

5. sin 110° 7. cos 40° 9. $\frac{1}{2}$ sin 140°
11. $\frac{1}{2}\cos 10x$ 13. cot 42° 15. $5\tan 8x$
17. $\sin 2A = -\frac{24}{25}$, $\cos 2A = \frac{7}{25}$, $\tan 2A = -\frac{24}{7}$
19. $\sin 2A = -\dfrac{4\sqrt{2}}{9}$, $\cos 2A = -\dfrac{7}{9}$, $\tan 2A = \dfrac{4\sqrt{2}}{7}$

EXERCISE 10-7 PAGE 214

1. $\dfrac{1}{2}, \dfrac{\sqrt{3}}{2}, \dfrac{1}{\sqrt{3}}$ 3. $\frac{1}{2}\sqrt{2+\sqrt{3}}, \frac{1}{2}\sqrt{2-\sqrt{3}}, 2+\sqrt{3}$
5. $\frac{1}{2}\sqrt{2-\sqrt{2}}, \frac{1}{2}\sqrt{2+\sqrt{2}}, \sqrt{2}-1$
7. $\frac{1}{2}\sqrt{2+\sqrt{3}},\ -\frac{1}{2}\sqrt{2-\sqrt{3}},\ -2-\sqrt{3}$ 9. cos 21°
11. tan 80° 13. tan 22° 15. $2\cos^2 35°$ 17. $\dfrac{1+\cos x}{2}$
19. $\dfrac{1-\cos x}{1+\cos x}$ 21. $\dfrac{1+\cos 10x}{2}$ 23. $\dfrac{3}{\sqrt{10}}, \dfrac{1}{\sqrt{10}}, 3$
25. $\dfrac{2}{\sqrt{6}}, -\dfrac{1}{\sqrt{3}}, -\sqrt{2}$
27. $\dfrac{\sqrt{50-5\sqrt{10}}}{10}, \dfrac{\sqrt{50+5\sqrt{10}}}{10}, \dfrac{\sqrt{10}-1}{3}$

EXERCISE 10-8 PAGE 217

1. sin 60° + sin 30° 3. $\frac{1}{2}$(cos 66° + cos 18°)
5. $\sin 4\theta + \sin 2\theta$ 7. $\frac{1}{2}$(sin 88° − sin 8°)
9. $\frac{1}{2}(\cos 7\theta + \cos\theta)$ 11. 2 sin 24° cos 8°
13. 2 cos 15° cos 10° 15. $-2\sin 3x \sin x$
17. $2\cos 7x \sin 2x$ 19. 2 sin 35° sin 10°
21. $\dfrac{2+\sqrt{3}}{4}$ 23. $\dfrac{\sqrt{2}+1}{4}$ 25. $-\dfrac{\sqrt{2}}{2}$

EXERCISE 11-1 PAGE 228

1. $-2+3i$ 3. $2+4i$ 5. 0 7. $x=3,\ y=4$
9. $x=-2,\ y=\frac{3}{2}$ 11. $x=0,\ y=\frac{3}{2}$ 13. $11+8i$

15. $-5 - 6i$ 17. $10 - 3i$ 19. $5 + 2i$ 21. $22i$
23. $12i\sqrt{2}$ 25. $6 + 13i$ 27. 52 29. $-6 - 10i$
31. $7 + 9i$ 33. $1 - 5i$ 35. $\frac{16}{25} + \frac{12}{25}i$
37. $\frac{11}{37} - \frac{8}{37}i$ 39. $4 + 3i$ 41. $-i$

EXERCISE 11-2 PAGE 233

1. $2 - 3i$ 3. $0 - 2i$ 5. $-2 - 6i$ 7. $-4 + 3i$
9. $\sqrt{2}(\cos 225° + i \sin 225°)$ 11. $4(\cos 0° + i \sin 0°)$
13. $4(\cos 330° + i \sin 330°)$ 15. $3(\cos 270° + i \sin 270°)$
17. $2(\cos 90° + i \sin 90°)$ 19. $6(\cos 300° + i \sin 300°)$
21. $2\sqrt{7}(\cos 240° + i \sin 240°)$ 23. $\sqrt{5}(\cos 27° + i \sin 27°)$
25. $\sqrt{3} + i$ 27. $-\sqrt{2} + i\sqrt{2}$
29. $3i$ 31. $-9i$ 33. $-6.69 - 7.43i$

EXERCISE 11-3 PAGE 236

1. $6i$ 3. $-9 + 9i\sqrt{3}$ 5. $27i$ 7. -2
9. $\frac{7}{2} - \frac{7i\sqrt{3}}{2}$ 11. $2\sqrt{2}(\cos 75° + i \sin 75°)$
13. $20\sqrt{2}(\cos 135° + i \sin 135°)$ 15. $8\sqrt{2}(\cos 315° + i \sin 315°)$
17. $2(\cos 30° + i \sin 30°)$ 19. $\frac{\sqrt{2}}{2}(\cos 345° + i \sin 345°)$

EXERCISE 11-4 PAGE 241

1. $4(\cos 70° + i \sin 70°)$ 3. $64(\cos 60° + i \sin 60°)$
5. $\cos 225° + i \sin 225°$ 7. $4(\cos 180° + i \sin 180°)$
9. $32(\cos 240° + i \sin 240°)$ 11. $1024(\cos 330° + i \sin 330°)$
13. $2(\cos 21° + i \sin 21°)$, $2(\cos 201° + i \sin 201°)$
15. $4(\cos 49° + i \sin 49°)$, $4(\cos 169° + i \sin 169°)$, $4(\cos 289° + i \sin 289°)$
17. $2[\cos (21° + k \cdot 90°) + i \sin (21° + k \cdot 90°)]$ where $k = 0, 1, 2, 3$
19. $\sqrt[6]{2}(\cos 15° + i \sin 15°)$, $\sqrt[6]{2}(\cos 135° + i \sin 135°)$, $\sqrt[6]{2}(\cos 255° + i \sin 255°)$
21. $\sqrt[4]{8}[\cos (67.5° + k \cdot 90°) + i \sin (67.5° + k \cdot 90°)]$, $k = 0, 1, 2, 3$
23. $\cos (54° + k \cdot 72°) + i \sin (54° + k \cdot 72°)$, where $k = 0, 1, 2, 3, 4$
25. $-32\sqrt{3} + 32i$ 27. i 29. $1024i$
31. $0.6840 + 1.879i$, $-1.970 - 0.3472i$, $1.286 - 1.532i$
33. $-0.493 + 1.45i$, $-1.01 - 1.15i$, $1.50 - 0.299i$

35. $\frac{\sqrt{2}}{2}(1+i), \frac{\sqrt{2}}{2}(-1+i), \frac{\sqrt{2}}{2}(-1-i), \frac{\sqrt{2}}{2}(1-i)$

37. $\cos(22\tfrac{1}{2}° + k \cdot 90°) + i \sin(22\tfrac{1}{2}° + k \cdot 90°), k = 0, 1, 2, 3$

EXERCISE 12-1 PAGE 244

1. $\{-\sqrt{5}, \sqrt{5}\}$
3. $\{-4i, 4i\}$
5. $\left\{-\frac{b}{a}i, \frac{b}{a}i\right\}$
7. $\{1, -2\}$
9. $\{1, 8\}$
11. $\{\frac{1}{2}, 1\}$
13. $\{-\frac{2}{9}, \frac{2}{9}\}$
15. $\{-\frac{4}{7}, \frac{4}{7}\}$
17. $\{\frac{1}{2}, -3\}$
19. $\{-\frac{1}{9}, -\frac{3}{2}\}$
21. $\{-\frac{1}{4}, \frac{2}{3}\}$
23. $\{-a, b\}$
25. $\left\{-b, \frac{1}{a}\right\}$

EXERCISE 12-2 PAGE 247

1. $\{-3, -1\}$
3. $\{-4, 5\}$
5. $\{-6, 5\}$
7. $\{\frac{1}{2}, 2\}$
9. $\{1, 1\}$
11. $\left\{\frac{-2-\sqrt{2}}{2}, \frac{-2+\sqrt{2}}{2}\right\}$
13. $\{-3-2i, -3+2i\}$
15. $\left\{\frac{2-i\sqrt{6}}{3}, \frac{2+i\sqrt{6}}{3}\right\}$
17. $\{-4, -2\}$
19. $\{-\frac{1}{2}, 2\}$
21. $\{1, -\frac{5}{3}\}$
23. $\left\{\frac{-5-\sqrt{17}}{4}, \frac{-5+\sqrt{17}}{4}\right\}$
25. $\left\{\frac{9-\sqrt{17}}{8}, \frac{9+\sqrt{17}}{8}\right\}$
27. $\left\{\frac{5-\sqrt{17}}{8}, \frac{5+\sqrt{17}}{8}\right\}$
29. $\left\{\frac{5-\sqrt{7}}{9}, \frac{5+\sqrt{7}}{9}\right\}$
31. $\left\{\frac{-1-3i}{2}, \frac{-1+3i}{2}\right\}$
33. $\left\{\frac{-6-\sqrt{14}}{5}, \frac{-6+\sqrt{14}}{5}\right\}$
35. $\left\{\frac{8-i\sqrt{6}}{5}, \frac{8+i\sqrt{6}}{5}\right\}$
37. $\left\{\frac{8-i\sqrt{14}}{13}, \frac{8+i\sqrt{14}}{13}\right\}$
39. $\left\{\frac{7-i\sqrt{11}}{10}, \frac{7+i\sqrt{11}}{10}\right\}$
41. $\left\{\left(\frac{1}{m}, 1\right)\right\}$

EXERCISE 12-3 PAGE 249

1. 5, 11 and $-5, -11$
3. 12, 14 and $-14, -12$
5. (*a*) 40 yd by 120 yd or 60 yd by 80 yd, (*b*) 50 yd by 100 yd
7. 9 in. by 12 in.
9. 200 mph and 700 mph

11. (*a*) 50 yd by 225 yd or 75 yd by 150 yd, (*b*) 100 yd by 150 yd
13. After 12 min and 20 min
15. (*a*) 336 ft
(*b*) 4 sec
17. 10 sec

EXERCISE 12-4 PAGE 253

1. $\{-3, -2, 2, 3\}$
3. $\{-\frac{1}{2}, -\sqrt{2}, \frac{1}{2}, \sqrt{2}\}$
5. $\left\{-\frac{2\sqrt{3}}{3}, -i\sqrt{3}, \frac{2\sqrt{3}}{3}, i\sqrt{3}\right\}$
7. $\{4\}$
9. $\{27, 64\}$
11. $\{-1, 2\}$
13. $\{-1, \frac{4}{5}\}$
15. $\left\{1, 4, \frac{5-\sqrt{37}}{2}, \frac{5+\sqrt{37}}{2}\right\}$
17. $\left\{-1, 4, \frac{3-\sqrt{13}}{2}, \frac{3+\sqrt{13}}{2}\right\}$
19. $\{\frac{1}{3}, \frac{6}{5}\}$
21. $\{-\frac{3}{7}\}$
23. $\{-\frac{3}{5}, 3\}$
25. $\{-1, 2, -1-\sqrt{6}, -1+\sqrt{6}\}$

EXERCISE 12-5 PAGE 255

1. $\{-3, 3\}$
3. $\{1, 3\}$
5. $\{-3\}$
7. $\varnothing$
9. $\{9\}$
11. $\varnothing$
13. $\{-1, 7\}$
15. $\{2\}$
17. $\{5\}$
19. $\{4\}$
21. $\{2\}$
23. $\{2, \frac{5}{2}\}$
25. $\{-1, 11\}$
27. $\{-2, 1\}$

EXERCISE 12-6 PAGE 259

1. Real, unequal, rational; 6, 5
3. Real, unequal, irrational; $-\frac{1}{2}, -\frac{7}{2}$
5. Real, equal, rational; $-3, \frac{9}{4}$
7. Imaginary, unequal; $0, \frac{11}{5}$
9. Real, unequal; irrational; $0, -\frac{4}{7}$
11. Real, unequal, irrational; $-\frac{4}{3}\sqrt{3}, \frac{1}{3}\sqrt{6}$
13. Real, unequal, irrational; $-\frac{1}{2}\sqrt{10}, -3$
15. $x^2 + 7x + 12 = 0$
17. $9x^2 - 6x - 8 = 0$
19. $10x^2 - 7x - 12 = 0$
21. $x^2 + 25 = 0$
23. $x^2 - 12x + 30 = 0$
25. $x^2 - 2\sqrt{7}x + 52 = 0$
27. $k = \pm 2$
29. $k = -11$
31. $k = -\frac{1}{2}$
33. $k = -18$
35. $k = 8, k = 0$
37. $k = 4$
39. $k = 7$

EXERCISE 12-7 PAGE 265

1. min $(-2, 1)$
3. max $(4, 19)$
5. min $(\frac{3}{4}, -\frac{1}{8})$
7. max $(\frac{1}{2}, -\frac{5}{2})$
9. max $(\frac{1}{3}, \frac{8}{3})$
11. 60 rd by 60 rd
13. 50 rd by 200 rd
15. 45

EXERCISE 12-8 PAGE 270

Solutions obtained graphically should differ, at most, only slightly from the results given here.

1. $\{(4.8, -3.8), (0.2, 0.8)\}$ **3.** $\{(3, 4), (-4, -3)\}$
5. $\{(-2.3, 1.6), (-2.3, -1.6), (2.3, 1.6), (2.3, -1.6)\}$
7. $\{(-3.2, 1.4), (-3.2, -1.4), (3.2, 1.4), (3.2, -1.4)\}$
9. $\{(3, 2.5), (-3, -2.5)\}$ **11.** $\{(0, -4), (3.5, 2), (-3.5, 2)\}$

EXERCISE 12-9 PAGE 274

1. $\{(-6, 5), (-6, -5), (6, -5), (6, 5)\}$
3. $\{(\sqrt{5}, -\sqrt{11}), (-\sqrt{5}, -\sqrt{11}), (-\sqrt{5}, \sqrt{11}), (\sqrt{5}, \sqrt{11})\}$
5. $\{(-i, 2i), (-i, -2i), (i, 2i), (i, -2i)\}$
7. $\{(-i\sqrt{3}, \sqrt{5}), (-i\sqrt{3}, -\sqrt{5}), (i\sqrt{3}, \sqrt{5}), (i\sqrt{3}, -\sqrt{5})\}$
9. $\{(-2, 1), (-2, -1), (2, -1), (2, 1)\}$ **11.** $\{(2, -1), (-4, 11)\}$
13. $\{(2 + i, 3 - i), (2 - i, 3 + i)\}$ **15.** $\{(2, 4), (-\frac{4}{5}, -\frac{22}{5})\}$
17. $\{(2, 1), (-2, -1), (\sqrt{3}, 0), (-\sqrt{3}, 0)\}$
19. $\{(\sqrt{2}, -3\sqrt{2}), (-\sqrt{2}, 3\sqrt{2})\}$
21. $\{(0, i), (0, -i), (2i, i), (-2i, -i)\}$
23. $\{(\sqrt{15}, 0), (-\sqrt{15}, 0), (1, -2), (-1, 2)\}$
25. $\left\{(3, -1), (-3, 1), \left(\frac{i\sqrt{34}}{2}, \frac{3i\sqrt{34}}{2}\right), \left(-\frac{i\sqrt{34}}{2}, -\frac{3i\sqrt{34}}{2}\right)\right\}$
27. $\left\{(3, -2), (-3, 2), \left(\frac{8}{\sqrt{7}}, \frac{-4}{\sqrt{7}}\right), \left(\frac{-8}{\sqrt{7}}, \frac{4}{\sqrt{7}}\right)\right\}$
29. $\{(-3, -1), (-1, -3), (3 - 2i, 3 + 2i), (3 + 2i, 3 - 2i)\}$
31. $\{(-4, -2), (-2, -4), (-8, -6), (-6, -8)\}$

EXERCISE 13-2 PAGE 282

1. $\{x > 3\}$ **3.** $\{x > 4\}$ **5.** $\{x < -\frac{1}{2}\}$
7. $\{-2 < x < 2\}$ **9.** $\{-5 \leq x \leq 5\}$ **11.** $\{-1 < x < 3\}$
13. $\{-2 < x < 2.5\}$ **15.** $\{1 < x < 3.5\}$
17. $\{x < 0 \text{ or } x > 1\}$ **19.** $\{x < -4 \text{ or } x > 4\}$
21. $\{x < 0\}$

EXERCISE 13-4 PAGE 287

1. $\{-1 < x < 3\}$ **3.** $\{-1 < x < 2\}$
5. No solution, or the empty set $\varnothing$ **7.** $\{1 < x < 5\}$
9. $\{2 < x < 7\}$ **11.** $\{-\frac{9}{5} \leq x \leq 3\}$

13. The angular region below the line $2x - y - 1 = 0$ and above the line $x - 8y + 1 = 0$, as shown in the figure below:

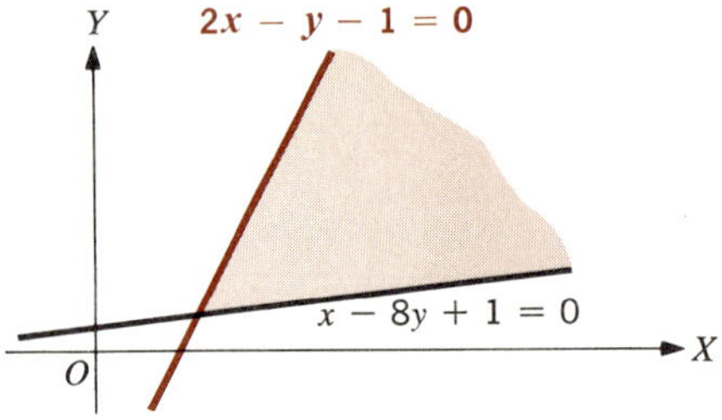

15. The region below the line $2x - y = 2$ and above the line $2x + y = 2$
17. The region above the line $4x + 3y - 6 = 0$
19. The angular region below the line $3x + 4y + 14 = 0$ and also below the line $2x - 5y - 6 = 0$
21. No solution
23. The region inside the circle $x^2 + y^2 = 16$ and below the line $y - 3x + 9 = 0$
25. The region inside the circle $x^2 + y^2 = 16$ and inside the parabola $x^2 - 4y = 0$

EXERCISE 14-1 PAGE 291

1. $1, 3, 5, 7;\quad a_{20} = 39$
3. $-2, 4, -8, 16;\quad a_{20} = 2^{20}$
5. $\frac{1}{2}, \frac{1}{6}, \frac{1}{12}, \frac{1}{20};\quad a_{20} = \frac{1}{420}$
7. $-1, \frac{1}{2}, -\frac{1}{3}, \frac{1}{4};\quad a_{20} = \frac{1}{20}$
9. $1, \frac{4}{5}, \frac{8}{10}, \frac{16}{17};\quad a_{20} = \dfrac{2^{20}}{401}$
11. 3, 9, 21, 45
13. $5, \frac{1}{5}, 5, \frac{1}{5}$
15. $-2, -1, -\frac{1}{3}, -\frac{1}{12}$

EXERCISE 14-2 PAGE 294

1. 22, 92
3. 20, 408.5
5. $-111, -816$
7. $38x + 23y,\ 260x + 143y$
9. 1470
11. 1785
13. 4, 7, 10, 13, 16, 19
15. $-43, -32, -21, -10, 1, 12, 23, 34$
19. $n = 15,\ S_n = 330$
21. $a_1 = -4,\ a_n = 21$
23. $n = 5,\ a_1 = 14$
25. $n = 4,\ a_1 = 14$
27. $a_1 = -9,\ n = 4$ and $a_1 = 7,\ n = 12$
29. 176 ft, 576 ft
31. $5,900
33. $6,405
35. $2,530

EXERCISE 14-3 PAGE 299

1. 729, 1093
3. −128, −84
5. $\frac{4}{9}, \frac{484}{9}$
7. $\frac{1}{3}, \frac{547}{3}$
9. 162, 54, 18, 6, 2
11. 8, −4, 2, −1, $\frac{1}{2}$, $-\frac{1}{4}$
13. $n = 9$, $S_n = 513$
15. $a_1 = \frac{1}{3}$, $S_n = \frac{364}{3}$
17. $r = \frac{2}{3}$, $S_n = \frac{65}{64}$
19. $n = 3$, $a_n = \frac{12}{5}$
21. $r = \frac{1}{4}$, $a_n = 1$ and $r = -\frac{5}{4}$, $a_n = 25$
23. $651.85
25. 256N
27. 59.05%, 7
29. $252\frac{16}{27}$ ft
31. The first proposition pays $5,800 for the fifth year and $24,500 for the 5-year period. The second proposition pays $5,856.40 for the fifth year and $24,420.40 for the 5-year period.

EXERCISE 14-4 PAGE 303

1. 1
3. $\frac{10}{3}$
5. $-\frac{25}{8}$
7. 10
9. $\frac{2}{3}$
11. $\frac{43}{99}$
13. $\frac{17}{33}$
15. $\frac{1043}{333}$
17. $\frac{431}{99}$
19. $\frac{1}{7}$
21. 2, $1\frac{511}{512}$
23. 1080 ft

EXERCISE 15-1 PAGE 306

1. 16, 12
3. 100, 48
5. 30,000
7. 1024
9. 120, 120
11. 210
13. 144
15. 420
17. 72, 144
19. 30,240; 17,280; 8640

EXERCISE 15-2 PAGE 312

1. 7, 210, 840, 5040
5. 5040; 40,320; 604,800
7. 60
9. 5040
11. 4320, 720
13. 34,560
15. 210; 3360; 34,650
17. 90,090
19. 13,860
21. 28,800; 2880
23. 1440, 3600

EXERCISE 15-3 PAGE 316

1. 465; 2600; 10,626; 1287
3. 2380
5. 36, 84
7. 35
9. 126, 256
11. 1960
13. 20,592
15. 800
17. 42
19. 70
21. 792

EXERCISE 15-4 PAGE 322

1. $\frac{1}{5}, \frac{4}{5}$ **3.** $\frac{1}{4}, \frac{1}{2}$ **5.** $\frac{1}{32}, \frac{5}{32}, \frac{5}{16}$
7. $\frac{1}{5}, \frac{2}{5}$ **9.** $\frac{4}{32}, \frac{2}{33}, \frac{5}{11}$ **11.** $\frac{10}{21}, \frac{121}{126}, \frac{125}{126}$
13. $\frac{1}{221}, \frac{1}{17}, \frac{4}{17}$ **15.** $\frac{1}{14}, \frac{1}{4}$ **17.** $\frac{1}{126}$

EXERCISE 15-5 PAGE 326

1. $\frac{5}{6}$ **3.** $\frac{1}{6}, \frac{5}{18}, \frac{5}{18}$ **5.** $\frac{1}{2197}, \frac{27}{2197}, \frac{1}{64}$
7. $\frac{1}{3}, \frac{1}{6}, \frac{1}{3}$ **9.** $\frac{21}{506}, \frac{63}{253}, \frac{175}{2024}$ **11.** $\frac{19}{30}$
13. $\frac{1}{462}, \frac{1}{462}$ **15.** $\frac{3}{4}, \frac{3}{16}$ **17.** $\frac{17}{28}, \frac{11}{18}$
19. $\frac{3}{4}$, \$375 **21.** \$3.60, \$4.29

EXERCISE 15-6 PAGE 329

1. $\frac{3}{8}, \frac{1}{4}, \frac{5}{16}$ **3.** $\frac{81}{2401}, \frac{864}{2401}, \frac{1024}{2401}$
5. 0.2401, 0.4116, 0.3483 **7.** 2, 1

EXERCISE 16-1 PAGE 336

1. $8a^3 - 12a^2y + 6ay^2 - y^3$
3. $a^5 + 10a^4x + 40a^3x^2 + 80a^2x^3 + 80ax^4 + 32x^5$
5. $243m^5 - 405m^4n^2 + 270m^3n^4 - 90m^2n^6 + 15mn^8 - n^{10}$
7. $a^8 - 8a^7x + 28a^6x^2 - 56a^5x^3 + 70a^4x^4 - 56a^3x^5 + 28a^2x^6 - 8ax^7 + x^8$
9. $x^8 + 8x^7y + 28x^6y^2 + 56x^4y^3 + 70x^4y^4 + 56x^3y^5 + 28x^2y^6 + 8xy^7 + y^8$
11. $\dfrac{a^5}{b^6} - \dfrac{6a^4}{b^4} + \dfrac{15a^2}{b^2} - 20 + \dfrac{15b^2}{a^2} - \dfrac{6b^4}{a^4} + \dfrac{b^6}{a^6}$
13. $\dfrac{x^5}{243} + \dfrac{5x^4y}{162} + \dfrac{5x^3y^2}{54} + \dfrac{5x^2y^3}{36} + \dfrac{5xy^4}{48} + \dfrac{y^5}{32}$
15. $\dfrac{a^{12}}{4096} + \dfrac{3a^9}{256} + \dfrac{15a^6}{64} + \dfrac{5a^3}{2} + 15 + \dfrac{48}{a^3} + \dfrac{64}{a^6}$
17. $35a^4x^3$ **19.** $330x^4$ **21.** $7920a^8$
23. $\dfrac{7a^5b^5}{216}$ **25.** $\dfrac{231a^6b^6}{16}$ **27.** $90{,}720x^8y^4$
29. 84 **31.** $-715x^3$ **33.** $-252x^5$
35. 1.172 **37.** 0.834 **39.** 0.851

EXERCISE 16-2 PAGE 339

1. $1 - x + x^2 - x^3 + \cdots$ **3.** $1 + y + y^2 + y^3 + \cdots$
5. $1 - 3x + 6x^2 - 10x^3 + \cdots$
7. $1 - \frac{9x}{2} + \frac{135x^2}{8} - \frac{945x^3}{16} + \cdots$
9. $1 + \frac{16x}{3} + \frac{224x^2}{9} + \frac{8960x^3}{81} + \cdots$
11. $a^{-3} + 6a^{-4}x + 24a^{-5}x^2 + 80a^{-6}x^3 + \cdots$
13. $2^{-11/4}\left(8 + x - \frac{3x^2}{16} + \frac{7x^3}{128} - \cdots\right)$
15. $2^{-11/3}\left(8 + \frac{8y}{3} + \frac{10y^2}{9} + \frac{40y^3}{81} + \cdots\right)$
17. 2.96 **19.** 1.01 **21.** 0.50 **23.** 0.97

EXERCISE 17-1 PAGE 349

1. 8 **3.** 13 **5.** -13 **7.** -17 **9.** -50
11. -2 **13.** 47 **15.** -8 **17.** -5
39. $x^3 - 7x + 6 = 0$ **41.** $x^3 + 6x^2 + 11x + 6 = 0$
43. $x^3 + x^2 - 12x = 0$ **45.** $3x^3 - x^2 - 3x + 1 = 0$
47. $x^3 - 2x^2 - 3x + 6 = 0$

EXERCISE 17-2 PAGE 353

1. $x^2 - 2x + 11, 2$ **3.** $3x^2 + 9x + 14, 15$
5. $2x^2 - 4x + 13, 7$ **7.** $5x^2 - 12x + 39, -145$
9. $x^3 - x^2 + 2x - 7, 24$ **11.** $2x^3 + x^2 - 3x + 10, -18$
13. $x^4 + x^3 + 2x^2 + 3x + 6, 16$
15. $x^4 - 2x^3 + 3x^2 - 12x + 48, -190$
17. $x^4 - x^3 + x^2 - x + 1, 0$ **19.** $x^4 - 2x^3 + 4x^2 - 8x + 16, 0$
21. $2x^3 - 4x^2 - 2x - 2, -5$ **23.** $4x^3 - 2x^2 + 4x - 4, 5$
25. $8x^2 - 4x + 8, -9$ **27.** $-63, -3, -3$ **29.** 8, 18, 388

EXERCISE 17-3 PAGE 358

1. $-i, 4$ **3.** $3 - i, 2$
5. $-\sqrt{7}, -2 \pm 2i$ **7.** $-1 - i\sqrt{3}, \pm\sqrt{2}$
9. $-3 \pm i\sqrt{2}, -i\sqrt{6}$ **11.** $-i, \pm 2i$
13. $x^3 - 5x^2 + x + 75 = 0$ **15.** $x^4 - 10x^3 + 27x^2 + 20x - 58 = 0$

17. $x^4 - 12x^3 + 77x^2 - 260x + 500 = 0$
19. $x^4 - 10x^3 + 42x^2 - 82x + 65 = 0$
25. All real; four real, two imaginary; two real, four imaginary; six imaginary

EXERCISE 17-4 PAGE 362

1. $2, 2, -3$
3. $3, 2 \pm i$
5. $\frac{1}{4}, \pm 2i$
7. $-\frac{1}{2}, -\frac{1}{2}, \frac{1}{2}$
9. $-\frac{2}{3}, -1 \pm 2i$
11. 2
13. $-\frac{1}{3}, \frac{1}{2}, \pm i$
15. $-\frac{1}{2}, -2, -2, -2$
17. $-2, -\frac{1}{3}, \frac{1 \pm \sqrt{2}}{2}$
19. $-3, -2, -2, -1, -1$
21. $\frac{3}{2}$
23. $(x - 1)(3x + 4)(2x + 1)$
25. $(2x - 1)(x^2 + 4x + 11)$
27. $(x + 1)(x - 1)(x - 2)^2$
29. $(x + 1)^3(x - 2)^2$

EXERCISE 17-6 PAGE 369

1. 1.51
3. 1.70
5. 2.04
7. 0.33
9. 1.08
11. 1.38
13. $-1.11, 1.25, 2.86$
15. $-0.11, 2.25, 3.86$
17. $-0.25, 1.45, 2.80$
19. $-0.47, 1.12, 2.35$
21. 1.45
23. 2.07

EXERCISE 18-1 PAGE 383

1. $\frac{\pi}{2}, 1, -1$
3. $6\pi, 1, -1$
5. $\frac{2\pi}{5}, \frac{1}{2}, -\frac{1}{2}$
7. $\frac{\pi}{2}$
9. $\frac{4\pi}{3}$
11. $\frac{2\pi}{7}$

EXERCISE 19-1 PAGE 388

1. $\left\{\frac{2\pi}{3}, \frac{4\pi}{3}\right\}$
3. $\left\{\frac{\pi}{4}, \frac{5\pi}{4}\right\}$
5. $\left\{\frac{5\pi}{6}, \frac{7\pi}{6}\right\}$
7. $\left\{\frac{\pi}{4}, \frac{7\pi}{4}\right\}$
9. $\left\{0, \frac{\pi}{2}, \pi, \frac{3\pi}{2}\right\}$
11. $\left\{\frac{\pi}{8}, \frac{7\pi}{8}, \frac{9\pi}{8}, \frac{15\pi}{8}\right\}$
13. $\{141° 4'\}$
15. $\{28° 58', 331° 2'\}$
17. $\{33° 41', 213° 41'\}$
19. $\left\{\frac{\pi}{6}, \frac{5\pi}{6}, \frac{7\pi}{6}, \frac{11\pi}{6}\right\}$
21. $\left\{\frac{\pi}{3}, \frac{2\pi}{3}, \frac{4\pi}{3}, \frac{5\pi}{3}\right\}$
23. $\left\{\frac{\pi}{2}, \frac{3\pi}{2}\right\}$
25. $\left\{\frac{\pi}{3}, \frac{2\pi}{3}\right\}$

27. $\left\{\frac{\pi}{3}, \pi, \frac{5\pi}{3}\right\}$ 29. $\left\{\frac{7\pi}{6}, \frac{11\pi}{6}\right\}$

31. $\{75°\ 31', 180°, 284°\ 29'\}$

33. $\left\{\frac{\pi}{2}\right\}$

35. $\{56°\ 19', 135°, 236°\ 19', 315°\}$

37. $\{18°\ 26', 135°, 198°\ 26', 315°\}$

39. $\{221°\ 49', 318°\ 11'\}$

41. $\{116°\ 00', 224°\ 00'\}$

EXERCISE 19-2 PAGE 392

1. $\left\{\frac{\pi}{6}, \frac{5\pi}{6}, \frac{3\pi}{2}\right\}$ 3. $\left\{\frac{\pi}{6}, \frac{5\pi}{6}, \frac{3\pi}{2}\right\}$ 5. No solution

7. $\left\{\frac{\pi}{3}, \frac{5\pi}{3}\right\}$ 9. $\left\{\frac{3\pi}{2}\right\}$ 11. No solution

13. $\left\{\frac{7\pi}{6}, \frac{11\pi}{6}\right\}$ 15. $\{90°, 216°\ 52'\}$

17. $\{57°\ 40', 159°\ 12'\}$ 19. $\{0°, 250°\ 32'\}$

21. $\left\{0, \frac{\pi}{2}, \pi, \frac{3\pi}{2}\right\}$ 23. $\left\{\frac{\pi}{4}, \frac{\pi}{3}, \frac{3\pi}{4}, \frac{5\pi}{4}, \frac{5\pi}{3}, \frac{7\pi}{4}\right\}$

25. $\left\{0, \frac{\pi}{3}, \frac{\pi}{2}, \frac{3\pi}{2}, \frac{5\pi}{3}\right\}$ 27. $\{14°\ 2', 90°, 194°\ 2', 270°\}$

29. $\left\{\frac{\pi}{6}, \frac{\pi}{2}, \frac{5\pi}{6}, \frac{3\pi}{2}\right\}$ 31. $\left\{\frac{\pi}{2}\right\}$

33. $\left\{\frac{\pi}{6}, \frac{\pi}{2}, \frac{5\pi}{6}, \frac{7\pi}{6}, \frac{3\pi}{2}, \frac{11\pi}{6}\right\}$ 35. $\left\{\frac{\pi}{4}, \frac{3\pi}{4}\right\}$

37. $\left\{\frac{\pi}{4}, \frac{3\pi}{4}, \frac{5\pi}{4}, \frac{7\pi}{4}\right\}$ 39. $\left\{\frac{\pi}{6}, \frac{5\pi}{6}, \frac{7\pi}{6}, \frac{11\pi}{6}\right\}$

41. $\left\{\frac{\pi}{12}, \frac{5\pi}{12}\right\}$ 43. $90°, 216°\ 52'$

45. $\{63°\ 26', 135°, 243°\ 26', 315°\}$

EXERCISE 19-3 PAGE 398

1. $\frac{\pi}{2}$ 3. $\frac{\pi}{4}$ 5. π 7. $\frac{\pi}{6}$

9. $\frac{\pi}{3}$ 11. $\frac{2\pi}{3}$ 13. $\frac{\pi}{6}$ 15. $-\frac{\pi}{6}$

17. 1.25 19. $\pi - 0.50$ 21. -1.44 23. $-\frac{1}{\sqrt{3}}$

25. $\dfrac{\sqrt{5}}{3}$ 27. $-\frac{1}{3}$ 29. $\dfrac{\sqrt{7}}{4}$ 31. $\dfrac{\pi}{4}$

33. $\frac{7}{25}$ 35. $-\frac{7}{9}$ 37. $\dfrac{\sqrt{6}}{6}$ 39. 0.7317

41. -0.2131

EXERCISE 19-4 PAGE 400

11. $\frac{16}{65}$ 13. $-\frac{33}{56}$ 15. 0

17. $y\sqrt{1-x^2} - x\sqrt{1-y^2}$ 19. $-\sqrt{1-x^2}$ 21. $x = \dfrac{1}{\sqrt{2}}$

23. $x = \dfrac{1}{\sqrt{5}}$ 25. $x = \frac{1}{6}$ 27. $x = \frac{1}{2}$

29. $x = \pm\dfrac{1}{\sqrt{2}}$ 31. $\{-1 \le x \le 1\}$

33. $\{-1 < x < 1\}$ 35. $\{-1 < x < 1\}$

37. $\{x \in R\}$ 39. $\{x \in R\}$

41. $\{x \in R\}$ 43. $\{x \in R\}$

45. $\{x \ge 0\}$ 47. $\{-1 \le x \le 1\}$

EXERCISE 20-1 PAGE 405

1. $\log_2 32 = 5$ 3. $\log_6 \frac{1}{6} = -1$ 5. $\log_8 1 = 0$
7. $\log_{16} 2 = \frac{1}{4}$ 9. $\log_{10} 1000 = 3$ 11. $\log_{3/4} \frac{64}{27} = -3$
13. $2^0 = 1$ 15. $5^2 = 25$ 17. $2^4 = 16$
19. $(\frac{1}{2})^{-3} = 8$ 21. $(\frac{1}{5})^{-3} = 125$ 23. $7^{-2} = \frac{1}{49}$
25. $x = 1000$ 27. $x = \frac{1}{11}$ 29. $x = 1$
31. $x = \frac{27}{8}$ 33. $x = 27$ 35. $a > 0$
37. $a = \frac{1}{25}$ 39. $a = \frac{32}{243}$ 41. $y = 4$
43. $y = -2$ 45. $y = -4$ 47. $y = -3$

EXERCISE 20-2 PAGE 408

1. $\log 251 + \log 46 + \log 18$ 3. $\frac{1}{2}\log 24 + 3\log 35$
5. $\log 49 + \log 65 - \log 71 - \log 86$
7. $\frac{1}{2}(\log 87 + \log 28 - \log 15)$
9. $\log 58 + \frac{1}{3}\log 28 - \log 34 - \frac{1}{2}\log 56$

11. $\log \dfrac{xy}{z}$ 13. $\log \pi r^2 h$ 15. $\log \dfrac{52^{1/2}}{5^{2/3} \cdot 3^{-4}}$

17. $\log \sqrt{\frac{(82)(44)(21)}{798}}$ **19.** 0.9542 **21.** 1.7323

23. 0.4184 **25.** 0.4643

EXERCISE 20-3 PAGE 415

1. 1 **3.** 2 **5.** 2 **7.** 1 **9.** 4
11. -2 **13.** -4 **15.** 0 **17.** -2 **19.** 1
21. 8 **23.** 0.7574 **25.** 4.7597 **27.** 9.9096 − 10
29. 8.9294 − 10 **31.** 2.8261 **33.** 6.8921 − 10
35. 5.8785 **37.** 0.4972 **39.** 3.8798
41. 9.9000 − 10 **43.** 4.9255 **45.** 7.6731 − 10
47. 2.7674 **49.** 1.0004 **51.** 2.0103
53. 6.0484 − 10 **55.** 5.8722 − 10 **57.** 2.7058
59. 5.5373 **61.** 3.9592 − 10 **63.** 8.8902

EXERCISE 20-4 PAGE 417

1. 126 **3.** 31.7 **5.** 0.611 **7.** 270,000
9. 0.00104 **11.** 22.24 **13.** 4082 **15.** 0.9346
17. 0.007265 **19.** 0.1294 **21.** 0.0001775 **23.** 514.1
25. 39.85 **27.** 516,400 **29.** 3,495,000

EXERCISE 20-5 PAGE 419

1. 4.293 **3.** 0.5510 **5.** 0.1200 **7.** 6173
9. 85.68 **11.** 738.8 **13.** 10.72 **15.** 7.390
17. 36.51 **19.** 2.007 **21.** 0.4637 **23.** 4.057
25. 0.2735 **27.** 0.04048 **29.** 3.132×10^{-8} **31.** 0.2058
33. 23.32 **35.** 6.080

EXERCISE 20-6 PAGE 421

1. $x = 3$ **3.** $x = 5$ **5.** $x = 2$
7. $x = -1$ **9.** $x = 0$ **11.** $x = 1.233$
13. $x = 0.8857$ **15.** $x = 2.014$ **17.** $x = 0.8414$
19. $x = 0$ **21.** $x = 0.04257$ **23.** $x = -1.857$
25. $x = 18.80$ **27.** $x = -1.366$

EXERCISE 20-7 PAGE 423

1. 4.605 **3.** -0.7382 **5.** 4.701 **7.** -1.265
9. 2.124 **11.** -3.194 **13.** 1.100 **15.** 0.3465

EXERCISE 20-8 PAGE 424

1. $9.6470 - 10$ **3.** $9.8895 - 10$ **5.** $9.7143 - 10$
7. $9.9993 - 10$ **9.** 0.0005 **11.** $59° 12'$
13. $43° 1'$ **15.** $34° 39'$ **17.** $5° 10'$
19. $45° 23'$

EXERCISE 21-1 PAGE 430

1. $C = 70°, \quad a = 61, \quad b = 76$ **3.** $B = 81°, \quad a = 1.9, \quad c = 3.5$
5. $A = 53°, \quad b = 8.5, \quad c = 23$
7. $A = 58° 50', \quad a = 334, \quad c = 369$
9. $C = 57° 30', \quad b = 0.177, \quad c = 0.359$
11. $A = 35° 3', \quad b = 1122, \quad c = 1271$
13. $A = 47° 42', \quad a = 16.62, \quad b = 22.36$
15. $C = 69° 38', \quad a = 1225, \quad c = 1778$
17. 31 ft **19.** 4900 ft, 6500 ft, 4000 ft
21. Current 1.6 mph, speed 2.9 mph
23. 260 mph, 300 mph **25.** 239 lb, 325 lb

EXERCISE 21-2 PAGE 436

1. $B = 36°, \quad C = 44°, \quad c = 28$ **3.** No solution
5. $B = 43°, \quad C = 85°, \quad c = 37$
7. $A = 107°, \quad B = 45°, \quad a = 4.9$
$A' = 17°, \quad B' = 135°, \quad a' = 1.5$
9. No solution
11. $A = 27° 30', \quad B = 51° 30', \quad a = 109$
13. $A = 39° 9', \quad C = 89° 28', \quad c = 3988$
15. $A = 86° 27', \quad C = 53° 30', \quad a = 199.0$
$A' = 13° 27', \quad C' = 126° 30', \quad a' = 46.40$
17. 372 ft **19.** N 46° E, 3.0 mph **21.** 144°, 590 mph

EXERCISE 21-3 PAGE 440

The unknown angles in problems 1 through 15 are results obtained without the use of $A + B + C = 180°$. Hence in some cases the sum of the results for the three angles is not exactly 180°.

1. $c = 3.2, A = 48°, B = 80°$
3. $b = 20, A = 33°, C = 51°$
5. $a = 15, B = 28°, C = 117°$
7. $a = 0.61, B = 19°, C = 30°$
9. $A = 41°, B = 79°, C = 61°$
11. $A = 53°, B = 71°, C = 56°$
13. $A = 36° 50', B = 50° 00', C = 93° 20'$
15. $A = 36° 50', B = 49° 20', C = 94° 00'$
17. 22.5 ft
19. 100 lb

EXERCISE 21-4 PAGE 444

1. 38
3. 430
5. 3600
7. 1970
9. 14
11. Two solutions: 233, 1840
13. 170
15. 0.2172

EXERCISE 22-1 PAGE 449

1. $A = 34°, \quad B = 87°, \quad c = 27$
3. $B = 26°, \quad C = 21°, \quad a = 140$
5. $A = 92°, \quad C = 42°, \quad b = 45$
7. $B = 55° 10', \quad C = 33° 10' \quad b = 110$
9. $A = 55° 30', \quad C = 86° 00', \quad b = 93.6$
11. $B = 64° 13', \quad C = 30° 54', \quad a = 7.821$
13. $A = 39° 29', \quad B = 21° 57', \quad c = 0.4707$
15. 970 lb
17. 15, 36
19. 40°, 220 mph

EXERCISE 22-2 PAGE 452

1. $A = 20°, \quad B = 32°, \quad C = 128°$
3. $A = 38°, \quad B = 45°, \quad C = 97°$
5. $A = 51°, \quad B = 39°, \quad C = 90°$
7. $A = 43° 20', \quad B = 50° 40', \quad C = 28° 00'$
9. $A = 35° 10', \quad B = 50° 40', \quad C = 94° 10'$
11. $A = 117° 49', \quad B = 39° 27', \quad C = 22° 44'$
13. 89° 20', 127° 40', 143° 00'
15. 84° or 96°

EXERCISE 23-1 PAGE 458

1. $\begin{bmatrix} -1 & 8 \\ 8 & 5 \end{bmatrix}$ 3. $\begin{bmatrix} 3 & -2 \\ 0 & -1 \end{bmatrix}$ 5. $\begin{bmatrix} 10 & 14 \\ 0 & 26 \end{bmatrix}$

7. $\begin{bmatrix} -5 & -2 & 1 \\ 9 & 10 & 11 \end{bmatrix}$ 9. $\begin{bmatrix} -2 & 2 & -1 \\ -8 & -2 & -10 \\ 20 & 10 & 28 \end{bmatrix}$

11. $\begin{bmatrix} 3 & 9 & 15 & 18 \\ 2 & 6 & 10 & 12 \\ 2 & 3 & 5 & 6 \end{bmatrix}$ 13. $\begin{bmatrix} 22 & -13 & 7 \\ 31 & -17 & 13 \\ 5 & -4 & 5 \end{bmatrix}$

EXERCISE 23-2 PAGE 461

1. -57 3. 100 5. -5
7. $x = -2, y = 3$ 9. $x = -\frac{3}{5}, y = -\frac{11}{5}$ 11. $x = -\frac{1}{2}, y = \frac{5}{2}$
13. $x = \frac{29c}{8a}, y = \frac{7c}{8b}$ 15. $x = 2, y = -\frac{2}{a}$ 17. 40
19. 47 21. 63 23. $x = 2$
25. $x = -2$ or 6 27. $x = -\frac{8}{3}$

EXERCISE 23-3 PAGE 468

11. $x = 3$ 13. $x = -2, 0, 2$

EXERCISE 23-4 PAGE 473

1. -96 3. 29 5. -26 7. 406 9. -36
11. 1083 13. 27 15. 9 17. 23.5 19. 19

EXERCISE 23-5 PAGE 480

1. $\{(1, 3, -2)\}$ 3. $\{(1, 3, -2)\}$ 5. $\{(3, 0, 2)\}$
7. $\{(1, -1, 1, -1)\}$ 9. $\{(1, -\frac{2}{7}, \frac{2}{7}, -\frac{1}{7})\}$
11. $\{(\frac{7}{3}, \frac{2}{3}, 1, -1)\}$ 13. $\{(-2, 3, \frac{3}{2}, -\frac{1}{2})\}$

INDEX

INDEX